Recent Advances in Biological Membrane Studies

Structure and Biogenesis, Oxidation and Energetics

NATO ASI Series
Advanced Science Institutes Series

A series presenting the results of activities sponsored by the NATO Science Committee, which aims at the dissemination of advanced scientific and technological knowledge, with a view to strengthening links between scientific communities.

The series is published by an international board of publishers in conjunction with the NATO Scientific Affairs Division

A	Life Sciences	Plenum Publishing Corporation
B	Physics	New York and London
C	Mathematical and Physical Sciences	D. Reidel Publishing Company Dordrecht, Boston, and Lancaster
D	Behavioral and Social Sciences	Martinus Nijhoff Publishers
E	Engineering and Materials Sciences	The Hague, Boston, and Lancaster
F	Computer and Systems Sciences	Springer-Verlag
G	Ecological Sciences	Berlin, Heidelberg, New York, and Tokyo

Recent Volumes in this Series

Volume 86—Wheat Growth and Modelling
edited by W. Day and R. K. Atkin

Volume 87—Industrial Aspects of Biochemistry and Genetics
edited by N. Gurdal Alaeddinoglu, Arnold L. Demain, and Giancarlo Lancini

Volume 88—Radiolabeled Cellular Blood Elements
edited by M. L. Thakur

Volume 89—Sensory Perception and Transduction in Aneural Organisms
edited by Giuliano Colombetti, Francesco Lenci, and Pill-Soon Song

Volume 90—Liver, Nutrition, and Bile Acids
edited by G. Galli and E. Bosisio

Volume 91—Recent Advances in Biological Membrane Studies: Structure and Biogenesis, Oxidation and Energetics
edited by Lester Packer

Volume 92—Evolutionary Relationships among Rodents: A Multidisciplinary Analysis
edited by W. Patrick Luckett and Jean-Louis Hartenberger

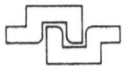

Series A: Life Sciences

Recent Advances in Biological Membrane Studies

Structure and Biogenesis, Oxidation and Energetics

Edited by

Lester Packer

University of California, Berkeley
Berkeley, California

Springer Science+Business Media, LLC

Proceedings of a NATO Advanced Study Institute on
New Developments and Methods in Membrane Research
and Biological Energy Transduction,
held August 16–29, 1984,
on the Island of Spetsai, Greece

Library of Congress Cataloging in Publication Data

NATO Advanced Study Institute on New Developments and Methods in Membrane Research and Biological Energy Transduction (1984: Nisos Spétsai, Greece)
Recent advances in biological membrane studies.

(NATO ASI series. Series A, Life sciences; v. 91)
"Proceedings of a NATO Advanced Study Institute on New Developments and Methods in Membrane Research and Biological Energy Transduction, held August 16–29, 1984, on the island of Spetsai, Greece"—T.p. verso.
"Published in cooperation with NATO Scientific Affairs Division."
Includes bibliographies and index.
1. Membranes (Biology)—Congresses. 2. Biological transport, Active—Congresses. 3. Bioenergetics—Congresses. I. Packer, Lester. II. North Atlantic Treaty Organization. Scientific Affairs Division. III. Title. IV. Series. ['DNLM: 1. Cell Membrane—congresses. QH601 N2795r 1984]
QH601.N375 1984   574.87'5   85-9481
ISBN 978-1-4684-4981-5   ISBN 978-1-4684-4979-2 (eBook)
DOI 10.1007/978-1-4684-4979-2

© Springer Science+Business Media New York 1985
Originally published by Plenum Press, New York in 1985
Softcover reprint of the hardcover 1st edition 1985

PREFACE

A NATO Advanced Study Institute on "New Developments and
Methods in Membrane Research and Biological Energy Transduction"
was held in order to consider some of the most recent developments
in membrane research methodologies and results, with particular
emphasis on studies of biological energy transduction. The partic-
ipants in the Institute dealt with three general areas of membrane
study: membrane structure (with emphasis on lipid and protein
components), membrane component assembly (with particular emphasis
on mitochondria and chloroplasts), and the specialized functions of
certain membrane systems. This last area included discussions of
topics such as drug transformation, the role of membrane electron
transport in the generation of oxygen radicals, the effect of oxygen
radicals on cellular homeostasis and on the structure, organization
and function of the acetylcholine receptor. Lectures and posters
were concerned with two central questions: what is the function of
membrane structure in energy transduction and how can energy transduction be effectively measured and assessed? This text presents
the content of the major lectures and important posters presented
during the Institute's program. In issuing this book, the editor
hopes to convey the proceedings of the Institute to a larger audi-
ence and to offer a comprehensive account of those developments
in membrane research that were considered on the Island of Spetsai
between August 16 and August 29, 1984.

L. Packer
Berkeley, California
February 1985

CONTENTS

III. ENERGETICS

MEMBRANE STRUCTURE : NEUTRON DIFFRACTION AND SMALL ANGLE

SCATTERING STUDIES

Giuseppe Zaccaï

Institut Laue-Langevin
156X
38042 Grenoble, Cedex, France

RADIATION SCATTERING BY ONE MOLECULE

Consider the molecule in Figure 1a in a beam of radiation of wavelength λ. The radiation wave scattered by the molecule in a direction making an angle 2θ with the incident beam is the sum of the waves scattered by all the atoms j in the molecule :

$$F(Q) = \sum_j b_j \exp i\, Q \cdot r_j \qquad (1)$$

where b_j is the scattering amplitude of atom j, r_j is its position relative to an arbitrary origin and Q is the scattering vector

$$Q = k_1 - k_0$$

$$|Q| = (4\pi \sin\theta)/\lambda \qquad (1a)$$

where k_0 is the incident wave vector and k_1 is the scattered wave vector.

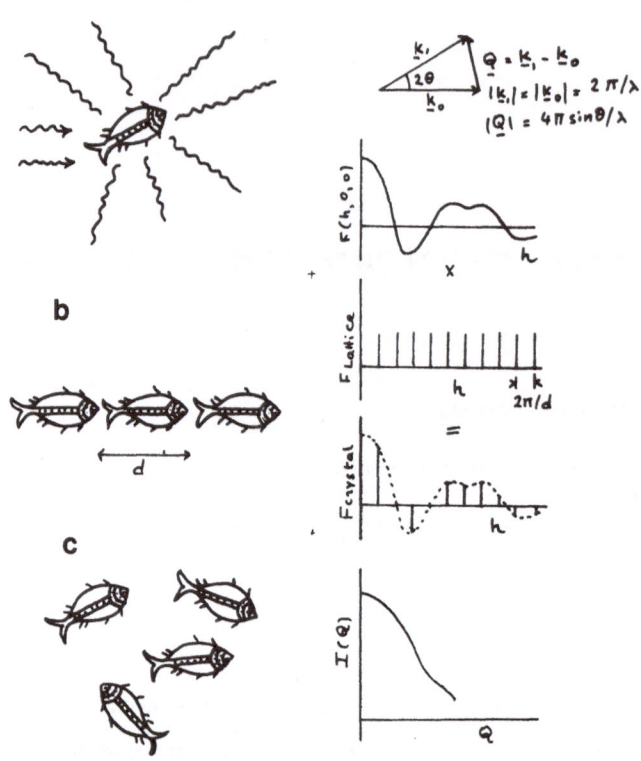

Figure 1. Structures in real space (on the left) and their Fourier
transforms (on the right).
(a) a single molecule : F(h,O,O) is one of the components of
vector F(Q) (e.g. the real part) for a given direction in
Q : (h,O,O).
(b) a one-dimensional crystal of spacing d.
(c) Identical molecules with no correlation between them.

There is a one to one correspondence between the function
F(Q), called the Fourier transform after the mathematician who
proposed it, and the arrangement of atoms in the molecule
described by b_j, r_j (its structure). If we knew the Fourier
transform of a molecule we would know its structure. It is not

possible, however, to measure $F(\underline{Q})$ completely, for several reasons. First, detectors measure intensity $|F(\underline{Q})|^2$ and not amplitude. Second, the radiation scattered by one molecule is too weak to observe. Third, there are lower and upper limits to the $\underline{Q}$ that can be observed, due to both the angular geometry of the experiment and the wavelength of the radiation.

The techniques of crystallography and small angle scattering have been developed to tackle these difficulties (1,2,3).

RESOLUTION

There is a reciprocal relationship between real space, where $\underline{r}_j$ is measured, and $\underline{Q}$ space (or Fourier space or "reciprocal" space) where $\underline{Q}$ is measured. A lattice of spacing d in real space has a Fourier transform of spacing $2\pi/d$ in reciprocal space (Figure 1b). The resolution of the experiment is the smallest separation that can be measured in real space ; it is $\simeq 2\pi/Q_{max}$ where Q_{max} is the maximum Q measured.

RADIATION SCATTERED BY MANY IDENTICAL MOLECULES

Figure 1b and 1c show the two extreme ways in which we can have many identical molecules in a sample : a crystal in which they are related to each other by the symmetry of a lattice, and complete disorder in which there is no correlation betweem molecules.

Radiation scattered by the crystal will result from inter-ference within each molecule and interference between different molecules. It is a useful and fortunate result that the two sets of interference can be separated and the Fourier transform of the crystal is simply the product of the Fourier transform of the lattice and that of the molecule (Figure 1b).

$$F_{cryst.}(\underline{Q}) = F_{latt.}(\underline{Q}) \times F(\underline{Q}) \qquad (2)$$

$F_{latt.}(Q)$ is itself a lattice in Q space, so that $F_{cryst.}(Q)$ has "reflections" at positions which are given by the lattice but of intensities which are given by the value of $F(Q)$ at that point (Figure 1b).

In the disordered sample (e.g. molecules in solution), there is no interference between waves from different molecules, the scattered radiation is given by $I(Q)$ (note that Q is not a vector; because of the disorder, all directions are equivalent).

$$I(Q) = \sum_{m} |F(\underline{Q})|^2_m \qquad (3)$$

where the sum is over $|F(Q)|^2_m$ for all orientation of the molecule ($F(Q)$ is different for each orientation, m). A usual form for $I(Q)$ is shown in Figure 1c. Note that it falls very rapidly with Q, i.e. high resolution information is destroyed by the disorder.

In the small Q limit, $I(Q)$ is described by the Guinier approximation:

$$I(Q) = I(0) \exp{-\frac{1}{3} R_G^2 Q^2} \qquad (4)$$

$$I(0) = \frac{c N_A}{M}\left(\sum_{j} b_j\right)^2 \qquad (5)$$

where c is mass concentration of molecules, M is molecular weight and N_A is Avogadro's number

$$R_G^2 = \sum_j b_j r_j^2 \Big/ \sum_j b_j \qquad (6)$$

where r_j is the distance of atom j to the centre of mass of the b_j distribution.

By analogy with mechanical moments of inertia, R_G is the radius of gyration of the b_j distribution.

Small angle scattering in the simplest applications yields two parameters : I(0), R_G ; the first is related to the total mass and the second to the distribution of mass in each molecule. Equations (5) and (6) are for molecules in a vacuum. Biological molecules, like the fish of Figure 1, are unhappy out of solution. The appropriate equations for solutions are

$$I(0) \propto \left| \sum_j \left(b_j - \rho^\circ v_j \right) \right|^2 \qquad (7)$$

and

$$R_G^2 = \sum_j (b_j - \rho^\circ v_j) r_j^2 \Big/ \sum_j (b_j - \rho^\circ v_j) \qquad (8)$$

where ρ° is the scattering length density of the solvent and v_j is the effective volume of atom j. The term $(b_j - \rho^\circ v_j)$ is called the excess scattering length of atom j over the solvent or the contrast length of atom j.

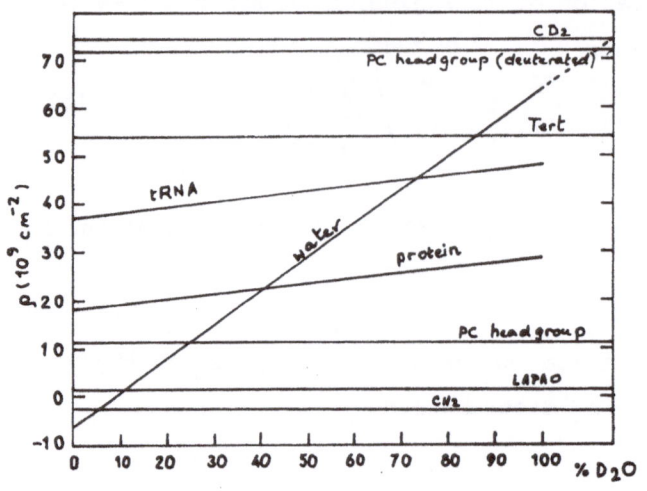

Figure 2. The scattering length density $\rho = (\Sigma b \text{ in Volume})/(\text{Volume})$ of different substances as a function of D_2O % in an $H_2O:D_2O$ mixed solvent. PC is phosphatidyl choline. Tert is a tertiary mixture of deuterated and H-detergent molecules which has homogeneous scattering density (22). LAPAO is the detergent of reference (23). The lines for tRNA and protein are not constant because of the exchange of labile H in the molecules. Contrast is proportional to $\rho(\text{substance}) - \rho(\text{solvent})$.

ISORMOPHOUS REPLACEMENT AND LABELLING

Structure determination is simplified in both small angle scattering and crystallography if parts of the molecule in the solvent are labelled so that they scatter differently (1,3). H and D (^{2}H) have very different b values for neutrons (-0.37421×10^{-12} cm and $+0.667 \times 10^{-12}$ cm, respectively). Isotopic replacement of H by D in a molecule (or H_2O by D_2O in the solvent) changes its neutron scattering pattern ($F(Q)$) and is a very powerful approach to solving its structure.

Contrast variation in solution scattering experiments is an approach where by varying ρ° (e.g. by changing the $D_2O:H_2O$ ratio in the solvent) the contrast of different parts of a macro-molecular complex (e.g. a membrane protein in a detergent micelle) is changed. Different parts of the complex can then be observed separately. There are sophisticated approaches to the analysis of contrast variation data, but the principles involved are simple (review in 3). Figure 2 shows the scattering densities of different molecules as a function of D_2O percentage in the solvent. For example, a measurement of a tRNA-protein complex in $\approx$ 70 % D_2O will essentially observe the protein moiety (tRNA contrast is $\approx$ 0); a measurement of a membrane protein in detergent, in the D_2O percentage which matches the detergent scattering density, will essentially observe the protein alone. Contrast variation has also been applied in crystallography at low resolution (4).

The detailed crystallographic analysis of isomorphous replacement of H by D, could be quite complex depending on the symmetry of the lattice, the number of sites and how much is already known about the structure (1).

LIPID BILAYERS

Extracted membrane lipids and related synthetic lipids spontaneously form a variety of structures when mixed with water, depending on water content, lipid composition and temperature. Many of these structures have been characterised by X-ray diffraction and phase diagrams have been drawn (5). Neutron diffraction studies have contributed greater detail in the structure of the lamellar phases L_α and L_β (one-dimensional crystals : of lipid bilayers with water in-between). L_α is a liquid crystalline phase with fluid or disordered hydrocarbon chains which give a broad 4.5 Å reflection in diffraction, L_β has gel-like ordered hydrocarbon chains which give a sharp 4.15 Å reflection in diffraction. There is evidence that lipids in biological membranes have a structure similar to the L_α phase (6).

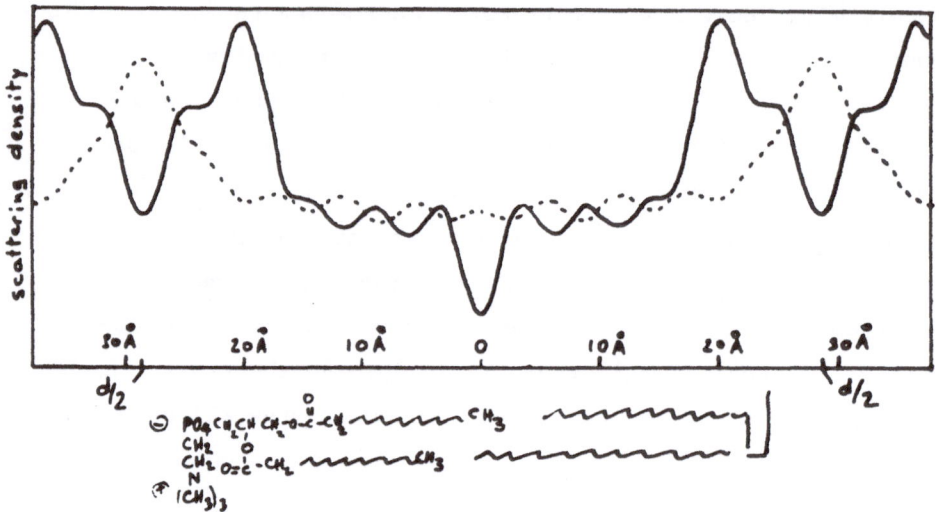

Figure 3. Scattering density profile of a dipalmitoyl phosphatidyl choline (DPPC) lipid bilayer at low water content in the gel phase. The dotted profile is the scattering density of the water distribution (hydration) (8). Below the profile, the molecule of DPPC shows the essential features which are maintained in the gel and lipid crystalline phases : headgroup parallel to the bilayer plane, out-of-step acyl chains.

The one dimensional scattering density profile of a lipid bilayer is shown in Figure 3. $H_2O:D_2O$ exchange has helped in the interpretation of the diffraction data and determined the hydration of the lipid headgroups (7). Specific chemical exchange of (CH_2) by (CD_2) in different segments of the hydrocarbon chains has located these labels in the profile and allowed a detailed description of the time averaged structure of both chains in 1,2 dipalmitoyl phosphatidyl choline (DPPC) and 1,2 dipalmitoyl phosphatidyl ethanolamine (DPPE) (8,9,10). Both in the L_α and L_β phases and for both headgroups the chains are not equivalent in their profile projection. Experiments on synthetic 1,3 phospholipids, on the other hand, have shown the chains to be equivalent, suggesting that their conformation in the bilayer is determined sterically by their attachment to the glycerol backbone and its orientation (11).

Similar experiments with D labels in the headgroup of DPPC have shown its time average orientation to be in the plane of the bilayer in both L_α and L_β phases (9). The neutron results correlate well and complement results from Deuterium Magnetic Resonance on the same systems (10).

Neutron diffraction experiments with specific deuteration have also been done to characterise interactions of other components such as cholesterol with lipid bilayers (12).

RETINAL ROD OUTER SEGMENT (ROS) DISK MEMBRANES

ROS disks are stacked naturally to form a one dimensional crystal with two membranes per unit cell (13). Neutron diffraction studies using $H_2O:D_2O$ exchange and a contrast variation approach have shown that most of the rhodopsin molecule is embedded in the bilayer (14). Increased protein density on the cytoplasmic side upon bleaching suggested an interpretation involving motion of the protein (14). It is now known, however, that bleaching starts an enzyme cascade on the cytoplasmic side (13) and what was probably observed in the neutron experiment was the beginning of this process with the interaction of another protein with rhodopsin. It is an interesting illustration of how care must be exercised in the interpretation of structural data when the biochemistry of a system is not fully known.

PURPLE MEMBRANE OF H. HALOBIUM

Purple membrane is naturally crystalline in two dimensions. The structure of bacteriorhodopsin, the only protein it contains, has been solved to 7 Å by electron microscopy (15). Its function as a

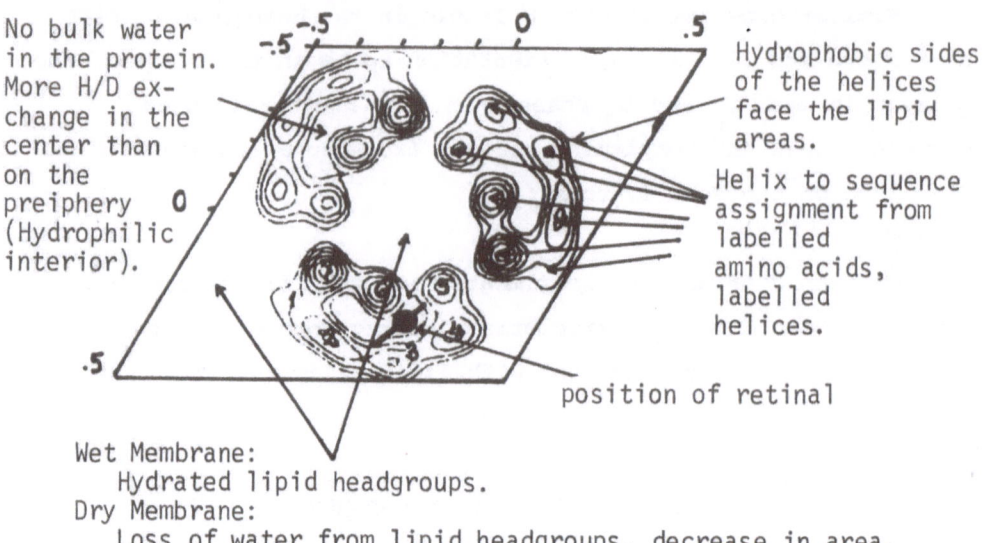

No bulk water in the protein. More H/D exchange in the center than on the preiphery (Hydrophilic interior).

Hydrophobic sides of the helices face the lipid areas.

Helix to sequence assignment from labelled amino acids, labelled helices.

position of retinal

Wet Membrane:
  Hydrated lipid headgroups.
Dry Membrane:
  Loss of water from lipid headgroups, decrease in area.
  Protein structure and lattice are maintained.

Figure 4. A unit cell of purple membrane showing three symmetry related molecule of bacteriorhodopsin. The seven α-helices are numbered conventionally and the information derived from neutron diffraction experiments is shown.

proton pump has been studied extensively, and it was the first membrane protein to be sequenced (16). Considerable complementary structural information on this membrane has been obtained from neutron diffraction. It is summarised in Figure 4. $H_2O:D_2O$ exchange has shown that in "wet" conditions hydration is predominantly around lipid headgroups, the protein does not have bulk water associated with it (17). Upon drying, the protein structure is maintained, but the lipids are dehydrated and their area in the plane of the membrane is reduced (18). Specifically deuterated amino acids have been incorporated in the structure by biosynthesis to act as labels (19,20). Diffraction experiment on these samples show that the protein α-helices are embedded in the bilayer with their hydrophobic faces towards the center of the molecule away from the lipid (19). Sequence to helix assignments have also been suggested from these data (20). The position and orientation of the retinal in the projection has been determined, by using a sample where deuterated retinal was incorporated into the membrane biosynthetically (21). A study is under way to locate two sequence helices in the structure (Popot, Trewhella, Zaccaï and Engelman, in preparation). It represented a preparative "tour-de-force". The molecule was split then reconstituted from deuterated and H-segments and recrystallised into its native two-dimensional lattice with lipid. Recent neutron diffraction studies have characterized the low resolution structure and hydration of the membrane at liquid nitrogen temperature, where the optical cycle of the retinal can be trapped in different states (Zaccaï, in preparation).

SOLUBILIZED PROTEINS

To successfully separate the radii of gyration of two components of a complex by contrast variation, it is not sufficient that the mean scattering density of one of the components match the solvent. It should also be homogeneous to the resolution required. There is a very detailed study of rhodopsin micelle complexes by SANS and contrast variation, in which a lot of effort was spent in ensuring that the detergent was homogeneous in scattering density (22). The R_G of rhodopsin was measured and the geometry of its interaction with the detergent was determined.

The molecular weight of membrane proteins is often a difficult parameter to determine, and the number of subunits in several membrane proteins remain controversial. Provided a pure system can be solubilized in detergent or lipid, a SANS experiment with contrast variation is a straightforward molecular weight determination of the system in the given conditions. If the radius

Vol. 109, No. 2, 1982 BIOCHEMICAL AND BIOPHYSICAL RESEARCH COMMUNICATIONS

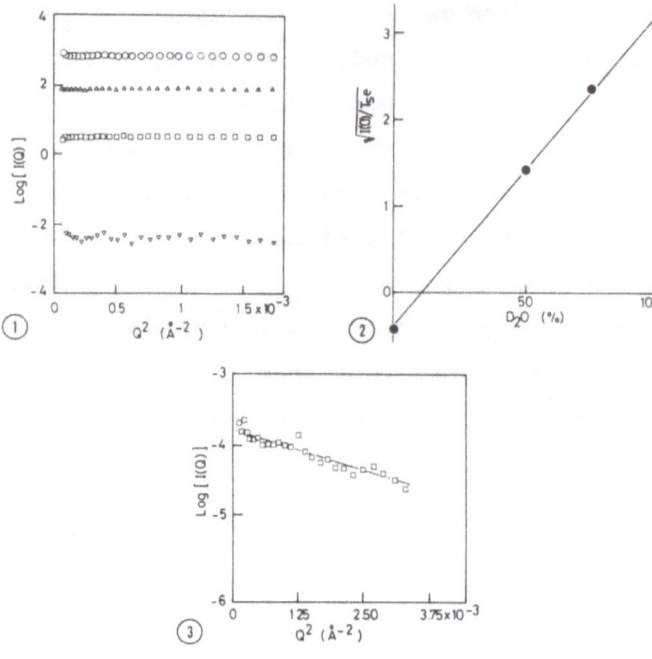

Figure 1 Guinier plots of scattering curves of pure LAPAO micelles. The detergent concentration was 4% (w/v). The pathlength was 0.200 cm for the three upper profiles and 0.100 cm for the lower one. The solvent composition was : 100% D_2O (O-), 75% D_2O (Δ-), 50% D_2O (□-) and H_2O (∇-).

Figure 2 Variation of zero angle scattering from LAPAO micelles with solvent deuteration. The intersection with the horizontal axis (I(0) = 0) gives the solvent composition whose scattering density is the same as the average scattering density of the micelle i.e. 10% D_2O. Detergent concentration : 4%.

Figure 3 Typical Guinier profile of LAPAO-carrier protein complex (protein liganded to carboxylatractyloside) in 10% D_2O. For this solvent composition, the contribution of detergent to I(0) was null. The protein concentration was 2.19 mg/ml and the LAPAO concentration 4.94%.

474

Figure 5. The set of Figures from the Block et al. study (23) of the ATP/ADP carrier protein showing the measurement of protein detergent micelles in 10 % D_2O where there is no contribution of the detergent to I(O), from which the molecular weight is determined.

of gyration is not required with precision, the other component (detergent, lipid) need not be homogeneous in scattering density. The method was applied successfully to the ATP/ADP transport protein solubilized in excess detergent (23) (Figure 5), and approaches using deuterated lipids are now under development.

CONCLUSIONS

Crystalline systems, either natural or artifically prepared have the advantage of providing information to relatively high resolution. Ultimately, structures to atomic resolution could only be obtained from good quality single crystals. In the case of lipid bilayers, however, the time average location of deuterium labels was determined to $\simeq 1$ Å from data of lower resolution by making certain assumptions. Similarly, the position and orientation of retinal in purple membrane was fixed with relatively high precision by assuming it has a rod-shape. Knowledge of the sequence and deuterium labelling of certain aminoacids also increase the effective resolution of the structure of bacteriorhodopsin beyond the resolution of the diffraction data. In the examples given above, deuterium labelling was done successfully by both chemical and biosynthetic means. It is a very powerful approach.

Contrast variation is a very low resolution approach but also very powerful in separating protein components from lipids or detergent. It is currently being used to help determine the structure of single crystals of matrix porin in detergent micelle (M. Zulauf, private communication).

Studies in solution have the significant advantages that the material need not be (indeed, should not be) ordered in any way, small amounts of protein are required ($\simeq 1$ mg) and different solvent conditions could be explored.

The present account is only an introduction and brief survey of the type of experiments which have been done by neutron diffraction and scattering on membrane systems. The original papers should be consulted for more detailed descriptions. A symposium on Neutrons in Biology was held at Brookhaven National Laboratory in 1982. The proceedings (24) are a good summary of projects in this field.

REFERENCES

1. C. R. Cantor and P.R. Schimmel, Techniques for the Study of Biological Structure and Function, <u>Biophysical Chemistry Part II</u>, W.H. Freeman and Company, San Francisco (1980).

2. A. Guinier and G. Fournet, <u>Small Angle Scattering of X-Rays</u>, Wiley, New York (1955).

3. G. Zaccaï and B. Jacrot, <u>Ann. Rev. Biophys. Bioeng.</u> 12, 139, (1983).

4. J. T. Finch, A. Lewit-Bentley, G. A. Bentley, M. Roth and P. A. Timmins, <u>Phil. Trans. Roy. Soc.</u> B290, 635 (1980).

5. A. Tardieu, V. Luzzati and F. C. Reman, <u>J. Mol. Biol.</u> 75, 711-733 (1973).

6. F. O. Schmitt, R. S. Bear, G. L. Clark, <u>Radiobiology</u> 25, 131-151 (1935).

7. D. L. Worcester in <u>Biological Membranes</u> (D. Chapman and D.F.H. Wallach, eds.) Academic Press, London (1976).

8. G. Büldt and J. Seelig, <u>Biochemistry</u> 19, 6170-6175 (1980).

9. G. Büldt, H. U. Gally, J. Seelig and G. Zaccaï, <u>J. Mol. Biol.</u> 134, 673-691 (1979).

10. G. Zaccaï, G. Büldt, A. Seelig and J. Seelig, <u>J. Mol. Biol.</u> 134, 693-706 (1979).

11. G. Büldt and G. H. De Haas, <u>J. Mol. Biol.</u> 158, 55-71 (1982).

12. D. L. Worcester and N. P. Franks, <u>J. Mol. Biol.</u> 100, 359-378 (1976).

13. E. A. Dratz and P. A. Hargrave <u>TIBS,</u> April 1983.

14. H. Saibil, M. Chabre and D. Worcester, <u>Nature</u> 262, 266-270 (1976).

15. R. Henderson and P. N. T. Unwin, <u>Nature</u> <u>257</u>, 28–32 (1975).

16. W. Stoeckenius and R. A. Bogomolni, <u>Ann. Rev. Biochem.</u> <u>52</u>, 587–616.

17. G. Zaccaï and D. Gilmore, <u>J. Mol. Biol.</u> <u>132</u>, 181–191 (1979).

18. P. K. Rogan and G. Zaccaï, <u>J. Mol. Biol.</u> <u>145</u>, 281–283 (1981).

19. D. M. Engelman and G. Zaccaï, <u>Proc. Natl. Acad. Sci. USA</u> <u>77</u>, 5894–5898 (1980).

20. J. Trewhella, S. Anderson, R. Fox, E. Gogol, S. Khan, G. Zaccaï and D. M. Engelman, <u>Biophys. J.</u> <u>42</u>, 233–241 (1983).

21. J. S. Jubb, D. L. Worcester, H. L. Crepi and G. Zaccaï, <u>EMBO J.</u> <u>3</u>, 1455–1461 (1984).

22. H. B. Osborne, C. Sardet, M. Michel-Villaz and M. Chabre, <u>J. Mol. Biol.</u> <u>123</u>, 177–206 (1978).

23. M.R. Block, G. Zaccaï, G. J. M. Lanquin and P. V. Vignais, <u>B.B.R.C.</u> <u>109</u>, 471–477 (1982).

24. <u>"Neutron in Biology"</u> Basic Life Sciences Vol. 27. (B. P. Schoenborn, ed.) Plenum Press, N.Y. (1984).

THE NATURE OF PROTEINS IN MEMBRANES

R.J.P. Williams

Inorganic Chemistry Laboratory
University of Oxford
South Parks Road, Oxford OX1 3QR

The first part of this article concerns a central theme which I have been pursuing for many years. It is the demonstration that the secret of biological systems rests in the connection

Chemical Composition → Structure → Mobility → Function

It has been usual to omit <u>mobility</u> but this is, in my view, a mistake. We then need to uncover four different features of a protein

(1) By analysis: its composition and sequence.
(2) By X-ray diffraction: its structure in a particular trapped state in a crystal. The dynamics of the structure in the crystal can add some useful knowledge, but relatively little because of the short time constants observed (10^{-12} sec).
(3) The mobility of the structure in solution as determined by NMR largely but with help from many spectroscopic tools. In essence this says that X-ray structures are unlikely to provide a unique answer to the states (plural) available to the protein.
(4) Function implies that there may be states not open to direct inspection under (2) and (3) but which may be inferred from (2) and (3) plus a knowledge of the activity of the protein.

The application of this approach to <u>membrane</u> proteins is hindered at almost every point. Sequence determination is difficult due to the hydrophobic stretches and DNA sequences are not sufficient on their own since proteins are often chemically modified. X-ray diffraction methods will fail when adequate crystals can not be prepared. NMR is not easily used with big molecules >20,000 daltons. Finally function is difficult to describe when the process of isolation could be critical (see Changeux in this volume). In this event and in order to uncover principles we must start by referring to a very simple protein. Here we choose cytochrome c, which performs a function on the surface of the mitochondrial membrane (1). It also has to move from one membrane to another, i.e. from the inner to the outer mitochondrial membrane, and along the inner membrane. After examining its properties we shall look at a second protein glycophorin which illustrates some other features of proteins which span membranes.

CYTOCHROME C

The crystal structures of several cytochromes c are known. I shall assume that these structures are correct in detail in the crystals which have been studied. We can then observe that

(a) The fold is independent of many different kinds of amino-acid substitution. It is also independent of salt concentration in the crystal.

(b) The position of internal side-chains shows only small changes from molecule to molecule where there are two different molecules in one unit cell or from one crystal structure to another.

(c) The position of side-chains on the surface of the protein shows a great deal of variation. In effect the surface has no fixed "structure" in the sense that the fold has.

(d) The internal structure changes slightly on change of oxidation state. Since the function of cytochrome c is to carry electrons this has obvious importance.

(e) The dynamics of the molecule are very restricted internally.

The study of proteins in solution by NMR allows a comparison to be made with the crystal structure. It may be difficult for many readers to follow the NMR description without some guidance so that a brief aside is made first to NMR methods (see reference (2)).

NMR Methods

Figure 1 shows part of a proton NMR spectrum of cytochrome c over a very narrow energy range which covers absorption by methyl groups. It will be seen that below the spectrum on a stick diagram many of the lines in the spectrum are assigned to individual methyl groups from amino-acid side chains indicated by their sequence numbers. The first aim of NMR studies is this assignment of the peaks and usually this problem is very difficult even at 10,000 molecular weight. Reference (2) deals with the assignment methods. Now once the assignment has been made the spectrum is understood at the fingerprint level. We then know the way in which the fold of the protein has affected the positions and other properties of the lines, due to each amino acid, so that the fingerprint is of structural parameters. Using these data thoroughly it is possible to show that the crystal structure can be taken over into the solution state. NMR can then be used to follow changes in solution for all parts of the molecule when we can follow the effect of very many parameters such as changes in spectral position or intensity of a line with temperature, pH, salt, binding agents, and so on and with man-made manipulations in the manner in which the spectrum is recorded. The latter allows exchange processes, relaxation rates, NH/ND exchanges, internal parameters of structures associated with individual residues (coupling constants) or external interactions between residues (NOE effects) to be followed. While crystal structure studies give an impression of a rigid molecule NMR gives the impression that there are a variety of motions within proteins covering a huge time-scale range. In fact it is the very fact that the crystal constrains motion that allows X-ray diffraction to be done while it is the very fact that solutions allow tumbling motions that allow NMR spectral studies. The time dependent nature of spectroscopy makes it not so good for structural analysis but exceedingly useful for studies of motion while the absence of time dependence in diffraction allows a good description of structure but a poor one of general motion. It is essential to study both ground state structure and time dependence of the structure in order to understand proteins so that NMR and X-ray diffraction are complementary techniques. Structures done by NMR will remain blurred since the molecules we describe here are blurred by motion in solution. The more rigid the molecule the better the correlation between the methods but the more dull the molecule.

Returning to Fig.1, the individual shifts, coupling constants, nuclear Overhauser enhancements of relaxation times and so on of assigned lines can be related to structural parameters obtained from crystal structures. Doing this for cytochrome c̲ has shown that
(1) The overall fold is closely the same in solution and in the crystal
(2) The side-chain positions are only altered in certain regions near the N- and C-termini and especially in regions under the heme i.e. on the methionine-80 side.
(3) The surface can not be given a structure
(4) The change in structure on change of oxidation state in solution has points in common with that seen in crystals but it is found that it is relayed all over the molecule (this is expected for a non-rigid body)
(5) Amino-acid substitutions can affect parts of the protein far from the position of substitution
(6) The biggest differences between crystal and solution structures arises from close contact distortions in crystals.

In addition NMR methods show that there are regions of mobility and regions of much less mobility, Fig.2. Of the aromatic residues

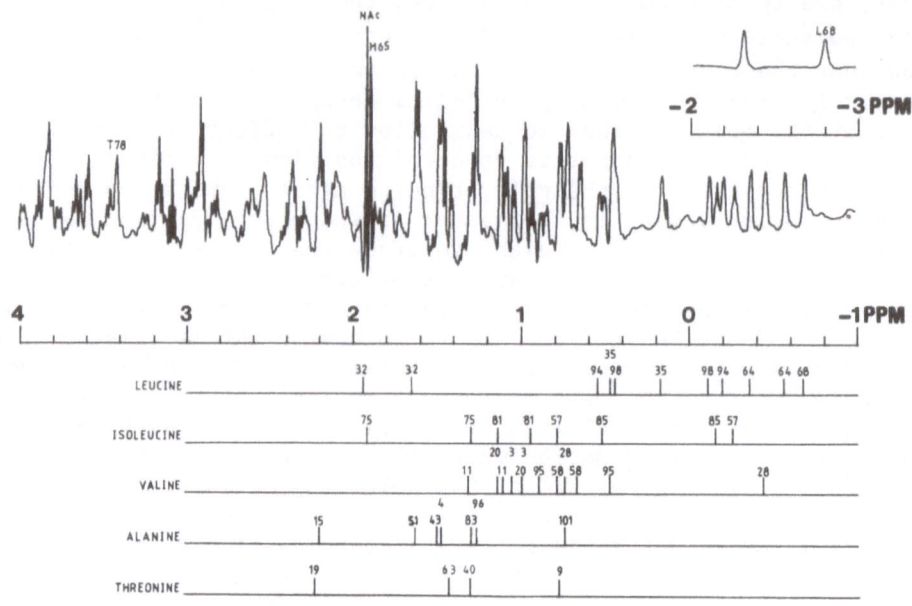

Fig. 1. Methyl region of the proton NMR spectrum of cytochrome c showing the assignments.

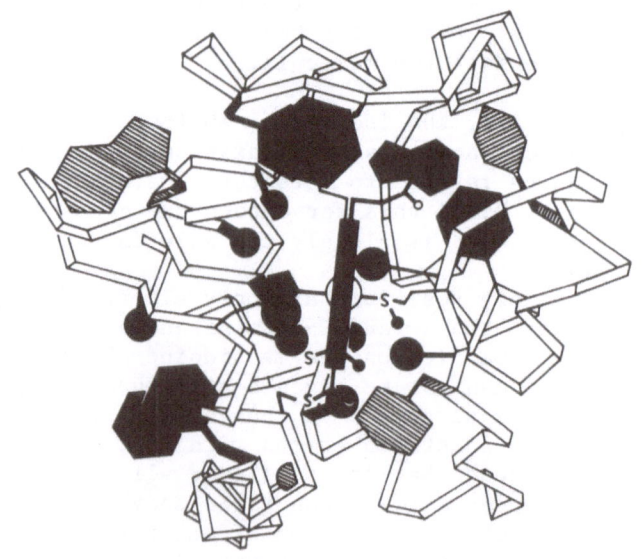

Fig. 2. Mobility map for cytochrome c. The filled-in regions are relatively immobile, the cross-hatched are the more mobile and there are three regions of intermediate mobility at the terminii and underneath the heme.

several flip rapidly, e.g. Phe-82 and Tyr-74, but other aromatic residues, 10, 46 and 48, flip slowly. Most valines and leucines flip very slowly e.g. around the heme. Many of these motions, when they occur, require considerable relaxation of the frame of the molecule which therefore must be breathing in some parts more than in others. Of the pieces of the protein those under the heme breath most extensively apart from regions close to the termini. This is in keeping with the observation that the methionine bound to the heme, Met-80, has peculiar properties in the Fe(III) state. The Met-80 to Fe(III) bond is easily broken. Again this rupture can be followed by NMR, magnetic susceptibility or even optical methods. Moreover we can relate this directly to a reaction since the Met-80 can be replaced on the minute timscale by cyanide. NMR can follow the protein through such changes. The relationship

     Structure - Mobility - Function

should now be obvious.

BINDING SURFACE OF CYTOCHROME-C

Let us now assume that cytochrome $\underline{c}$ is understood as far as its internal structure and internal motions are concerned. We are left with the description of the surface which can not be described by structure since in solution it is floppy and crystal structures are multiply degenerate. The inside was usefully described by one structure - the outside is an ensemble of states. We must reduce the level of demand on the word "structure" asking first what we need to know about the surface. Clearly the answer is we need to know where things bind. This is a question about the potential (free) energy wells which the surface generates for different kinds of molecules e.g. cations, anions, hydrophobic molecules. Using NMR we can generate potential energy maps by employing probe molecules. We will then have derived three series of maps for cytochrome $\underline{c}$ in solution.
   (1) Internal maps of structures of the Fe(III) and Fe(II) molecules mainly obtained from X-ray diffraction and confirmed by NMR
   (2) Internal maps of mobilities illustrating time scales from 10^{-13} sec down to 100sec mainly from NMR but located in the maps under (1)
   (3) Surface maps showing binding energies all over the surface for spherical negative ions, spherical positive ions and spherical molecules such as $[Cr(AcAc)_3]^0$, again from NMR but located on maps through (1).
The maps of binding under (3) have been generated for the first time for cytochrome $\underline{c}$, Fig.3. We find six sites for $[Fe(CN)_6]^{3-}$. No wonder the kinetics of electron transfer reactions are a puzzle.

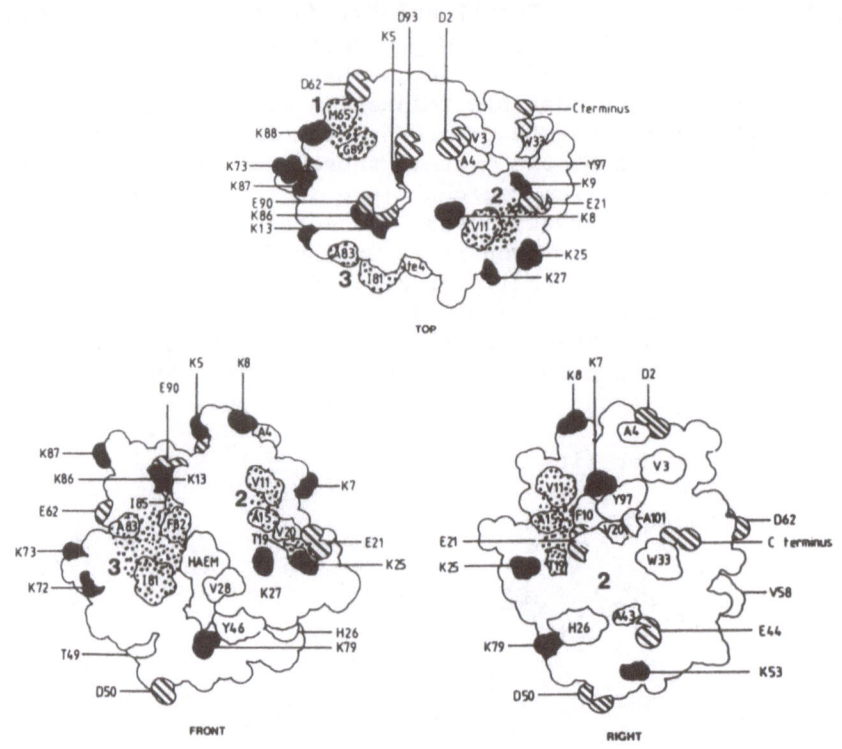

Fig. 3.   Maps of the binding locations (dotted regions) for
$[Cr(CN)_6]^{3-}$ or $[Fe(CN)_6]^{3-}$ on the surface of cytochrome c.
Other points designated are acidic and basic side-chains.

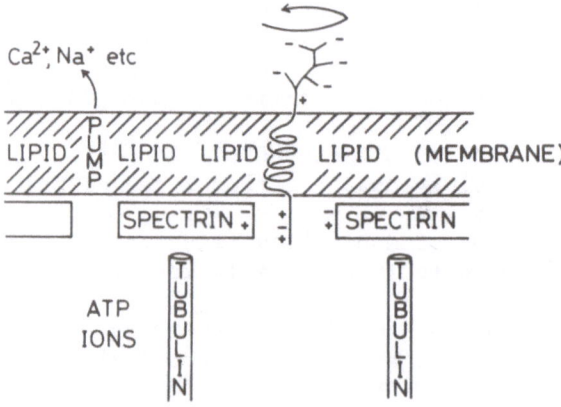

Fig.4.   An outline picture for glycophorin, shown by the helix
cross-membrane linkage, and its connection to the cell
apparatus. We consider that many trans-membrane proteins
work through an adjustment of the helices in the membrane.

THE FUNCTIONS OF CYTOCHROME C

Cytochrome c is an electron transfer protein. We need to understand electron transfer in the complexes which cytochrome c forms with its donor and acceptor partners. Secondly cytochrome c acts as an electron shuttle moving from centre to centre. We need to understand how cytochrome c moves and how it selects its binding partners.

Tackling binding first we have seen that there are many potential energy wells on the surface of cytochrome c so that it resembles a mosaic patch-work. Each patch can not be represented by a molecular structure since this hides its dynamic character. The binding is a statistical thermodynamic property not a static one. Selectivity now does not arise in the same way as the selectivity of small molecule binding to proteins. In the latter the small molecule fits into a hole provided by the protein, lock and key fit, and even if there is induced fit the major selective effects are generated by repulsion or loss of solvation. The positive interactions need not be selective to such a great degree. By way of contrast the fitting of mobile surfaces can not be like this but arises instead through the large number of contacts with attractive interactions. Where there are large numbers of contacts reciprocal matching of mosaic patches will give selectivity. Thus we believe that cytochrome c finds its protein partners through patch matching.

Patch matching of open surfaces has a great advantage over close fitting. It can be fast. Thus the on-rate of mobile bodies into partnership is much faster than the fitting of a die into a mould. Cytochrome c must bind rapidly so that good binding strength is possible concommitant with a fast off rate.

Now cytochrome c has many patches of the same charge type placed quite close together. Thus when it meets a surface of opposite charge type it can bind it in many ways e.g. $[Fe(CN)_6]^{3-}$ or negative phospholipids. It follows that cytochrome c can roll on a negatively charge surface diffusing from site to site without leaving the surface just as $[Fe(CN)_3]^{3-}$ can roll on the positively charged surface of cytochrome c. It can undergo two-dimensional diffusion on such a surface much as DNA-binding proteins (histones) can roll along DNA. Thus inspection of structure and mobility reveals how the electron transfer activity is limited in its expression by binding traps.

ELECTRON TRANSFER IN COMPLEXES

It is fundamental to electron transfer that if it is to be very fast then
(a) the distance between centres must be very short. This is

not possible for any protein such as cytochrome c since it has to use its surface to provide a <u>selective</u> binding region. Electron transfer rate is then limited in all biological systems but it is made optimally fast by (b) and (c).

(b) Little conformation change should occur with redox state change.

(c) Since some conformation change must occur on redox state change, the iron changes its properties, the redox centre must have easy mobility to allow relaxation to the required "intermediate" geometry of fast electron transfer.

We have seen how very well cytochrome c generates the necessary features described under (b) and (c) given (a). In essence we can understand this protein. There is small internal mobility and large external mobility which are respectively the requirements for fast electron transfer in complexes and fast complex formation or diffusion on a surface.

Now we have seen how one protein can function while connected to the surface of a membrane we need to make one or two further observations before we consider a protein which penetrates a membrane and acts as a communicating unit between two aqueous phases. It is important to recognise that the surface of a protein can behave in a quite distinct way so that as well as forming the kind of association seen for cytochrome c it can form a tightly bound unit of very specific structure. Biological organisation requires both this type of association and dynamic mosaic patch matching since local recognition with tight binding gives a firm framework structure while more general use of a surface and loose binding give the mobile units of a machine. Firm protein/protein binding is illustrated next with cytochrome c as an antigenic protein.

ANTIGENIC SURFACES

We have described elsewhere (3) our reasons for supposing that the antigenic properties of proteins are associated with the more mobile segments. In the case of cytochrome c we pointed mainly to the parts of the protein around Ile 57 and Tyr 74 a region which has a fast flipping ring, has a varying NOE distance between groups, differs from crystal to solution, and has a temperature dependent conformation. We have found since that the N- and C-terminii are both mobile and antigenic. The idea then is that the antibody seeks to find the more structured regions of a protein immediately underlying the surface by distorting a protein surface structure. This type of binding is intrinsically of higher selectivity locally than that described above and in order to function must be of very considerable binding strength. Both on and off rate constants are likely to be slow. The antigenic binding properties are then totally different from the biological mobile assembly properties both in the nature of the side-chains used and the rates of exchange. Recently further data on viral proteins have confirmed our views (4).

MEMBRANE PROTEINS

We may not have good information about most membrane proteins for some time and perhaps we shall have to be satisfied with blurred pictures. Maybe blurred pictures are of the essence of many such proteins. I arrive at these positional statements in the following way. There is good evidence that the proteins which span membranes do so through the formation of helical stretches. The helical stretches give rise to the communication network between the inside and the outside of the cell whether this communication is by mechanical or chemical transduction. All such tranductions require motion of a much more considerable kind than that seen in cytochrome c. The motion is likely to be the twisting and sliding of the helices relative to one another and relative to the lipids. I have shown that in water it is the helical portions of relatively loose proteins which allow transduction e.g. in haemoglobin and calmodulin Table I. Now the membrane proteins are exposed to three phases, inside aqueous, outside aqueous and membrane lipid. What forces could maintain such molecules in rigid positions over periods of time given that they must relax during activity? It seems most likely that the assembly is again based on dynamic mosaic patch matching so that function can be rapidly expressed. Glycophorin is an example (5).

We have described glycophorin as follows, Fig.4. The protein portion in the outside aqueous phase is a heavily glycosylated pep-tide chain carrying many negative charges. Its function is not known but it will occupy a large volume due to electrostatic repul-sion. The volume occupied will be constrained by the fact that it is held in the membrane so that the outer negatively charged unit is pulling on the transmembrane section of the protein and then on the portion in the aqueous phase. Any adjustment of the outer peptide then gives a change of force which is transmitted by the trans- mem-brane helical rod portion to the inside. The inside portion then relays the message to the network of cellular proteins. If this description is true then glycophorin, like many another trans-membrane protein, will act as a signalling device. It will signal the interaction of the spread-out, outside,negative probe portion with changing mechanical and electrical fields and it can even signal a change in specific chemical interactions. The essence of the device is the relative position of the helical transmembrane section to the surfaces of the membrane. It is like a float on the end of a fishing line. A very similar description applies to recep-tors such as the epidermal growth factor receptor (6).

Table I. Proteins using Helix/Helix Changes

Protein	Device
Calmodulin, Troponin	Transduction of Ca^{2+} current to a mechanical relay
Haemoglobin	Allosteric communication between subunits
Insulin	Small helix changes on binding
Cro-repressor	Movement and Binding to DNA
Proton ATP-synthase	Transduction of H^{+} current to chemical reaction
Acetylcholine Receptor	Transduction of transmitter (hormone) binding to gating
Ion Channels	Electric Field switch of helical dipoles (Edmonds gating)
Kinases (and P-450?)	Helix/helix re-orientation to give hinge-bending and reaction site protection
Glycophorin	Motion of a single helix up and down through a membrane (entropic signal)
Growth Factor Receptor	Motion of a single helix up and down through a membrane (chemically induced external conformation change).

N.B. Helix/helix motions may well be general to contractile devices as well as signalling and energy transduction. The general idea is that an energy input to a very local region can be transmitted through relative rotation of pairs or larger numbers of helices to generate mechanical changes at a distance or to open and shut channels. The energy input can be from light, ion or other binding or chemical reaction and can be made reversible or irreversible as required.

Now if we put together in a more complex unit several such helices all of which can slip up and down then we shall have created a central channel (of any size we wish) which can be modulated by the relative motion of the helices. A little thought allows the construction of models for the passage of ions, or whatever molecules one wishes, upon the receipt of a message more strongly linked to one then the others of the bundle of helices.

It would be quite wrong to presume that all membrane proteins have a blurred structure of course. If we consider those parts of the organisation which are involved in electron transfer only then they should be relatively rigid, like cytochrome <u>c</u>. Again the parts of a membrane protein which are involved in enzyme action and have a high specificity for small molecules must be relativly rigid. It is the transducing apparatus which must have the greatest mobility.

SUMMARY

This paper has set out to show that the very nature of membrane proteins as transducers may make their study extremely difficult. A transducing device must respond rapidly. Rapid response demands that the ground state lies close in energy to many other states so that the protein concerned has a time-fluctuating structure. Getting a picture of one structure (by X-ray diffraction) helps to picture the other related structures by NMR even if the picture is of low resolution but we shall have to be very careful with the interpretation since some important structure could be a considerably displaced form of the molecule. It should be remembered that proteins which are carriers or enzymes have to be relatively rigid to keep the specificity of action over long periods of time but that other proteins such as calmodulins, histones, glycophorin, glucagon and so on have a very different lability in that they control and manipulate mobile organisation. Such proteins must have high mobility – and this is now known to be so. In water the high mobility is generated by helical proteins with a high content of charged amino-acids. In membranes mobility will arise from helical proteins of a high hydrophobic content. By putting together helical membrane segments which can move laterally relative to one another (with simultaneous rotation) and helical aqueous segments on both sides of the membrane a complex connection between events external and internal to the cell can be generated. Because of the lability of the proteins a great variety of states will be possible. In such a way highly refined localised signalling is generated. The signalling can be purely mechanical as for glycophorin or a more complex motion giving energy transduction (ATP-synthetase) or ion conduction paths, the acetylcholine receptor channel.

REFERENCES

1. G.R. Moore, Z-X. Huang, C.G.S. Eley, H.A. Barker, G. Williams, M.N. Robinson and R.J.P. Williams, Electron Transfer in Biology, Faraday Disc. Chem. Soc., 74:311 (1982).

2. G.R. Moore, R.G. Ratcliffe and R.J.P. Williams, NMR and the Biochemist, Essays in Biochemistry, 19:143 (1983).

3. G.R. Moore and R.J.P. Williams, Antigenic Differences amongst Cytochromes c, Europ. J. Biochem., 103:543 (1980).

4. E. Westhof, D. Altschuh, D. Moras, A.C. Bloomer, A. Mondragon, A. Klug and M.H.V. Van Regenmortel, Correlation between Segmental Mobility and Antigenic Properties, Nature 311:123 (1984).

5. N.R. Egmond, R.J.P. Williams, E.J. Webb and D.A. Rees, NMR Studies of Glycophorin, Europ. J Biochem., 97:73 (1979).

6. T. Hunter, The Epidermal Growth Factor Receptor Gene and its Product, Nature 31:414 (1984).

28

PHOSPHATIDYLCHOLINE TRANSFER PROTEIN:

MODEL FOR LIPID-PROTEIN INTERACTIONS AND PROBE FOR MEMBRANE RESEARCH

Karel W.A. Wirtz, Anton J.W.G. Visser[*],
Jos A.F. Op den Kamp, Ben Roelofsen and
L.L.M. van Deenen

Laboratory of Biochemistry
State University of Utrecht
Utrecht, The Netherlands

[*]Department of Biochemistry
 Agricultural University
 Wageningen, The Netherlands

INTRODUCTION

Nature has provided for proteins that enable the constituent lipids to leave the physical constraints of the membrane structure. By their mode of action, these proteins facilitate *in vitro* the transfer of monomer lipid molecules between membranes. Lipid transferring proteins have been identified for phospholipids (for reviews, see Refs. 1 and 2), cholesterol[3] and glycolipids[4,5]. The physiological role of these proteins is still a matter of conjecture. In this regard, it is of interest that the transfer of ganglioside G_{M2} between membranes is facilitated by a specific protein, whose actual function is to present this ganglioside to hexosaminidase A as a water-soluble activator/lipid complex[6].

To date, phospholipid transfer proteins with different specificities have been purified to homogeneity from mammalian tissues[7-14], plants[15-16], and yeast[17]. In this paper, we will discuss some recent studies on the phosphatidylcholine-specific transfer protein (PC-TP)[†] and its application in membrane research.

[†]Abbreviations: PC-TP, phosphatidylcholine transfer protein; PC, phosphatidylcholine; PA, phosphatidic acid; PE, phosphatidyl-ethanolamine; 1-PnA-PC, 1-parinaroyl-2-palmitoyl-*sn*-glycero-3-phosphocholine; 2-PnA-PC, 1-acyl-2-parinaroyl-*sn*-glycero-3-phospho-choline; diPnA-PC, 1,2-diparinaroyl-*sn*-glycero-3-phosphocholine.

One of the major sources of PC-TP is the cytosolic fraction from bovine liver. A 4200-fold purification yields a homogeneous protein which, stored in 50% glycerol at -10°C, is stable for years[18]. Upon purification this protein contains one molecule of non-covalently bound PC, which can be exchanged for PC from vesicles when PC-TP interacts with these vesicles[19]. This exchange reaction at the vesicle interface forms the basis for PC-TP acting as a specific carrier of PC between membranes, as shown in experiments with phospholipid monolayers[20]. Release of PC into the interface does not appear to be coupled to binding of PC from that interface, as PC-TP also mediates a net transfer of PC to interfaces devoid of PC. Net transfer as part of a continuous process has been detected by applying electron spin resonance spectroscopy to a mixture of donor vesicles consisting of spin-labeled PC/PA (75:25, mol %) and acceptor vesicles of PE/PA (81:19, mol %)[21]. In the absence of PC-TP, the spin-spin exchange-broadened spectrum (spectrum B, Fig. 1), did not change with time, indicating a lack of spontaneous redistribution of spin-labeled PC between the vesicles. Addition of PC-TP, however, gave rise to the time-dependent appearance of a three-line spectrum superimposed on the exchange-broadened spectrum

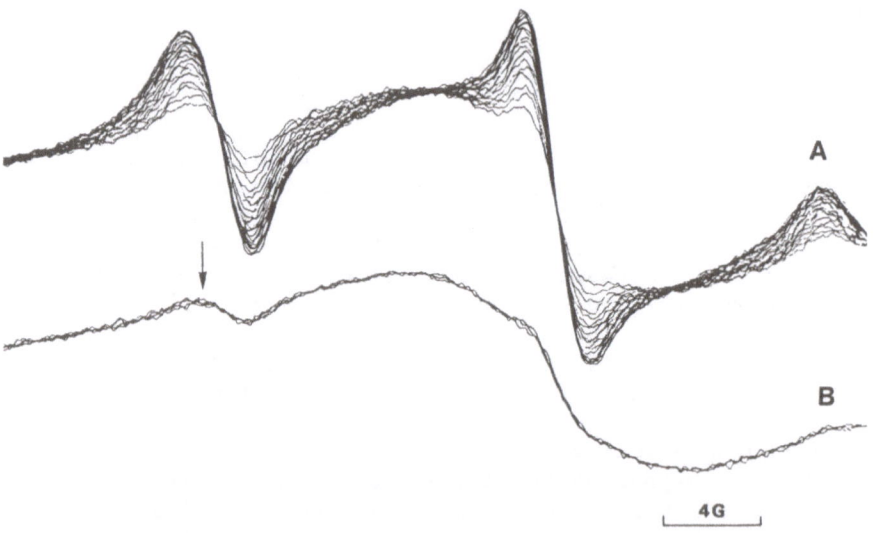

Fig. 1. Changes in the electron spin resonance spectra of spin-labeled PC vesicles on addition of non-labeled vesicles and PC-TP. Spectra A: Spin-labeled vesicles (6.3 nmol spin-labeled PC-PA, 75:25, mol %) were mixed with non-labeled vesicles (57 nmol PE-PA, 81:19, mol %) and PC-TP (15 μg); Spectra B: as in A, without PC-TP. For details see Ref. 21.

(spectrum A, Fig. 1). This three-lined spectrum resulted from the PC-TP—catalyzed insertion of spin-labeled PC from the donor into the acceptor vesicles. These results are consistent with results from similar experiments where the PC-TP—mediated net transfer of fluorescently labeled PC from quenched donor vesicles to acceptor vesicles composed of either PA or PE/PA (80:20, mol %), was measured[22].

The specificity of PC-TP for PC suggests that the protein exposes at the membrane interface a recognition site that interacts with the phosphorylcholine head group. This interaction has been explored in detail by measuring the transfer of PC-analogues modified in the polar head group[19]. The experiments demonstrated that transfer is inhibited when (i) the distance between phosphate moiety and quaternary nitrogen is altered, and (ii) a methyl group on the quaternary nitrogen is removed or substituted by an ethyl or propyl group. The recognition site has been further explored by treating PC-TP with the α-carbonyl reagents 2,3-butanedione and phenylglyoxal, known to specifically modify Arg-residues[23]. Both reagents completely inactivated the transfer protein by specific rapid modification of three of the ten Arg-residues. In the presence of negatively charged vesicles, butanedione rapidly modified two Arg-residues without a significant loss of activity (Fig. 2). This strongly suggests that the vesicle interface protects one Arg-

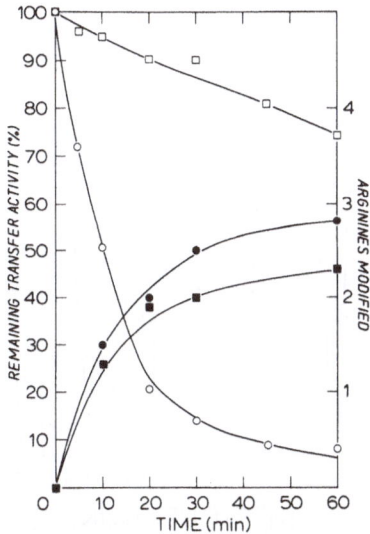

Fig. 2. Inactivation of PC-TP by 2,3-butanedione (30 mM) in the presence (□) and absence (○) of vesicles consisting of PC-PA (80:20, mol %), and the corresponding number of Arg residues modified (■,●). For details, see Ref. 23.

residue against modification essential for transfer activity. This Arg-residue may be part of the recognition site where it interacts with the phosphate moiety of the phosphorylcholine head group. Similar kind of interactions have been shown to be essential for those enzymes acting on phosphate-containing substrates[24-27].

In addition to a recognition site, PC-TP has a lipid binding site that accommodates the PC-molecule. At present it is not known how these two site cooperate in order to enable the protein to extract and bind a PC-molecule. As a first step in gaining knowledge on these sites, Moonen et $al.$[28] and Akeroyd et $al.$[29] elucidated the primary structure. The protein consists of a single polypeptide chain of 213 amino acid residues and has N-acetyl methionine as a blocked N-terminus and threonine as the C-terminus. It contains two disulfide bridges at Cys^{17}-Cys^{63} and Cys^{93}-Cys^{207}. Examination of the structure reveals the presence of very hydrophobic peptides of which the most prominent are Val^{98}-Val-Tyr-Trp-Gln-Val^{103}, Val^{171}-Phe-Met-Tyr-Tyr-Phe^{176} and Trp^{186}-Val-Ile-Asn-Trp-Ala-Ala^{192}. A more systematic analysis by the method of Rose[30] gave the distribution of hydrophobicity along the chain (Fig. 3). In particular, the location of the maxima in the C-terminal half corresponding with the peptides Val^{171}-Phe^{176} and Trp^{186}-Ala^{192}, is striking where these maxima are alternated by minima of zero hydrophobicity. In addition to the hydrophobicity profile, the α-helix, β-strand and β-turn

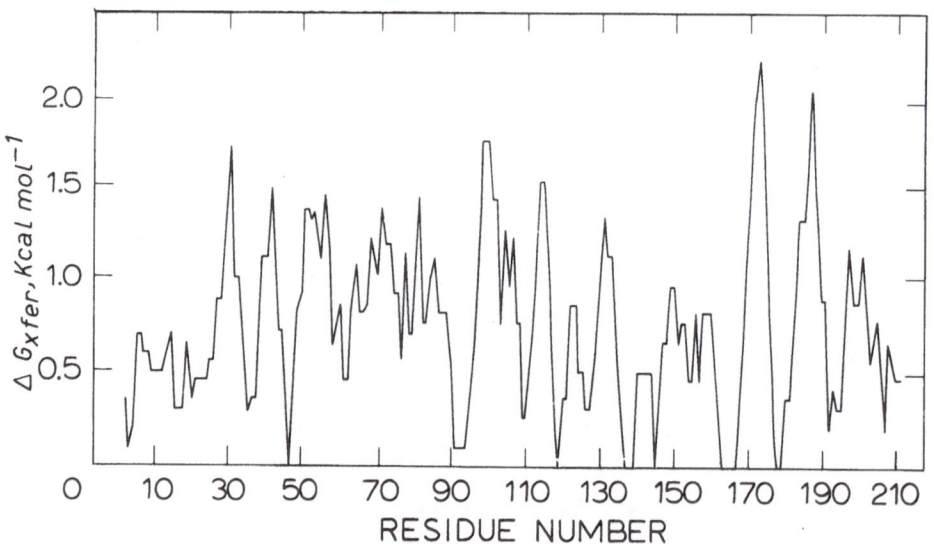

Fig. 3. Smoothed hydrophobicity profile for PC-TP. ΔG_{xfer} indicates the free energy of transfer from water into organic solvents. The profile was smoothed by taking a five-point moving average along the polypeptide chain. For details, see Ref. 31.

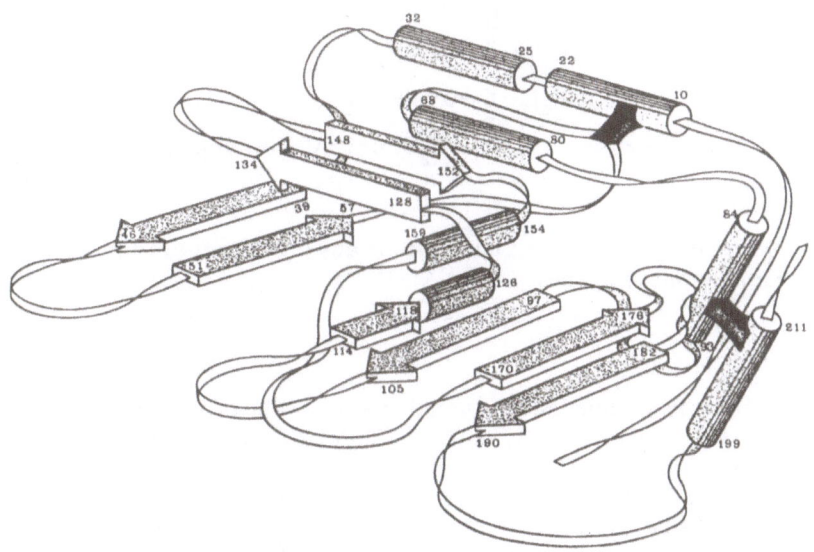

Fig. 4. A tentative folding model of PC-TP. Cylinders indicate
α-helix and arrows β-strand. For details, see Refs. 31 and 36.

structures were also derived from the primary structure. Details
of these preductions have been recently presented[31]. Based on
these predictive methods, a speculative folding model of PC-TP was
developed (Fig. 4). This model has taken some very recent concepts into consideration, where the apolar core is formed by anti-
parallel β-sheets shielded by α-helical structures at the surface
of the protein[32-35]. Inasmuch as reduction of the disulfide bonds
abolishes transfer activity it is important to note that the bond
at Cys^{17}-Cys^{63} restricts the arrangement of the secondary structure
at the N-terminus; the bond at Cys^{93}-Cys^{207} may be important for
the alignment of the hydrophobic and amphiphilic β-strand peptide
Thr^{97}-Tyr^{105} close to the anti-parallel β-sheet (residues 170-176/
182-190). For a more detailed discussion of the model, see Ref.
36.

Lipid Binding Site

    The binding site of PC-TP has been investigated with static
and time-resolved fluorescence spectroscopy by use of PC containing
cis-parinaric acid (1-PnA-PC, 2-PnA-PC, diPnA-PC)[37]. This fluorescent polyene probe (see Fig. 5) has very attractive molecular and
spectroscopic features for studying lipid-protein interactions[38-40].
One of these features is that vesicles prepared of pure PnA-PC do
not fluoresce as a result of self-quenching. Titration of these
vesicles with PC-TP gives rise to an increase of fluorescence in-

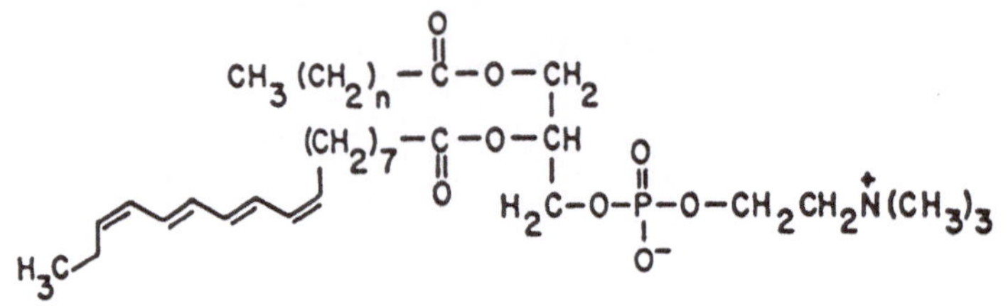

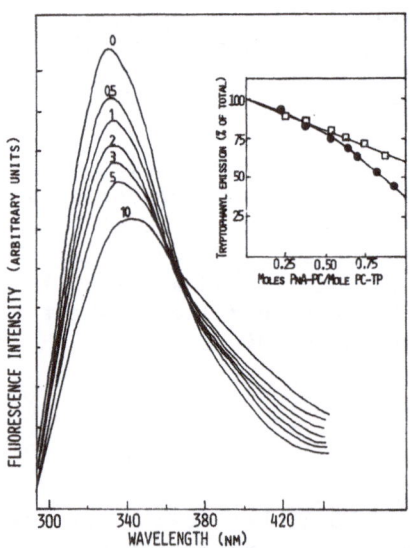

Fig. 5. The chemical structure of 1-acyl-2-*cis*-parinaroyl-*sn*-glycero-3-phosphocholine.

tensity, indicating that self-quenching was abolished due to binding of PnA-PC to the protein. By this criterium, both 1-PnA-PC, 2-PnA-PC and diPnA-PC were shown to bind to PC-TP. It is of great interest that diPnA-PC in PC-TP has double the fluorescence intensity of 1- and 2-PnA-PC. This indicates that self-quenching of diPnA-PC on PC-TP does not occur. Self-quenching due to an interaction of the *sn*-1- and *sn*-2-parinaroyl chain was observed when diPnA-PC was present in an excess egg PC-vesicles.

Fig. 6. Fluorescence emission spectrum of PC-TP in the presence of 2-PnA-PC. The PC-TP concentration was 0.5×10^{-6} M. Excitation was at 255 nm. The numbers refer to the nanomoles of vesicle 2-PnA-PC added.
(Insert) Moles of PnA-PC/PC-TP complex formed are plotted against the residual tryptophanyl fluorescence. 2-PnA-PC (□); diPnA-PC (●). For details, see Ref. 37.

Binding of the PnA-PC derivatives to PC-TP was also determined by monitoring the endogenous tryptophanyl fluorescence spectrum. Incorporation of 2-PnA-PC into PC-TP resulted in a decrease of tryptophanyl fluorescence intensity (Fig. 6). This reflects a radiation-less energy transfer between tryptophan residues and PnA-PC bound to PC-TP. From the plot of tryptophanyl fluorescence intensity against the moles of 2-PnA-PC per mole of PC-TP (see insert to Fig. 6), it appears that there is a linear relationship between tryptophanyl fluorescence quenching and the concentration of 2-PnA-PC on the protein. At the theoretical 1 mole of PnA-PC per mole of PC-TP, quenching would amount to 40%. Addition of 1-PnA-PC had essentially the same effect on the tryptophanyl fluorescence. Quenching is not additive as binding of diPnA-PC to PC-TP resulted in a maximal quenching of about 60% (see insert to Fig. 6).

The absorption spectrum of 2-PnA-PC bound to PC-TP is shown in Fig. 7. For comparison, we have also included the spectrum of 2-PnA-PC in ethanol. It is evident that the maxima of the vibronic bands of the bound chromophore are slightly red-shifted. The spectra of 1- and diPnA-PC bound to PC-TP displayed identical characteristics. The dynamic fluorescence behaviour of PnA-PC in PC-TP was analyzed by determining the fluorescence anisotropy decay. To this end, the initial anisotropy was generated by exciting the parinaroyl-fluorophore with polarized laser pulses at 305 nm (for details of the procedure, see Ref. 41). As a representative

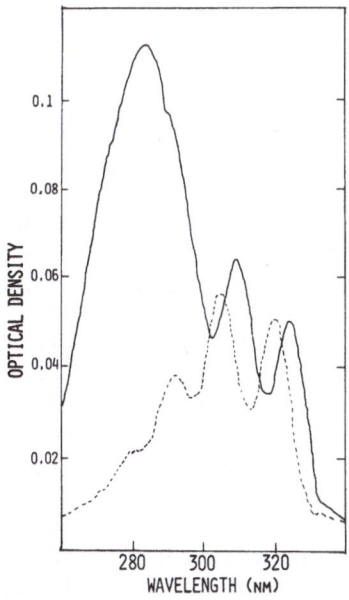

Fig. 7. Absorption spectrum of 2-PnA-PC bound to PC-TP (——) and in ethanol (---). For details, see Ref. 37.

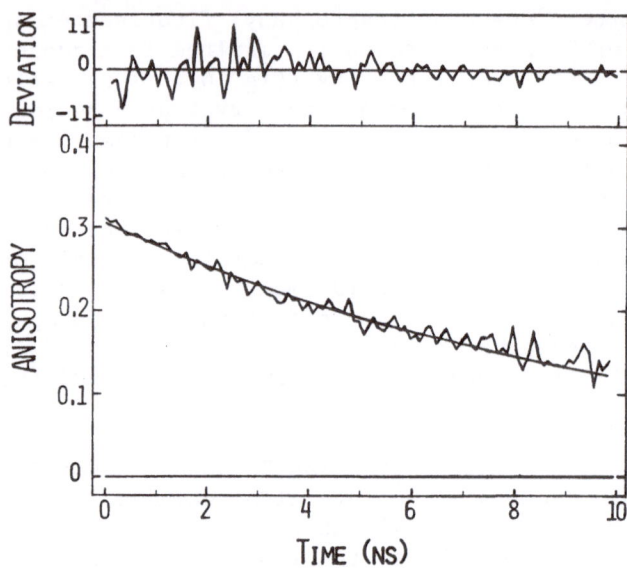

Fig. 8. Fluorescence anisotropy decay of 2-PnA-PC bound to PC-TP.
The experimental anisotropy r(t) is represented by the noisy curve;
the smooth line is the fitted decay function. For details, see
Refs. 37 and 42.

example the anisotropy decay pattern of 2-PnA-PC bound to PC-TP is
shown in Fig. 8. This decay can be fitted to a single exponential
function in the time range of observation (*i.e.*, 10 ns) and yields
a rotational correlation time of 11 ns. A similar behaviour was
observed for 1-PnA-PC and diPnA-PC in PC-TP yielding rotational
correlation times of 26 ns and 15 ns, respectively. The different
rotational correlation times lead to the following conclusions:
  i) independent of the shape of the protein, the 1- and 2-parinaroyl
     chains cannot be aligned in one binding site;
 ii) the correlation time of 11 ns for the 2-parinaroyl chain is
     characteristic for that of PC-TP (M.W. 25,000) assuming the
     protein is spherical;
iii) the correlation time of 26 ns for the 1-parinaroyl chain indi-
     cates that PC-TP is not spherical but elongated.
The model that fits the data, is given in Fig. 9. The long correlation time of 26 ns for 1-PnA-PC strongly suggests that the linear
polyene is oriented parallel to the main symmetry axis (for theoretical details, see Ref. 42). The short correlation time of 11 ns for
2-PnA-PC implies that the fluorophore is more or less parallel to
the short semi-axis and thus perpendicular to the other acyl chains.
In support of this arrangement, a correlation time of 15 ns was
observed for diPnA-PC, where this time agrees very well with the
harmonic mean of the correlation times of 1- and 2-PnA-PC. This
strongly suggests that the two chromophores are independently photo-

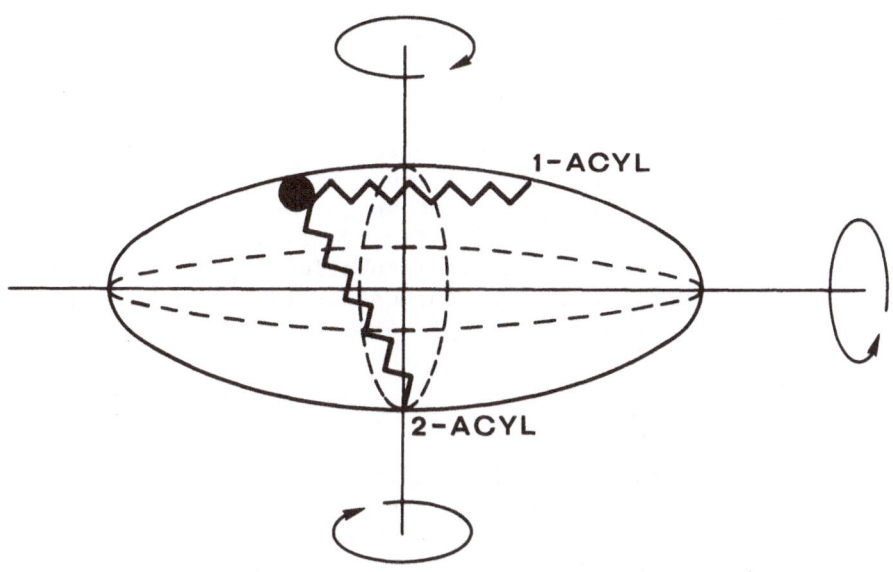

Fig. 9. Tentative orientation of the 1-acyl- and 2-acyl chain of PC bound to PC-TP.

selected and behave as isolated probes.

The organization of the lipid binding site has been partially elucidated by use of PC carrying photolabile 4-azido-2-nitrophenoxy- or m-diazirinophenoxy moieties on the *sn*-2-fatty acyl chain[43,44]. This photolabeling approach has provided strong evidence that the 2-acyl chain is alligned along the peptide Val^{171}-Phe^{176} (see Fig. 4). In addition to this peptide, an extensive coupling to Tyr^{54} was observed. Arguments against this residue being part of the actual binding site, have been put forward[44].

Applications in Membrane Research

In the transfer of PC between membranes PC-TP exchanges PC present in the outer leaflet without perturbing the membrane structure. This has made PC-TP a valuable tool in establishing the size of the PC-pool in the outer leaflet and the rate at which PC-TP equilibrates between the outer and inner leaflet. So it has been demonstrated that approximately 70% of PC in single bilayer vesicles is present in the outer leaflet (*i.e.*, available for exchange), and that the estimated half-life of transbilayer movement is in excess of 4-11 days[45-47]. Incorporation of glycophorin into PC-vesicles greatly enhanced the rate of transbilayer movement[48]. It can be seen from Fig. 10, that 67% of PC was exchangeable in the absence of glycophorin, whereas 90% was exchangeable in the presence of this protein. It was estimated that the transbilayer movement in glycophorin-containing vesicles had a half-time of less than 1 hr.

In recent studies, PC-TP has been used to gain knowledge on the organization of PC in intact erythrocytes[49]. In these experiments erythrocytes were incubated with rat liver microsomes containing [^{14}C]PC. The amount of erythrocyte PC which had exchange with microsomal PC ($C_{e,t}$), was calculated from determining the radio-activity in the microsomal PC before incubation ($R_{e,o}$), the radio-activity in the erythrocytes after incubation ($R_{e,t}$) and the absolute amount of microsomal PC (C_m). Then the following expression holds:

$$\frac{R_{m,o} - R_{e,t}}{C_m} = \frac{R_{e,t}}{C_{e,t}}$$

where the left term represents the specific activity of PC in the microsomes after the incubation, and the right term the specific activity of the pool of erythrocyte PC exchanged. In general, the amount of erythrocyte PC exchanged ($C_{e,t}$), is expressed as a per-centage of the total amount. The time dependence of PC-TP—mediated exchange of PC between microsomes and intact human and rat ery-throcytes is presented in Fig. 11. After 2 hrs of incubation, a maximum was reached where about 75% of PC in the human erythrocyte was exchanged (Fig. 11A). This pool represents the outer leaflet and does not appear to equilibrate PC in the inner leaflet. In rat

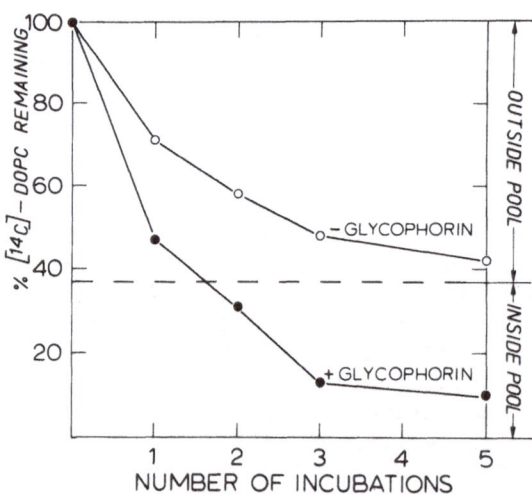

Fig. 10. PC-TP—mediated transfer of [^{14}C]dioleoyl-PC from uni-lamellar vesicles to multilamellar liposomes. Vesicles contained [^{14}C]PC with (●) and without (○) glycophorin. For details see Ref. 48.

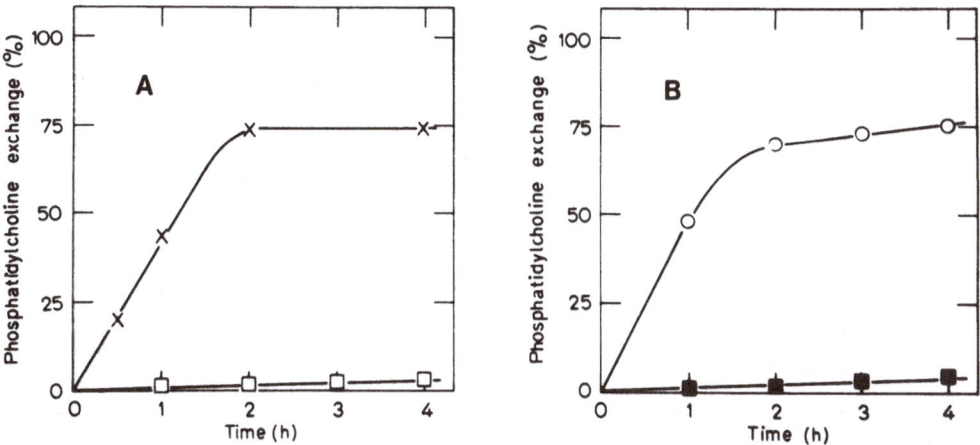

Fig. 11. Exchange of PC between erythrocytes and rat liver microsomes. The amount of PC exchanged in the absence (□,■) and presence (x,○) of PC-TP is plotted as a percentage of total erythrocyte
PC. Human (A) and rat (B) erythrocytes. For details, see Ref. 49.

erythrocytes, PC behaves differently where one observes a fast-
exchangeable pool of 50–60% of the PC present (i.e., the outer
leaflet) and a slowly exchangeable pool (i.e., the inner leaflet)
(Fig. 11B). From this biphasic curve, it can be estimated that PC
equilibrates between inner and outer leaflet with a half-time of
7 hrs. These results are in excellent agreement with the data
obtained with phospholipases[50],[51].

Similar experiments have been performed with homozygous
reversible sickle cells, where these erythrocytes have a normal
discoid shape under room air but adopt the sickled shape under
nitrogen atmosphere[52]. The discoid cells behave as normal erythrocytes having 76% of the PC available for exchange. Interestingly,
incubation of the deoxygenated sickled cells results in an essentially complete exchange of the entire PC-pool. The half-time of
transbilayer movement of PC in these cells has been determined by
introducing [^{14}C]PC from rat liver microsomes into the outer leaflet
followed by measuring the appearance of this label in the inner
leaflet. To this end, erythrocytes labeled with [^{14}C]PC were incubated for different periods of time, treated with phospholipase A_2
and the specific radioactivity of [^{14}C]lysoPC determined. The
decrease of this specific radioactivity with time is a measure for
the rate of transbilayer movement of [^{14}C]PC. From Fig. 12A it can
be learned that the half-time of transbilayer movement is 14 hrs for

the oxygenated discoid cells and 3.5 hrs for the deoxygenated sickled cells. The increased transbilayer mobility of PC in the sickled cells is immediately restored to the normal low rate upon reoxygenation of the cells, indicating a complete reversibility of this phenomenon (Fig. 12B).

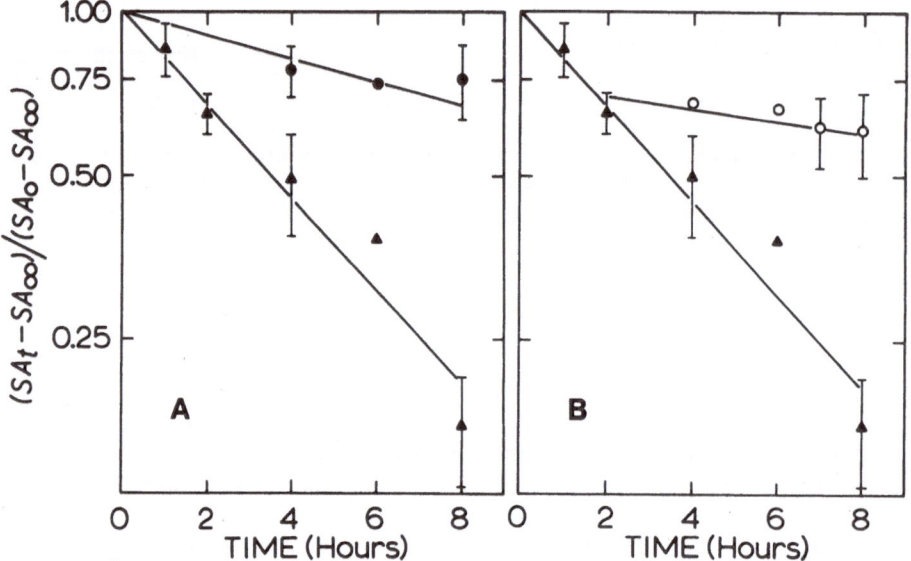

Fig. 12. Transbilayer movement of PC in oxygenated, deoxygenated and reoxygenated homozygous reversible sickle cells. Relative specific radioactivity (SA) of PC in the outer leaflet, calculated as $(SA_t-SA_\infty)/(SA_0-SA_\infty)$ is plotted semilogarithmically against time: A, oxygenated (●) and deoxygenated (▲) cells; B, deoxygenated (▲) and reoxygenated (○) cells. For details, see Ref. 52.

Inherent to the mode of action of PC-TP one has the possibility to replace the natural PC in biological membranes by PC of distinct fatty acid composition[53-56]. By this manipulation we have observed that, for example, the human erythrocytes tolerate only limited changes in the molecular species of PC[53,56]. Both an increase in the degree of saturation and unsaturation affected the permeability barrier finally leading to hemolysis.

REFERENCES

1. K.W.A. Wirtz, in: "Lipid-Protein Interactions" (Jost, P.C. and Griffith, O.H., Eds.), Vol. 1, pp. 151-233, Wiley-Interscience, New York (1982).
2. J.C. Kader, D. Douady and P. Mazliak, in: "Phospholipids" (Hawthorne, J.N. and Ansell, G.B., Eds.), pp. 279-311. Elsevier Biomedical Press, Amsterdam (1982).
3. J.M. Trzaskos and J.L. Gaylor, Biochim. Biophys. Acta 751, 52-65 (1983).
4. R.J. Metz and N.S. Radin, J. Biol. Chem. 257, 12901-12907 (1982).
5. A. Abe, K. Yamada and T. Sasaki, Biochem. Biophys. Res. Commun. 104, 1386-1393 (1982).
6. E. Conzelmann, J. Burg, G. Stephan and K. Sandhoff, Eur. J. Biochem. 123, 455-464 (1982).
7. H.H. Kamp, K.W.A. Wirtz and L.L.M. van Deenen, Biochim. Biophys. Acta 318, 313-325 (1973).
8. B.J.H.M. Poorthuis, T.P. van der Krift, T. Teerlink, R. Akeroyd, K.Y. Hostetler and K.W.A. Wirtz, Biochim. Biophys. Acta 600, 376-386 (1980).
9. G.M. Helmkamp, M.S. Harvey, K.W.A. Wirtz and L.L.M. van Deenen, J. Biol. Chem. 249, 6382-6389 (1974).
10. P.R. DiCorleto, J.B. Warach and D.B. Zilversmit, J. Biol. Chem. 254, 7795-7802 (1979).
11. R.J. Read and J.D. Funkhouser, Biochim. Biophys. Acta 752, 118-126 (1983).
12. B. Bloj and D.B. Zilversmit, J. Biol. Chem. 252, 1613-1619 (1977).
13. R.C. Crain and D.B. Zilversmit, Biochemistry 19, 1433-1439 (1980).
14. E.V. Dyatlovitskaya, N.G. Timofeeva and L.D. Bergelson, Eur. J. Biochem. 82, 463-471 (1978).
15. D. Douady, M. Grosbois, F. Guerbette and J.C. Kader, Biochim. Biophys. Acta 710, 143-153 (1982).
16. J.C. Kader, M. Julienne and C. Vergnolle, Eur. J. Biochem. 139, 411-416 (1984).
17. G. Daum and F. Paltauf, Biochim. Biophys. Acta 794, 385-391 (1984).
18. J. Westerman, H.H. Kamp and K.W.A. Wirtz, Methods Enzymol. 98, 581-586 (1983).
19. H.H. Kamp, K.W.A. Wirtz, P.R. Baer, A.J. Slotboom, A.F. Rosenthal, F. Paltauf and L.L.M. van Deenen, Biochemistry 16, 1310-1316 (1977).
20. R.A. Demel, K.W.A. Wirtz, H.H. Kamp, W.S.M. Geurts van Kessel and L.L.M. van Deenen, Nature New Biol. 246, 102-105 (1973).
21. K.W.A. Wirtz, P.F. Devaux and A. Bienvenue, Biochemistry 19, 3395-3399 (1980).
22. J.W. Nichols and R.E. Pagano, J. Biol. Chem. 258, 5368-5371 (1983).

23. R. Akeroyd, L.G. Lange, J. Westerman and K.W.A. Wirtz, _Eur. J. Biochem._ 121, 77–81 (1981).
24. L.G. Lange, J.F. Riordan and B.L. Vallee, _Biochemistry_ 13, 4361–4370 (1974).
25. F.J.M. Daemen and J.F. Riordan, _Biochemistry_ 13, 2865–2871 (1974).
26. E. Morkin, I.L. Flink and S.K. Naberjee, _J. Biol. Chem._ 254, 12647–12652 (1979).
27. J.J. Schrijen, W.A.H.M. Luyben, J.J.H.H.M. De Pont and S.L. Bonting, _Biochim. Biophys. Acta_ 597, 331–344 (1980).
28. P. Moonen, R. Akeroyd, J. Westerman, W.C. Puijk, P. Smits and K.W.A. Wirtz, _Eur. J. Biochem._ 106, 279–290 (1980).
29. R. Akeroyd, P. Moonen, W.C. Puijk and K.W.A. Wirtz, _Eur. J. Biochem._ 114, 385–391 (1981).
30. G.D. Rose, _Nature_ 272, 586–590 (1978).
31. R. Akeroyd, J.A. Lenstra, J. Westerman, G. Vriend, K.W.A. Wirtz and L.L.M. van Deenen, _Eur. J. Biochem._ 121, 391–394 (1982).
32. M.J.R. Sternberg and J.M. Thornton, _Nature_ 271, 15–20 (1978).
33. G.E. Schulz and R.H. Schirmer, _in_: "Principles of Protein Structure" (Cantor, C.R., Ed.), pp. 66–165, Springer-Verlag, Heidelberg (1979).
34. M.I. Kanehisa and T.Y. Tsong, _Biopolymers_ 19, 1617–1628 (1980).
35. G.D. Rose and S. Roy, _Proc. Natl. Acad. Sci. U.S.A._ 77, 4643–4647 (1980).
36. K.W.A. Wirtz, T. Teerlink and R. Akeroyd, _in_: "Enzymes of Biological Membranes" (Martonosi, A.N., Ed.), Vol. 2, pp. 111–138, Plenum Press, New York (1985).
37. T.A. Berkhout, A.J.W.G. Visser and K.W.A. Wirtz, _Biochemistry_ 23, 1505–1513 (1984).
38. L.A. Sklar, B.S. Hudson and R.D. Simoni, _Biochemistry_ 16, 5100–5108 (1977).
39. L.A. Sklar, B.S. Hudson, M. Peterson and J. Diamond, _Biochemistry_ 16, 813–819 (1977).
40. P.K. Wolber and B.S. Hudson, _Biochemistry_ 20, 2800–2810 (1981).
41. A. van Hoek, J. Vervoort and A.J.W.G. Visser, _J. Biochem. Biophys. Methods_ 7, 243–254 (1983).
42. A.J.W.G. Visser and K.W.A. Wirtz, _in_: "Excited State Probes in Biochemistry and Biology" (Szabo, A.G. and Mazotti, L., Eds.), Plenum Press, New York, in press (1985).
43. P. Moonen, H.P. Haagsman, L.L.M. van Deenen and K.W.A. Wirtz, _Eur. J. Biochem._ 99, 439–445 (1979).
44. J. Westerman, K.W.A. Wirtz, T. Berkhout, L.L.M. van Deenen, R. Radhakrishnan and H.G. Khorana, _Eur. J. Biochem._ 132, 441–449 (1983).
45. L.W. Johnson and D.B. Zilversmit, _Biochim. Biophys. Acta_ 375, 165–175 (1975).
46. J.E. Rothman and E.A. Dawidowicz, _Biochemistry_ 14, 2809–2816 (1975).
47. B. de Kruijff and K.W.A.Wirtz, _Biochim. Biophys. Acta_ 468,

318-326 (1977).

48. B. de Kruijff, E.J.J. van Zoelen and L.L.M. van Deenen, Bio-
     chim. Biophys. Acta 509, 537-542 (1978).

49. G. van Meer, B.J.H.M. Poorthuis, K.W.A. Wirtz, J.A.F. Op den
     Kamp and L.L.M. van Deenen, Eur. J. Biochem. 103, 283-288
     (1980).

50. A.J. Verkleij, R.F.A. Zwaal, B. Roelofsen, P. Comfurius, D.
     Kastelijn and L.L.M. van Deenen, Biochim. Biophys. Acta 323,
     178-193 (1973).

51. W. Renooij, L.M.G. van Golde, R.F.A. Zwaal and L.L.M. van
     Deenen, Eur. J. Biochem. 61, 53-58 (1976).

52. P.F.H. Franck, D.T.Y. Chiu, J.A.F. Op den Kamp, B. Lubin,
     L.L.M. van Deenen and B. Roelofsen, J. Biol. Chem. 258, 8435-
     8442 (1983).

53. L.G. Lange, G. van Meer, J.A.F. Op den Kamp and L.L.M. van
     Deenen, Eur. J. Biochem. 110, 115-121 (1980).

54. S.N. Mathur, I. Simon, B.R. Lokesh and A.A. Spector, Biochim.
     Biophys. Acta 751, 401-411 (1983).

55. P. North and S. Fleischer, Biochim. Biophys. Acta 772, 65-76
     (1984).

56. F.A. Kuypers, B. Roelofsen, J.A.F. Op den Kamp and L.L.M. van
     Deenen, Biochim. Biophys. Acta 769, 337-347 (1984).

PHOSPHATIDYLINOSITOL KINASE, A KEY ENZYME OF PHOSPHATIDYLINOSITOL METABOLISM: ITS ROLE IN AN INTRACELLULAR SECOND MESSENGER SYSTEM

Zafiroula Georgoussi and Ludwig M. G. Heilmeyer, Jr.

The National Hellenic Research Foundation, 48 Vas Constantinou Ave. 11635 Athens, Greece, and Institut für Physiologische Chemie, Lehrstuhl 1, Ruhr-Universität Bochum, Universitätsstr. 150, 4630 Bochum, West-Germany

It is now well established that many cell agonists, hormones or neurotransmitters called "first messengers" provoke intra-cellulary a concentration change of a "second messenger" like cyclic AMP, cyclic GMP or Ca^{2+} (for review: Dumont et al., 1981), (Fig. 1). An increase of these second messengers leads to saturation of specific receptor proteins which can be either regulatory proteins or domains of enzymes or in case of Ca^{2+}, specific Ca^{2+} binding proteins, e. g. calmodulin.

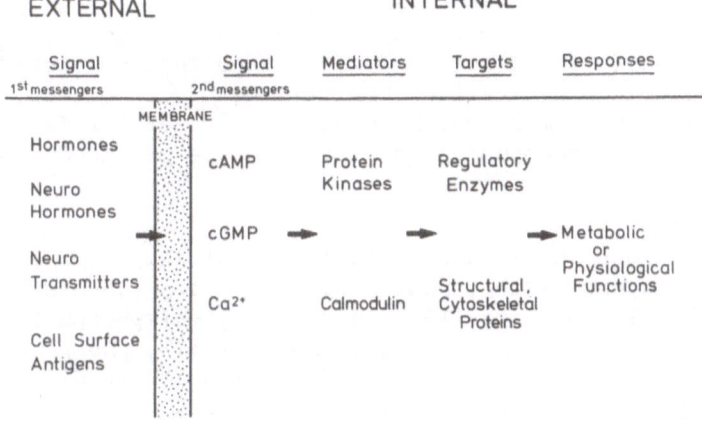

Figure 1: Control of Cellular Processes by Phosphorylation

This process triggers comformational changes which constitute a signal. It is transmitted to modifiers like protein kinases. They can be specific enzymes which phosphorylate target proteins at serine or threonine residues. Phosphorylation of these proteins changes their physicochemical properties and/or their enzymatic activities which results in an alteration of the physiological or metabolic state of a cell.

Originally the Hokins (1958) observed that upon binding to its specific receptors many agonists which very often employ Ca^{2+} as second messenger additionally increase phosphatidylinositol metabolism in cell membranes. Three metabolites of phosphatidylinositol have the characteristics of second messengers, namely: diacylglycerol (DG), inositoltrisphosphate (IP_3) and arachidonic acid (Berridge, 1984). The two first metabolites exhibits characteristic properties of classical second messengers such as cAMP. Their production is initiated upon specific agonist-receptor interaction. They are produced rapidly and act at very low concentrations; there exist specific reaction sequences for removing these messengers, once the external signal is withdrawn. From arachidonic acid substances like prostaglandins, thromboxanes and leucotrienes are synthesized which are considered intercellular rather than intracellular messengers.

Figure 2 shows the cyclic reaction sequence which is involved in the formation of these second messengers via the polyphosphoinositides and the resynthesis of phosphatidylinositol as well as the site of action of these messengers. Ligand-receptor interaction activates probably a phosphodiesterase that splits PIP_2 to IP_3 and DG (Michell, 1982). DG activates the Ca^{2+} and phospholipid dependent protein kinase C which can phosphorylate specific cytosolic proteins (Nishizuka, 1984; Michell, 1983). IP_3 is binding to membranes of the endoplasmic reticulum which causes the release of stored Ca^{2+} (Berridge, 1983; Streb et al., 1983). Thus, Ca^{2+} can be considered as a "third messenger". IP_3 is dephosphorylated and inactivated by phosphomonoesterases, first IP_2 is formed then IP and finally I. DG is inactivated by rephosphorylation to PA (Michell and Kirk, 1981; Berridge, 1981) and then converted to the cytosine nucleotide derivative (CDP-DG). Both metabolites, CDP-DG and I, combine to form PI which serves as a reservoir for the formation of these second messengers. They are produced from the

---

Abbreviations used: phosphatidylinositol (PI), phosphatidyl-inositol-4-phosphate (PIP), phosphatidylinositol-4,5-bisphosphate (PIP_2), inositol-phosphate (IP), inositol-1,4-bisphosphate (IP_2), inositol-1,4,5-trisphosphate (IP_3), inositol (I), phosphatidic acid (PA), diacylglycerol (DG), cytidine diphosphate diacyl-glycerol (CDP-DG), sarcoplasmic reticulum (SR).

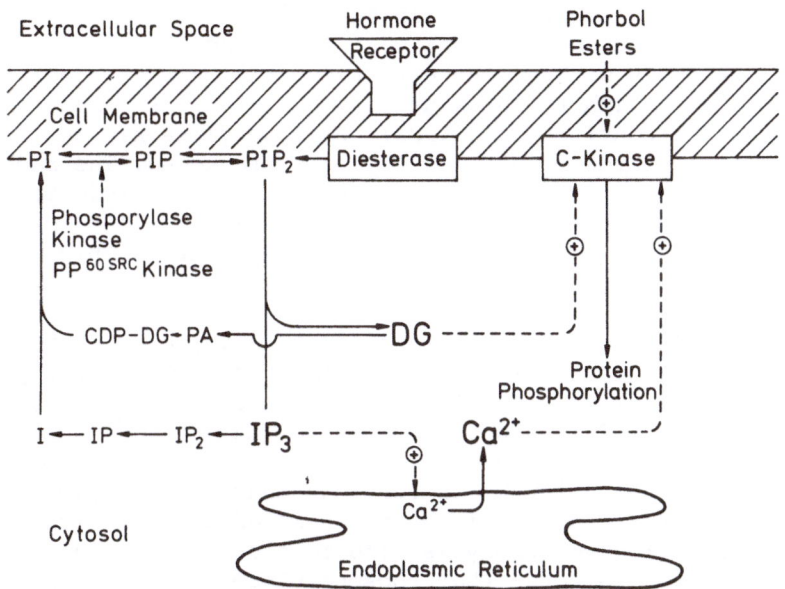

Figure 2: Model for the role of polyphosphoinositides in receptor activation

highly phosphorylated phospholipid, PIP_2. Two phospholipid kinases are involved in this cascade, phosphatidylinositol kinase and phosphatidylinositolphosphate kinase which act consecutively on PI and PIP respectively. A rate limiting step is catalysed by phosphatidylinositol kinase which will be described in more detail below.

Aside from these specific actions of the two second messengers, IP_3 and DG, one can observe the following effects on the metabolism of different cells upon ligand-receptor interaction: ·
1. An increase of PA in biological membranes can lead to a lateral lipid phase separation. This effect can be the explanation that PA has the capability of a Ca^{2+} ionophore (Michell et al., 1977; Putney et al., 1980; Putney, 1981; Lapetina, 1982).
2. PA and DG are substrates of lipases which generate arachidonic acid (Lapetina et al., 1981; Bell et al., 1979).
3. The synthesis of these anionic phospholipids, especially of PIP and PIP_2 can lead to an increased Ca^{2+} binding since polyphosphoinositides have a higher affinity for Ca^{2+} than the non phosphorylated PI. It can influence the transport of Ca^{2+} or the surface activity in certain areas of the membranes (Michell, 1982).
4. Changes of the phospholipid composition especially in relation to the charge can influence the activity of membrane bound enzymes either directly or indirectly due to changes in the transmembranal potential, membrane fluidity, or ion binding (Farese, 1983).

To understand the regulation of PIP formation the PI-kinase

must be known. It has been shown in several tissues to be a membrane bound enzyme located in plasma as well as in endoplasmic reticulum membranes. Recently we have shown that surprisingly a soluble cytoplasmic enzyme, phosphorylase kinase, stimulates PIP formation in SR-membranes (Varsanyi et al., 1983). Phosphorylase kinase is a multifunctional enzyme regulating glycogen breakdown. The rabbit muscle holoenzyme, first isolated by Krebs et al. in 1964 is a large phosphoprotein of molecular weight $1.2 - 1.3 \times 10^6$ (Hayakawa et al., 1973; Cohen, 1973). It is composed of four tightly associated subunits, designated α, β, γ and δ.

Localisation of a phosphorylase kinase antigen in SR-membranes (Gröschel-Stewart et al., 1978; Jennissen et al., 1979) the presence of a phosphorylase kinase activity in SR membranes of I-strain mice (Varsanyi et al., 1978), the stimulation of the SR Ca^{2+}-transport ATPase activity by phosphorylase kinase (Hörl et al., 1978; Hörl and Heilmeyer, 1978) all were indications that this enzyme plays a role in the Ca^{2+}-transport into the SR in addition to its function as a key enzyme of glycogenolysis. The molecular basis became apparent as it was detected that phosphorylase kinase phosphorylates phosphatidylinositol which is associated with the SR Ca^{2+}-transport ATPase (Varsanyi et al., 1983). From these observations an obvious question that arises is: what is the relation of phosphatidylinositol kinase to phosphorylase kinase.

To study this relationship an assay for the phosphatidylinositol kinase has to be developed. This activity determination can be based on the fact that phosphatidylinositol which is tightly associated with the Ca^{2+}-transport ATPase can be phosphorylated and precipitated with trichloroacetic acid.

Figure 3 shows that incubation of the Ca^{2+}-transport ATPase in absence of Ca^{2+} leads to incorporation of trichloroacetic acid precipitable radioactivity as PIP which is catalysed by an endogenous phosphatidylinositol kinase. Addition of purified phosphorylase kinase enhances the rate as well as the amount of the incorporated radioactivity ca. 2 fold. μM Ca^{2+} leads to decomposition of the formed PIP due to the presence of phosphomono and/or diesterase activity.

Figure 4 shows that a short trypsin treatment of the Ca^{2+}-transport ATPase inactivates almost completely the endogenous phosphatidylinositol kinase activity. Employing this trypsin treated Ca^{2+}-transport ATPase as substrate the system can be reconstituted: PIP is formed again upon exogenous phosphorylase kinase addition.

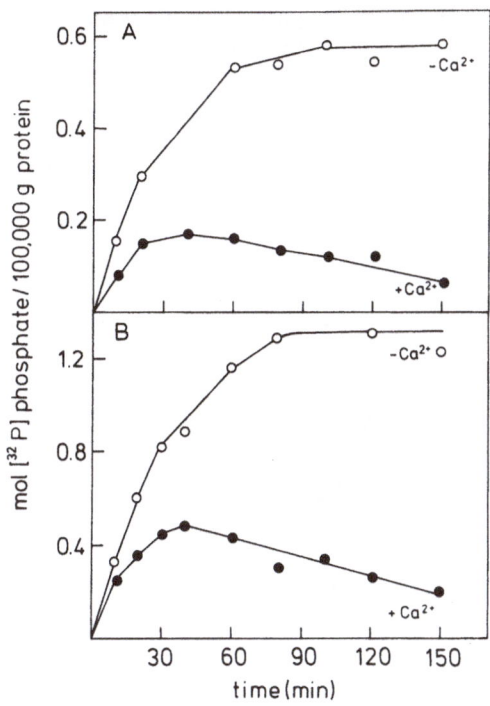

Figure 3: PIP formation in the isolated native Ca^{2+}-transport ATPase

Radioactivity incorporation was catalysed by endogenously (A) or exogenously added (B) phosphatidylinositol kinase (= phosphorylase kinase) in presence of 0.5 μM (+ Ca^{2+}) or 1.6 nM (- Ca^{2+}) free Ca^{2+}-concentrations.

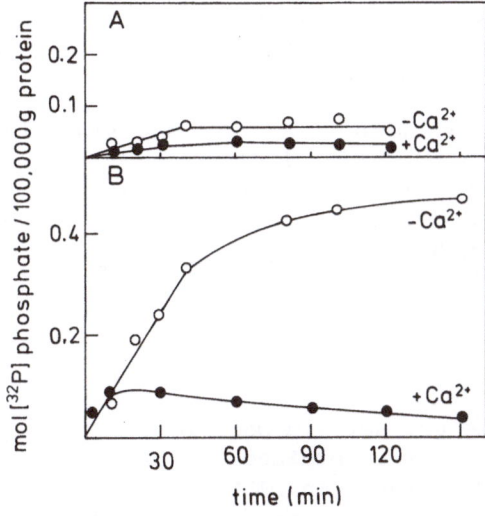

Figure 4: PIP formation in the trypsin treated Ca^{2+}-transport ATPase

Phosphorylation was carried out in absence (A) and presence (B) of purified phosphatidylinositol kinase in presence of 1.6 nM (- Ca^{2+}) and 0.5 μM (+ Ca^{2+}) free Ca^{2+}-concentrations.

Purified phosphorylase kinase solely catalyses PIP formation (Fig. 5) as does the endogenous kinase present in the purified Ca^{2+}-transport ATPase (Fig. 5, I and II). There is no difference either using the native or trypsin treated ATPase as substrate (Fig. 5, II and IV). Diacylglycerol kinase additionally produces PA when the native ATPase is employed as substrate and exclusively PA when the trypsinized material is used (Fig. 5, III). This new assay system can now be employed to follow the enrichment of phos-

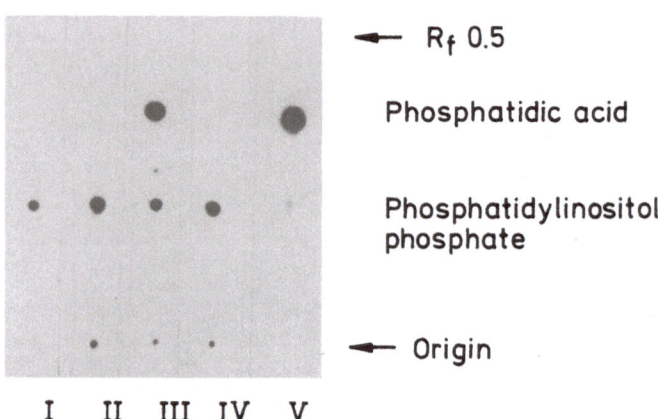

←— R_f 0.5

Phosphatidic acid

Phosphatidylinositol phosphate

←— Origin

I   II   III   IV   V

Figure 5: Identification of phosphorylated products
Phosphorylation was carried out under the conditions described in
Fig. 1 and 2. The formed products were extracted and identified
by thin layer chromatography. I. PIP formation is catalysed by the
endogenous phosphatidylinositol kinase (Fig. 1 A). II. Additionally
300 µg/ml phosphorylase kinase was added (Fig. 1 B). III. Diacyl-
glycerol kinase was added. IV and V. The trypsin treated Ca^{2+}-
transport ATPase was employed as substrate and phosphorylase kinase
or diacylglycerol kinase were added, respectively.

phatidylinositol kinase from a crude extract. In parallel, the purification of phosphorylase kinase was determined. Gelfiltration over Sepharose 4 B as well as chromatography on DEAE-cellulose could not separate phosphatidylinositol kinase from phosphorylase kinase activity. To further explore a resolution of phosphatidyl-inositol kinase from phosphorylase kinase the purified phosphorylase kinase was applied onto a hydroxyapatite column equilibrated with buffer containing 1 M KCl.

As shown in Figure 6 the main phosphatidylinositol kinase activity does indeed coelute with the main A_2 activity of phosphorylase kinase (Kilimann and Heilmeyer, 1982, a and b). Therefore one might assume that phosphorylase kinase is identical to phosphatidylinositol kinase. This conclusion is further supported by the parallel purification of both enzymatic activities from a crude muscle extract, the same molecular weight as demonstrated by gelfiltration and the same charge as concluded from the identical elution pattern from DEAE cellulose. In contrast a diacylglycerol

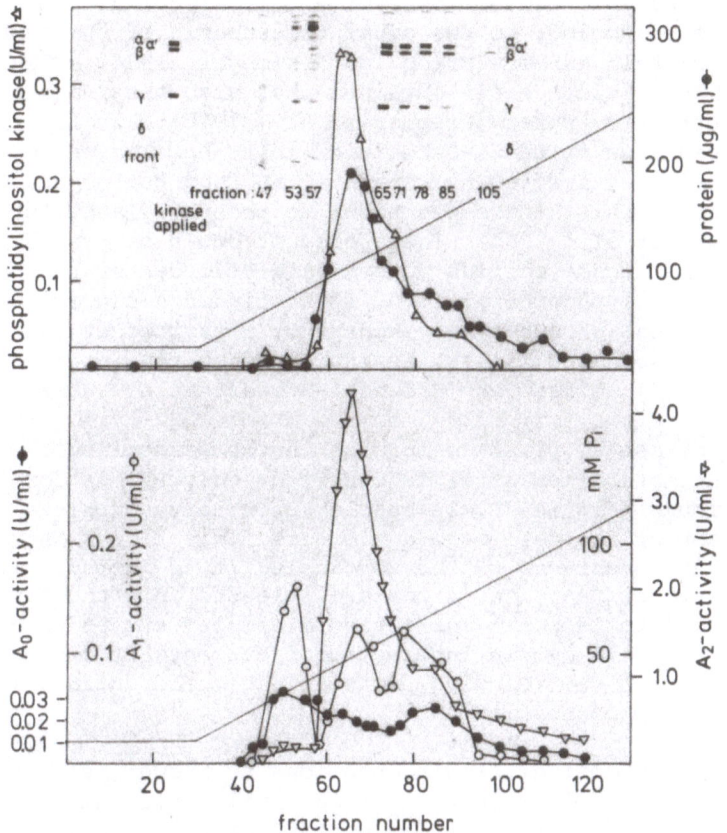

Figure 6: Hydroxyapatite chromatography of phosphatidylinositol kinase and phosphorylase kinase
Hydroxyapatite (0.6 x 10 cm) was equilibrated with 50 mM sodium glycerol 2-phosphate, 2 mM EGTA, 2 mM dithioerythritol, 10 mM KH_2PO_4, 1 M KCl, pH 6.8. Phosphorylase kinase purified by DEAE-cellulose chromatography was dialysed against the same buffer for 5 h at 0 °C. The protein was eluted with a linear gradient from 10 to 200 mM KH_2PO_4. The insets show gel electrophoretic patterns of the fractions indicated.

kinase which can be enriched by DEAE-cellulose chromatography and gelfiltration over Sepharose 4 B is easily separable from phosphorylase kinase.

It seems to be a discrepancy that a protein kinase, phosphorylase kinase, could be a phospholipid kinase too, i. e. phosphatidylinositol kinase. However, it is also known that phosphorylase kinase is a multifunctional protein kinase. Dickneite et al. (1978) first showed that more than one catalytic site might be present in phosphorylase kinase which accepts troponin T and phosphorylase b as substrates, respectively. It seems now clear that the γ-subunit carries a catalytic center (Skuster et al., 1980) and that this subunit shows homology to the catalytic subunit of the cAMP dependent protein kinase (Crabb and Heilmeyer, 1984, a; Reimann et al., 1984). Evidence was obtained that also the β-subunit contains a catalytic center (Fischer et al., 1978). Finally the NH_2-terminal sequence of the α-subunit exhibits homologies to the pp60[v-src] kinase, a tyrosine kinase (Crabb and Heilmeyer, 1984, b). Very recently, this kinase was shown to phosphorylate also PI and DG (Sugimoto et al., 1984). Therefore, it could be possible that phosphorylase kinase carries also a catalytic center for phosphatidylinositol phosphorylation. These findings support the speculation that phosphorylase kinase or a part of it can be also a membrane associated kinase, as indicated by the intracellular histochemical localisation (Gröschel-Stewart et al., 1978).

In contrast to plasma membranes in the sarcoplasmic reticulum a different metabolism of PI is observed; only PIP is formed, no PIP_2 was found (Figure 7). In muscle, upon nerve impulses depolarisation of the cell membrane occurs. This signal is transmitted to the membranes of the SR causing it to increase its permeability for Ca^{2+}. Ca^{2+} ions are released into the sarcoplasm and are bound to troponin initiating a complex set of reactions that lead to actin-myosin interaction i.e. contraction. Ca^{2+} binds also to specific Ca^{2+} binding proteins e.g. calmodulin which can cause protein phosphorylation.

One of the most difficult questions to answer is, however, what role the PIP plays in the sarcoplasmic reticulum. It could be involved in regulation of Ca^{2+} uptake or Ca^{2+} release. Indeed, as already mentioned, PIP enhances the maximal activity of the Ca^{2+} transport ATPase ~ 2 fold and increases its affinity for Ca^{2+} ~ 7 fold (Varsanyi et al., 1983). Similar observations that the Ca^{2+}-transport ATPase activity might depend upon changes in membranes polyphosphoinositides were reported by Buckley and Hawthorne, 1972, and Penniston, 1982. However, more evidence should be provided that agonist dependent phosphatidylinositol metabolism does indeed induce or play a functional change in the permeability of Ca^{2+} across membranes (Figure 7).

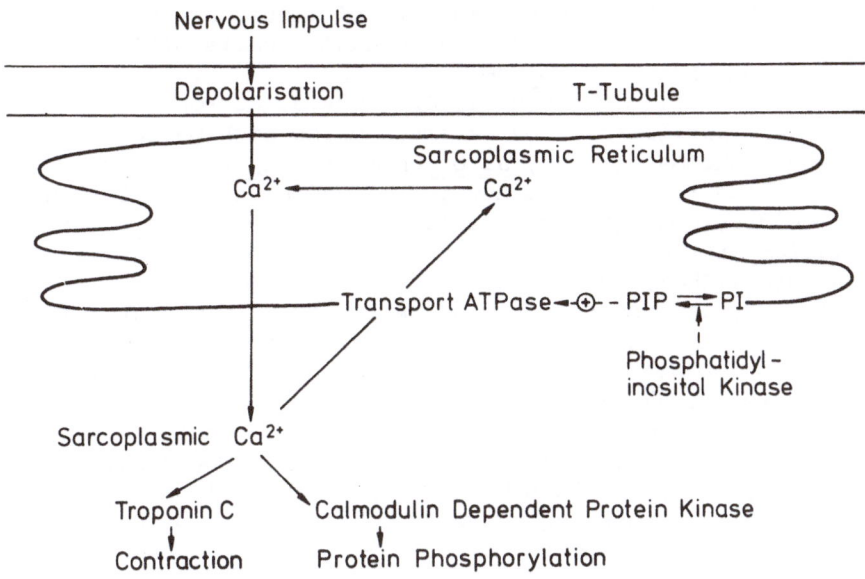

Figure 7: Model for the role of PIP in skeletal muscle

ACKNOWLEDGEMENTS

This work was supported by grants from the Deutsche Forschungs-gemeinschaft and the Fonds der Chemie. Z. Georgoussi is a recipient of a N.H.R.F. scholarship. The help in typing this manuscript by Ulrike Rosenbaum is gratefully acknowledged.

REFERENCES

Bell, R.L., Kennerly, D.A., Stanford, N. and Majerus, P.W., 1979, Diglyceride lipase: a pathway for arachidonate release from human platelets, Proc. Natl. Acad. Sci. USA, 76:3238.

Berridge, M.J., 1981, Phosphatidylinositol hydrolysis: a multi-functional transducing mechanism, Molecular and Cellular Endocrinology, 24:115.

Berridge, M.J., 1983, Rapid accumulation of inositol trisphosphate reveals that agonists hydrolyse polyphosphoinositides instead of phosphatidylinositol, Biochem. J., 212:849.

Berridge, M.J. and Irvine, R.F., 1984, in: "Cyclitols and Ino-sitides", Bleasdale, J., Eichberg, J. and Hauser, G., Humana Press.

Berridge, M.J., 1984, Inositol trisphosphate and diacylglycerol as second messengers, Biochem. J., 220:345.

Buckley, J.T. and Hawthorne, J.N., 1972, Erythrocyte membrane poly-phosphoinositide metabolism and the regulation of Ca^{2+} binding, J. Biol. Chem., 247:7218.

Cohen, P., 1973, The subunit structure of rabbit skeletal muscle
phosphorylase kinase and the molecular basis of its activation
reaction, Eur. J. Biochem., 34:1.

Crabb, J.W. and Heilmeyer,L.M.G., Jr. 1984 a, Micropreparative
protein purification by reversed-phase high-performance
liquid chromatography, J. of Chromatography, 226:129.

Crabb, J.W. and Heilmeyer, L.M.G., Jr. 1984 b, High performance
liquid chromatography purification and structural characterization of the subunits of rabbit muscle phosphorylase kinase,
J. Biol. Chem., 259:6346.

Dickneite, G., Jennissen, H.P. and Heilmeyer, L.M.G., Jr., 1978,
Differentiation of two catalytic sites on phosphorylase
kinase for phosphorylase b and troponin T phosphorylation,
FEBS lett., 87:297.

Dumont, J.E., Greengard, P. and Robinson, A.G., 1981, "Advances in
Cyclic Nucleotide Research", Raven Press, New York.

Farese, R.V., 1983, The phosphatidate-phosphoinositide Cycle: An
intracellular messenger system in the action of hormones
and neurotransmitters, Metabolism, 32:628.

Fischer, E.H., Alaba, I.O., Brautigan, D.L.,Malencik, D.A.,
Moeschler, H.I., Picton, C. and Procingwong, S., 1978,
"Versatility of Proteins" (C.H. Li. ed), Academic Press,
New York, 133.

Gröschel-Stewart, U., Jennissen, H.P., Heilmeyer, L.M.G., Jr. and
Varsanyi, M., 1978, Localization of Ca^{2+}-dependent protein
kinases in various tissues by the immunofluorescent technique,
Int. J. Peptide Protein Res., 12:177.

Hayakawa, T., Perkins, J.P. and Krebs, E.G., 1973, Studies on the
subunit structure of rabbit skeletal muscle phosphorylase
kinase, Biochemistry, 12:574.

Hokin, M.R. and Hokin, L.E., 1958, Enzyme secretion and the
incorporation of ^{32}P into phospholipids of pancreas slices,
J. Biol. Chem., 203:967.

Hokin, L.E. and Hokin, M.R., 1958, Phosphoinositides and protein
secretion in pancreas slices, J. Biol. Chem., 233:805.

Hörl, W.H., Jennissen, H.P. and Heilmeyer, L.M.G., Jr., 1978,
Evidence for the participation of a Ca^{2+} dependent protein
kinase and a protein phosphatase in the regulation of the
Ca^{2+} transport ATPase of the sarcoplasmic reticulum.
I. Effect of inhibitors of the Ca^{2+} dependent protein kinase
and protein phosphatase, Biochemistry, 17:759.

Hörl, W.H. and Heilmeyer, L.M.G., Jr., 1978, Evidence for the
participation of a Ca^{2+} dependent protein kinase and
protein phosphatase in the regulation of the Ca^{2+} transport
ATPase of the sarcoplasmic reticulum, II. Effect of phosphorylase kinase and phosphorylase phosphatase, Biochemistry, 17:766.

Jennissen, H.P., Veh, R.W., Petersen, J.K.H. and Neubauer, H.P.,
1979, Immunological studies of phosphorylase kinase on the
subunit level, Hoppe Seyler´s Z. Physiol. Chem., 360:293.

Kilimann, M.W. and Heilmeyer, L.M.G., Jr., 1982, Multiple Activities on phosphorylase kinase, 1. Characterization of three partial activities by their response to Ca^{2+}, Mg^{2+}, Mn^{2+}, pH and NH_4Cl and effect of activation by phosphorylation and proteolysis, Biochemistry, 21:1727.

Kilimann, M.W. and Heilmeyer, L.M.G., Jr. 1982, Multiple activities on phosphorylase kinase, 2. Different specificities towards the protein substrates, phosphorylase b, troponin and phosphorylase kinase, Biochemistry, 21:1735.

Krebs, E.G., Love, D.S., Bratvold, G.E., Trayser, K.A., Meyer, W.L. and Fischer, E.H., 1964, Purification and properties of rabbit skeletal muscle phosphorylase b kinase, Biochemistry, 3:1022.

Lapetina, E.G., Billah, M.M., Cuatrecasas, P., 1981, The initial action of thrombin on platelets. Conversion of phosphatidyl-inositol to phosphatidic acid preceding the production of arachidonic acid, J. Biol. Chem., 256:5037.

Lapetina, E.G., 1982, Regulation of arachidonic acid production: role of phospholipase C and A_2, Trends Pharmacol. Sci., 3:115.

Michell, R.H., Jafferji, S.S. and Jones, L.M., 1977, The possible involvement of phosphatidylinositol breakdown in the mechanism of stimulus-response coupling at receptors which control cell-surface calcium gates, Adv. Exp. Med., 83:447.

Michell, R.H., Kirk, C.J., 1981, Why is phosphatidylinositol degraded in response to stimulation of certain receptors? Trends Pharmacol. Sci., 2:86.

Michell, R.H., 1982, Inositol phospholipids and cell calcium, Cell Calcium, 3:285.

Michell, R.H., 1982, Is phosphatidylinositol really out of the calcium gate? Nature, 296:492.

Michell, R.H., 1983, Ca^{2+} and protein kinase C: two synergistic cellular signals, Trends Biochem. Sci., 8:263.

Nishizuka, Y., 1984, The role of protein kinase C in cell surface signal transduction and tumor promotion, Nature, 308:693.

Penniston, J.T., 1982, Plasma Membrane Ca^{2+}-Pumping ATPases, in: "Transport ATPases", Carafoli, E. and Scarpa, A. ed., The New York Academy of Sciences, New York.

Putney, J.W., Weiss, S.J., Van de Walle, C.M. and Haddes, R., 1980, Is phosphatidic acid a calcium ionophore under neurohumoral control, Nature, 284:344.

Putney, J.W., 1981, Recent hypotheses regarding the phosphatidyl-inositol effect, Life Sciences, 29:1183.

Reimann, E.M., Titani, K., Ericsson, L.H., Wade, R.D., Fischer, E.H., Walsh, K.A., 1984, Homology of the -subunit of phosphorylase b kinase with cAMP-dependet protein kinase, Biochemistry, in press.

Skuster, J.R., Chan, J.K.F. and Graves, D.I., 1980, Isolation and properties of the catalytically active subunit of phosphorylase b kinase, J. Biol. Chem., 255:2203.

Sugimoto, Y., Whitman, M., Cantley, C.L. and Erikson, R.L., 1984, Evidence that the Rous sarcoma virus transforming gene product phosphorylates phosphatidylinositol and diacyl-glycerol, Prod. Natl. Acad. Sci., USA, 81:2117.

Streb, H., Irvine, R.F., Berridge, M.J. and Schulz, I., 1983, Release of Ca^{2+} from a nonmitochondrial intracellular store in pancreatic acinar cells by inositol-1,4-5-trisphosphate, Nature, 306:67.

Varsanyi, M., Gröschel-Stewart, U. and Heilmeyer, L.M.G., Jr., 1978, Characterization of a Ca^{2+}-dependent protein kinase in skeletal muscle membranes of I-strain and wild-type mice, Eur. J. Biochem., 87:331.

Varsanyi, M., and Heilmeyer, L.M.G., Jr., 1981, Phosphorylation of the 100.000 M_R Ca^{2+} transport ATPase by Ca^{2+} or cyclic AMP dependent and independent protein kinases, FEBS Lett., 131:223.

Varsanyi, M., Tölle, H.G., Heilmeyer, L.M.G., Jr., Dawson, R.M.C. and Irvine, R.F., 1983, Activation of sarcoplasmic reticular Ca^{2+} transport ATPase by phosphorylation of an associated phosphatidylinositol, The EMBO Journal, 2:1543.

ELECTRON PARAMAGNETIC RESONANCE IN BIOLOGY:

SPIN TRAPPING

Alexandre T. Quintanilha

Department of Physiology-Anatomy
University of California
Berkeley, CA 94720, U.S.A.
and
Departamento de Biofisica
Instituto Abel Salazar
Universidade do Porto
Porto, Portugal

INTRODUCTION

Discovered in 1945 by Zavoisky[1] electron paramagnetic resonance (EPR) has since been applied to a large number of research areas. It is a spectroscopic method using frequencies in the microwave region (i.e. 10^9 up to 10^{11}hz) and wavelengths from approximately 10^{-3} to 10^{-1} meters; it is limited to the detection of unpaired electrons. Unpaired electrons are present in free radicals, triplet electronic states and transition and rare earth ions. The sensitivity of the method allows radical concentrations of 10 nM to be detected.

In the same way as in the case of other forms of electromagnetic spectroscopy the experimental information in the EPR spectrum is contained in the location and amplitude of the absorption bands.

THE RESONANCE PHENOMENON

An atom or molecule which contains unpaired electrons has non-zero spin and therefore possesses a corresponding magnetic moment. A single free electron possesses a spin of 1/2. The magnetic moment of an electron can be written as

$$\mu = g\beta M_s$$

where g, the spectroscopic splitting factor has a value which is a function of the electron's environment; β, the Bohr magneton is a constant; and M_s, the angular momentum quantum number, can take the values of +1/2 and -1/2 for the case of a free electron.

In the absence of an external magnetic field, free electrons in either of the two states, +1/2 or -1/2 have equal energy. In the presence of an external magnetic field H the free electrons will occupy two distinct energy levels. The lower energy state will have its spin magnetic moment aligned in the direction of the external applied magnetic field. The difference in energy between the two levels is linearly proportional to the applied magnetic field, and is given by

$$\Delta E = g\beta H$$

Transitions between the two energy levels can be induced by supplying the electron system with energy, usually in the form of electromagnetic radiation. If ν is the frequency of the radiation, then transitions will occur when

$$h\nu = g\beta H$$

which represents the resonance condition (h is Planck's constant). Most EPR measurements are made keeping the frequency constant and varying the applied magnetic field. The most popular frequencies used have been (in the microwave region) at approximately 9 GHz (X-band), 23 GHz (K-band) and 35 GHz (Q-band). For a g value of roughly 2 (the case for a free electron) and a frequency of 9 GHz, the magnetic field required for resonance will be 3,200 gauss which is readily attainable with classical electromagnets.

Transitions between the two energy levels result in absorption of the electromagnetic radiation supplied to system. For a single spin system, the population of the lower energy level (n_{lower}) is only slightly greater than the population of the higher energy level (n_{upper}). These populations are governed by the Boltzmann distribution

$$\frac{n_{upper}}{n_{lower}} = - \; exp(\Delta E/kT)$$

where k is the Boltzmann constant and T the absolute

temperature. At 300 degrees K and with a magnetic field of 3,200 gauss the above ratio will be equal to 0.998.

A detectable absorbtion requires that this population difference be maintained. While it is clear that the radiation field tends to equalize these populations, thermal equilibrium will always tend to keep $n_{lower} > n_{upper}$ Experimentally, EPR measurements are usually performed with radiation intensities that are too small to effectively disturb the Boltzmann distribution of the electron populations in the two energy levels. Clearly, the difference in the two populations can be made larger by either lowering the temperature T or increasing ΔE by increasing H.

g-FACTORS

Each electron, in a particular ion, is characterized by both its spin quantum number and its orbital angular momentum quantum number. These are combined into a total angular momentum characterized by a different quantum number determined by the rules of quantum mechanics.

The interaction giving rise to this total angular momentum is known as the spin-orbit coupling. This coupling results in the electron usually having a different effective magnetic moment from that of a free electron and therefore also a different g-value. This means that for a given frequency, unpaired electrons associated with different ions or radicals will resonate at different values of the externally applied magnetic field H. In the case of radicals, the differences are small but they nevertheless can provide valuable information about their structures.

The spin-orbit coupling of an electron depends on the orientation of the molecule with respect to the applied magnetic field. It follows that g is in fact a tensor. However, in the case of free electrons associated with molecules in solution, where there is rapid tumbling, g will take on a time-averaged value.

HYPERFINE SPLITTING

If the unpaired electron is close enough to sense the magnetic moment of a particular nucleus then the electron will not only be under the effect of the externally applied magnetic field, H, but also the effective magnetic field of the nucleus, Hn. The resonance condition should then be modified to read

$$h\nu = g\beta \,|H + H_n|$$

where, of course, H and Hn are both vectors. However, the nuclear moment is also quantized and can take on 2I + 1 discreet values, where I is the nuclear spin quantum number. Therefore, for a particular value of there will be 2I + 1 values of the externally applied magnetic field for which we will observe the resonance condition and a corresponding absorption of electromagnetic radiation.

In a hydrogen atom the nucleus has a spin I = 1/2, meaning that there are two possible orientations for the proton spin (+ 1/2 and - 1/2). Four energy levels for the unpaired electron will result (fig. 1).

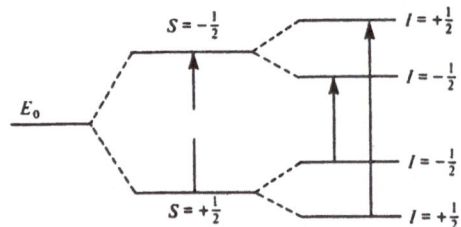

Fig. 1:   Energy Levels of an Electron Interacting
          with a Proton

Since quantum mechanics does not allow the nuclear spin quantum number to change during an electronic transition we are left with two allowed transitions which are indicated by the arrows (fig.1). The spectrum of the hydrogen atom will therefore consist of two lines of equal intensity. The separation between these lines (measured in gauss or in millitesla) is referred to as the hyperfine splitting constant a.

The hyperfine constant is sensitive to the dielectric constant of the medium in which the unpaired electron is located and is therefore a useful parameter to study partition distributions of the paramagnetic ions under study. If the electron interacts with more than one nucleus and each nucleus has a different magnetic moment the spectra obtained can be very complex and difficult to interpret. Of particular interest to us in this chapter are unpaired electrons centered on nitrogen where I = 1. In this case we expect the spectrum to be a triplet of identical lines (fig. 2).

If the electron "sees" both a nitrogen and a proton then the the three lines will be split into six. If it "sees" a nitrogen and two non-equivalen protons it will give rise to 12 lines, and so on.

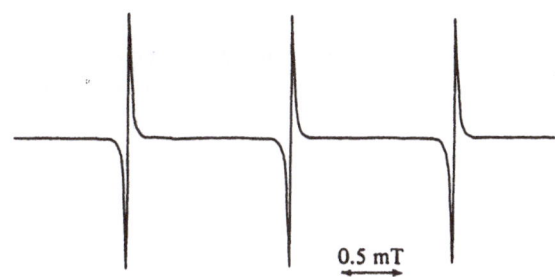

Fig. 2: Spectrum of the NO· Radical

BIOLOGICAL STUDIES

Excluding the methods of electron nuclear double resonance (ENDOR) and electron double resonance (ELDOR), EPR studies in biology can be divided into three broad areas:

1) The study of specific paramagnetic ions (eg. Fe, Mn, Cu, Co) important in a multitude of enzymatic and catalytic functions.

2) The use of spin labels to study primarily physical and structural aspects of biological systems.

3) The use of spin traps to identify short lived and highly reactive radicals generated in biological systems.

I will concentrate on the third area only.

SPIN TRAPPING

Many biologically important mechanisms are known to involve free radicals. The majority of radicals have a very short lifetime. Until 1968 the direct detection by EPR of transient radicals was restricted to studies using flow systems, which unfortunately require relatively large amounts of material.

This problem was solved by using suitable radical scavengers which react with radical intermediants to form relatively long-lived radicals. These radical scavengers are referred to as spin traps [2-4].

The most well known spin traps are

PBN: α-phenyl N-tert-butylnitrone
POBN: α-(4-pyridyl-1-oxide) N-tert-butylnitrone
DMPO: 5,5-dimethyl-1-pyrroline-N-oxide
MNP or NtB: 2-methyl-2-nitroso propane (fig. 3)

$$(CH_3)_3C-N\overset{O}{} + R\bullet \longrightarrow (CH_3)_3C-\overset{O\bullet}{N}-R$$

Fig. 3:   Reaction of MNP with a Radical R·

When a radical reacts with one of these traps (fig. 3) the resulting radical shows a spectrum characterized by a 1:1:1 splitting due to the interaction of the unpaired electron with the nitrogen nucleus, which has a nuclear spin quantum number of 1. This triplet is split by the coupling of the unpaired electron with magnetic nuclei on the α- and β-carbon atoms (fig. 4).

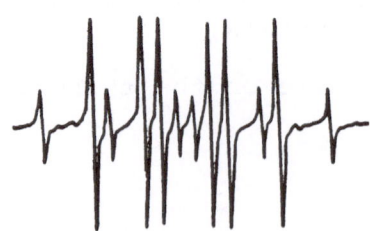

Fig. 4:   Characteristic Spectrum of MNP-R · Adduct

The traps PBN and POBN have the advantage that the protons of the t-butyl groups hardly interact with the unpaired electron. The MNP trap has an advantage over the others in that the radical fragment R is closer to the radical center (NO·) and thus contributes more to the splitting pattern. However, MNP is very light-sensitive and therefore less useful than the other traps.

The apparent rate constants for the reaction between different radicals and these traps can vary by several orders of magnitude (table I). The water solubility of these traps also varies a great deal.

Once the stable radical-adduct has been formed it may be isolated and its spectrum determined in a variety of solvents.

This allows for the determination of several hyperfine splitting constants that can be compared to literature values when attempting to identify the radical under study (table II). We should stress that an unambiguous identification of the radical can only be achieved using mass spectroscopy.

Table I.   Apparent Rate Constants ($M^{-1} s^{-1}$) of the Spin Trapping Reactions

Radical	DMPO	PBN	MNP
$O_2^{-\cdot}$	10		
HO·	3.4×10^9	1.0×10^9	
RO·	5.0×10^8	5.5×10^6	1.5×10^6
$R\overset{\cdot}{C}H_2$	2.5×10^6	1.3×10^5	9.0×10^6
$R\underset{}{\overset{H}{C}}R'$	4.2×10^5	6.5×10^4	6.1×10^6
$R\underset{\cdot}{\overset{R'}{C}}R''$		1.0×10^4	3.3

Table II. Hyperfine Splitting Constants For a Representative Selection of Radical-Adducts (in Benzene)

Type	Radical	PBN		DMPO	
		a_N	a_H	a_N	a_H
ROO·	t-BuOO·	13.40	1.57		
	CumylOO·	13.55	1.82		
	HOO·	13.28	2.25		
RO·	t-BuO·	14.21	1.83	13.17	7.93
	Et-O·	14.01	2.01	13.22	6.96
	Me-O·	13.76	2.00	13.58	7.61
	HO·	14.48	2.17	14.83	14.83
R·	Cumyl·	14.25	2.19		
	Et·	14.32	3.22	14.20	20.49
	Me·	14.81	3.47	14.31	20.52
	Benzyl·	13.78	2.01	14.16	20.66
	Phenyl·	14.41	2.21	13.79	19.22

ACKNOWLEDGEMENT

        I wish to thank Dr. Ohara Augusto for helping me put together tables I and II.

REFERENCES

    1. E. Zavoiski, J. Phys. U.S.S.R. Vol. 9, p. 211 (1945).

    2. M. J. Perkins, in "Essays in Free Radical Chemistry" p.97, Special Publication No.24, Chemical Society London, 1970.

    3. E. G. Janzen, C. A. Evans, and E. R. Davis, in "Organic Free Radicals" p. 433 (ACS Symposium Series No. 69) American Chemical Society, 1978.

    4. G. M. Rosen and E. J. Rauckman "Spin Trapping of Superoxide and Hydroxyl Radicals" in Methods of Enzymology (ed. S. P. Colowick and N. O. Kaplan) Vol. 105, p. 198, 1984.

STRUCTURE OF ENVELOPED VIRUSES AND THEIR INTERACTION WITH CELL SURFACES

Frank R. Landsberger

The Rockefeller University
New York, New York, U.S.A.

Viruses possessing a lipid-protein membrane are collectively referred to as enveloped viruses. Because of the ability to vary independently, the lipid and protein composition of the virus particles, they may be considered as model biological membranes. As such they offer a unique opportunity to study lipid-protein interactions as well as membrane-membrane interactions.

Enveloped Virus Structure:

Enveloped viruses such as the influenza and parainfluenza viruses assemble by budding at the cell surface during which they acquire a lipid bilayer. (Choppin and Compans, 1975; Compans and Choppin, 1975). Influenza virus has associated with its viral envelope the HA glycoprotein which possesses hemagglutinating activity and is responsible for virus binding to the host cell surface and the neuraminidase protein (Choppin and Compans, 1975). Sendai virus, a parainfluenza virus, when grown in MDBK cells contains large amounts of a glycoprotein designated F_o, which is absent in egg-grown Sendai virus (Scheid and Choppin, 1977). MDBK-grown Sendai virus, lacks in cell-fusing and hemolysis activity as well as in infectivity. The F_o precursor in MDBK-grown Sendai virions can in vitro be cleaved to yield $F_{1,2}$. Egg-grown Sendai virions also possess $F_{1,2}$. Virions containing the cleaved F protein have hemolysis and cell-cell fusion activity and are also infectious. In addition, Sendai virus has a hemagglutinin glycoprotein. A major nonglycosylated protein (M-protein) is associated with the inner surface of the lipid bilayer of each of these viruses (Compans and Dimmock, 1969; Bachi et al., 1969; Compans et al., 1970; White et al., 1970).

The unusual feature of all the enveloped viruses examined to date is that the rigidity of the viral lipid bilayer is considerably greater than that of the host plasma membrane (Landsberger et al., 1971, 1973; Sefton and Gaffney, 1974; Landsberger and Compans, 1976; Barenholz et al., 1976; Moore et al., 1976a,b). [Under the conditions used, the detected ESR signal is due to spin label in the plasma membrane (Landsberger and Compans, 1976).] Membranes prepared from lipids extracted from vesicular stomatitis virus (VSV), a rhabdovirus, are more fluid than the viral lipid membrane (Altstiel et al., 1975; Altstiel and Landsberger, 1981a). Since the lipid composition of enveloped viruses is determined largely by that of the host cell plasma membrane (Klenk and Choppin, 1969, 1970a,b; McSharry and Wagner, 1971; Choppin et al., 1971, 1972), the spin label electron spin resonance (ESR) data indicate that membrane-associated viral proteins markedly increase the rigidity of the viral envelope (Landsberger and Compans, 1976: Sefton and Gaffney, 1974; Altstiel and Landsberger, 1981a). Several lines of evidence have suggested that the M-protein plays a role in the observed enhanced rigidity of the viral lipid bilayer.

Virus-membrane interactions:

    1. Attachment of viruses to cells: The process of infection of susceptible cells starts with the adsorption of virus particles to the receptors on the plasma membrane of a cell (Howe and Lee, 1970; Dales, 1973; Lonberg-Holm and Philipson, 1974). Adsorption to receptors on host cells appears to be analogous to adsorption to erythrocytes (Bachi et al., 1977).

    Spin label ESR techniques have been used to probe the change in the structure of the lipid phase of a plasma membrane following attachment of viruses or lectins (Lyles and Landsberger, 1976, 1977; Landsberger and Altstiel, 1981). Upon agglutination of avian or amphibian erythrocytes by Sendai virus grown in MDBK cells, and influenza virus and by the lectins concanavalin A and wheat germ agglutinin, the erythrocyte bilayer becomes more fluid as assayed by five different spin labels (Lyles and Landsberger, 1976, 1978). This structural change upon virus adsorption is complete at approximately 1 to 80 particles per cell (Lyles and Landsberger, 1976). Virus or lectin attachment produces no detectable changes in the lipid bilayer of mammalian erythrocytes. Avian, but not mammalian, erythrocytes contain microtubules associated with the plasma membrane (Fawcett and Witebsky, 1964). Colchicine and vinblastine, which can disrupt microtubules but do not affect the fluidity of the membrane lipid phase by themselves (Lyles and Landsberger, 1976; Altstiel and Landsberger, 1977), inhibit the fluidity increase that accompanies the addition of virus to avian erythrocytes. Thus, binding of viruses or lectins to receptors on avian erythrocyte surfaces results in a change in the membrane bilayer through a system that resembles microtubules in its drug sensitivity. These results have important implications for the initial events in virus infection. They suggest that the structure of the host cell plasma membrane is altered by virus attachment, as perhaps a preparatory step to penetration.

The change in bilayer fluidity of chicken erythrocytes upon attachment of virus or lectins requires cross-linking of receptor molecules (Landsberger et al., 1978; Lyles and Landsberger, 1978). A water-soluble hemagglutinin fragment of influenza virus (Brand and Skehel, 1978) attaches to the cell surface but does not change the bilayer fluidity. In contrast, a multivalent aggregate ("rosettes") of influenza hemagglutinin (Scheid and Choppin, 1973) increases the fluidity of the chicken erythrocyte membrane (Lyles and Landsberger, 1978). Experiments with succinyl-con A (Gunther et al., 1973), which cannot agglutinate erythrocytes, show no change in plasma membrane fluidity (Lyles and Landsberger, 1978). However, the addition of antibody to con A to succinyl-con A-pretreated cells increases erythrocyte bilayer fluidity.

We have also investigated the binding of VSV, a rhabdovirus, to BHK cells (Altstiel and Landsberger, 1981). The ESR spectra of a fatty acid spin label incorporated into the plasma membrane of intact BHK cells indicate that the plasma membrane bilayer becomes rapidly more rigid after addition of VSV. This change appears also to involve microtubules in that it can be inhibited by colchicine. The monomeric water-soluble G_s form of VSV G-protein does not cause a change in the plasma membrane fluidity. Upon addition of anti-G antibody to cells pretreated with G_s, the cell membrane bilayer becomes more rigid.

These data are strikingly similar to those obtained with influenza and Sendai virus attachment to avian erythrocytes. Both support the hypothesis that agglutination-induced fluidity changes involve lateral rearrangement of receptors in the plane of the membrane. The fact that VSV attachment to BHK cells causes a decrease in bilayer fluidity whereas influenza and Sendai virus attachment to avian erythrocytes causes an increase probably reflects differences in lipid composition and lipid-protein interactions in the two systems.

Experiments with fluorescently-labeled VSV support the suggestion that VSV attachment to the BHK cell surface results in a lateral rearrangement of the receptors (Altstiel and Landsberger, 1981).

Using [35S]-methionine labeled virus, the binding of VSV to BHK and L cells appears to be positively cooperative, indicating that the binding of a few virions to the cell surface makes the binding of subsequent particles easier. It appears that binding of VSV to the cell surface causes a lateral rearrangement of plasma membrane components requiring the multivalency of the virus particle (i.e., a VSV particle has many G-proteins) and involving microtubules. This lateral rearrangement of membrane receptors could facilitate the binding of additional virus particles.

2. Ionic double layer at membrane surfaces: For virus penetration of a cell to occur, the negatively charged surfaces of enveloped viruses and cells (Brinton and Lauffer, 1958; Sherbet, 1978) must be brought together. By a variety of experiments, the surface potentials or charge on membranes have been measured (e.g., McLaughlin and Harary, 1976; Gaffney and Mich, 1976). According to the ionic double layer theory, a charged surface in an ionic medium attracts counterions which will attract their own counterions. These electrostatic attractions are balanced by the thermal motion of the ions. Thus the double layer consists of the negatively charged membrane surface and the distribution of counterions (Verwey and Overbeek, 1948). This theory predicts that at a distance dependent on the ionic strength of the medium the charge on a membrane will be effectively screened and the surface will appear to be electrically neutral.

We have investigated properties of the ionic double layer at the surface of negatively charged lipid vesicles and VSV particles using spin label ESR methods (Landsberger and Altstiel, 1980). The isotropic hyperfine coupling constant A_{iso} (a parameter which can be calculated from the ESR spectrum) is proportional to the local electrical field (Griffith et al., 1974; Griffith and Jost, 1976). We synthesized a series of four phospholipid spin labels with a nitroxide moiety bound to the lipid phosphate by spacer groups of varying lengths. Using $NiCl_2$ it was determined that the spin labels in the lipid vesicles and in VSV are located on the exterior surface of the lipid bilayer. Using the nitroxide-induced broadening of the ^{1}H NMR linewidth of the lecithin choline methyl group, the average separation of the nitroxide moiety from the lecithin phosphate was compared with direct measurement from CPK models. The results indicate that the distance of the nitroxide moiety from the surface of the lipid bilayer is determined by the extended length of the spacer group.

The agreement between the experimental ESR data and the predictions of the ionic double-layer theory is excellent. The ionic strength of the medium calculated from the ESR data is that actually used in the experiment. At physiological ion concentrations, the double layer is relatively compact. With decreasing ionic strength, the double layer broadens resulting in less effective screening of the charge on the membrane surface, suggesting that with decreased ionic strength the ability of a virus particle to bind to a cell is decreased. Calculation of the repulsive energy, due to the interactions between the ionic double layers at the surfaces of the cell and virus, indicates a square root dependence on the ionic strength.

The efficiency of attachment of VSV, as measured by efficiency of infection, was indeed found to decrease with decreasing ionic strength, and to have the same square root of the ionic strength dependence as the repulsive energy. Thus, it appears that the ionic double layer at the surface of membranes is a factor in enveloped virus attachment to cells.

3. <u>Sendai virus-induced hemolysis and fusion with erythrocyte membranes</u>: In the intact erythrocyte, a phosphatidylcholine derivative spin label (i.e., PC12) resides in a more rigid environment than the corresponding phosphatidylethanolamine (PE12) spin label (Tanaka and Ohnishi, 1976; Lyles and Landsberger, 1977). [For both spin labels, the nitroxide moiety is on the twelfth carbon of the number 2 fatty acyl chain.] Upon Sendai virus-induced hemolysis, the region probed by PC12 becomes more fluid while that of PE12 becomes more rigid, such that the end-point spectra of the two labels are identical (Lyles and Landsberger, 1977). These changes reflect the amount of hemoglobin released. In our studies, 90% hemolysis, corresponding to an almost complete structural change, occurs at a virus-to-cell ratio of about 160 to 1600 (Lyles and Landsberger, 1977).

The kinetics of fusion of the Sendai virus envelope with erythrocyte membranes were studied by spin label ESR methods (Lyles and Landsberger, 1979). The virion envelope was labeled with PCn or PE12. As the virus envelope fuses with the erythrocyte membrane, the viral lipids mix with the cellular lipid bilayer and the ESR spectrum changes from that characteristic of the virus to that of the erythrocyte. The fusion process appears to be a first-order reaction which is independent of virus concentration with a half-time of 7 min at 37° C.

The kinetics of the viral hemolytic activity were also investigated, using spin label ESR methods, in terms of the structural change in the erythrocyte membrane induced by hemolysis. At sub-maximum levels of hemolysis, the half-time of the hemolytic reaction is similar to that of envelope fusion. These results suggest that envelope fusion is the limiting step in virus-induced hemolysis. Early harvest egg-grown Sendai virus has very low hemolytic activity but fuses with erythrocyte membranes as efficiently as late harvest egg-grown Sendai virus, indicating that fusion of the viral envelope with the erythrocyte membrane does not require the presence of full hemolytic activity. On the other hand, fusion of viral and erythrocyte membranes is a prerequisite for hemolysis.

Kuroda et al., (1980) used Sendai virus spin labeled with PC12 at a density of about 10-15% of the total phospholipid. Fusion with erythrocytes was detected as an increase in the spectrum amplitude resulting from spin-spin interactions. Their results differ from ours in that they are dependent on the amount of virus used and in that they observe larger spectral changes when Sendai virus is fused with intact erythrocytes than with ghosts. Several reasons could account for these discrepancies. Their data is not corrected for the hemolysis induced structural change (Lyles and Landsberger, 1979). This may explain their observation that the fusion kinetics of Sendai virus with intact erythrocytes and ghosts are different, which we found to be the same. The measured spectral splitting $(2A'_{zz})$ is linear in the amount of virus fused (Lyles and Landsberger, 1979); the increased peak height due to decreased spin-spin interaction is not linear in terms of fused virus. The extent of fusion measured by our approach is consistent with biochemical evidence (c.f. Lyles and Landsberger, 1979).

4.   Sendai virus fusion with liposomes: It has been of interest to determine whether there is a protein receptor in the cellular plasma membrane for the Sendai virus F1,2 protein which is involved in fusion of the virus particle with the plasma membrane. Using electron microscopy methods, Haywood (1974, 1975) showed that Sendai virus could fuse with protein-free liposomes containing phosphatidylcholine, phosphatidylethanolamine, cholesterol, and bovine gangliosides, where the gangliosides serve as virus attachment sites.

Since it was not possible from these experiments to determine the kinetics of fusion of Sendai virus with the liposomes, spin label ESR methods were used to examine the rate and extent of Sendai virus fusion with liposomes composed of sphingomyelin, phosphatidylcholine, cholesterol, phosphatidylethanolamine, and brain gangliosides (Landsberger et al., 1981). A significant fraction (40-50%) of egg-grown Sendai virions fuses with these liposomes. The fusion reaction does not absolutely require the presence of host cell proteins. No fusion was observed between liposomes and MDBK-grown Sendai virions which lack fusion activity. The fusion process appears to follow the form of a first-order reaction which is independent of virus concentration. The half-time of the reaction is about 55 min.

The qualitative similarity in the kinetics of Sendai virus fusion with liposomes and erythrocytes suggests that liposomes are appropriate models for the study of viral and cell plasma membranes. The rate and extent of fusion of Sendai virus with liposomes is dependent on the composition of the liposomes but no absolute lipid requirements were observed. The presence of gangliosides as virus binding sites in the liposomes does not appear to be a prerequisite for fusion with Sendai virus. In the ESR experiments, the binding step may be replaced by close contact between the virions and the liposomal surfaces. In the more physiological situation, where before a virus is bound to the cell the mean distance between virus particle and cell is comparatively large, virus binding to the cell surface may lengthen the residency time of a virus particle sufficiently to permit fusion to occur.

5. Structural changes in the host plasma membrane during the replicative cycle of VSV: We have examined changes in the lipid bilayer of the plasma membrane of intact BHK cells as a function of the time after infection with VSV (Landsberger and Altstiel, 1980; Altstiel and Landsberger, 1982b). Initially, there is an increase in the plasma membrane bilayer rigidity due to binding of the virus to the cell surface. At approximately four hours post-infection, there is also an increase in the plasma membrane rigidity in comparison with uninfected cells. Comparison with the growth curve of VSV progeny particles suggests that this increase in plasma membrane bilayer rigidity may be due to the insertion of viral proteins into the plasma membrane. This result is consistent with the observation that the lipid bilayer of the mature virus particle is more rigid than that of the plasma membrane.

6. Interaction of interferon with the plasma membrane: Interferons are proteins which are excreted by cells after exposure to either infectious or inactivated virus or to double-stranded RNA. When added to uninfected cells, interferons inhibit the replication of a wide variety of viruses (c.f. reviews by Chany, 1976; Gresser, 1977; Friedman, 1977; Stewart, 1979). It has been suggested that interferon has an effect similar to hormones on the cell membrane (Baron and Buckler, 1963; Grollman et al., 1977).

We have used spin label methods to investigate the structural changes in the plasma membrane lipid bilayer of interferon-treated HeLa-S3 tumor cells (Pfeffer et al., 1981). Treatment for 1 hr causes an increase in bilayer rigidity which disappears within 6 hr. By 24 hr after the beginning of interferon treatment, there is another increase in plasma membrane rigidity which is observed for at least two days. These effects appear to be species specific. Mouse interferon treated L-cells have an increased plasma membrane bilayer rigidity. However, human interferon-treated L-cells or mouse interferon-treated HeLa cells are not detectably changed in their plasma membrane. The dose dependence of the interferon effect on the plasma membrane lipid bilayer has been examined, and it appears to follow the dose dependence of the biological effects of interferon. We have found that human interferon also changes the lipid bilayer fluidity of intact human erythrocytes, supporting the suggestion that interferon can act at the plasma membrane.

Treatment of human HeLa-S3 cells with human B interferon alters lipid metabolism by increasing cholesterol synthesis 60% at 24 hr after the beginning of treatment, and 450% at 48hr (Pfeffer et. al.). In contrast, the (^{14}C)acetate labeling of sphingomyelin, phosphatidylethanolamine, and triglycerides shows no change as a result of interferon treatments, while the labeling of phosphatidylcholine was moderately increased.

The apparent biphasic nature of the interferon effect on membrane rigidity may reflect two separate phenomena: an early stage which may be related to the attachment of interferon to its cell surface receptors, and a late stage which may be associated with the fully developed phenotype of interferon-treated cells characterized by an altered cytoskeleton structure and decreased rate of cell proliferation and other changes.

ACKNOWLEDGEMENTS

This work has been supported by Research Grants GM31790 from the National Institutes of General Medical Sciences and PCM8409213 from the National Science Foundation. F.R.L. is an Andrew W. Mellon Foundation Fellow. The author wishes to acknowledge Linda M. Forsythe and Jane I. Epstein for their help in the preparation of this manuscript.

REFERENCES

Altstiel, L.D., and Landsberger, F.R. (1977), Nature 269, 70.
Altstiel, L.D., and Landsberger, F.R. (1981), J. Virol. 39, 82.
Altstiel, L.D., and Landsberger, F.R. (1981a), Virology, 115, 1.
Altstiel, L.D., Landsberger, F.R., and Compans, R.W. (1975), Abs. Ann. Meeting ASM, p.238.
Bachi, T., Deas, J.E., and Howe, C. (1977), in "Cell Surface Reviews," G. Poste and G.L. Nicholson, eds., 2, p. 83, North-Holland, the Netherlands
Bachi, T., Gerhard, W., Lindenmann, J., and Muhlethaler, K. (1969), J. Virol. 4, 769.
Barenholz, R., Moore, N.F., and Wagner, R.R. (1976), Biochemistry 15, 3563.
Baron, S., and Buckler, C.E. (1963), Science 141, 1061.
Chany, C. (1976), Biomedicine 24, 148.
Choppin, P.W., Compans, R.W., Scheid, A., McSharry, J.J., and Lazarowitz, S.G. (1972), in "Membrane Research," C.F. Fox, ed., Academic Press, New York.
Choppin, P.W., Klenk, H.D., Compans, R.W., and Caliguiri, L.A. (1971), Perspect. Virol. 7, 127.
Compans, R.W., and Dimmock, N.J. (1969), Virology 39, 499.

Compans, R.W., Klenk, H.D., Caliguiri, L.A., and Choppin, P.W. (1970), Virology 42, 880.

Dales, S. (1973), Bact. Rev. 37,103.

Fawcett. D.W., and Witebsky, F. (1964), Z. Zellforsch. Mikrosk. Anat. 62, 785.

Friedman, R.S. (1977). Bact. Rev. 41, 543.

Gaffney, B.J., and Mich, R.J. (1976), J. Amer. Chem. Soc. 98, 3044.

Gresser, I. (1977), Cellular Immun. 34, 406.

Griffith, O.H., Dehlinger, P.J., and Van, S.P. (1974), J. Mem. Biol. 15 159.

Griffith, O.H., and Jost, P.C. (1976), in "Spin Labeling Theory and Applications," L.J. Berliner, ed., p. 453, Academic Press, New York.

Grollman, E.F., Lee, G., Ramos, S., Lazo, P.S., Kaback, H.R., Friedman, R. and Kohn, L.D. (1978). Cancer Res. 38, 4172-4185.

Gunther, G.R., Wang, J.L., Yahara, I., Cunningham, B.A., and Edelman, G.M. (1973), Proc. Natl. Acad. Sci. USA 70, 1012.

Haywood, A.M. (1974), J. Mol. Biol. 83, 427.

Haywood, A.M. (1975), in "Negative Strand Viruses," R.D. Barry and B.W.J. Mahy, eds., Vol. II, p.923, Academic Press, London.

Howe, C., and Lee (1970), Adv. Virus Res. 17, 1.

Klenk, H.D., and Choppin, P.W. (1969), Virology 38, 255.

Klenk, H.D., and Choppin, P.W. (1970a), Virology 40, 939.

Klenk, H.D., and Choppin, P.W. (1970b), Proc. Natl. Acad. Sci. USA 66, 57.

Kuroda, K., Maeda, R., Ohnishi, S.-I. (1980), Proc. Natl. Acad. Sci. USA 77, 804.

Landsberger, F.R., and Altstiel, L.D. (1980), Ann. N.Y. Acad. Sci. 348, 419.

Landsberger, F.R., and Compans, R.W. (1976), Biochemistry 15, 2356.

Landsberger, F.R., Compans, R.W., Choppin, P.W., and Lenard, J. (1973), Biochemistry 12, 4498.

Landsberger, F.R., Greenberg, N., and Altstiel, L.D. (1981), in "Replication of Negative Strand Viruses." D.H.L. Bishop and R.W. Compans, eds., Elsevier. p. 517.

Landsberger, F.R., Lenard, J., Paxton, J., and Compans, R.W. (1971), Proc. Natl. Acad. Sci. USA 68, 2579.

Landsberger, F.R., Lyles, D.S., and Choppin, P.W. (1978), in "Negative Strand Viruses and the Host Cell," B.W.J. Mahy and R.D. Barry eds., p. 787, Academic Press, London.

Lonberg-Holm, K. and Phillipson, L. (1974), in "Monographs in Virology," J.L. Melnick, ed., 5, Karger, Basel.

Lyles, D.S., and Landsberger, F.R. (1976), Proc. Natl. Acad. Sci. USA 73, 3497.

Lyles, D.S., and Landsberger, F.R. (1977), Proc. Natl. Acad. Sci. USA 74, 1918.

Lyles, D.S., and Landsberger, F.R. (1978), Virology 88, 25.

Lyles, D.S., and Landsberger, F.R. (1979), Biochemistry 18, 5088.

McLaughlin, S., and Harary, H. (1976), Biochemistry 15, 1941.

McSharry, J.J., and Wagner, R.R. (1971), J. Virol. 7, 412.

Moore, N.F., Barneholz, Y., McAllister, P.E., and Wagner, R.R. (1976a), J. Virol. 19, 275.

Moore, N.F., Barneholz, Y., and Wagner, R.R. (1976b), J. Virol. 19, 126.

Pfeffer, L.M., Landsberger, F.R., and Tamm, I. (1981), J. Interferon Res., 1, 613.

Scheid, A., and Choppin, P.W. (1973), J. Virol. 11, 263.

Scheid, A., and Choppin, P.W. (1977), Virology 80, 54.

Sefton, B.M., and Gaffney, B.J. (1974), J. Mol. Biol. 90, 343.

Sherbet, G.V. (1978), "The Biophysical Characterization of the Cell Surface," Academic Press, New York.

Stewart, W.E. (1979), "The Interferon System," Springer, Austria.

Tanaka, K.-I., and Ohnishi, S.-I. (1976), Biochim. Biophys. Acta 426, 218.

Verwey, E.J.W., and Overbeek, J.T.G. (1948), "Theory of Lyophobic Colloids," Elsevier Publishing Co., Amsterdam.

White, D.O., Tayler, J.M., Haslam, E.A., and Hampson, A.W. (1970), in "The Biology of Large RNA Viruses," R.D. Barry and B.W.J. Mahy, eds., p. 602, Academic Press, London.

MODIFICATION AND SPIN LABELING STUDIES OF MEMBRANE BIOENERGETICS

Lester Packer and Rolf J. Mehlhorn

Department of Physiology and Anatomy
University of California
Berkeley, CA 94720

INTRODUCTION

We are using chemical modification techniques, alone and in combination with nitroxide spin probes, to study biological energy converters, emphasizing light-energized systems with relatively simple structures. The principal research objective is to elucidate the molecular mechanisms of energy conversion by photocatalysts. Bacteriorhodopsin (BR) has been the primary focus of our research program; but recently we initiated some studies with bacterial chromatophores, which contain only photosystem I, but not the water-splitting photosystem II of higher plants.

Bacteriorhodopsin is a relatively low molecular weight protein found in the purple membranes of the halophilic bacterium Halobacterium halobium. The function of this protein in the bacterium is known to be the active pumping of protons from the cytoplasm to the external aqueous phase, using light energy at wavelengths of about 570nm. Unlike the pigments that are found in higher plants, bacteriorhodopsin acts as a direct photoelectrical device, without the participation of electron-transport proteins (1). This simplicity of action as well as the remarkable stability of BR in isolated purple membranes has generated intense, worldwide interest in how bacteriorhodopsin achieves its energy conversion. Research in other laboratories has yielded the primary amino acid sequence (2), electron density data to a resolution of seven angstroms (3), and information about the position of the light-absorbing moiety, retinal (4-6). Studies in our laboratory have been concerned primarily with identifying specific amino acids that actively participate in BR function.

Our basic strategy for studying the roles of specific amino acids in bacteriorhodopsin consists of altering the structure of BR with chemical reagents of high specificity to learn how the various amino acids contribute to photon capture and the subsequent proton pumping. Among the functions that we routinely assay with spectrophotometric and laser flash photolysis techniques are BR absorption spectra and the kinetics of the photocycle both in isolated purple membranes and in sealed envelope vesicles derived from the bacteria. We also perform proton pumping assays with sealed membrane preparations, including lipid vesicles that have been reconstituted with modified purple membranes. There are multiple appearances of all the amino acids that occur in BR and it is therefore difficult to label only a single one and to assign a functional significance to it. Thus, most of our studies have analyzed the effects of multiple labeling under various modification conditions and have sought to characterize the spatial distribution of labeled sites within the protein by structural analyses. This approach has provided evidence for the roles of lysines (7-10), arginines (10-12), tyrosines (9,13-15), tryptophans (14,16) and carboxyl (9,10,18,19) residues in the activity of the protein (20). We have shown that some of these amino acids are in close proximity to the retinal chromophore, whereas others are involved directly or indirectly in conducting the proton across the membrane.

We have also experimented with retinal-deficient mutant-derived "white membranes" (21) which can be converted to spectroscopically normal purple membranes upon addition of retinal (22). These preparations provide novel research opportunities for us because they can be chemically modified when the normal retinal binding site is not occupied. White membranes also have quite different physical properties than purple membranes (23).

SPIN PROBE METHODS

We have developed the use of spin labels for accurately measuring energy transduction phenomena involving biological membranes, including internal aqueous volume, pH and electrical potential (24-28), and membrane surface potential (29). These tools have been successfully applied to four systems: halobacterial membrane vesicles (24,26,28,30), isolated and reconstituted purple membranes (31) and bacterial chromatophores (33). Our technique consists of observing electron spin resonance spectra of nitroxide stable free radicals within membrane enclosed domains and inferring the concentrations of probes from these spectra. Among the probes we use for this purpose are weakly polar nitroxides whose intracellular signals are representative of cell volumes, spin-labeled amines and weak acids, whose intracellular spectra are indicative of transmembrane pH gradients, and spin-labeled hydrophobic ions which provide information about

SPIN PROBES FOR BIOENERGETICS STUDIES

PARAMETER	NAME	STRUCTURE	COMMENTS
Volume	TK	O-N⟩=O	a, b
	2N3	O-N⟩O	a, b
Δ pH	TA (R=R'=H) MTA (R=H,R'=Me) DMTA (R=R'=Me)	O-N⟩—N⟨R R'	c, d, pK_a=9.0 c, d, pK_a=9.1 c, d, pK_a=8.6
	Tempmorpholine	O-N⟩—CH₂CH₂—N⟨O	c, d, pK_a=7.7
	TC	O-N⟩—C-OH	c, e, pK_a=4.8
$\Delta\Psi$	Kϕ	O-N⟩—CH₂—P⁺(C₆H₅)₃	f
$\Delta\Psi_0$	CAT_n	O-N⟩—N⁺(CH₃)₃—$(CH_2)_{n-1}$—CH₃	g
	AN_n	O-N⟩—O-P(O)(O⁻)-O-$(CH_2)_{n-1}$—CH₃	h

TK = Tempone, TA = Tempamine, TC = Tempcarboxylate

a) high permeability ($t_{1/2} < 0.1Gs$)
b) narrow intrinsic line width W_o= 0.4G
c) high permeability for unprotonated species ($t_{1/2} < 0.1s$)
d) intrinsic line width W_o= 1.6G
e) intrinsic line width W_o= 1.5G
f) slow permeability ($t_{1/2} \sim 1m$)
g) slow permeability as ion pairs in halobacterial cell envelope
   vesicles ($t_{1/2}$ = 8m for n=8); impermeable in erythrocytes.
h) slow permeability in erythrocytes ($t_{1/2} \sim 10m$)

77

electrical potentials. Since all these parameters are measured under nearly identical conditions the results of our analyses are considerably more accurate than commonly used indirect assays of electrochemical potentials, such as measurements of radiolabeled tracers in flow dialysis experiments or studies with fluorescent dyes based on self quenching effects.

The essence of our method is to treat membrane preparations with a combination of nitroxide spin labels and paramagnetic ions complexes. The paramagnetic ions are not seen directly in the ESR spectra, but they cause a broadening of the nitroxide signal which is linearly dependent on the concentration of added paramagnetic ion complexes. By choosing multivalent ionic complexes, we ensure that these complexes do not cross biological membranes and thus only the nitroxides that are located outside a cell are broadened. With sufficient concentrations of a paramagnetic complex, line broadening is so effective that only the intracellular nitroxides are observed in the spectra; thus we can directly infer intracellular concentrations of probes under these conditions. We have tested a considerable variety of paramagnetic ion complexes and have identified several that are suitable for studies of bioenergetic phenomena. Among the most effective complexes are ferricyanide and manganese-EDTA. In most situations these agents are effective at concentrations of less than 100mM, a concentration that does not cause serious osmotic problems with cells or membrane vesicles.

Often it is of interest to perform time-resolved volume studies and this requires spin probes that are capable of crossing membranes at a rapid rate. We have examined the membrane permeability of a diverse group of probes with a rapid mixing apparatus and have found that a sufficient condition for rapid membrane permeability is that nitroxides contain fewer than two hydrogen bonding residues within the molecule. The most useful nitroxide we have characterized is Tempone. This molecule has the desirable characteristics of narrow intrinsic line widths, rapid membrane permeability and sufficient water solubility to yield sharply defined aqueous spectral lines in most membrane systems.

Proton gradients are common among energy-transducing membranes and their measurement can provide essential information for understanding energy conversion mechanisms. Spin probes also lend themselves to such measurements with nitroxide-labeled amines and carboxylic acids. The former accumulate in acidic aqueous compartments whereas the latter concentrate in alkaline compartments. Measured concentration gradients of these spin labeled species can be related to the magnitudes of pH gradients. A major advantage of spin labels over other probes for this purpose is that the spin labels give clearly distinct spectra depending on whether they are located in polar or nonpolar environments. This property allows us to determine the extent of membrane binding of the probes. Since the proportion of membrane-bound probes can be quite large for some molecules the correction for this effect can be crucial for an accurate determination of pH gradients.

Electrical events appear to play fundamental roles in most biological energy conversion processes; accordingly the accurate measurement of electrical transmembrane potentials is of great interest. The two major approaches for measuring membrane potentials are the use of microelectrodes and the application of probes that can cross membranes as charged species. Although many monovalent ions appear to exhibit some membrane permeability, the most useful ions for bioenergetic studies appear to be the so-called hydrophobic ions. These ions are surrounded by covalently-linked hydrophobic groups which confer significant lipid solubility on the ions and thus enable them to cross membranes at relatively rapid rates. The classical hydrophobic ion that has enjoyed extensive application to bioenergetics studies is tetra-phenyl phosphonium. The time resolution of the phosphonium ions does not permit the measurement of some interesting transient phenomena; however equilibrium potentials are adequately determined. The correction for membrane binding of the phosphonium probes is substantial. We have shown that its neglect can lead to overestimates of potentials of more than 60 millivolts (24). Such an error can have enormous effects on testing theories concerned with the mechanism of energy coupling between electrochemical and chemical potentials, e.g., Mitchell's chemiosmotic hypothesis.

We have been intrigued by the possibility that other ions whose charge is highly delocalized might exhibit superior permeability characterisics. To explore this possibility we have synthesized a series of spin labeled pyridinium ions and have studied their permeability characteristics (34). The results of these studies are that charge delocalization does not confer the desired potential-sensitivity upon these probes, although they do appear to have high membrane permeability. This apparent paradox has been ascribed to the inference that these probes cross membranes as ion pairs. Apparently, the structure of the aromatic rings associated with the pyridinium salts is sufficiently planar to allow the formation of uncharged complexes which easily cross membranes, but which are, of course, unaffected by electrical gradients. In light of these findings we have continued to rely upon the phosphonium spin labels for measuring electrical membrane potentials.

CHEMICAL MODIFICATION ON AND SPIN PROBE RESULTS FOR BACTERIORHODOPSIN

The table on the following page summarizes what we have learned from chemical modification studies of bacteriorhodopsin.

Table: Data on BR from chemical modification and spin-probe studies
----

Lysines residues do not participate in the pathway of proton conduction.

Three tyrosines are proton acceptors/donors: at 26 and 64, as revealed by iodination and within the 72-248 proteolytic fragment as shown by nitration

Reprotonation requires the participation of ion pairing between carboxyl and arginine groups.

At least one carboxyl group is deeply imbedded within the hydrophobic core of BR (spin labeling of carboxyls)

Essential groups: Schiff base[*], phenolate, carboxylate

Schiff base is protonated for BR_{570}, deprotonated for M_{412}[*]

The O_{640} intermediate is not related to the blue acid species

Proton transfer accompanies the rise and decay of the M_{412} intermediate (deuterium isotope effects cause a 5-6-fold slowdown of the rise and a 2-fold slowdown of the decay kinetics of M_{412})

Photocycle kinetics are regulated by the protonmotive force (ESR spin probe studies)

Intermolecular motion is not a prerequisite for function (crosslinking data)

Intermolecular interaction is not required for function (monomerization of BR in vesicles)[*]

Intramolecular conformational freedom is essential for function (crosslinking data)

Cytoplasmic tail is partially immobilized, required for full activity (spin labeling of carboxyls, proton/M_{412} stoichiometry)

Tryptophans interact with retinal (iodination)

Significance unclear (zero-length crosslinking alters pattern of proteolytic cleavage)
----
[*] work from other laboratories.

Apart from reversible protonation of the Schiff base attachment site of retinal to the epsilon amino group of lysine 216 (which has been demonstrated by work in other laboratories), we have shown that perhaps three tyrosine residues participate in the activitiy of bacteriorhodopsin as a proton pump. In addition to reversible protonation of tyrosyl groups, the participation of carboxyl residues is essential for photocycle activity. However, lysine groups, long considered as potential candidates for proton translocation, do not appear to be involved directly in proton translocation.

Short range molecular motion appears to be essential for activity because crosslinking inhibits the reprotonation phase of the photocycle, particularly, zero length crosslinking occuring between lysine and carboxyl residues. The latter observation has been exploited recently to show that selective patterns of proteolytic cleavage of bacteriorhodopsin can be accomplished following such intermolecular crosslinking.

SELECTED DETAILED RESULTS

Lysine Groups: There are seven lysine groups in bacteriorhodopsin. One of them, lysine 216, has its epsilon amino group covalently bound in a Schiff base linkage with the retinal chromophore. The other six groups were earlier thought to be favorable candidates for participating in the translocation of protons in a relay involving their uptake and release from the surfaces of the membranes and possibly even through the protein interior. This idea was tested very early in our studies using a wide variety of reagents that react with lysine groups to form derivatives with altered proton-binding characteristics. Among these, we used imidoesters of different sizes , which do not appreciably change charge, but do change the pK_a's of reacted lysines (35,36). If the charge of the lysine group is unmodified but a bulky substituent is reacted with it, an average of 5.5 of the 6 available lysine residues can be modified (35). Under these conditions no change in photocycle kinetics or proton pumping activity is observed. These finding indicate that lysine groups per se are not involved directly in proton translocation. However, crosslinking of lysine groups with other lysine groups or carboxyl groups (see below) inhibits proton pumping in the second phase of the photocycle, i.e., reprotonation (36). Fluorescamine modification of lysines causes loss of the exciton coupling band in the CD spectrum, without inhibiting light-to-dark adaptation, indicating that this coupling between neighboring bacteriorhodopsin molecules is not required for adaptation.

Tyrosine modification: BR contains 11 tyrosines whose titratable hydroxyls are plausible candidates for conducting protons across the membrane. We have employed two strategies for modifying this amino acid: iodination of surface groups with the membrane-impermeable reagents glucose oxidase and lactoperoxidase in the presence of iodide ion (38,39), and nitration with tetranitromethane (39,40-45). The former treatment showed that the tyrosines most sensitive to alterations in their photocycle kinetics are located at or near the cytoplasmic surface of the purple membrane. Correlation of an altered pH-dependence of proton pumping activity with pK_a shifts of iodinated tyrosines has implicated tyrosines as donors/acceptors in the proton conduction pathway. Nitration caused blue-shifts in the absorption spectrum, suggesting that tyrosines play a role in determining the absorption spectrum of BR. Nitration also induced changes in the circular dichroism spectra of purple membranes that could be ascribed to a weakening of the interaction between neighboring BR molecules. When nitration was performed under illumination at low pH, we discovered a new light-dependent nitration reaction (45,46) having a considerable improvement in the specificity of the reaction - a single tyrosine residue could be modified under the reaction conditions. Nitration in the dark at alkaline pH caused one type of blue shift, whereas nitration in the light at acidic pH affected another tyrosine and caused a distinct blue shift. This new modification, taken together with results from other laboratories, suggests that at least 3 of the 11 tyrosine residues are essential for proton pumping activity (39).

Arginine modification: treatment of BR with 2,3-butanedione revealed that only 3 to 4 of the 7 available arginines in BR are accessible to this reagent (35,44). Modification of only two of the arginines caused a marked slowdown of the photocycle kinetics but without an inhibition of the rise time of the M412 intermediate. An analysis of the pH dependence of the photocycle kinetics revealed that arginines play a role in the rate of photocycling. The mechanism of arginine involvement in the photocycle kinetics could be as structure stabilizers by forming salt bridges with anionic amino acids, or, alternatively, arginines could serve as proton donors to the Schiff base nitrogen of retinal. The arginine residues affected by this chemical modification appear to be involved in the reprotonation phase of the photocycle.

Carboxyl modification: We have applied a water soluble and a hydrophobic carboxyl-activating reagent to BR to probe the occurrence of acidic amino acids in environments of different polarities (47-50). A hydrophobic quinoline, N-ethoxycarbonyl-2-ethoxy-1,2-dihydroquinoline (EEDQ), has been used to activate primarily carboxyls that are buried within the nonpolar interior of BR. Polar carboxyl groups in BR were activated with the water soluble reagent ethyldimethylpropylaminecarbodiimide (EDC). Activated carboxyls will react with available nucleophiles to form amide bonds. If lysine residues within BR are located in sufficient proximity to the activitated carboxyls they will react with them

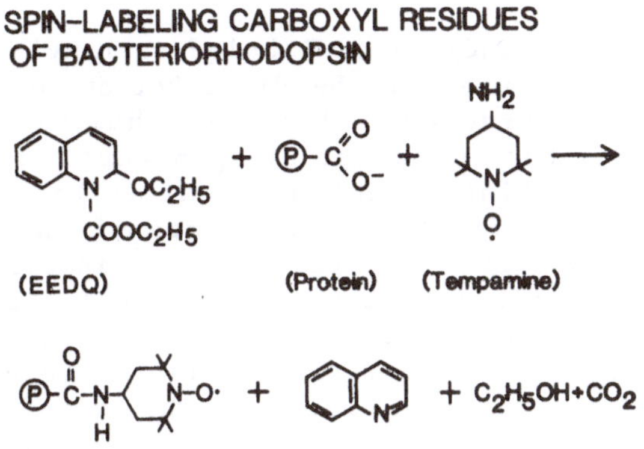

SPIN-LABELING CARBOXYL RESIDUES OF BACTERIORHODOPSIN

(EEDQ)   (Protein)   (Tempamine)

thus effecting intramolecular (and, perhaps, intermolecular) crosslinkages. This might be expected to reduce the capability of BR to undergo energy-linked conformational changes. Indeed, we have observed that crosslinked BR has altered photocycle kinetics and is substantially impaired in its ability to pump protons. When crosslinking is partially prevented by treatment with an exogenous nucleophile, e.g., glycine methyl ester (GME), less inhibition is observed. The choice of GME as a nucleophile in this reaction is predicated on the fact that the extent of reaction is easily determined with an amino acid analyzer. Other useful nucleophiles are aminoethylsulfonic acid (AES), whose negative charge ensures that the ionic properties of BR will be conserved during the modification, nitrotyrosinemethylester, whose pH-sensitive chromophore yields data on energy tranfer processes with retinal (49), and 4-amino,2,2,6,6-tetramethyl piperidine-N-oxyl (Tempamine), a spin labeled reporter group (51). The reaction scheme for the spin labeling is shown above.

In the presence of an excess of the nucleophile GME, up to 12 of the 23 carboxyl-containing amino acids of the protein can be chemically modified. Interestingly, up to 10 carboxyls can be modified without any perceptible loss of structural integrity.

Treatment of control and crosslinked preparations with proteolytic enzymes has enabled us to assign modified amino acids to specific BR fragments (50); this information in turn has helped in tentatively locating the carboxyls in the tertiary structure of BR. Future use of these modifications, in conjunction with proteolytic cleavage and reconstitution work, will also enhance the structural studies of bacteriorhodopsin by computer modeling techniques (53).

Because of the abundance of labeled carboxyls in BR, it has been difficult to learn about the distribution of reacted carboxyls within the BR structure. This issue has been tackled with some success using the technique of spin labeling (31). We have shown with spin labeling techniques that at least one carboxyl group is located in the hydrophobic core of the protein, at a distance of more than 16A from the membrane interface. Moreover, the distribution of spin labels was unexpectedly heavily weighted towards non-aqueous amino acids. Apart from the important conclusion that ionizable groups can reside within hydrophobic domains of BR, the spin-labeling study has laid the groundwork for analyzing the labeling pattern of many chemical modification procedures. This will enable us to perform more incisive chemical modification studies of proteins in the future.

The spin-labeled bacteriorhodopsin could be used to study proteolytic cleavage of the C-terminal tail of the protein as shown in Fig. 1, on the following page. While this relatively polar tail was attached to the protein it yielded a partially immobilized ESR spectrum, but upon treatment with trypsin the ESR spectra revealed a progressive increase in motional freedom of some spectral component, which was ascribed to the liberated tail in separate centrifugation experiments.

Retinal Chromophore: The retinal chromophore of bacteriorhodopsin has been shown in work from other laboratories to be essential for the proton pumping function of bacteriorhodopsin. The Schiff base nitrogen at the site of attachment of retinal to lysine 216 becomes deprotonated and protonated during the photoreaction cycle, suggesting that the residue may be one source of the protons pumped. Retinal also undergoes a cyclic change in the structure of its polyene chain, from all-trans to 13-cis configuration (M_{412} state), returning to the all-trans configuration upon completion of the photocycle (BR_{570} state).

The retinal chromophore of bacteriorhodopsin can be removed in the light, in the presence of hydroxylamine, yielding the colorless bacterioopsin. We have shown that the passive proton permeability across bacterioopsin is slightly enhanced and that this enhancement is specifically blocked by regenerating the chromophore with all-trans retinal. These studies suggest that retinal

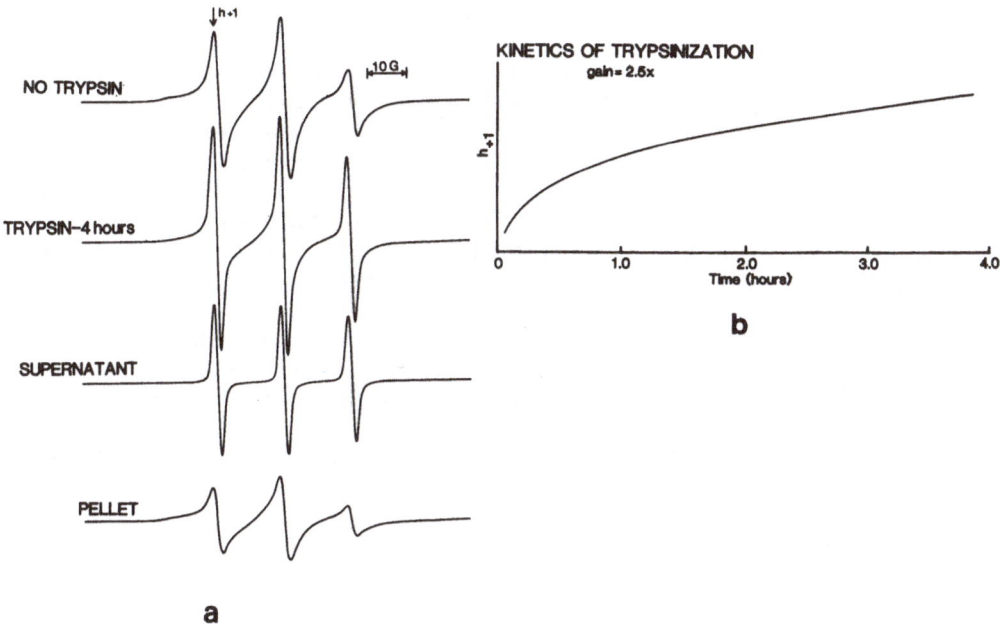

TRYPSIN TREATMENT OF CARBOXYL
SPIN-LABELED BACTERIORHODOPSIN

Fig. 1a) ESR spectra of carboxyl spin-labeled BR treated with
trypsin. b) Kinetic ESR line height trace showing the action of
trypsin on purple membranes (51,with permission from the authors).

may act as the gate for proton conduction.

The bacterioopsin apoprotein can be reconstituted with retinal analogues in order to probe the interactions of the protein
with the chromophore. In one collaborative effort we have been
preparing the way for mechanistic studies using photoaffinity
labeling with retinal analogues (52) by using a "mesityl" retinal.
Results have shown that bacteriorhodopsin reconstituted in this
way retains a photocycle, albeit greatly slowed, indicating that
the substitution is in the same place as the native retinal.

White membranes, obtained from retinal-deficient mutants of
halobacteria, have also been investigated. Our work has shown that
the structure of reconstituted bacteriorhodopsin in the white
membrane and its function are quite similar to that in native
purple membranes (i.e., primary sequence and pathway of the
photocycle are the same); however the lipid environment of the
protein is different. Consistent with this altered lipid
structure, the dependence of the photoreaction cycle and the
proton pumping process is affected differently in white and purple
membranes by physical factors such as temperature. These studies

85

indicate that the lipids interspersed between bacteriorhodopsin molecules in the purple membrane are important for modulating the activity of bacteriorhodopsin.

QELS studies of the size and surface charge of BR: We have collaborated with Bernard Arrio and. Pierre Volfin (University of Paris, Orsay) in utilizing laser doppler velocimetry (LDV) and quasi-elastic light scattering (QELS) to evaluate accurately and quantitatively changes in surface charge of purple membranes and purple membrane-containing vesicles (LDV technique) and the size distribution of purple membrane preparations (QELS technique). These studies have led to two important results (54):

(1) an absolute measurement of the surface charge of the purple membrane before and after chemical modification has been made, and light-induced changes in surface charge have been determined in suspensions of purple membranes. The method also appears promising for separating and characterizing reconstituted liposomes containing purple membranes with or without other catalysts as it is possible to directly visualize, and thus to separate, populations of these vesicles moving with different velocities in an electric field.

(2) routine analysis revealed a striking difference in the mobilities of purple and white membrane preparations. Purple membrane preparations migrated more slowly than white membrane preparations under applied electric fields, implying that the effective sizes of the former were larger than the latter. Negative staining microscopy in conjunction with QELS revealed that purple membrane preparations were aggregated or stacked whereas white membranes were smaller and unstacked. We were able to show that this was, in effect, due to low level proteolytic cleavage of the terminal tail. Subsequently we showed that controlled trypsin cleavage of the C-terminal tail produced a massive aggregation of purple membrane preparations. Under these conditions membranes were observed, by electron microscopy, to be regularly stacked upon one another. This is readily explained by the removal of the highly charged C-terminal tail of 21 amino acid residues having 5 carboxyl groups, which reduces the electrostatic relative to the hydrophobic forces between purple membrane sheets. When such aggregated preparations are mixed with lipids and treated under appropriate conditions to form liposomes, apparently these aggregates are teased apart.

These findings suggest that the low values of proton-to-M_{412} stoichiometry (or cations released per M_{412} stoichiometry), reported in the literature for isolated so-called "native purple membranes" as being lower than reconstituted liposomes are artifacts which result from this stacking/aggregation phenomenon. This effect needs to be considered in evaluating all earlier studies reporting stoichiometries of cation-to-photocycle activity by suspensions of purple membrane fragments. Our results are consistent and provide new insights into previous controversies in the literature about these stoichiometries.

Proton Pumping and the Sodium Proton Antiporter in Halobacterial Membrane Vesicles: ESR methods with isolated halobacterial membrane vesicles have proven to be extremely useful for characterizing the activities of bacteriorhodopsin and also halorhodopsin, a second retinal-containing protein in halobacteria, which appears to pump chloride ions. Using the methods described above, accurate changes in internal pH and changes as a function of external pH have been calculated. Since the proton pumping function in naturally oriented BR vesicles is such that protons are pumped from inside of the vesicles to the outside external medium, the probe of choice for light-induced pH studies is is a spin labeled weak acid, Tempcarboxylate (4-carboxy-2,2,6,6-tetramethyl piperidine-N-oxyl) whereas in reconstituted membrane vesicles made from isolated purple membranes and lipids, where the flux of light-induced proton translocation is from the outside to the inside, as well as for halorhodopsin vesicles, where protons passively follow the actively pumped chloride ions, the probe of choice has proven to be Tempamine (4-amino-2,2,6,6-tetramethyl piperidine-N-oxyl). ESR experiments with membrane vesicles in sodium chloride vs. potassium chloride media have permitted us to investigate and quantitate the action of the sodium proton antiporter driven by the proton translocation activity of bacteriorhodopsin. Fig. 2, on the following page, shows that illumination of BR envelope vesicles causes volume loss which is inhibited by addition of the uncoupler FCCP.

These ESR volume measurements have recently been applied to studies with membrane vesicles lacking bacteriorhodopsin but containing halorhodopsin and we have been able to prove unambiguously that volume changes do indeed occur accompanying the action of halorhodopsin, and that these volume increases are dependent on the presence of chloride (or bromide) anions (Mehlhorn, Packer, Schobert and Lanyi, manuscript in preparation).

Demonstration of the Equilibrium between the Photocycle of Bacteriorhodopsin and the Proton Electrochemical Potential Generated by Bacteriorhodopsin: Reconstituted bacteriorhodopsin liposomes have proven to be ideal for demonstrating, for the first time, that the proton electrochemical potential that can be generated by bacteriorhodopsin in a sealed membrane system is in equilibrium with the photoreaction cycle (46,55). By combining the ESR spin probe method for measuring volume, pH and membrane potentials with illumination studies of the formation and decay of the M_{412} intermediate of the photocycle, we have been able to show that the amount of the M intermediate formed during steady state illumination is diminished by reagents which collapse the membrane potential but not the pH gradient (Fig. 3, see over).

This indicated that the proton pumping activity is slowed by the development of a transmembrane electrical potential. Parallel

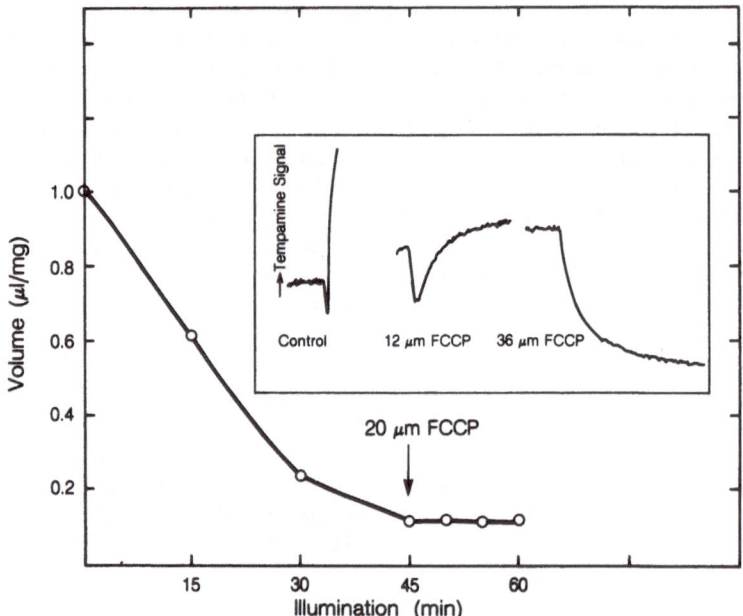

Fig. 2. ESR line heigths of envelope vesicles of S-9 halobacteria as a function of time of illumination and addition of the uncoupler FCCP. To avoid artifacts, the illumination was conducted in the absence of nitroxide and quenching agent; the volumes of the vesicles were assayed immediately afterwards.

studies with laser flash photolysis methods have shown that the transmembrane potential slows down the decay of the M_{412} intermediate but not its rise. These studies are being extended in our laboratory and other laboratories to assess the consequences of this type of regulation not only in reconstituted liposomes but in native membrane preparations and intact cells.

Surface potentials associated with the photocycle of purple membranes: In earlier studies we used our spin probe techniques for measuring surface potentials in conjunction with laser flash excitation of purple membranes in the ESR spectrometer to show that the excitation of BR causes purple membranes to become more negatively charged and that these charge alterations are consistent with the rate of formation of the M412 intermediate (Fig. 4).

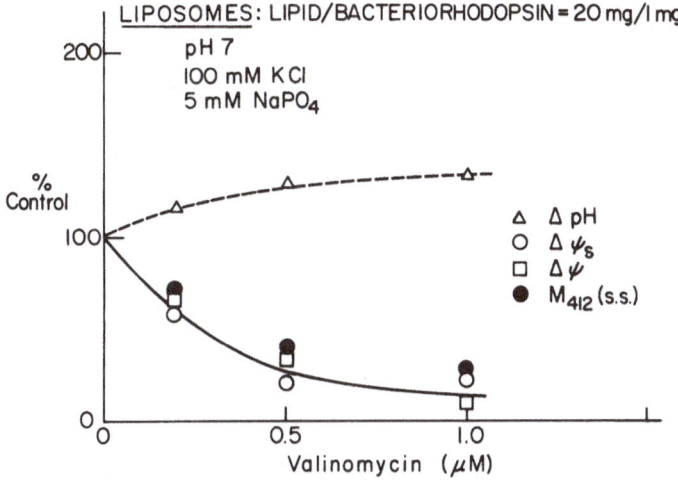

Fig 3. Effect of valinomycin, which collapses transmembrane electrical gradients when added to vesicles suspended in KCl, upon photocycle kinetics.

Subsequently, one of our collaborators on this project (C. Carmelli) pursued these studies in his own laboratory with optical probes having improved time-resolution characteristics and showed that surface charge changes preceded the appearance of the M412 intermediate. On the basis of these observations we have proposed that newly formed negative charges on the purple membrane provide a "sink" for protons that are subsequently released from the Schiff's base nitrogen when the photocycle reaches the M412 stage.

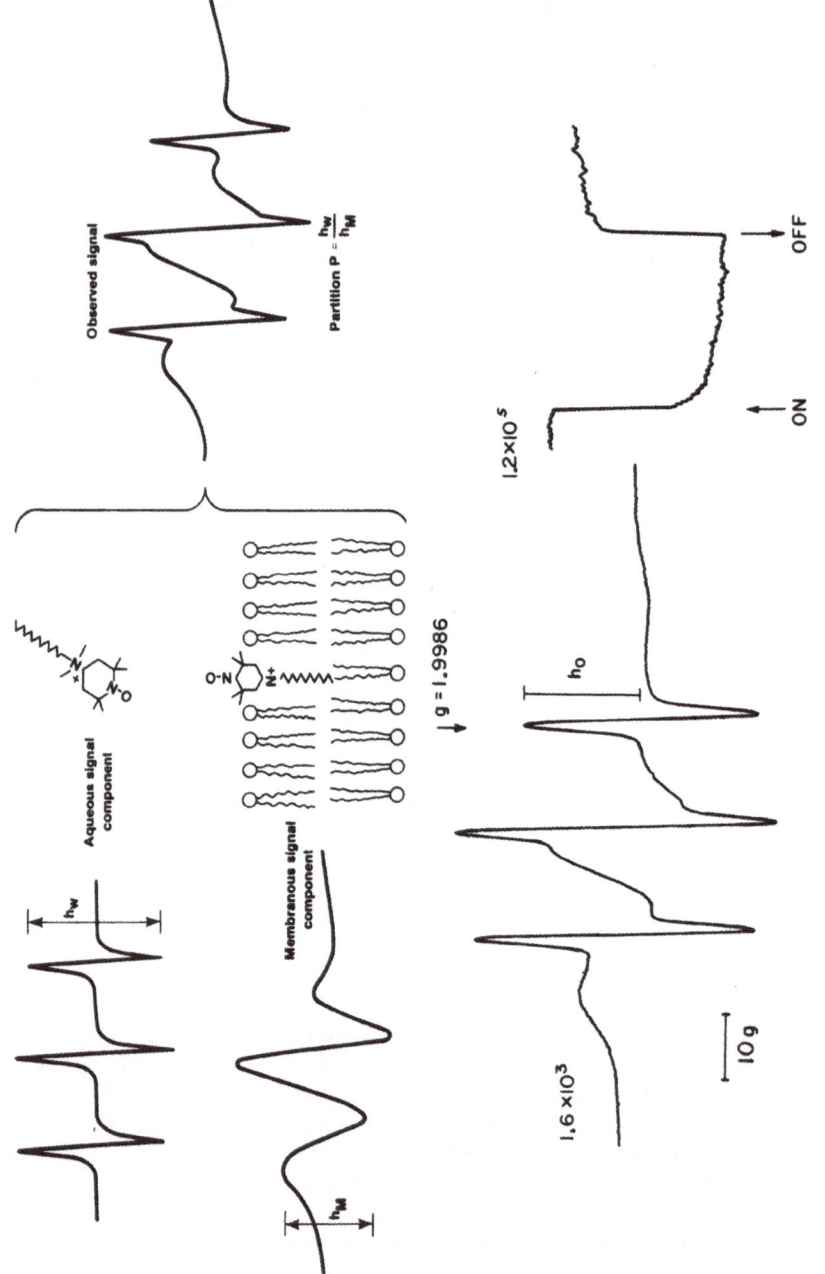

Fig 4. Surface potential changes measured in purple membrane suspensions with the spin-labeled cationic amphiphile CAT$_{12}$.

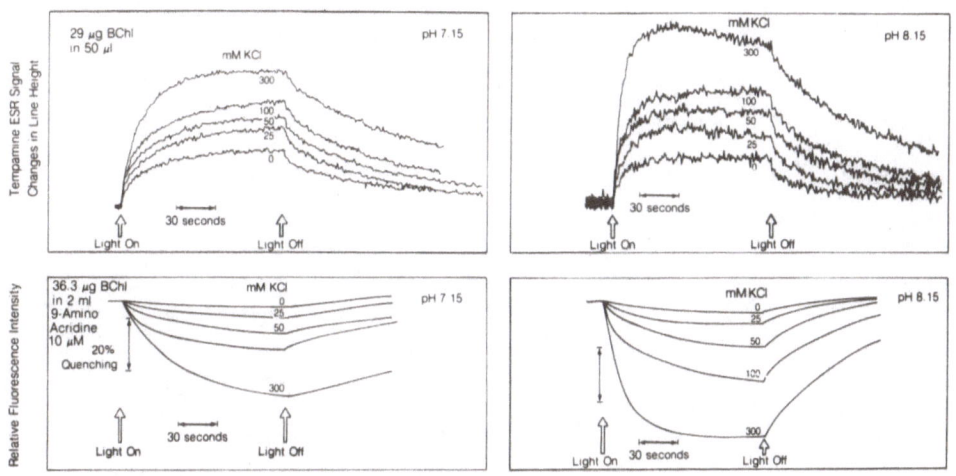

Fig. 5. Fluorescence vs. ESR probe responses to light-induced pH gradients in chromatophores.

Chromatophore studies: our recent studies with chromatophores have been concerned with measuring the pH gradients that are formed upon illumination (33). We have used nitroxide spin probe techniques to show that these pH gradients reach a magnitude slightly greater than two units when the membranes are bathed in media containing relatively high concentrations of potassium. Fig. 5 shows that fluorescence probes and spin probes give similar kinetic responses in chromatophores.

The spin probe technique has two significant advantages for measuring bioenergetics parameters over commonly-used alternative methods, when chemical reduction of the probes does not occur (e.g., in chromatophores). These are that volumes and pH measurements are made under virtually identical conditions, and that membrane binding artifacts can be eliminated. Fig. 6, on the next page shows how we used single light flashes to show that protons are released from the membranes in a progressive manner, thus ruling out a postulated "proton sink" within the membrane interface. Such potential wells had been invoked to suggest that measurements of electrochemical potentials in bulk aqueous media might underestimate the true potentials that could be generated within microdomains of energy-transducing membranes. The high sensitivity of the spin label method made these measurements feasible and makes it possible to detect pH gradients resulting from a single turnover of proton pumping proteins like bacteriorhodopsin.

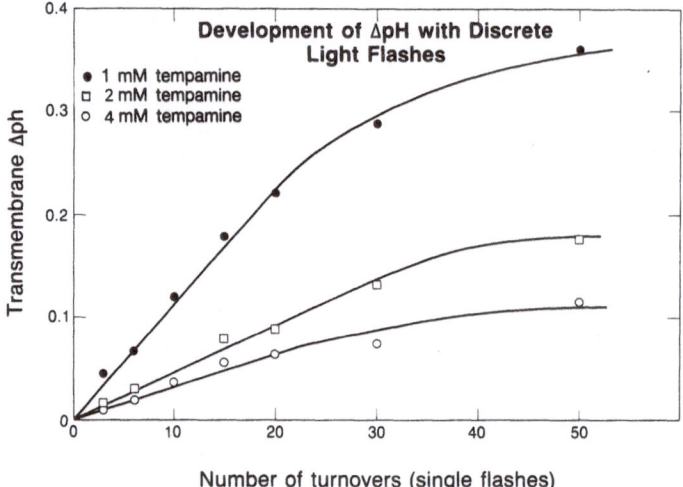

Fig. 6. Measurement of pH gradients resulting from a train of single light flashes in chromatophores of R. sphaeroides. The tempamine concentration was 1 mM (o), 2 mM (□), and 4 mM (•).

ACKNOWLEDGEMENTS

This work was supported by: The Office of Biological Energy Research, Division of Basic Energy Sciences, U.S. Department of Energy (Contract # DEAC03-76SF00098), The National Institute of Health (AG-04818), through the Department of Energy and a NASA Ames Research Centre / University of California, Berkeley interchange agreement.

REFERENCES

(1) Khorana, H.G., Gerber, G.E., Herlihy, W.C., Gray, C.P., Anderegg, R.J., Nihei, K., and Biemann, K. (1979) Amino acid sequence of bacteriorhodopsin. Proc. Natl. Acad. Sci. USA 76:5046-5050.
(2) Henderson, R., and Unwin, P.N.T. (1975) Three-dimensional model of purple membrane obtained by electron microscopy. Nature 257:28-32.

(3) Huang, K.S., Radhakrishman, R., Bailey, H., and Khorana, H.G. (1982) Orientation of retinal in bacteriorhodopsin as studied by cross-linking using a photosensitive analog of retinal. J. Biol. Chem. 257:13616-13623.

(4) King, G.I., Mowery, P.C., Stoeckenius, W., Crespi, H.L., and Schoenborn, B.P. (1980) Location of the chromophore in bacteriorhodopsin. Proc. Natl. Acad. Sci. USA 77:4726-4730.

(5) Jubb, J.S., Worcester, D.L., Crespi, H.L., and Zaccai, G. (1984) Retinal location in purple membranes of Halobacterium halobium: a neutron diffraction study of membranes labelled in vivo with deuterated retinal. EMBO J. 3:1455-1461.

(6) Konishi, T., and Packer, L. (1976) Light-dark conformational states in bacteriorhodopsin. Biochem. Biophys. Res. Comm. 72:1437-1442.

(7) Konishi, T., and Packer, L. (1976) Biochem. Biophys. Res. Comm. 72:1437-1442.

(8) Konishi, T., Tristram, S., and Packer, L. (1979) The effect of cross-linking on photocycling activity of bacteriorhodopsin. Photochem. and Photobiol. 29:353-358.

(9) Packer, L., Quintanilha, A.T., Carmeli, C., Sullivan, P.D., Scherrer, P., Tristram, S., Herz, J., Pfeifhofer, A., and Mehlhorn, R.J. (1981) Molecular aspects of light induced uptake and release of protons by purple membranes. Photochem. and Photobiol. 33:579-585.

(10) Packer, L., Tristram, S., Herz, J.M., Russell, C., and Borders, C.L. (1979) Chemical modification of purple membranes: role of arginine and carboxyllic acid residues in bacteriorhodopsin. FEBS Lett. 108:243-248.

(11) Lam, E., and Packer, L. (1981) Effect of fluorescamine modification of purple membranes on exciton coupling and light-to-dark adaptation. Biochem. Biophys. Res. Comm. 101:464-471.

(12) Tristram-Nagle, S., and Packer, L. (1981) Effects of arginine modification on the photocycle and proton pumping of bacteriorhodopsin. Biochem. Intl. 3:621-628.

(13) Scherrer, P., Packer, L., and Seltzer, S. (1981) Effect of iodination of the purple membrane on the photocycle of bacteriorhodopsin. Arch. Biochem. Biophys. 212:589-601.

(14) Katsura, T., Lam, E., Packer, L., and Seltzer, S. (1982) Light dependent modification of bacteriorhodopsin by tetranitromethane. Interaction of a tyrosine and a tryptophan residue with bound retinal. Biochem. Intl. 5:445-456.

(15) Packer, L., Scherrer, P., Yue, K.T., Varo, Gy., Ormos, P., Barabas, K., Der, A., and Keszthelyi, L. (1982) Electrical signals in the photocycle of tyrosine modified bacteriorhodopsin. Biochem. Intl. 5:437-443.

(16) Konishi, T., and Packer, L. (1977) Chemical modification of bacteriorhodopsin with N-bromosuccinimide. FEBS Lett. 79:369-373.

(17) Herz, J.M., and Packer, L. (1981) Structural involvement of carboxyl residues in the photocycle of bacteriorhodopsin. FEBS Lett. 131:158-164.

(18) Herz, J.M., Mehlhorn, R.J., and Packer, L. (1983) Topographic studies of spin-labeled bacteriorhodopsin: Evidence for buried carboxyl residues and immobilization of the C-terminal tail. J. Biol. Chem. 258:9899-9907.

(19) Herz, J., Hrabeta, E., and Packer, L. (1983) Evidence for a carboxyl group near the vicinity of the retinal chromophore in bacteriorhodopsin. Biochem. Biophys. Res. Comm. 114:872-881.

(20) Packer, L. (1983) Bacterial rhodopsin and halorhodopsin. In: The Function of Vitamin A: A Minisymposium Summary. Fed. Proc. 42:2741-2744.

(21) Lam, E., Fry, I.F., Packer, L., and Mukohata, Y. (1982) Comparison of the O_{640} photo-intermediate and acid-induced species in membrane patches from Halobacterium halobium S_9 and R_1mW strains. FEBS Lett. 146:106-110.

(22) Mukohata, Y., Sugiyama, Y., Kaji, Y., Usukura, J., and Yamada, E. (1981) Photochem. Photobiol. 33:593-600.

(23) Lam, E., Fry, I., Packer, L., and Mukohata, Y. (1982) Comparison of the O_{640} photo-intermediate and the acid-induced species in membrane patches from halobacterium halobium S9 and RlmW strains. FEBS Lett. 146:106-110.

(24) Mehlhorn, R.J., Candau, P., and Packer, L. (1982) Measurements of volumes and electrochemical gradients with spin probes in membrane vesicles. Methods Enzymol. 88:751-762.

(25) Mehlhorn, R.J. and Packer, L. (1981) Measurements of volumes and light-induced pH and electrical gradients in sealed membranes with spin probes. In: Photophysical Processes-Membrane Energization, G. Akoyunoglou, ed., Balaban International Science Services, 443-450.

(26) Mehlhorn, R.J. and Probst, I. (1982) Measurements of pH gradients with spin probes-Example of halobacterial cell envelopes. Methods Enzymol. 88:751-762.

(27) Candau, P., Mehlhorn, R.J., and Packer, L. (1982) ESR measurements of bioenergetics parameters inside intact cells of Anacystis nidulans. In: Photosynthetic Procaryotes: Cell Differentiation and Function, (L. Packer and G. Papageorgiou, eds.), New York, Elsevier Biomed Press pp.91-101.

(28) Mehlhorn, R.J. and Packer, L. (1983) Bioenergetics studies with spin probes. In: Biomembranes and Cell Function (Kummerow, F.A., Benga, G., and Holmes, L., eds.) Proc. N.Y. Acad. Sci. 414:180-189.

(29) Mehlhorn, R.J. and Packer, L. (1979) Membrane surface potential measurements with amphiphilic spin labels Methods Enzymol. 56:515-526.

(30) Kamo, N., Takeuichi, M., Hazemoto, N., and Kobatake, Y. (1983) Light-Induced Delta pH of Envelope vesicles containing Halorhodopsin Measured by Use of a Spin Probe. Arch. Biochem. Biophys. 221:514-525.

(31) Carmeli, C., Quintanilha, A., and Packer, L. (1980) Surface charge changes in purple membranes and the photoreaction cycle of bacteriorhodopsin. Proc. Natl. Acad. Sci. USA 77:4707-4711. (32) Packer, L., Quintanilha, A.T., and Mehlhorn, R.J. (1982) Application of chemical modification and spin labeling techniques to the study of energy conversion by bacteriorhodopsin. In: Proceedings of the Second International Conference on Water and Ions in Biological Systems. Bucharest Sept 6-11, 1982.

(33) Melandri, B.A., Mehlhorn, R.J., and Packer, L. (1984) Light-induced proton gradients and internal volumes in chromatophores of Rhodopseudomonas sphaeroides. Arch. Biochem. Biophys. 235: 97-105.

(34) Mehlhorn, R. J., Packer, L., Macey, R., Balaban, A. T., and Dragutan, I. Measurement of Transmembrane Proton Movements with Nitroxide Spin Probes, in press.

(35) Tristram-Nagle, S.A. (1981) The role of positively charged amino acids in bacteriorhodopsin. Doctoral Thesis, University of California, Berkeley.

(36) Herz, J.M. and Packer, L.(1981) Structural involvement of carboxyl residues in the photocycle of bacteriorhodopsin. FEBS Lett. 131:158-164.

(37) Lam, E. and Packer, L. (1981) Effect of fluorescamine modification of purple membranes on exciton coupling and light-to-dark adaptation. Biochem. Biophys. Res. Comm. 101:464-471.

(38) Scherrer, P., Packer, L., and Seltzer, S. (1981) Effect of iodination of the purple membrane on the photocycle of bacteriorhodopsin. Arch. Biochem. Biophys. 202:589-601.

(39) Iwasa, T., Takeda, K., Tokunaga, F., Scherrer, P.S., and Packer, L. (1982) The photoreaction of tyrosine-iodinated bacteriorhodopsin at low temperatures. Biosci. Rep. 2:948-958.

(40) Lam, E., Seltzer, S., Katsura, T., and Packer, L. (1983) Light-dependent nitration of bacteriorhodopsin. Arch. Biochem. Biophys. 227:321-328.

(41) Lam, E. and Packer, L. (1984) The role of tyrosine residues in functional and structural properties of bacteriorhodopsin. In: Proceedings of the Symposium on H+-ATPase Synthase: Structure, Function, Regulation, Bari, Italy, April 5-7, 1984, pp. 181-189.

(42) Lam, E., Pande, A., Callender, R., Hilinski, E.F., Rentsepis, P., and Packer, L. (1984) Spectroscopic characterization of nitrated purple membranes. Biochem. Intl. 8:217-224.

(43) Mukohata, Y., Watanabe, T., Lam, E., and Packer, L. (1984) Differential chemical modification of bacteriorhodopsin. Proceedings of the EMBO Workshop on Molecular Biology of Retinal Proteins, Rottach-Engern September 10-14, 1984.

(44) Tristram-Nagle, S. and Packer, L. (1981) Effects of arginine modification on the photocycle and proton pumping of bacteriorhodopsin. Biochem. Internatl. 3:621-628.

(45) Katsura, T., Lam, E., Packer, L., and Seltzer, S. (1982) Light-dependent modification of bacteriorhodopsin by tetranitromethane. Interaction of a tyrosine and a tryptophane residue with bound retinal. Biochem. Intl. 5:445-456.

(46) Quintanilha, A.T. (1984) Electrical effects and the control of the photocycle/ H+-pumping in bacteriorhodopsin. In: Hydrogen Ion Transport in Epithelia, (J. Forte, ed.) New York: Wiley Interscience pp. 35-45.

(47) Herz, J.M. and Packer, L. (1984) Carboxyl groups and the proton pump of bacteriorhodopsin. Biochemical Society Transactions 12:405-408.

(48) Hrabeta, E., Robinson, A.E., and Packer, L. (1984) Effect of specific carboxyl modifications on the blue acid species and O650 photocycle intermediate of bacteriorhodopsin.Biochem. Biophys. Res. Comm. 122:1110-1116.

(49) Herz, J., Hrabeta, E., and Packer, L. (1983) Evidence for a carboxyl group near the vicinity of the retinal chromophore in bacteriorhodopsin. Biochem. Biophys. Res. Comm. 114:872-881. (50) Wu-Chou, S., Robinson, A.E., Hrabeta, E., and Packer, L. (1984) Cross-linking of bacteriorhodopsin using specific carboxyl modifications and proteolytic cleavage. Biochem. Biophys. Res. Comm. 124:565-571.

(51) Herz, J.M., Mehlhorn, R.J., and Packer, L. (1983) Topographic studies of spin-labeled bacteriorhodopsin: Evidence for buried carboxyl residues and immobilization of the C-terminal tail. J. Biol. Chem. 258:9899-9907.

(52) Sonnewald, U., Seltzer, S., Robinson, A.E., and Packer, L. [Mesityl] bacteriorhodopsin. The properties of an analogue of the purple membrane containing [mesityl] retinal as the chromophore. Photochem. Photobiol. in press.

(53) Zwerling, H., Mehlhorn, R.J., Packer, L., and MacElroy, R. (1982) A computer technique for structural studies of bacteriorhodopsin. Methods Enzymol. 88:772-784.

(54) Packer, L., Arrio, B., Johannin, G., and Volfin, P. (1984) Surface charge of purple membranes measured by electrophoretic light scattering. Biochem. Biophys. Res. Comm. 122:252-258.

(55) Quintanilha, A. (1980) Control of the photocycle in bacteriorhodopsin by electrochemical gradients. FEBS Lett. 117:8-12.

SPECTROSCOPIC PROBES FOR CONFORMATIONAL TRANSIENTS OF MEMBRANE

PROTEINS

R. Wagner

University of Osnabrueck, Biophysics
Department of Biologie/Chemistry
D-4500 Osnabrueck, FRG

INTRODUCTION

Energy transduction in biological membranes is supported by
functionally distinct protein complexes embedded in the lipid
bilayer matrix. Various biophysical methods introduced to the
field of biomembranes yielded indispensible information about
structure and function of the lipid bilayers and their embedded
protein complexes. The demonstration of the lateral and rotational
mobility of membrane proteins (1,2) undoubtedly influenced the
current concepts of membrane structure and function to a large
extent. In principle there are three main methods which can yield
quantitative information on lateral and rotational mobility of
membrane proteins. These are: fluorescence, phosphorescence, and
absorption polarization techniques; fluorescence photobleaching
recovery; and saturation transfer in electron parametric resonance
(for reviews see 3,4). Here the use of phosphorescence (triplet)
probes for measuring the rotational diffusion of intrinsic and
extrinsic membrane proteins will be described. Besides yielding
information on internal fluctuations within the protein, this
technique enables the measurement of comparatively slow rotational
diffusion (> $\emptyset$.1 Pa s = 1 cp), i.e., in biological membranes. Thus
it extends the application range to time domains which are more
relevant to the biological action of enzymes.

In addition to this, under aerobic conditions the lifetime
of the spectroscopic triplet state depends mainly on the axes of
oxygen to the binding site(s) of the probe within the protein;
therefore it can be used to monitor the location of the covalently
attached dye molecule within the protein.

THEORETICAL BACKGROUND

Spectroscopic Properties of Triplet Probes

Measurement of the rotational diffusion in the μs and ms timescales by spectroscopic techniques requires a spectroscopic state having at least a lifetime in the same order of magnitude. As proposed earlier (3, 13) this can be achieved by using the long lived triplet state of probe molecules. They are superior to the usual intrinsic chromophores of proteins, namely tryptophan and tyrosin, for several reasons: 1) Absorbance and emissions at longer wavelength (less scattering), 2) Higher extinction coefficients and defined transition geometry, and 3) High quantum yield for triplet formation.

Triplet probes already have been successfully used to study the rotational diffusion of intrinsic and extrinsic membrane proteins (5,6,7,) as well as of large macromolecular complexes in solution (8,9). The most suitable triplet probes were heavy halogen derivatives of fluorescein (3,11), namely tetrabromo-fluorescein (eosin) and tetraiodo-fluorescein (erythrosin) (see Fig. 1, left). These probes were derivatized with reactive groups like isothiocyanat (-NCS) and iodoacetamide (-NH-CO-CH$_2$I), by which they can be covalently coupled to proteins in a rather specific way (1∅).

In the following, the principles of photoselection necessary to measure rotational motion and the spectroscopic properties of eosin will be summarized briefly (11,12,13,14).

By excitation of eosin with plane polarized light a population of molecules whose transition dipole for absorption is parallel or at a small angle to the E vector of the incident light are preferentially excited. This induces a preferentially (anisotropic) population of excited molecules. If a short (compared to random Brownion motion) excitation pulse is used, the initial absorption will be dichroic and the emission from the excited state will be polarized. As the anisotropic population becomes randomized again, the initial polarization of the emission and the dichroism will decay with a certain time constant (rotational correlation time). This is shown in Fig. 1, right, schematically.

The triplet state can be detected by monitoring either phosphorescence emission, emission of delayed fluorescence, or ground state depletion. Their lifetimes are all identical, about 1 ms at 2∅ ° C in the absence of oxygen (6,11). Characteristic parameters of eosin in dilute aqueous solutions (< 1∅$^{-6}$) are: ϕ_f = ∅.2, τ_f = ∅.9 ns, ϕ_{isc} = ∅.7, τ_p = 1.8 ms; where ϕ_f and τ_f are the fluorescence quantum yield and the lifetime of the prompt fluorescence, ϕ_{isc} is the quantum yield for intersystem crossing, and τ_p

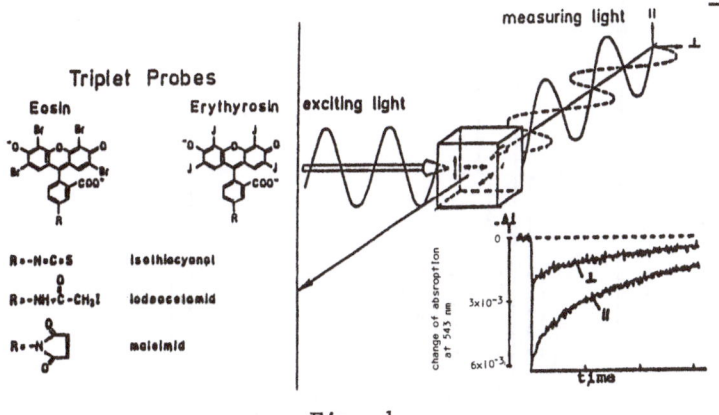

Fig. 1

is the phosphorescence lifetime (see Fig. 2). When the triplet
state is populated by flash excitation by plane polarized light
under photoselection, the absorption changes are dichroic. From
the timecourse of the dichroism the rotational motion of the probe
molecule can be investigated.

Analyses of the Absorbance Anisotopy Decay

The anisotropy parameter r(t) for the absorbance signal ΔA is
defined as

$$r(t) = \frac{\Delta A_{11}(t) - \Delta A_{\perp}(t)}{\Delta A_{11}(t) + 2\Delta A_{\perp}(t)} \qquad (1)$$

where ΔA_{11} and $\Delta A_{\perp}$ refer to the absorption changes obtained for
parallel and perpendicular polarizations between the exciting and
measuring light. r(t) depends only on rotational motion, whereas
the denominator, the total absorbance change, depends only on the
lifetime of the excited state, provided the excited state decays
with one component only (12). For an ideal absorber where the
angle between the oscillators for absorbance and measurement is
zero, the initial anisotropy will be $\emptyset.4$ (13). The theoretical
relationship between r(t) and the rotational diffusion has been
treated by several authors (12,14). In the case of an arbitrarily
shaped molecule, r(t) will decay as the sum of five exponentials

$$r(t) = \sum_{i=1}^{5} a_i \; \exp.(-t/\tau_i) \qquad (2)$$

the resolution of which is still beyond present-day experimental
resolution. The theory for the experimental situation where a

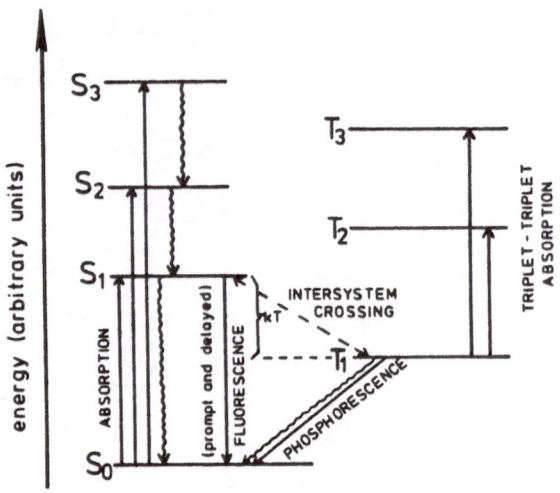

Fig. 2. Jablonski diagram of electronic energy levels for
E-Type Triplet probes. Note that the energy between
the first excited singulet state and the first triplet
state is in the order of magnitude for kT at room
temperature.

comparatively small probe molecule is bound to a larger molecule
or in an environment where its rotational mobility is restricted
has been published by several authors (12,15). In the case of the
covalent binding to a protein the decay of the absorption anisotropy
consists of at least two components: the motion of the probe
relative to the protein and the motion of the protein itself.
Usually the motion of the probe is faster by a few orders of
magnitude than that of the larger protein. This is shown in
Fig. 3, where the different time domains for rotational motion
of a protein bound probe (linear absorber) are drawn schematically.

Besides this, the situation can be more complicated if the
protein chain which covers the binding site of the probe can move
considerably in respect to the protein backbone (segmential motion).
Thus, in principle the decay of the absorption anisotropy can
be highly complicated. Nevertheless, in practice the analyses of
the data can be straightforward; since rotational motions of the
probe from different origins are in different time domains, they
can be considered to be independent and separated from each other.
This can be achieved for example by proper experimental design as
described in ref. 8.

Measurement of the Rotational Diffusion of Large Protein Complexes
in Solution

The use of triplet probes enables the measurement of large
oligomeric protein complexes in solution. From the measured
rotational correlation times, size and shape of these proteins can

100

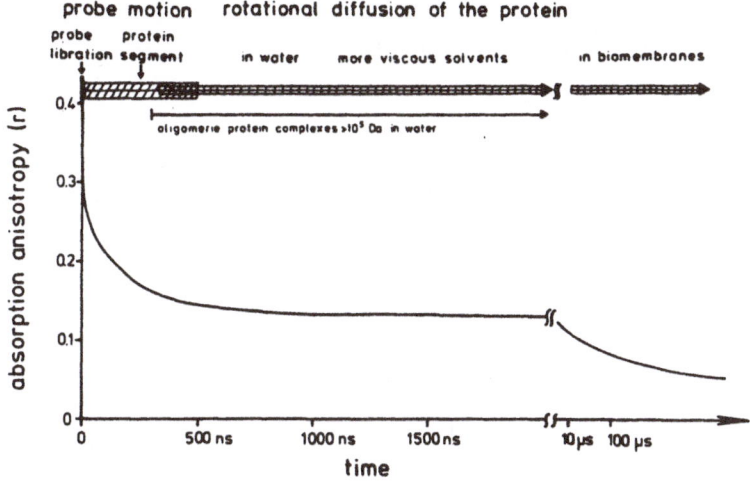

Fig. 3. Time domains for different rotational motions of a protein bound probe at 20 °C.

be deduced. Among the hydrodynamic properties of macromolecular complexes, the diffusion coefficient for rotational motion is more sensitive to the conformation of these complexes than translational diffusion and intrinsic viscosity and therefore useful in obtaining desired structural information. This information is of special interest, since the three-dimensional structure of large protein complexes cannot be determined by x-ray diffraction.

The first step in the analyses of the experimentally observed decay of the absorption anisotropy is the assignment of the decay components to the possible origins of rotational motion. Librational motion of the probe occurs at very short times (< 50 ns), so it will show up in virtually smaller initial absorption anisotropy $(r(\emptyset))$. Segmental motions within the protein and rotational diffusion of the protein can be discriminated experimentally by measuring the absorption anisotropy at different solvent viscosities. Only the rotational diffusion depends linearly on the viscosity. The components assigned to the rotational diffusion of the whole protein can be analyzed to obtain structural information as described briefly in the following.

The decay of the absorption anisotropy due to Brownian rotational motion of the macromolecular protein complex depends on the volume and shape of the macromolecule, the temperature and the viscosity of the solvent, and the orientation of the probe with respect to the macromolecule. In the general case of an arbitrarily shaped macromolecule five exponentials in the decay of the absorption anisotropy are expected. In the simplest case of a rigid sphere only one component will appear. However, so far almost all macromolecules exhibit at least one axis of approximate symmetry, so a maximum of three exponentials will appear in the decay of the absorption anisotropy.

$$r(t) = \sum_{i=1}^{3} a_i \exp.(-t/\tau_i) \tag{3}$$

where

$$\tau_1 = 1/6D_1, \quad \tau_2 = 1/5D_1 + D_3, \quad \tau_3 = 4D_1 + 2D_3$$

and for a rigid sphere

$$\tau_s = 1/6D_s \tag{4}$$

and

$$1/6D_s = V \eta / k T \tag{5}$$

where V is the volume, n the viscosity, k the Boltzmann constant, and T the temperature. D_3 is the diffusion coefficient for the rotation about the symmetry axis and D_1 is the diffusion coefficient of any axis perpendicular to it. To gain structural information from the experimentally observed rotational correlation times, a theoretical formalism is necessary. The rotational diffusion coefficients for "model macromolecules," as calculated from the hydrodynamics, are compared with the experimental observed results. Besides the classical ellipsoid approximation for macromolecular shapes of Perrin (16), recent improved calculations of hydrodynamic properties of several model structures are available (17). Table 1 shows, for example, a comparison of the diffusion coefficients for lateral and rotational motion for oligomeric protein complexes. It is obvious that only the diffusion coefficient for rotational motion is very sensitive to the spatial arrangement of the subunits. Therefore the measurement of the rotational diffusion provides a rather sensitive method to investigate the structure of large macromolecular complexes.

Measurement of the Rotational Diffusion of Proteins in Biological Membranes

Intrinsic proteins of biological membranes can be considered to be immersed in a one-dimensional lattice which determines their possible motions. Regarding rotational motion, only significant rotation about the axis parallel to the normal of the membrane plane will be possible. In addition to this, wobbling of parts of the protein or of the whole protein have to be considered. The former movement can be characterized by a diffusion constant parallel to the normal of the membrane plane and in the latter case by a wobbling diffusion constant and the degree of orientation constraint. Both cases have been given extensive theoretical treatment (3,18); here only a brief description will be given.

Table 1. Translational and Rotational Diffusion Coefficients of Oligomeric Subunit Structures

NO. OF SUBUNITS n	GEOMETRY	D_t/D_t^{sph}	D_r^{11}/D_r^{sph}	D_r^{33}/D_r^{sph}
1	Sphere	1.000	1.900	1.000
2	Dimer	0.725	0.264	0.564
3	Triangle	0.621	0.236	0.175
3	Linear	0.586	0.108	0.397
4	Square	0.550	0.159	0.119
4	Tetrahedron	0.564	0.165	0.165
4	Linear	0.500	0.056	0.306
5	Pentagon	0.491	0.111	0.084
5	Bipyramid	0.520	0.118	0.156
6	Hexagon	0.440	0.081	0.061
6	Octahedron	0.494	0.115	0.115
6	Trigonal prism	0.482	0.100	0.109
8	Cube	0.433	0.075	0.075

D_r^{33} corresponds to rotation around the axis of highest symmetry, and D_r^{11} is for any axis perpendicular to it.

The values are normalized to D_t^{sph}, D_r^{sph} for the spherical monomer.

Data from: J. Garcia de la Torre and V.A. Bloomfield (1981), Quarterly Reviews of Biophysics 14, 81-139

The absorption anisotropy for an intrinsic membrane protein will decay according to

$$r(t) = a_1 \exp.(-D_3 t) + a_2 \exp.(-4D_3 t) + a_3 \qquad (6)$$

where

$$a_1 = \frac{6}{5} (\sin^2\theta\cos^2\theta); \quad a_2 = \frac{3}{10} (\sin^4\theta); \quad a_3 = \frac{1}{10} (3\cos^2\theta-1)^2$$

This predicts that the absorption anisotropy decays to a finite time-independent value of a_3, from which the angle θ, the orientation of the chromophore to the normal of the membrane plane, can be calculated. This angle represents, in the case of portein bound extrinsic probes, only a mean value of the independent probe motion rather than a definite angle. If the experimentally observed decay of the absorption anisotropy does not fall to a time-independent value, this indicates that the protein under investigation is not a homogenous population, but a mixture of different aggregation states. As for isotropic rotational motion in solution, the anisotropic rotational motion of a protein in the membrane is determined by the

temperature, the vicosity of the membrane, and the size and shape
of the protein. In a good approximation the protein is treated
as a cylinder of the radius (a) and the height (h) with its
symmetry axis parallel to the membrane plane. This is described
by the following equation.

$$D_3 = k \, T / \, 4 \, \Pi \, a^2 h \, \eta \tag{7}$$

This equation implies the lipid bilayer to be a continuum, which
has been shown to be a valid approximation (20).

For extrinsic membrane proteins two cases have to be considered:
proteins which are specifically anchored via membrane counter
proteins like ATPases and proteins which are bound rather unspecifically to membrane lipids via electrostatic interactions. In the
first case the situation is quite similar to the one of intrinsic
membrane proteins. The overall rotational mobility of the extrinsic
part of the protein complex will be determined by the size and shape
of the intrinsic part, the temperature, and the viscosity of the
membrane, provided the viscosity ratio of the bulk phase and the
membrane is < 0.01 (19).

In contrast to this, extrinsic proteins bound to the membrane
lipids by electrostatic interactions are expected to be immobile
in the timewindow of rotational motion (> 50 ns to ms). Possible
fluctuations of their ionic attraction and repulsion (which can
possibly completely randomize the initial photoselected anisotropy)
should occur at much shorter times. So far no measurements of
this kind of membrane proteins have been reported.

Finally, an interesting application of the rotational diffusion
measurement should be pointed out. This is the estimation of free
local lateral diffusion (in contrast to long range lateral diffusion)
which might help in the understanding of electron transport processes
in membranes since parts of this may be mediated by local collisions
between their components. From the rotational diffusion coefficient
of a protein normal to the membrane plane, the lateral diffusion
coefficient for free local motion can be calculated according to (18)

$$D_{lat} = (\ln \eta \, / \eta_w - \gamma \,) \, a^2 D_3 \tag{8}$$

where γ is Euler's constant.

The measurement of rotational diffusion of a variety of intrinsic
membrane proteins with intrinsic chromophores (e.g. Cytochrom-C
oxidase) and covalently attached triplet probes (erythrosythe band
3 protein (33); SR-ATPase (34); BF1 (35); CF1 (36); and FNR (7))
provided valuable information on their roles in membrane functions.

In the following, the measurements of CF1 and FNR will be described in some detail.

Instrumentation and Practical Problems

The triplet state of the protein bound probe can be monitored either in absorbance (ground state depletion, triplet-triplet absorption) or emission (phosphorescence, delayed fluorescence). In both cases the most severe practical problem is the unavoidable flash burst of the excitation pulse which blindfolds the detection system at least for times < 50 ns. In the case of emission measurements, which are in principle superior at chromophore concentrations < 10^{-6} M in the measuring cuevette, this is accomplished by the very intense light burst of prompt fluorescence. Since the triplet-triplet transition is very weak and the quantum yield of phosphorescence is rather low, ground state depletion and delayed fluorescence, both intense transitions, are most suitable. We normally use the ground state depletion (no prompt fluorescence at this wavelength) for the detection of the triplet state. The principle instrumental set up for this measurement is shown in Fig. 4. For high time resolution we used a xenon flash lamp with tunable duration (100 us-1.5 ms) as measuring light (intensity up to $0.7W/cm^2$). The focused measuring light is passed through a filter of proper wavelength and a Glan Tomson prism, which can be rotated automatically by a stepping motor, and then through the measuring cuevette. For detection, a photodiode with 0.5 cm^2 area and a risetime (C_{90}) of 0.5 ns is used. This photodiode is protected from the scattered excitation light by a special cutoff filter with T = 70% at the measuring wavelength and A > 7 at the excitation wavelength. Excitation at right angle to the measuring beam is provided by a frequency-doubled, Q-switched Nd-YAG laser (intrinsically polarized) with a flash duration of 5 ns fwhm, the energy of which is adjusted to saturate no more than 30% of the probe molecules. The signal

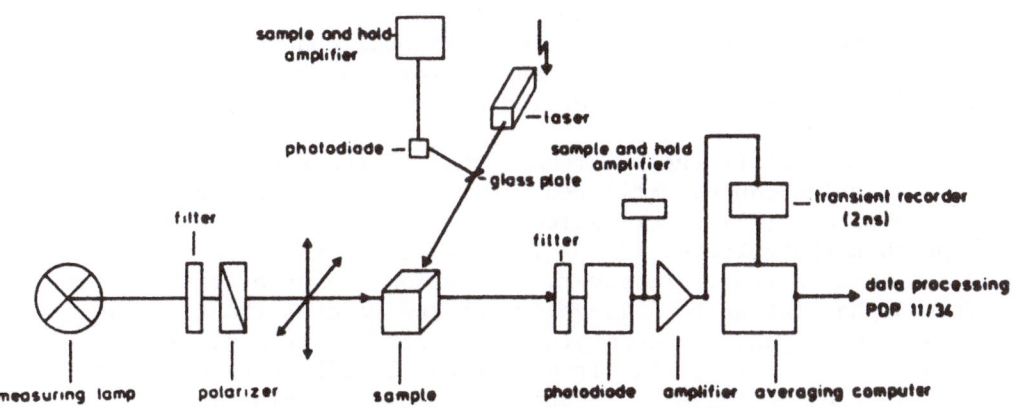

Fig. 4

output of the photodiode is amplified (100 times) and recorded on a Biomation 6500 transient recorder (max. time resolution 2 ns/address) digitally interfaced to a Tracor TN 1500 averaging computer and averaged in succession. The DC output of the photodiode and the energy of the excitation flash were recorded on two dual sample and hold amplifiers. The absorbance changes for parallel and perpendicular polarizations between excitation and measuring light were combined as described in equation (1) and analyzed by means of a Fourier Analyses for multiple exponential decay components for their amplitudes and realization times (32).

On the Role of Conformational Changes in the Chloroplast F1-ATPase (CF1) in Photophosphorylation

Photosynthetic ATP-synthesis in green plants is mediated by the membrane-bound, proton-translocating enzyme complex CF1-CF0. This oligomeric enzyme complex is composed of two portions, a hydrophobic intrinsic membrane protein CF0 and a peripherical part CF1; the peripherical part contains the catalytic site(s) and the F0 part conducts protons specifically across the membrane. It is widely accepted now that these two together act as a reversible ATP dependent H^+-pump as proposed by Mitchel (for reviews see ref. 22,23). The structures of the proton pumps are quite similar in all energy coupling membranes (22). In chloroplasts the CF0-CF1 complex is composed of eight different subunits, three in the F0 part and five in the F1 part. Although amino acid sequences of the enzymes from E. coli, mitochondria, and (in part) from chloroplasts are known (see ref. 22,23) and an increasing number of structural information has become available, the structure and the structural interplay of the subunits as well as the mechanism of the ATP-synthesis remain unknown.

Structural changes of the enzyme in the F0 and F1 parts seem to play an essential role during the transduction of the energy (22,23,24,25) in all coupling membranes. It is not known whether these structural changes, which have been demonstrated by different methods, reflect the regulation of the enzyme and/or the energy carrying step. Although these conformational changes are also essential parts in proposed mechanisms for phosphorylation (22,25), little information exists about the molecular level.

In chloroplasts the membrane bound CF1 shows only latent ATPase activity, which can be activated under electrochemical energization of the membrane (27). Activation of the enzyme is also necessary for photophosphorylation (28). The activation is enhanced by reducing agents (26). Moreover, activation increases the rate of photophosphorylation to a large extent and removes the time-lag in the onset of photophosphorylation, which now is controlled thermodynamically (27,29). It was shown (27,30) that activation of CF1 in the prescence of reducing agents is accompanied by the reduction of a disulfide bridge in the γ subunit, whether the enzyme is membrane bound or isolated (27). Therefore it seems as if this activation, which might be accompanied by changes in the ternary and quaternary structure of the enzyme, is a unique requirement for the synthesis/hydrolysis of ATP in CF1, and that the activation of the isolated enzyme with reducing agents closely mimics the situation in the membrane.

We measured the rotational diffusion of the eosin-NCS labeled CF1 before and after activation by DTT at 25 °C, to see whether this activation is accompanied by cross-conformational changes. The

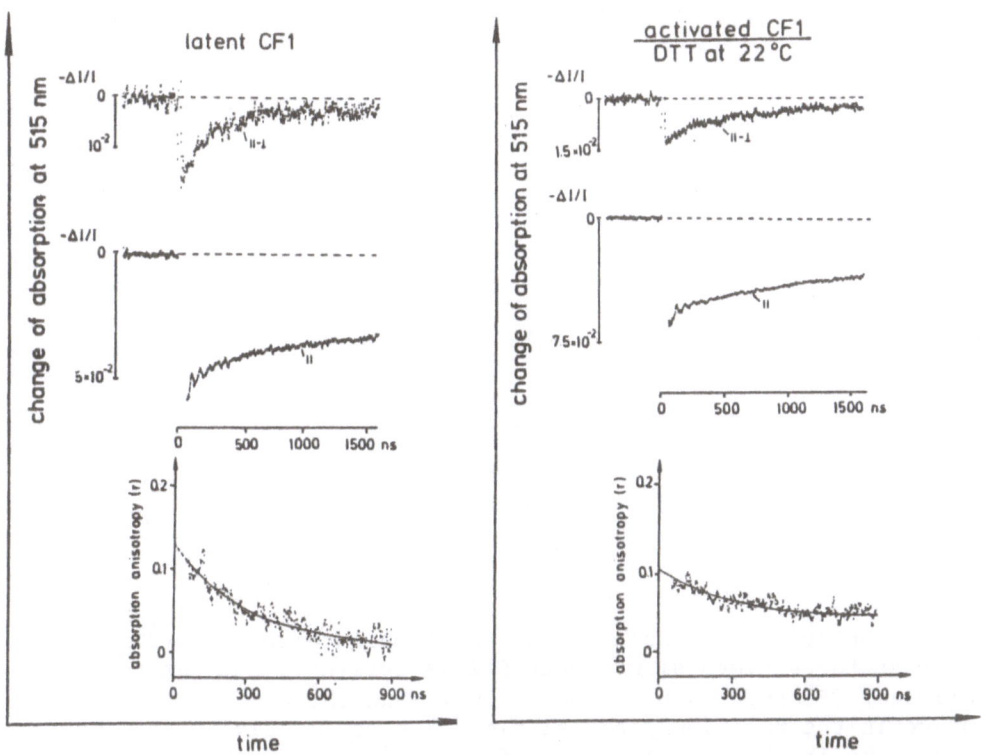

Fig. 5. Time course of the absorption changes of eosin-NCS bound to CF1. Upper traces: differences of the absorption changes obtained for parallel and perpendicular polarization of the exciting and measuring light. Middle traces: absorption changes for parallel polarization of exciting and measuring light. Lower traces: time course of the absorption anisotropy (r) as calculated (see eq. 1) from the upper and middle traces. The inserted lines show the fit of the data obtained by computer analyses.

enzyme bound eosin-NCS was used as probe in photoselection experiments as described in the preceding. Here the results of these measurements will be described briefly; they are published in detail elsewhere (9).

In Fig. 5 a typical set of experimental data for latent CF1 is shown. CF1 has been labeled (1mol eosin/ mol CF1) without impairing any of its activities. The label was bound exclusively in the α subunit of CF1.

From the rotational correlation time obtained for the latent enzyme (390 ns) we could exclude an overall spherical geometry (9), but see below. In Fig. 6 the time course of the absorption anisotropy for CF1 lacking the σ subunit and (σ,ε) subunits are shown and in Table 2 the rotational correlation times obtained for the different CF1 preparations are summarized.

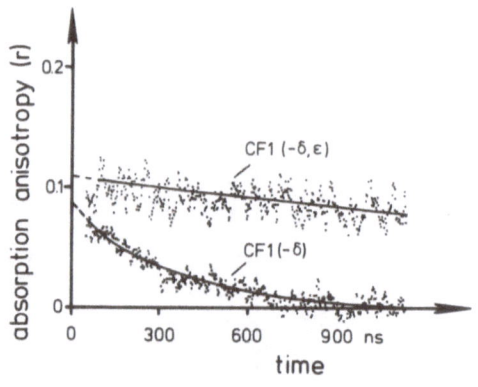

Fig. 6

As obvious from Table 2, the rotational correlation times are at least three times slower when CF1 is activated and the (σ,ε) deficient CF1 rotates 8.2 times slower than latent CF1. This change in the rotational correlation time could only be assigned to changes in the overall shape of CF1, since all other possible reasons (e.g., aggregation, change in hydration) could be ruled out experimentally. Although, as with other transport properties of macromolecules, it is not possible to get unambiguous shape information from the observed rotational correlation times, a range of possible structures compatible with the experimentally observed values can be deduced. Moreover, the results clearly show that activation of the enzyme and removal of the σ and ε subunits change the quaternary structure of the enzyme no matter what structural models are applied to the experimental data. To correlate

Table 2. Rotational Correlation Times and Amplitudes Obtained for Different CF_1 Samples and for EF_1

NO.	SAMPLE	τ_{r1} (ns)	a_1	τ_{r2} (ns)	a_2
1	CF_1 (latent)	388	0.098	145	0.032
2	CF_1 (latent) n = 11 cp, 60% v/v, Glycerol	4400	0.05	461	0.048
3	CF_1 (activated) DTT, 4 h, 25 C	1530	0.091	109	0.024
4	CF_1 (activated) DTT, 4 min. 63 C	1500	0.090	100	0.021
5	CF_1 (activated) 40 mM octylglucoside	1270	0.072	105	0.037
6	CF_1 (activation reversed) 1 ρM $CuCl_2$, 1 h	400	0.097	150	0.033
7	EF_1	1500	0.060	115	0.078
8	CF_1 ($-\delta$)	320	0.11	—	—
9	CF_1 ($-\delta\varepsilon$)	3200	0.08	—	—

the correlation times with structural information we applied a
theoretical formalism for oligomeric protein structures (17) rather
than the classical ellipsoid approximation (16) (which in the case
of $CF1-(\sigma,\varepsilon)$ would require that the longer axis of this CF1 be
equal to 620Å). This calculation showed that the large increase
in the rotational correlation times was due to a change in the
spatial arrangement of the subunits in CF1 upon activation or
removal of the σ and ε subunits. This rearrangement of the subunits
would not change the overall dimensions of the enzyme (increase of
the longer axis, 160Å, by 10%) but rather its shape (9).

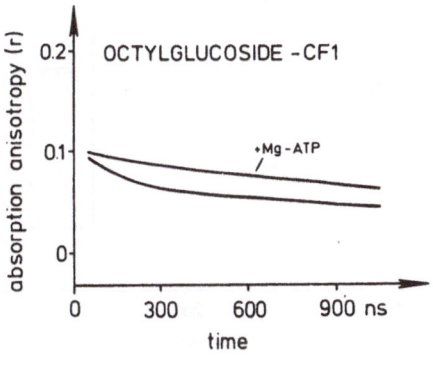

Fig. 7

It is worth noting that the binding of nucleotides did not change the overall structure of CF1, but binding of Mg^{2+}-ATP to activated CF1 largely affected segmential mobility of protein chains in the subunit of CF1. This is shown in Fig. 7. In the absence of Mg^{2+}-ATP the absorption anisotropy decayed with two exponentials; the faster one could be assigned to segmental motion of the protein chains which cover the eosin-NCS binding site in the subunit. This component disappeared completely after addition of Mg^{2+}-ATP, no matter which type of activation had been performed.

Movement of protein chains in the subunit of CF1 upon activation, both when membrane bound or in solution, could also be detected by using the triplet lifetime of the bound eosin-NCS as a measure of the proximity of the binding site to the bulk medium (37). This is shown in Fig. 8 for isolated latent and activated CF1. In the latent enzyme where the label is almost exclusively bound to the surface, the triplet state decayed faster (typically 10-30 us in air saturated solutions at 25 °C). But after activation with DTT at 25 °C most of the eosin molecules revealed a longer triplet lifetime (typically 120 us-250 us). Longer lifetime means hindered access of oxygen to the dye or deeper burring of the dye in the protein (37,38). Therefore we concluded that activation of the CF1 transfers protein chain segments in the subunit form the outside to the interior. As observed for the changes in the spatial arrangement of the subunits, the former changes in the α subunit are reversed when the enzyme is deactivated (9).

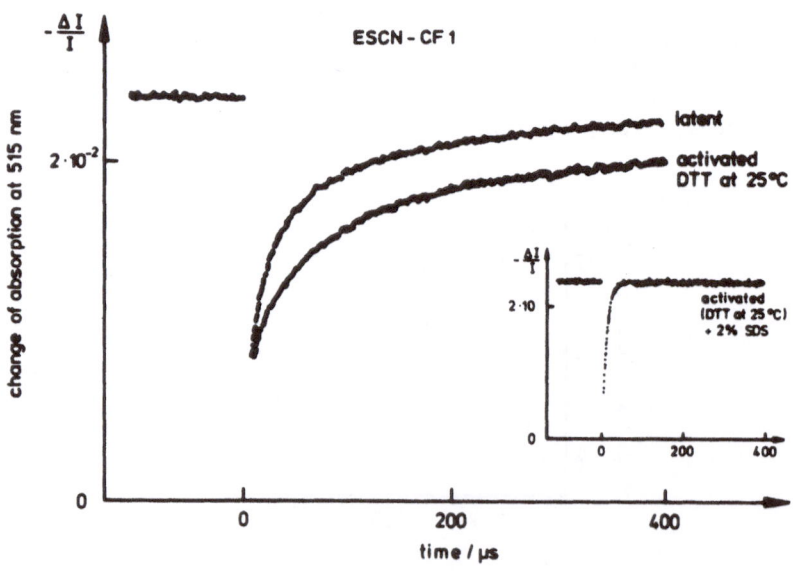

Fig. 8

In conclusion, from these experiments the conformational changes in isolated CF1 accompanying activation may be described as follows:

1. The reduction of a cystin in the γ subunit leads to a rearrangement of the subunits ("loosening or less dense packing"); this rearrangement seems to be the structural requirement for both ATP-hydrolysis and ATP-synthesis.

2. In addition to this, activation is accompanied by movement of protein chains in the α subunit of CF1 which are closely related to the binding of ATP. This conformational change, a local effect, might refect smaller changes in the ternary structure of the enzyme related to the regulation by binding or release of nucleotides.

Binding of CF1 to the Membrane and its Rotational Mobility

We measured the binding of eosin-NCS labeled CF1 to thylakoid membranes which were depleted from entire CF1 either by EDTA or by NaBr treatment (31,35). For this we used labeled CF1 preparations containing five subunits, four subunits (-σ), or three subunits -(σ,ε). The amount of rebound labeled CF1 could simply be monitored by measuring either the eosin absorbance or fluorescence in the reconstituted membranes. Only the 5-subunit CF1 was able to reconstitute photophosphorylation in depleted membranes. Labeled CF1 lacking the σ and ε subunits revealed the same extent of rebinding to NaBr depleted membranes. Moreover, the rebinding of these two subunits lacking CF1 preparations seemed to be specific to the CF∅ counterpart, since their binding was competitive to nonlabeled 5-subunit CF1 (R. Wagner, C.S. Andreo, and S. Engelbrecht, unpublished results).

The measurement of the rotational diffusion of membrane-bound, eosin-NCS labeled CF1 revealed a rather low rotational mobility of CF1 (35). The observed correlation times were also dependent on how thylakoid membranes had been depleted before reconstitution. In EDTA-depleted membranes a rotational correlation time of 1.5 ms was observed (35), whereas in NaBr-depleted membranes the correlation time was > 3 ms. The 4-subunit and 3-subunit CF1 revealed the same rotational correlation time as the 5-subunit CF1. From these results it is evident that neither the σ subunit nor the ε subunit is responsible for the binding of CF1 to its membrane counterpart CF∅. The whole CF∅-CF1 complex is rather immobilized in the thylakoid membrane. This indicates strong interaction of the enzyme with other membrane proteins or self-aggregation. In contrast to the CF∅-CF1 complex, the membrane-bound BF1 from Rhs. spheroides revealed in chromatophores from the same bacteria a rotational correlation time in the order of magnitude (16∅ us at 1∅ °C) expected for rather free rotational motion of the BF∅-BF1 complex in the membrane (36). It is worth noting that in chloroplasts energization of the membrane in the presence of ADP and P_i accelerated the rotational motion of CF1, or at least of those parts in the enzyme which covered the eosin binding site (35).

Rotational Mobility and Conformational Changes of Reconstituted Ferredoxin-NADP-Oxidoreductase

The complex formation between the water soluble redoxmediator, ferredoxin (Fd), and the terminal enzyme of the electron transport chain in chloroplasts, ferredoxin-NADP-oxidoreductase (FNR), in solution was studied by measuring the rotational diffusion of the eosin-NCS labeled enzyme. In addition to this the effect of NADP and DADPH binding and release on the conformation of the FNR was investigated by measuring the changes in the triplet lifetime of the bound probe (39). The results showed that a strong binary complex between FNR and Fd is formed only if either NADP or NADPH is bound to the enzyme. The results obtained from the measurements of the triplet lifetime of the bound probe showed that the binding of NADP and NADPH allosterically affected the ferredoxin binding domain in the enzyme (39). These results suggest that the enzyme is allosterically regulated by the binding and release of the substrates.

The eosin-NCS labeled ferredoxin-NADP-oxidoreductase (FNR) when membrane bound revealed a very high rotational mobility (correlation times less than 5 us) but in the presence of the water soluble redox mediator ferredoxin, this mobility is drastically reduced (correlation time 40 us) (7). This value is compatible with rotational diffusion of a 300 kDa protein in a viscosity of 1 Pa s (= 10 poise), i.e., an intrinsic membrane protein. We therefore concluded that ferredoxin links the FNR to a membrane protein complex of the electron transport chain, either the cytochrome b/f_6 complex or photosystem I. This ternary complex would live longer than required for the photoreduction of the FNR by photosystem I via ferredoxin (7).

ACKNOWLEDGMENTS

The author is very grateful to the cited colleagues N. Carrillo, C.S. Andreo, R. Casadio, and S. Engelbrecht for their collaboration. Special thanks are due to W. Junge for valuable discussion.

REFERENCES

1. Chapman, D., Gomez-Fernandez, J.C. and Goni, F.M. (1979) FEBS-Lett. 98, 211-223
2. Cherry, R.E. (1976) in Biol. Membranes, vol. 3 pp. 47-102 Chapman and Whalch eds.
3. Cherry, R.E. (1979) Biochim. Biophys, Acta 559, 289-327
4. Johansson L.B.A. and Lindblom, G. (1980) Quat. Rev. Biophys. 13, 63-118
5. Buerkli, A. and Cherry, R.E. (1980) Biochemistry 20, 138-145

6. Austin, R.H., Chan, S.S. and Jovin, T.M. (1979)
    Proc. Natl. Acad. Sci. U.S.A. 76, 5650-5654
7. Wagner, R., Carrillo, N., Junge, W. and Vallejos, R.H.
    (1982) Biochim. Biophys. Acta 680, 317-330
8. Wagner, R. and Junge, W. (1982) Biochemistry 21 1890-1899
9. Wagner, R., Engelbrecht, S. and Andreo, C.S. (1985)
    Europ. J. Biochem., in press
19. Means, G.E. and Feney, R.E. (1971) Chemical Modification of
    Proteins, Holden-Day, San Francisco
11. Moore, C., Boxer, D. and Garaland, P. (1979) FEBS-Lett.
    108, 161-166
12. Riggler, R. and Ehrenberg, M. (1973) Quat. Rev. Biophys.
    6, 139-199
13. Albrecht, C. (1961) J. Molec. Spectrosc. 6, 84-108
14. Belford, G.G., Belford, R.L. and Weber, G. (1972)
    Proc. Natl. Acad. Sci. U.S.A. 69, 1392-1393
15. Wahl. Ph. (1975) Chem. Phys. 7, 210-219
16. Perrin, F. (1934) J. Phys. Radium 5, 479-518
17. Garcia Bernal, J.M. and Garcia de la Torre, J. (1981)
    Bioploym. 20, 129-139
18. Kawato, S. and Kinoshita, K. (1981) Biophys. J. 36, 277-296
19. Saffman, P.G. and Delbrueck, M. (1975)
    Proc. Natl. Acad. Sci. U.S.A. 72, 3111-3113
20. Cherry, R.E. and Godfrey, R.E. (1981) Biophys. J. 36, 277-296
21. Mitchel, P. (1966) Bodmin Cornwall, Engl., Glynn Reas. Lab.
    pp. 1-192
22. Amzel, L.M. and Pederson, P. (1983) Ann. Rev. Biochem. 52,
    801-824
23. Cross, R.L. (1981) Ann. Rev. Biochem. 50, 681-714
24. Mc Carty, R.E. and Moroney, J.V. (1984) in The Enzymes of
    Biological Membranes, (Martonosi, A. ed.) 2nd Ed. pp. 383-413
25. Boyer, P.D., Cross, R.L. and Momsen, W. (1974)
    Proc. Natl. Acad. Sci. U.S.A. 70, 2837-2839
26. Backer-Grunwald, T and van Damm, K. (1974) Biochim. Biophys.
    Acta 347, 290-298
27. Ketchman, S.R., Davenport, J.W., Warnecke, K. and McCarty
    R.E. (1984) J. Biol. Chem. 259, 7286-7293
28. Junge, W., Rumberg, B. and Schroeder, H. (1970)
    Europ. J. Biochem 14, 575-581
29. Davenport, J.W. and McCarty, R.E. (1981) J. Biol. Chem, 256,
    8947-8954
30. Arana, J.L. and Vallejos, R.H. (1982) J. Biol. Chem, 257,
    1125-1127
31. Andreo, C.S., Patrie, W.J. and McCarty, R.E. (1982)
    J. Biol. Chem. 257, 9968-9975
32. Provencher, S. (1976) Biophys, J. 16, 27-41
33. Nigg, E.A. and Cherry, R.E. (1980)
    Proc. Natl. Acad. Sci. U.S.A. 77, 4702-4706
34. Hofmann, W. Sarzala, M.G. and Chapman, D. (1979)
    Proc. Natl. Acad. Sci. U.S.A. 76, 3860-3864

35. Wagner, R. and Junge, W. (1980) FEBS-Lett. 114, 327-333
36. Casadio, R. and Wagner, R., Biochim. Biophys. Acta, in press
37. Wagner, R., Carrillo, N., Junge, W. and Vallejos, R.H. (1981) FEBS-Lett. 136, 208-212
38. Wagner, R., Andreo, C.S. and Junge, W. (1983) Biochim. Biophys, Acta 723, 123-127
39. Wagner, R., Carrillo, N., Junge, W. and Vallejos, R.H. (1981) FEBS-Lett. 131, 335-340

DETERMINATIONS OF FUNCTIONAL HETEROGENEITY

IN PROTEOLIPOSOMES

Duncan H. Bell

Botany Department
University of Glasgow
Glasgow   G12 8QQ
United Kingdom

INTRODUCTION

In this chapter the role of membranes and membrane proteins in the organization and maintenance of biological function are discussed. Reconstitution experiments in the study of membrane proteins are considered along with a discussion of physical and functional heterogeneity as it pertains to these proteoliposome preparations. In this context, a novel dual-entrapped-probe method for the analysis of this heterogeneity is detailed. Specific reference is made to the application of the dual-probe approach to membrane proteins which generate pH gradients. The chapter concludes with a brief discussion of further applications of the technique and its limitations.

The Role of Membranes and Membrane Proteins

Living organisms have developed levels of organization which permit them to function. At the cellular level, the main unit of differentiation between compartments or organelles and the cytoplasm is a phospholipid membrane. These membranes are a mosaic of bilayer regions of phospholipids and integral membrane proteins (Hargreaves and Deamer, 1978). Whereas the membranes serve primarily as semi-permeable barriers to the flow of molecules, the proteins have several functions, namely: the facilitation of the passive transport of materials, the active transport of molecules, the synthesis of important compounds, and the mediation of bioenergetic processes. It should be remembered of course that any particular protein may perform several of these functions.

Isolation and Purification of Membrane Proteins

Although membrane proteins play key roles in the biochemistry of cells, it has been only relatively recently that techniques have been employed to study their activity in detail. There are several reasons for this lag behind the water-soluble enzymes. (i) By their very nature, membrane proteins require a membrane to have activity. Thus upon extraction and purification with the techniques then available, the resultant purified protein was usually completely inactive. (ii) The membrane proteins often have an asymmetry to their structure. This means that the primary amino acid sequence and secondary structure of the protein confer on it a preferred orientation when present in a phospholipid membrane. Thus for many membrane proteins, a thorough analysis of their activity requires that the molecules be re-inserted in a non-random fashion. (iii) Concomitant with this problem is the ancilliary problem of recording the activity of the inner space of an organelle or vesicle. Frequently one of the end results of the protein's activity is a transmembrane gradient. That is to say, a difference in concentration between the "outside" and "inside of the organelle. (iv) Membrane proteins are notoriously difficult to crystallize. X-ray crystallographic studies of the three-dimensional structure are therefore impossible with most membrane proteins.

Given the above-mentioned difficulties in working with membrane proteins, the development of techniques for their isolation and reconstitution into artificial membranes by Racker and his associates is a great achievement. The first report of the use of reconstitution was with a mitochondrial ATPase complex inserted into asolectin vesicles by Kagawa and Racker (1971). Since that time the list of proteins successfully studied with this method has grown remarkably.

The significant number of enzymes studied by reconstitution is not to suggest that the reincorporation of these enzymes and the assay of their activity is in any way straightforward or routine. The section below will detail the difficulties in the approach.

An article by Racker in the Methods in Enzymology series (1979) outlines the problems faced by an investigator who wishes to isolate a functional membrane protein and reconstitute it into proteoliposomes. Space does not permit a full recounting of the difficulties presented by this method. It is perhaps sufficient to note however, that the successful isolation of membrane proteins can depend on factors in addition to those normally required for the isolation of water-soluble enzymes. These factors are primarily the detergent used and the presence or absence of substrate during the isolation (Racker, 1979).

Reconstitution of Membrane Proteins

But the isolation of the protein is only the first step in the
studying of it. Each protein seems to have its own requirements for
maximum activity. The major variables are the phospholipid composition of the artificial membranes, the technique used to make the
vesicles, and the method for the incorporation of the protein into
the proteoliposome. Once again, the article by Racker (1979)
should be consulted for details.

Properties of Vesicles and Proteoliposomes

Before considering the properties of proteoliposomes, it may be
instructive to consider the properties of the vesicles themselves
(Szoka and Papahadjopoulos, 1981). The major characteristics are:
vesicle size and distribution, diffusion rates of substances
through the membranes, and the chemical composition of the vesicles.

Vesicle sizes can range from the small ones made by sonication
(200-500 angstroms diameter; Watts et al., 1978) to very large ones
(3000-10000 angstroms) made by the extrusion of multilamellar dispersions through polycarbonate filters (Olson et al., 1979; Rigaud
et al., 1983) or the slow hydration of a phospholipid mixture
(Darzon et al., 1980). Important in these preparations is the distribution of sizes. The smallest vesicles (sonication) appear to
be the most uniform. The extent of multilamellar vesicles is also
minimized by sonication. Many studies have been conducted on the
permeability of ions and substrates of interest. As might be expected, the permeabilities are greatest for uncharged species and
least for those with charges (Jain and Wagner, 1980).

HETEROGENEITY IN PROTEOLIPOSOMES

There are many sources of heterogeneity in proteoliposomes.
Some have been noted already in that they are related to the physical characteristics of the phospholipid vesicle even before the
protein is inserted. These are the vesicle size, the permeability
of the membranes and aging qualities of the vesicle.

The action of inserting a protein into these membranes will
probably affect the above properties and introduce some additional
sources of heterogeneity. Of these, inactive protein is no doubt
the most serious offender. The observation that different enzyme
preparations have different rates when measured per weight of protein is explained by having proteins with variable activities or a
varying proportion that are active. The second alternative very
likely predominates if the proteins are subjected to sonication as
part of the preparation technique (Racker, 1979). Also, certain

proteins can be inactive if inserted into the proteoliposomes in a
monomeric rather than dimeric configuration (Robinson and Capaldi,
1977). A second source of heterogeneity is the protein orientation
and applies to those proteins which generate or respond to a transmembrane gradient. It is well known that proteins can, under
certain conditions, orient themselves randomly in the membrane.
This problem is probably greatest in large liposomes where the surface curvature might play less of a role in determining the orientation (Van Dijck et al., 1981). A third source of variability
is in the number of proteins per vesicle (Casey et al., 1984). It
is not uncommon to have preparations where a significant proportion
of the vesicles have no protein inserted (Madden et al., 1983). A
final type of heterogeneity is the permeability of the vesicles.
Inasmuch as the membrane proteins are isolated by extraction with
agents which disrupt the membrane (detergents), to the extent that
the detergents are not thoroughly removed, they can affect the
permeability of some of the vesicles.

Attempts to eliminate the heterogeneity are logical consequences
of the problems themselves. Vesicles without protein inserted can
be removed by density centrifugation (Casey et al., 1984) or Seph-
adex columns (Madden et al., 1983). The other sources of heterogeneity such as inactive protein and membrane leakiness are not as
easily dealt with because functional qualities are involved instead
of physical differences such as charge or density.

In some cases corrections for heterogeneity are straightforward
or the assay can be designed to look only at those vesicles that are
operating properly. For example, if one only wants H^+/O ratios from
cytochrome oxidase vesicles, the substrates can be added to the
outside only. Then only those proteins that have the correct orientation will be studied (Nicholls et al., 1980; Wrigglesworth and
Nicholls, 1979). Also, if one side of the protein is sensitive to
inhibitors, they can be added to the outside or trapped inside to
select for proteins in one configuration or the other (Drachev et
al., 1976). In the case of bacteriorhodopsin proteoliposomes, it is
possible to determine the ratio of the orientations by cleaving the
exposed ends of the protein with papain or chymotrypsin (Rigaud et
al., 1983). Finally, if the protein is of sufficient size to be
visualized with the electron microscope, corrections can be made for
sidedness (Van Dijck et al., 1981). It should be emphasized however
that even in these cases, heterogeneity still exists.

But why is heterogeneity such a problem and why does the careful investigator need to be concerned about it? The answer lies in
the reason for doing reconstitution assays in the first place — the
functioning of the enzyme under study. An assay mixture composed of
fully-active proteoliposomes and fully-inactive proteoliposomes is
fundamentally different from a fairly homogeneous mixture of parti-

ally active proteoliposomes. To take an extreme example, if one is studying a membrane protein which takes up protons while hydrolyzing ATP, determinations of the external pH cannot distinguish whether a small proportion of the vesicles are taking up most of the hydrogen ions or all of the vesicles are operating at almost the same efficiency. The implications of the two alternatives for pH gradient formation, internal buffering effects, mobilities of counterions, etc. are very great.

The common neglect of the problem of heterogeneity has its origins in the techniques used to study protein activity. With the possible exception of ^{31}P-NMR (Trandinh et al., 1981; Van Dijck et al., 1981), the instrumental signals recorded are averages for the entire reaction mixture. To take a more specific example, the monitoring of pH gradients, one can look at changes in the bulk (outside) pH (Oesterhelt and Stoeckenius, 1973); the uptake of radioactive (Ramos et al., 1979), fluorescent (Rottenberg, 1979), or ESR-susceptible (Mehlhorn and Probst, 1982) amines; or the optical changes of entrapped pH-sensitive probes (Scarpa, 1979; Bell et al., 1983). Even in the latter case of probes trapped exclusively within vesicles, the signal is the sum of all the partial signals from the active and inactive vesicles.

DUAL-ENTRAPPED-PROBE TECHNIQUE

In this section will be described the theoretical background to a simple technique to measure the parameters of enzyme activity in a way which accounts for heterogeneity.

General Theory

To explain how the dual-probe technique functions, it is best to take as a model a system with a given degree of heterogeneity and to see what the response of each probe would be. In practice however, the reverse holds — namely that the responses of the probes are determined and the degree of heterogeneity calculated.

Figures 1 and 2 give hypothetical examples of some responses of entrapped probes to various conditions. As can be seen, those probe pairs which are satisfactory in this approach are those which give different absolute determinations of the average condition (C1 ≠ C2). One advantage of this method is that the probes need not be of the same type or respond in the same way. Fluorescence probes can be combined with probes that change their absorption or ESR signal or anything else. The only criteria are that their responses not be mirror images of each other or a direct proportionality (see Fig. 2. for examples). Under these circumstances, the two probes would always show the same average condition (C1≈C2) and be useless

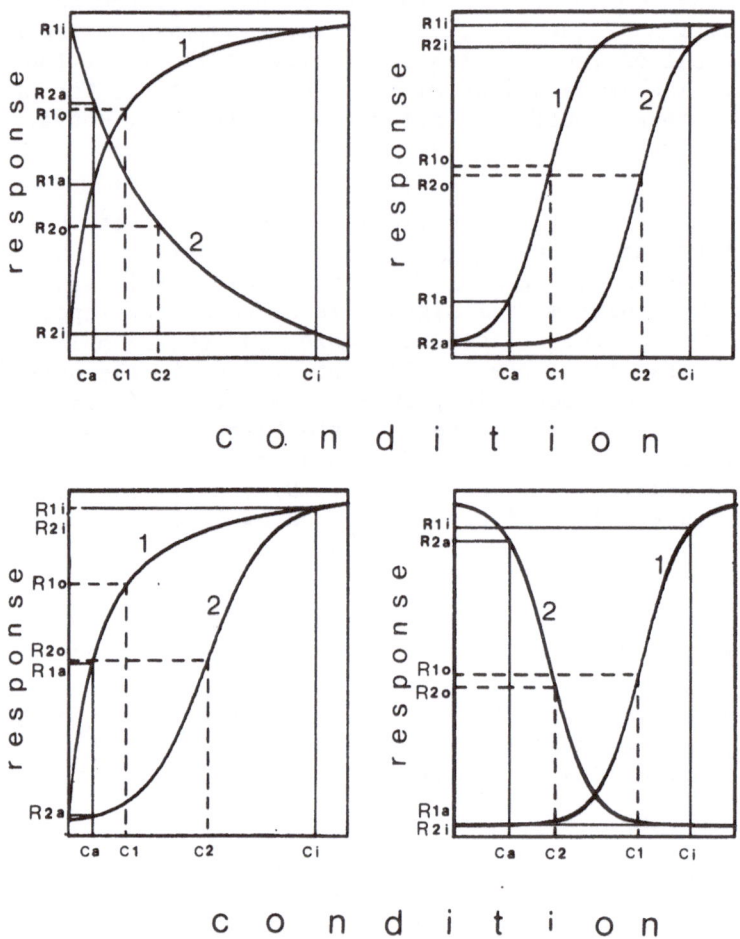

Fig. 1. Responses of hypothetical probes in studies with dual-
entrapped probes. For this figure, it has been assumed
that 50% of the vesicles are active and 50% inactive. C_i
is both the starting internal condition (pH, $\Delta\Psi$, ΔpH, etc.)
and the condition if all of the vesicles are unresponsive.
C_a is the condition if all of the vesicles are active.
C_1 and C_2 are the conditions that would appear to be
measured by the two probes. R_{1a},R_{1i}, R_{2a},R_{2i}, are the
responses expected from probes 1 and 2 when the vesicles
are fully active or fully inactive. R_{1o} and R_{2o} are the
responses expected to be observed if 50% of the vesicles
are active. Note that for each probe : $R_o = 0.5R_a + 0.5R_i$.

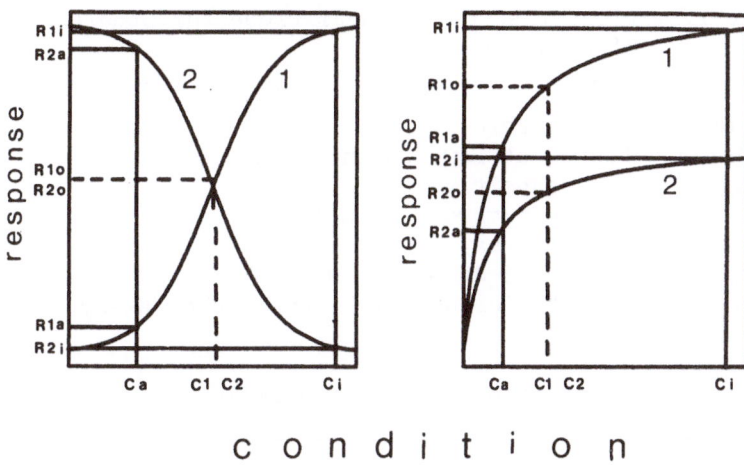

Fig. 2.   Responses of hypothetical probes in studies with dual-
          entrapped probes. Unlike figure 1, with these relation-
          ships, C1 equals C2. Thus for all degrees of heterogen-
          eity, the observed conditions will appear to be identical
          and it is impossible to determine the active fraction.

for determinations of heterogeneity.

     But how are the responses of the two probes used in determining
the degree of heterogeneity and the true activity of the active
vesicles? For each probe, there will be an unique relationship
between the condition and response. This will be determined experimentally, usually by trapping the probes in the vesicles and artificially imposing the condition desired (internal pH, membrane potential, etc.) and measuring the response. The experiment is then conducted and the response of the probes monitored. To the extent that
the internal conditions at some time are not the same, the responses
of the probes will indicate different average internal conditions.
This information can be inserted in the equations [1] and [2]:

$$R1o = [R1a \cdot F_{act}] + [R1i \cdot (1 - F_{act})] \tag{1}$$

$$R2o = [R2a \cdot F_{act}] + [R2i \cdot (1 - F_{act})] \tag{2}$$

where $R1o$ and $R2o$ are the observed responses of the two entrapped
probes in the system under investigation. In each equation the left
and right bracketed terms are the contributions to the response
from the active and inactive vesicles respectively. F_{act} is the
fraction of the vesicles which are active and $R1a$, $R1i$, $R2a$, $R2i$ are
the responses of the two probes when the vesicles are all active or
inactive. It should be clear at this point that with the know-

Table 1.   Properties of an Ideal Probe for Studies
            of Heterogeneity

Membrane impermeable   (highly charged)
Easily detected at low concentrations
Response very sensitive
Unreactive with membrane components
Unreactive with assay components
Not easily reduced or oxidized
No spectral interference with membrane or assay
      components
Stable
Easily removed from outside of vesicles

---

ledge of the given response at any condition with each probe (see
Fig. 1), equations [1] and [2] have two unknowns, the active fraction and the condition of the active vesicles, and can be solved to
give unique values.

The properties of an ideal probe for analyses as described here
are outlined in Table 1. Except for electrochromic probes which are
inserted across and within the membrane, the probes must be water-
soluble and thus able to be trapped inside the proteoliposomes without leaking out during the assay. The probes would also be highly
sensitive to the quantity being measured and highly detectible. As
regards the latter, the ideal probe would be detected without interference from the reconstituted proteins or chromophores. In addition, the ideal probe would be unreactive to the components necessary for the assay such as reductants or oxidants and just as important, stable with respect to photochemistry. The ideal probe
would be easily removed from the outside of the vesicle and have no
effect on the physical properties (permeability, fluidity, etc.) of
the vesicle membrane. Finally, the probe molecules should not
react with each other at the concentrations required to have probe
molecules in every vesicle and to be detected.

Additional considerations become important when designing ideal
probe pairs. They must respond with different sensitivities to the
condition being studied, such as having different pK's or Stern-
Volmer constants. They must not, however, be so different that one
is only sensitive in a range outside of those to be encountered.
The two probes must not interact with each other — either chemically
or photochemically. Also the responses of the probes must not inter-
fere with one another and be measured independently.

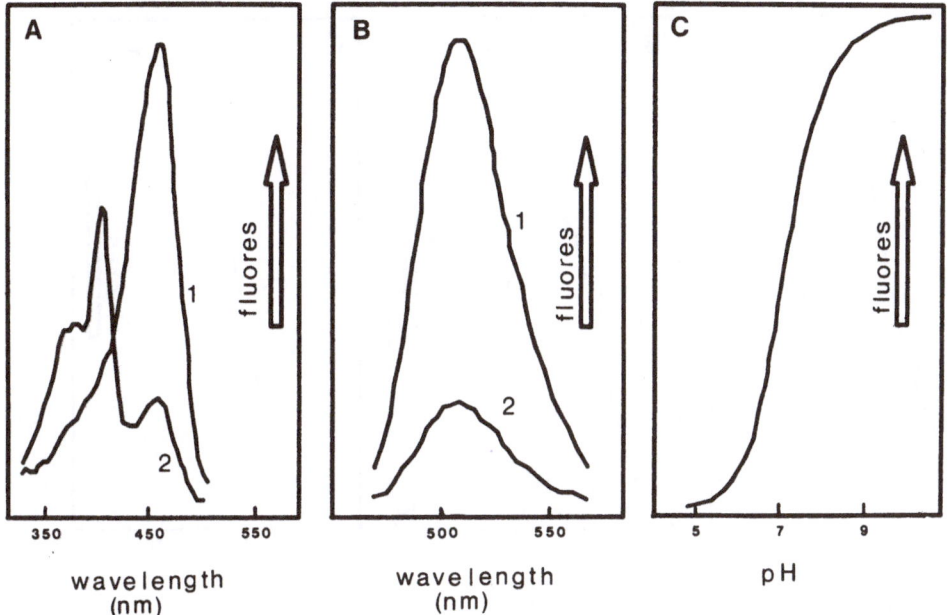

Fig. 3.   Properties of Pyranine. Part A. Excitation spectrum: emission detected at 510 nm. Traces 1 and 2 at pH's 8.3 and 6.6. Part B. Emission spectrum: excitation at 460 nm. Traces 1 and 2 at pH's 8.3 and 6.6. Part C. pH titration of Pyranine fluorescence. Excitation at 460 nm and emission at 510 nm.

Application of the Dual-Probe Approach to Bacteriorhodopsin Proteoliposomes

Bacteriorhodopsin is a photoactive protein which exists in sheets of purple membrane and can be isolated from the halophylic bacterium, Halobacterium halobium. When reconstituted into proteoliposomes, it has been shown to function as a light-driven proton pump. That is to say, bacteriorhodopsin can induce a pH gradient across a proteoliposome in the presence of light (Oesterhelt and Stoeckenius, 1973).

One approach to study these proteoliposomes has been the entrapment of a pH-indicating fluorescent probe, 8-hydroxy-1,3,6-pyrenetrisulfonic acid (Pyranine; Clement and Gould, 1981; Gould and Bell, 1981). As Fig. 3 shows, the fluorescence of Pyranine is sensitive to pH in a range (pK=7.2) of biological interest.

More recently, Bell et al., (1983) introduced a naphthalene

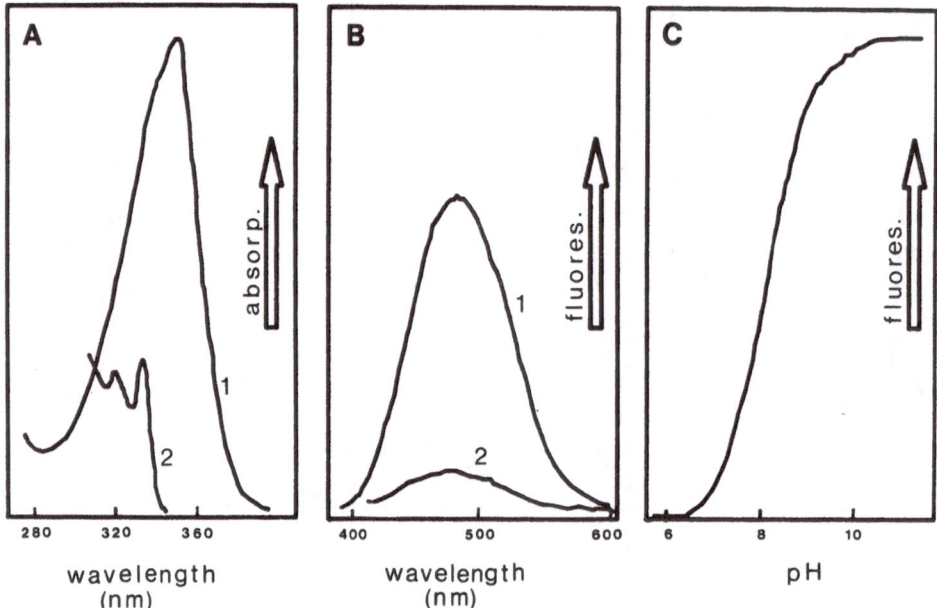

Fig. 4.   Properties of Naps. Part A. Absorption spectrum. Traces 1 and 2 at pH's 10.0 and 4.0. Part B. Emission spectrum: excitation at 350 nm. Traces 1 and 2 at pH's 10.0 and 4.0. Part C. pH titration of Naps fluorescence. Excitation at 337 nm and emission at 510 nm.

derivative (1-hydroxy-6,8-naphthalenedisulfonic acid; Naps) as a similar type of pH-indicating probe. Like Pyranine, it is negatively-charged and has a suitable pK (8.2; see Fig. 4). The pKs of both of the probes are not highly sensitive to entrapment inside vesicles or proteoliposomes, and change by less than 0.2 units when enclosed in vesicles (Bell et al., 1983; Clement and Gould, 1981).

Figure 5 confirms that the probes do in fact monitor the internal pH. In proteoliposomes made with bacteriorhodopsin and entrapped Pyranine, the fluorescence decreases upon light activation, indicating a decrease in the internal pH.

As mentioned above, it is possible to "cotrap" the two probes within the vesicles. Upon illumination, the probe with the higher pK (Naps) usually records a higher internal pH than the Pyranine (see Table 2). This result is consistent with the expectation if there is heterogeneity in the final pH's reached at steady-state (compare C1 and C2 in Fig. 1). One can use these values to determine the active fraction and the steady-state levels in the active proteoliposomes (Table 2) as described in general above. To solve

124

Table 2.   Determinations of Functional Heterogeneity in
            Proteoliposomes formed with Bacteriorhodopsin

Concentration of Gramicidin [μM]	Initial pH	Observed Steady-State pH		Calculated pH	Active Fraction
		Naps	Pyranine		
0.0	8.35	7.85	7.72	7.5	0.68
5.4	8.35	8.20	8.08	7.6	0.19
10.7	8.35	8.30	8.20	7.65	0.07

Asolectin vesicles contained 60 mM Naps and 2 mM Pyranine. Steady-state pH determinations were done after 2-3 min. of constant illumination.

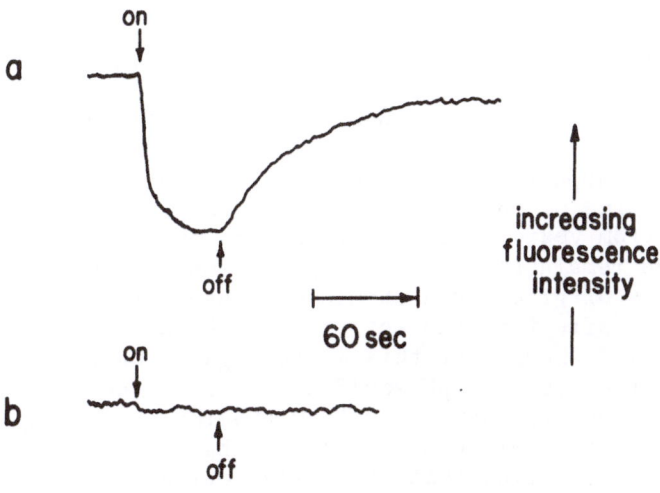

Fig. 5.   Light-induced changes in the fluorescence of Pyranine
           trapped inside bacteriorhodopsin proteoliposomes. Trace
           a, control. Trace b, control plus 10 μg/ml gramicidin.

this specific case, the usual strategy is to use the equation which describes the relationship between pH and fluorescence intensity for each probe. This comes from the equation

$$pH = pK + \log ([A^-] / [AH])$$ [3]

where A^- is the ionized (and fluorescent) form of the probe and AH is the unionized form. This is rearranged to

$$F1 = k [A^-] = 1 + \frac{1}{1 + 10^{(pH - pK)}}$$ [4]

where k is an arbitrary proportionality constant which accounts for the optical arrangement, total probe concentration, photomultiplier sensitivity etc. This equation is used in conjunction with equations [1] and [2] to solve for Fact and the internal pH of the active vesicles. In practice, one arbitrarily selects a trial internal pH and solves for F_{act} with each probe separately. The values will be different but will suggest whether to increase or decrease the trial pH. One can systematically move the trial pH until the F_{act}'s determined by each probe are identical. Ideally, a short computer program can be written to solve these equations with the only inputs needed being the external pH, the pK's of the two probes, and the internal pH determined with each probe.

One word of warning is required here though. Up to this point, the reasoning has been circular. It has been assumed that the proteoliposomes are heterogeneous and a technique used to confirm it. It is necessary to test these predictions by unequivocally introducing additional heterogeneity into the preparation and determining that the approach is still valid. This has been done by increasing the number of inactive vesicles with the addition of gramicidin. The mode of action of this uncoupler has been well characterized (Biegel and Gould, 1981). Unlike various other proton-carrying ionophores (FCCP, amines, etc.) gramicidin molecules form stable pores or channels across membranes. These channels do not diffuse from vesicle to vesicle but become insterted into a membrane, and remain there. The implication for these studies is that additions of gramicidin should decrease the active fraction (F_{act}) without affecting the activity of those vesicles that lack a pore. Table 2 shows that this is the case. On the basis of each probe alone, the average ΔpH would appear to be decreased to 0.05-0.15 units in the presence of 10.7 μM gramicidin. Using the dual-probe technique however, one calculates that there are still a small fraction of the vesicles that are fully active.

Thus this dual-probe method seems to distinguish between changing levels of pH as a result of heterogeneity and a true decrease in the activity of all the vesicles.

126

Dual-Probe Approach in Other Systems

Although this new method has been clearly shown to function
in bacteriorhodopsin proteoliposomes for the determination of
pH gradients, it probably can be applied to other systems as well.
The best candidates would appear to be determinations of membrane
potentials. At this point two approaches seem possible, the first
is the uptake of an ion which quenches the fluorescence of an entrapped probe and the second is the monitoring of membrane-
trapped (electrochromic) probes.

The theory behind the first approach is not unlike the determination of ΔpH. For these types of studies, one could envision a
trapped probe which has a response which is sensitive to the internal
concentration of a cation. To measure $\Delta\Psi$, the permeability of the
membrane to the cation would need to be sufficient to allow the
cation to redistribute across the membrane in response to a charge
asymmetry. It should be remembered that this approach is different
from the uptake of amines in that the moving molecule (the cation)
is not the probe itself but rather a change in the cation's concentration causes a change in the response of the entrapped probes.
Perhaps the two best applications which incorporate this theory would
be the quenching of arsenazo molecules by Ca^{2+} in the presence of
the calcium ionophore A23187 (Scarpa, 1979) and the heavy-atom
quenching by ions such as Cs^+ or Tl^+ in the presence of valinomycin
(Karper et al.,1983).

As mentioned above, one could also use the dual-probe technique
in vesicles with electrochromic probes imbedded in the membrane
(Bashford and Smith, 1979). Once again the key factor in the use of
the probes is that both of the probes must be inserted into each of
the vesicles and not diffuse between vesicles.

Limitations of the Dual-Probe Technique

The limitations of the dual-probe approach need to be considered
in light of its usefulness in correcting for heterogeneity in reconstituted systems. The principle assumption of two fractions, one
completely active and one completely inactive is a consequence of
having only two probes as markers. Whereas the presence of only
two types of proteoliposomes is unlikely for a large number of reasons, the values obtained certainly give much better approximations
than an assumption of homogeneity. Clearly greater resolution
would be affected by the co-entrapment of three or four probes in
each vesicle but the difficulty of analysis and of finding probes
with the right characteristics is probably and exponential function
of the number of probes!

In one respect the use of the dual probe technique is unequivocal - if the preparation is functionally homogeneous, the probes will show it. This is demonstrated in Fig. 6, where vesicles containing Pyranine and Naps are given a pulse of KOH and the change of the internal pH monitored. At all times the internal pH determined by each probe is identical. Therefore the dual-probe method would seem to be a useful tool in the routine assay of reconstituted membrane proteins. Alterations of the preparative steps would be checked against the determinations of F_{act} until the preparation is truly homogeneous.

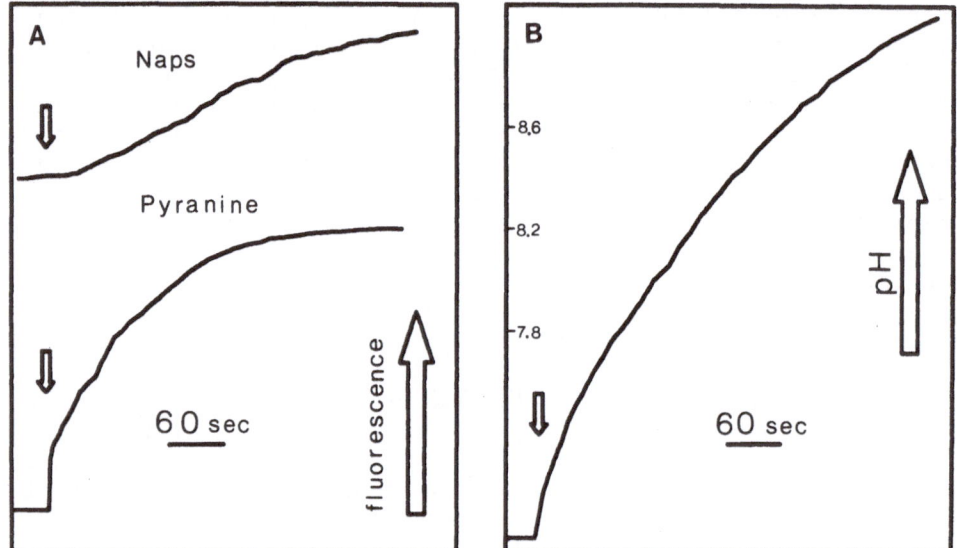

Fig. 6. Changes in the fluorescence intensity of asolectin vesicles with entrapped Pyranine and Naps after a pulse of KOH (at the arrow). Part A. Simultaneous fluorescence changes. Part B. Internal pH. Both probes gave values on the same line at all times (see Bell et al., 1983).

CONCLUSION

This paper has presented a powerful technique to study the functional heterogeneity of reconstituted systems. It is hoped that this brief introduction into the problems associated with heterogeneity will stimulate further research into entrapped probes and probe pairs by the readers.

REFERENCES

Bashford, C. L., and Smith, J. C., 1979, The use of optical probes to monitor membrane potential, Methods Enzymol., 55:569.

Bell, D. H., Patterson, L. K., and Gould, J. M., Transmembrane pH gradients and functional heterogeneity in reconstituted vesicle systems, Biochim. Biophys. Acta, 725:368.

Biegel, C. M., and Gould, J. M., 1981, Kinetics of hydrogen ion diffusion across phospholipid vesicle membranes, Biochemistry, 20:3474.

Casey, R. P., O'Shea, P. S., Chappell, J. B., and Azzi, A., 1984, A quantitative characterisation of H^+ translocation by cytochrome c oxidase vesicles, Biochim. Biophys. Acta, 765:30.

Clement, N. R., and Gould, J. M., 1981, Pyranine (8-hydroxy-1,3,6-pyrenetrisulfonate) as a probe of internal aqueous hydrogen ion concentration in phospholipid vesicles, Biochemistry, 20:1534.

Darszon, A., Vandenberg, C. A., Schonfeld, M., Ellisman, M. H., Spitzer, N. C., and Montal, M., 1980, Reassembly of protein-lipid complexes into large bilayer vesicles: Perspectives for membrane reconstitution, Proc. Natl. Acad. Sci. U.S.A., 77:23.

Drachev, L. A., Frolov, V. N., Kaulen, A. D., Liberman, E. A., Obstroumov, S. A., Plakunova, V. G., Semenov, A. Yu., and Skulachev, V. P., Reconstitution of biological molecular generators of electric current. Bacteriorhodopsin, J. Biol. Chem., 251:7059.

Gould, J. M., and Bell, D. H., 1981, Hydrogen ion permeability in reconstituted vesicle systems, in: "Energy Coupling in Photosynthesis," Selman, B. R. and Selman-Reimer, S. eds., Elsevier, New York.

Hargreaves, W. R., and Deamer, D. W., 1978, Origin and early evolution of bilayer membranes, in: "Light Transducing Membranes," D. W. Deamer ed., Academic Press, New York.

Jain, K. M., and Wagner, R. C., 1980, "Introduction to Biological Membranes," Wiley and Sons, New York.

Kagawa, Y., and Racker, E., 1971, Partial resolution of the enzymes catalyzing oxidative phosphorylation. XXV. Reconstitution of vesicles catalyzing $^{32}P_i$-adenosine triphosphate exchange, J. Biol. Chem., 246:5477.

Karpen, J. W., Sachs, A. B., Cash, D. J., Pasquale, E. B., and Hess, G. P., 1983, Direct spectrophotometric detection of cation flux in membrane vesicles: Stopped-flow measurements of acetylcholinereceptor-mediated ion flux, Anal. Biochem., 135:83.

Mehlhorn, R. J., and Probst, I., 1982, Light-induced pH gradients measured with spin-labeled amine and carboxylic acid probes: Application to Halobacterium halobium cell envelope vesicles, Methods Enzymol., 88:334.

Nicholls, P., Hildebrandt, V., and Wrigglesworth, J. M., 1980, Orientation and reactivity of cytochrome aa_3 heme groups in proteoliposomes, Arch. Biochem. Biophys., 204:533.

Oesterhelt, D., and Stoeckenius, W., 1973, Functions of a new photoreceptor membrane, Proc. Natl. Acad. Sci. U.S.A., 70:2853.

Olson, F., Hunt, C. A., Szoka, F. C., Vail, W. J., and Papahadjopoulos, D., 1979, Preparation of liposomes of defined size distribution by extrusion through polycarbonate membranes, Biochim. Biophys. Acta, 557:9

Racker, E., 1979, Reconstitution of membrane proteins, Methods Enzymol., 55:699.

Ramos, S., Schuldiner, S., and Kaback, H. R., 1979, The use of flow dialysis for determinations of ΔpH and active transport, Methods Enzymol., 55:680.

Rigaud, J.L., Bluzat, A., and Buschlen, S., 1983, Incorporation of bacteriorhodopsin into large unilamellar liposomes by reverse phase evaporation, Biochem. Biophys. Res. Comm., 111:373.

Robinson, N. C., and Capaldi, R. A., 1977, Interaction of detergents with cytochrome c oxidase. Biochemistry, 16:375.

Rottenberg, H., 1979, The measurement of membrane potential and ΔpH in cells, organelles, and vesicles, Methods Enzymol., 55:547.

Scarpa, A., 1979, Measurements of cation transport with metallochromic indicators, Methods Enzymol., 56:301.

Szoka, F., and Papahadjopoulos, D., 1981, Liposomes, preparation and characterization, in: "Liposomes: From Physical Structure to Therapeutic Applications," C. G. Knight, Ed., Elsevier New York.

Trandinh, S., Prigent, Y., Lacapere, J.-J., and Gary-Bobo, C., An NMR method for the study of proton transport across phospholipid vesicles, Biochem. Biophys. Res. Comm., 99:429.

Van Dijck, P. W. M., Nicolay, K., Leunissen-Bijvelt, J., Van Dam, K., and Kaptein, R., 1981, 31-P-NMR and freeze-fracture electron microscopic studies on reconstituted bacteriorhodopsin vesicles, Eur. J. Biochem., 117:639.

Watts, A., Marsh, D., and Knowles, P.F., 1978, Characterization of dimyristoylphosphatidylcholine vesicles and their dimensional changes through the phase transition: Molecular control of membrane morphology, Biochemistry, 17:179.

Wrigglesworth, J. M., and Nicholls, P., 1979, Turnover and vectorial properties of cytochrome c oxidase in reconstituted vesicles, Biochim. Biophys. Acta, 547:36.

ON THE MODE OF ACTION OF THE DELTA-ENDOTOXIN

OF BACILLUS THURINGIENSIS

Maja Huber-Lukač

Institute of Microbiology, ETH-Zürich
Haldeneggsteig 4, CH-8092 Zürich, Switzerland

INTRODUCTION

The interest in microbial insect control has increased over the years because the application of chemical insecticides has become more and more problematic. Several important pest insects have acquired resistance to chemical insecticides. Furthermore, the action of these insecticides generally is not limited to the pest insects. Often they also affect useful invertebrates as well as some vertebrates. Thus, they are hazardous both to the environment and public health.

Insect pathogenic bacteria represent one alternative to control certain pest insects specifically. The most important microbial product applied today is Bacillus thuringiensis. Up to date no resistance to the B.thuringiensis preparations has been observed. They excel in a defined host spectrum, which is restricted to only a few pest insect species, and therefore they present no harmful impact on environment.

The species B.thuringiensis is divided into 21 serotypes based on antigenic differences of the flagellae. The biochemical differences between strains within the same serotype are used for further division into varieties. Strains belonging to serotypes 1-13 and 15-21 are toxic to more than 100 insect species of the order Lepidoptera. However, the biological activity of any one variety is limited to only a few specific lepidopterous larvae. Some susceptible insects are listed in table 1. In contrast to the lepidopteran-specific varieties, B.thuringiensis var. israelensis, serotype 14, is primarily toxic to certain insects of the order Diptera. Such insects include those of the vectors of onchocerciasis and malaria, Simulium damnosum and Anopheles stephensi.

Table 1. Important insects susceptible to <u>Bacillus thuringiensis</u>, together with some host crops (Lüthy and Ebersold, 1981a).

Insect species	Crops
Pieris brassicae (large white butterfly)	vegetables
Trichoplusia ni (cabbage looper)	vegetables, tobacco
Plutella maculipennis (diamond black moth)	vegetables
Heliothis virescens (tobacco budworm)	tobacco, soybean
Manduca sexta (tobacco hornworm)	tobacco
Choristoneura fumiferana (spruce budworm)	forest
Malacosoma disstria (tent caterpillar)	forest
Lymantria dispar (gypsy moth)	forest
Anagasta kuehniella (Mediterranean flour moth)	stored flour
Ephestia cautella (almond moth)	stored grain
Anopheles stephensi	vector of malaria
Simulium damnosum	vector of onchocerciasis

The insecticidal activity of <u>B.thuringiensis</u> is primarily due to a proteinaceous crystal, often bipyramidal in shape (Fig. 1), which is produced by the bacterium during sporulation. This crystal, or parasporal body, can make up 20-30% of the dry weight of a sporulating cell. It is released at the last phase of sporulation when the cell wall is lysed.

The crystal protein genes of different <u>B.thuringiensis</u> varieties are located either on plasmids (var. <u>kurstaki</u>, <u>israelensis</u>) or on chromosomal DNA (<u>dendrolimus</u>) or on both (<u>thuringiensis</u>, strain berliner 1715) (Klier et al., 1982).

For further background information on <u>B.thuringiensis</u> the reader is referred to reviews by Bulla et al. (1980) and Lüthy and Ebersold (1981a).

THE CRYSTAL PROTEIN

The parasporal bodies of most varieties are composed of dimeric subunits which are linked by disulfide bonds to form the rigid crystal structure. Based on X-ray diffraction data, Holmes and Monro (1965) calculated a molecular weight (MW) of 230,000 D for the dimer of the variety <u>thuringiensis</u>. The MW of the monomers of different lepidopteran-active serotypes range from 130,000 to 145,000 D (Huber et al., 1981). As a rule two identical monomers form the dimer. The protein crystals can be dissolved with retention of toxicity by use of buffer solutions with pH > 8 containing reducing agents.

Figure 1. Parasporal bodies of <u>Bacillus</u> <u>thuringiensis</u> var. <u>thuringiensis</u> (micrograph supplied by H.R. Ebersold).

The parasporal bodies as well as the dissolved subunits are protoxins. They are not toxic upon injection into the hemocoel of host larvae nor do they lyse tissue culture cells from susceptible insects. Proteolytic activation is required to yield toxic polypeptides. Under natural conditions the activation step takes place in the intestine of the insect larvae by gut juice proteases following uptake of the crystals along with food. <u>In</u> <u>vitro</u> digestion of the protoxin with trypsin leads rapidly to a fragment with MW 70,000 D which is degraded further to a more stable polypeptide with MW 55,000 D. Both biologically active fragments are referred to as delta-endotoxin.

Although a few authors have also reported toxicity of proteolytic products with MW < 10,000 D (Fast and Angus, 1970) most others agree that biological activity is lost below a MW of 30,000 D (Bulla et al., 1981; Tojo and Aizawa, 1983). The delta-endotoxin is inactivated by organic solvents such as methanol, ethanol, and acetone, as well as by denaturing agents like $HgCl_2$, trichloro acetic acid, sodium dodecyl sulfate, and heat. pH >11 also leads to rapid loss of toxicity (Nishiitsutsuji-Uwo et al., 1977). Although various authors have found concentrations of carbohydrates ranging from 0.5-12% to be associated with the crystal protein, no correlation between sugar content and biological activity has been detected (Nickerson, 1980).

The mosquito-active protein crystal of B.thuringiensis var. israelensis differs in several respects from the lepidopteran-active ones. (1) It consists of 3 subunits with MW of 130,000, 73,000, and 25,000 D. (2) It is surrounded by a coat-like, alkali-resistent envelope (Huber and Lüthy, 1981). (3) 40% of the protein crystal, mainly the 25'000 D component, is soluble in buffer solutions with pH 10.5 without reducing agents. (4) This protein requires no proteolytic activation for in vitro toxicity to Aedes albopictus cells (Thomas and Ellar, 1983a).

EFFECTS OF THE DELTA-ENDOTOXIN

When susceptible lepidopterous larvae ingest protein crystals or activated delta-endotoxin the first symptom of poisoning, cessation of feeding, appears within 2-10 min. Subsequently, the larvae become sluggish, some insect species are paralysed, and get diarrhoea (Heimpel and Angus, 1959). Larvae which have taken up lethal doses of toxin can nevertheless remain alive for several days.

Histopathological investigations reveal that the midgut epithelial cells of these insects are damaged within 10 min after toxin administration. Drastic changes in both the columnar and goblet cells can be observed by light and electron microscopy (Lüthy and Ebersold, 1981b). The cells swell, form spherical cytoplasmic protrusions, disintegrate from the tissue and often lyse. Parallel with these alterations, the endoplasmatic reticulum, particularly near Golgi complexes, and mitochondria undergo severe vacuolization (Ebersold et al., 1977). In addition, the number of lytic vacuoles in the cytoplasm increases significantly. The same effects are caused by the delta-endotoxin in vitro as was demonstrated with tissue culture cells of Choristoneura fumiferana (Fig. 2 and 3), Trichoplusia ni (Nishiitsutsuji-Uwo et al., 1977), and Bombyx mori (Percy and Fast, 1983).

The delta-endotoxin of the variety israelensis affects midgut cells of sensitive dipteran larvae in the same way in vivo: Rapid rounding up of the cells followed by sequential swelling, blebbing of membrane vesicles from the exterior, and finally cell lysis. Analogous effects were observed in vitro with cells of Aedes albopictus (Thomas and Ellar, 1983a). The ultrastructural changes of the cell organelles were found to be identical to the ones reported for the lepidopteran-active delta-endotoxins (Charles, 1983).

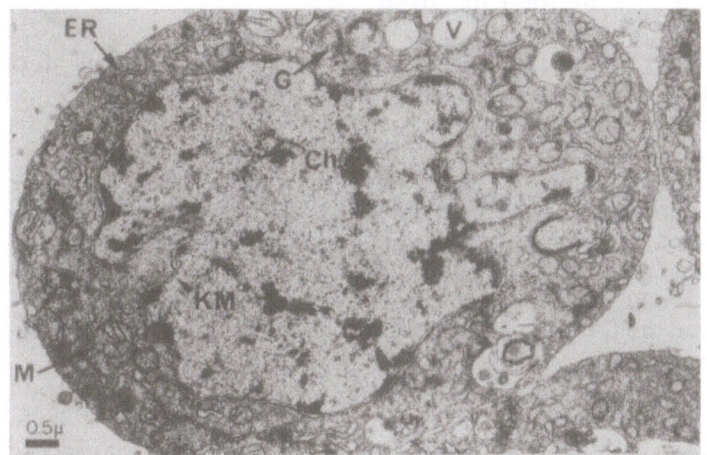

Figure 2. Thin section of an untreated <u>Choristoneura</u> <u>fumiferana</u> Cf-1
cell. Ch: chromatine, ER: endoplasmatic reticulum,
G: Golgi's apparatus, KM: nuclear membrane, M: mitochon-
drium, V: autolytic vacuole. From Geiser (1979).

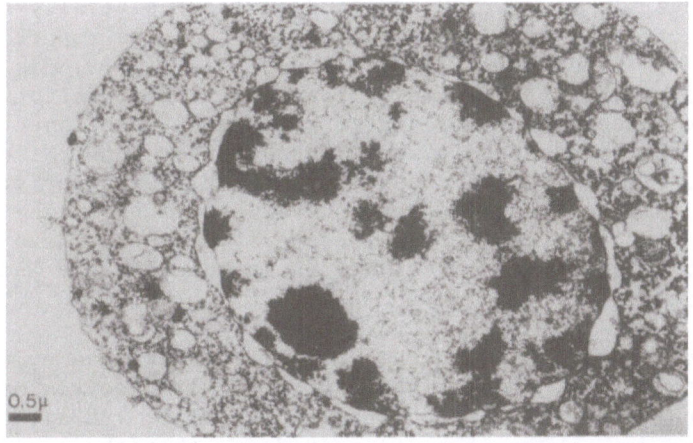

Figure 3. Thin section of a Cf-1 cell incubated with 10 µg/ml delta-
endotoxin of the variety <u>thuringiensis</u> for 30 min at 23°C.
From Geiser (1979).

SPECIFICITY OF THE DELTA-ENDOTOXIN

Though the lepidopteran-active and mosquito-active delta-endotoxins display similar effects on susceptible cells they differ
significantly in their specificities and will be discussed separately.
The different specificities of the toxins are certainly due to
differences in primary and tertiary structures of the delta-endotoxins (Zalunin et al., 1983). Besides, several properties of the
host insects may account for divergent susceptibilities. It is
probable that factors like pH of the insect guts, their redox potential, the presence of specific proteases, the composition of the gut
epithelium and other insect specific qualities decide on the sensitivity of the host.

The action of lepidopteran-active delta-endotoxins in vivo seems
to be confined to midgut epithelial cells of sensitive insects. In
vitro not only columnar and goblet cells are affected but also hemo-
cytes and ovary cells of susceptible insects were found to respond to
the toxin, however, more slowly and less extensively (Nishiitsutsuji-
Uwo et al., 1980). There exist contradictory results on the action of
lepidopteran-active delta-endotoxins on mammalian cells. Prasad and
Shetna (1974) found that the protoxin of the variety thuringiensis
inhibited growth of Yoshida ascites sarcoma in vivo and in vitro.
However, these results should be interpreted with care since the
protoxin they used had been dissolved in 1M NaOH for 15 h. It is
likely that under these harsh conditions of dissolution the protoxin
was no longer in its native form and at least partially degraded.
This conjecture is justified by the fact that the reported antitumour
protein had a MW of only 13,000 D. Rao et al. (1979) discovered that
parasporal bodies of the same variety elicited inhibitory effects on
chemically induced fibrosarcoma in Swiss mice when injected subcutan-
eously at the site of the tumour. On the other hand, Nishiitsutsuji-
Uwo et al. (1980) could not confirm any antitumour activity of the
proteolytically activated delta-endotoxin on Yoshida ascites sarcoma
cells, on mouse sarcoma 180 ascites cells, on human HeLa or KB cells.
Therefore, the antitumour activity seems questionable and needs
confirmation.

The delta-endotoxin of the variety kurstaki had no effect on
human erythrocytes, mouse fibroblasts, pig lymphocytes, and cell
lines of chicken fibroblasts (Thomas and Ellar, 1983a).
Thus, up to date there is no firm evidence that the lepidopteran-
active delta-endotoxins affect any cells other than those of susceptible insects.

In contrast, the delta-endotoxin of the variety israelensis does
not only act on cells of dipteran larvae. Thomas and Ellar (1983a)
demonstrated that the alkali-soluble crystal protein also affects

several cultured cell lines from insensitive insect larvae as well as a variety of vertebrate cells. Mouse fibroblasts, primary pig lymphocytes, and three mouse epithelial carcinoma cell lines strongly responded to the delta-endotoxin. The toxin also lysed erythrocytes of human, horse, sheep, mouse, and rat within 3 h. The same authors showed that intravenous and subcutaneous injection of this delta-endotoxin to Balb/c mice at a dose of 15-30 µg of protein per gram body weight resulted in rapid paralysis followed by death within a few hours. The same delta-endotoxin was not toxic however when administered per os. Nevertheless field application of B.thuringiensis var. israelensis is not considered problematic since not dissolved parasporal bodies were never found to be toxic for vertebrates.

THE ACTION OF THE DELTA-ENDOTOXIN
IN COMPARISON WITH OTHER BACTERIAL PROTEIN TOXINS

Bacterial protein toxins may be classified into two general categories, cytolytic and cytotoxic. The first category consists of a variety of toxins whose function is the destruction of the permeability barrier of target cells, usually leading to lysis and cell death. Several of these membrane damaging toxins, like streptolysin O, streptolysin S, staphylococcal alpha-, beta-, and delta-toxins have been reviewed by Alouf (1977). On the other hand, cytotoxic toxins cause specific lesions in cell processes distinct from membrane permeability. Their effects may be on molecules associated with the plasma membrane or upon cytoplasmic components. Thus, cholera toxin, Escherichia coli heat-labile toxin, and pertussis toxin activate the adenylate cyclase at the inner surface of the plasma membrane, whereas diphtheria toxin, Pseudomonas toxins, and Shigella toxins, among others, act in the cytosol by modifying factors required for protein synthesis.

Recent results on the mode of action of both lepidopteran-active and mosquito-active delta-endotoxins support the hypothesis that they are cytolytic toxins. Ebersold et al. (1978) demonstrated that the selective permeability of midgut epithelial cells of P.brassicae larvae is destroyed by the delta-endotoxin of the variety thuringiensis. They showed that ruthenium red (MW 552) which cannot penetrate the membrane of intact cells is evenly distributed in microvilli of these cells 5 min after ingestion of the toxin by the larvae. Severe damage of cell membranes was also induced by the delta-endotoxin in vitro. Lactate dehydrogenase, an enzyme present in an unbound state in the cytoplasm, was detected in the supernatant 10 min following exposure of C.fumiferana cells to the toxin (Ebersold et al., 1980). After 1 h 80% of this enzyme had leaked out of the cells. Furthermore, p-nitrophenol phosphate, a substrate of the acid phosphatase, could enter the cells after toxin treatment. Within 10 min the activity of the lysosomal enzyme was found to increase by 50%.

These results demonstrate that the permeability barrier of the plasma membrane of sensitive cells is destroyed by the delta-endo-toxin. Earlier reports on changes in potassium ion permeability induced by the delta-endotoxin are consistent with this conclusion. The nerve-blocking effect of the toxin on ganglia of Periplaneta americana can also be attributed to a disturbed potassium ion regula-tion (Cooksey et al., 1969). However, it is not known yet whether the delta-endotoxin primarily affects potassium ion permeability in particular (Griego et al., 1979; Harvey and Wolfersberger, 1979) or causes a general breakdown of the permeability control which allows potassium ions as well as other molecules to leak out of the cells. Angus (1968) postulated that the delta-endotoxin acts as a ionophor since it exhibits morphological effects similar to valinomycin. On the other hand, Fast and Morrison (1972) disputed this presumption because they found no alteration in concentration of K+, Na+, Ca2+, and Mg2+ in the hemolymph during the first 10 min upon toxin adminis-tration in vivo. An increase in the K+-level could not be measured earlier than 20 min after force-feeding the larvae with toxin (Nishii-tsutsuji-Uwo and Endo, 1981).

Whilst the intracellular changes have not been observed earlier than 15 min after toxin administration, alterations in membrane permeability occur already in less than 5 min. Therefore, it may be assumed that the delta-endotoxin primarily acts on the permeability barrier of cells whereas all effects in the cytoplasm are secondary.

SPECULATIONS ON THE MOLECULAR MODE OF ACTION OF THE DELTA-ENDOTOXIN

According to Alouf (1977) cytolytic toxins can be classified into three groups according to the mechanism by which they modify membrane permeability. (1) Toxins with a general surfactant, deter-gent-like activity which solubilize components of the plasma membrane (streptolysin S, surfactin). (2) Toxins with enzymatic activity hydrolysing certain membrane components (S.aureus beta-toxin, C.per-fringens alpha-toxin) or toxins that activate cellular enzymes which induce self-destruction of the membrane bilayer (S.aureus delta-toxin). (3) Toxins binding to specific molecules of the membrane causing a transient or permanent rearrangement of the bilayer frame-work by formation of functional holes (streptolysin O, staphylococcal leucocidin). As a rule cytolytic toxins interact with membrane phospholipids or sterols whereas cytotoxic toxins interact with glycoproteins and/or gangliosides.

Studies on the molecular modes of action of B.thuringiensis delta-endotoxins are not advanced enough to allow a clear-cut assign-ment of the toxins to any of these 3 groups. One characteristic of all cytolytic toxins, with the exception of leukocidins, is their broad spectrum of activity. They lyse a variety of mammalian and other kinds of cells. While the alkali-solubilized delta-endotoxin of the variety israelensis exhibits similarly unspecific action, lepidopteran-active delta-endotoxins destroy apparently only cell membranes of a few specific insect species. Therefore, the molecular modes of action of these two types of toxins might well be different.

Thomas and Ellar (1983b) demonstrated that the dissolved israelensis toxin is efficiently neutralized in vivo and in vitro by preincubation with phospholipids extracted from A.albopictus cells. It is also inactivated by lipid dispersions of phosphatidyl choline, sphingomyelin, phosphatidyl ethanolamine, and to a lesser extent by phosphatidyl serine provided that these lipids contain unsaturated fatty acids and are embedded in an organized fluid bilayer. In contrast, preincubation of the delta-endotoxin with multilamellar liposomes containing phosphatidyl inositol, cardiolipin, cerebroside, or cholesterol had no effect on its biological activity. The toxin was neither found to stimulate endogenous phospholipase activity in the membrane of susceptible cells nor to possess lipase activity itself. Based on all these results the authors concluded that the israelensis toxin acts directly as a protein-surfactant to destroy plasma membrane integrity. However, their findings are also consistent with the hypothesis that the toxin causes a disturbance in the bilayer fluidity by interacting with certain membrane phospholipids leading to a loss of integrity of the osmotic barrier. Thus, the preliminary data on the mode of action of the israelensis toxin only speak against the possibility that it belongs to the second group of cytolytic toxins.

Even less is known about the mechanism of action of lepidopteran-active delta-endotoxins. Because of their high specificity there is little likelihood that they are endowed with a general surfactant activity. Up to date no proteolytic or lipolytic activity has been attributed to the lepidopteran-active delta-endotoxins, nor have they been found to activate any membrane associated enzymes. Furthermore, no membrane components have been postulated as toxin receptors until recently. The morphological modifications of cell membranes and cytoplasmic organelles induced by the delta-endotoxin, however, very much ressemble the effects of the thiol-activated cytolytic toxins (Ebersold et al., 1977; Alouf, 1977).

We conducted a series of preliminary experiments in order to evaluate whether lepidopteran-active delta-endotoxins bind to specific membrane lipids. The interaction of the delta-endotoxin of the variety thuringiensis with different lipids was investigated by several binding assays. First, the binding of the delta-endotoxin to midgut lipids of sensitive P.brassicae larvae was examined.

The lipids were extracted with chloroform-methanol (2:1, v/v) and were separated on silica gel plates by thin-layer chromatography. To prevent nonspecific toxin binding, the chromatograms were saturated with polyisobutylmethacrylate and bovine serum albumin (Brockhaus et al., 1983). Then, the chromatograms were overlayed with 125I-labelled delta-endotoxin. Subsequently, binding of the toxin to the lipids was analysed by autoradiography.

The 125I-delta-endotoxin of variety thuringiensis was found to bind to five lipids of the midgut extract of P.brassicae (Fig. 4,b). In order to characterize these lipid components different indicator reagents were applied. Some properties of the lipids are summarized in table 2. Orcinol was used to detect glycolipids, ninhydrin to trace primary amino groups, and phospray to indicate phospholipids. The three hydrophilic delta-endotoxin binding lipids were identified as phospholipids, one of which contains a primary amino group. Two of these phospholipids were hydrolysed by treatment with 0.5 M NaOH for 30 min at 37°C whereas the most polar phospholipid was found to be alkali-stable. The two apolar lipids could not be identified with the indicators used. Although one lipid was stained by orcinol it became atypically coloured. Both hydrophobic lipids were not affected by exposure to NaOH.

To exclude the possibility that binding of 125I-delta-endotoxin to the insect lipids is due to a change in structure of the protein induced by radioiodination, the interaction of native, unlabelled delta-endotoxin was studied in an indirect binding assay. Two different 125I-labelled monoclonal antibodies, which recognize only native delta-endotoxin (Huber-Lukač et al., 1983), were utilized to reveal binding of unlabelled toxin to P.brassicae lipids. This indirect test demonstrated that the native thuringiensis toxin binds to the same lipids as the 125I-labelled toxin (not shown). 125I-antibodies alone showed only weak unspecific affinity to one of the hydrophobic toxin binding lipids.

Table 2. Properties of midgut lipids of P.brassicae larvae binding delta-endotoxin. The lipids were separated by thin-layer chromatography using chloroform-methanol-water (55:45:10 by volume) as developing solvent.

R_f-values	alkali-stability	reaction with		
		orcin	ninhydrin	phospray
0.97	+	−	−	−
0.91	+	?	−	−
0.73	−	−	+	+
0.56	−	−	−	+
0.46	+	−	−	+

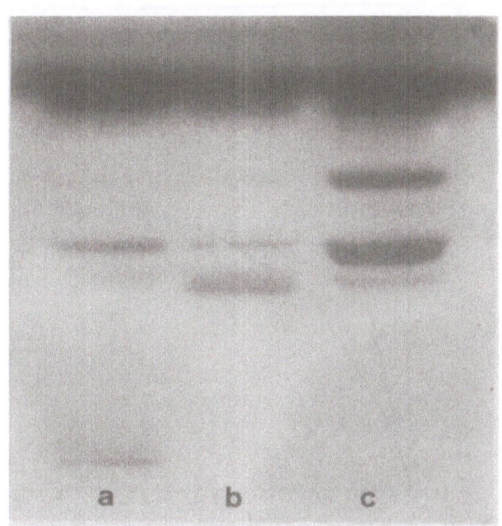

Figure 4. Binding of delta-endotoxin to midgut lipids of D.balteata
(a), P.brassicae (b), and S.littoralis (c) larvae. Thin-
layer chromatography of the lipids was carried out on
silica gel plates developed with chloroform-methanol-water
(55:45:10 by volume). Autoradiogram of chromatogram after
incubation with 125I-labelled delta-endotoxin.

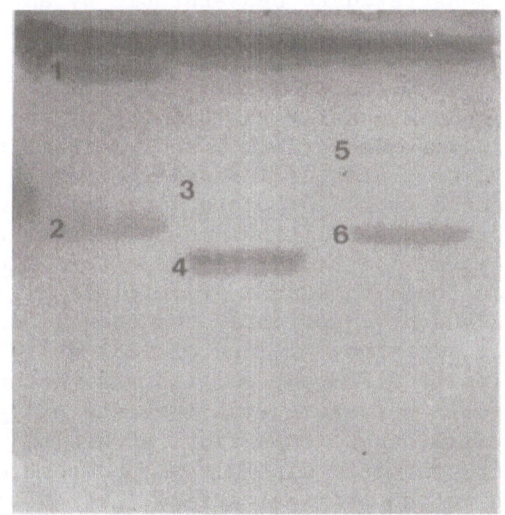

Figure 5. Binding of delta-endotoxin to cholesterol (1), phosphatidyl
choline (2), phosphatidyl inositol (3), sphingomyelin (4),
phosphatidyl ethanolamine (5), and dipalmitoyl phosphatidyl
choline (6). Thin-layer chromatography was performed as described in Fig. 4. Autoradiogram of chromatogram after incubation with 125I-labelled delta-endotoxin.

141

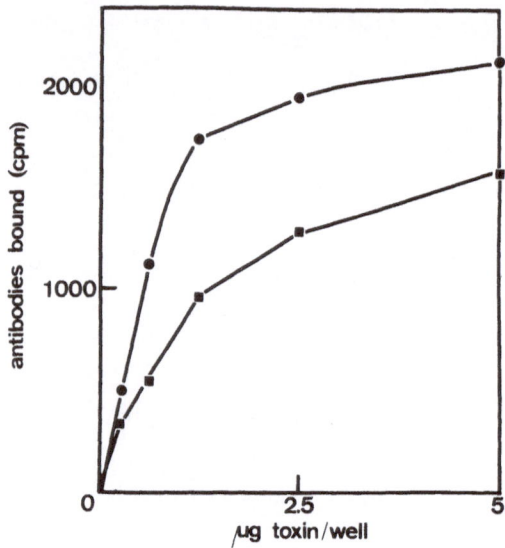

Figure 6. Binding of native, unlabelled delta-endotoxin to midgut
            lipids of <u>P.brassicae</u> as evaluated by a solid phase radio-
            immunoassay using 125I-labelled monoclonal antibodies
            94.2 (■) and 44.1 (●) (Huber-Lukač et al., 1983).

Furthermore, the binding of the delta-endotoxin to midgut lipids of
<u>P.brassicae</u> was demonstrated to be saturable by an indirect solid
phase radioimmunoassay using the 125I-labelled monoclonal antibodies
(Fig. 6).

        In order to examine whether the binding of the delta-endotoxin
is limited to lipids of susceptible insect larvae, the binding of the
<u>thuringiensis</u> toxin to lipids of insensitve insects was analysed,
using lipids extracted from <u>Spodoptera littoralis</u> and <u>Diabrotica
balteata</u>. Like <u>P.brassicae</u> <u>S.littoralis</u> belongs to the order <u>Lepidop-
tera</u> whilst <u>D.balteata</u> is a member of the order <u>Coleoptera</u>. The
delta-endotoxin was found to bind to five lipids from both of these
insects (Fig. 4, a and c), apparently the same lipids as of <u>P.brassi-
cae</u> (Fig. 4, b). In addition, binding of 125I-delta-endotoxin to
various other lipids of different origins was tested. It was demonstrated that the toxin interacts with cholesterol, phosphatidyl
choline, sphingomyelin, and phosphatidyl ethanolamine from vertebrates as well as with synthetic dipalmitoyl phosphatidyl choline (Fig.
5). A weak binding to phosphatidyl serine (not shown) but no binding
to phosphatidyl inositol from soybeans was observed (Fig. 5, 3).

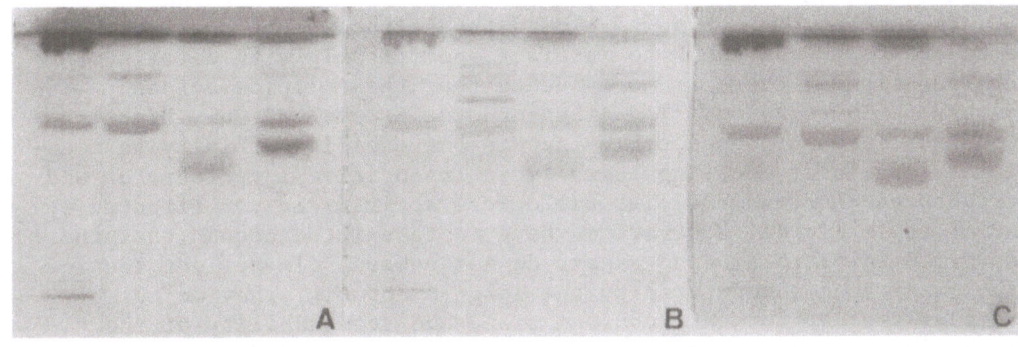

Figure 7. Binding of <u>thuringiensis</u> delta-endotoxin (A) and protoxin (B) as well as of <u>israelensis</u> toxin (C) to lipids of <u>D.balteata</u> (1), <u>S.littoralis</u> (2), <u>P.brassicae</u> (3), and to cholesterol, phosphatidyl ethanolamine, phosphatidyl choline, and sphingomyelin (4). Thin-layer chromatography was carried out as described in Fig. 4. Autoradiogram of chromatograms after incubation with 125I-labelled toxins.

When ^{125}I-labelled <u>thuringiensis</u> delta-endotoxin was replaced in the binding assay by the protoxin of the same variety or by the mosquito-active <u>israelensis</u> toxin only slight differences in the reaction pattern were encountered (Fig. 7, A–C). Despite their completely different host spectrum, both the <u>thuringiensis</u> protoxin and the <u>israelensis</u> toxin interacted with the same lipids as the activated delta-endotoxin. They showed additional affinity to a few more lipid components which were identical for both toxins and therefore cannot be responsible for the different specificity.

However, while the <u>israelensis</u> toxin was reported to be neutralized by a variety of phospholipids in the form of liposomes (Thomas and Ellar, 1983b) the <u>thuringiensis</u> delta-endotoxin was not. Preincubation of the lepidopteran-active toxin with multilamellar liposomes of <u>P.brassicae</u> and <u>D.balteata</u> lipids in large excess did not reduce the biological activity of the toxin. The same was true if the delta-endotoxin was force-fed together with lipid vesicles of cholesterol, sphingomyelin, or phosphatidyl choline to larvae of <u>P.brassicae</u>.

In summary, it can be concluded that although the thuringiensis delta-endotoxin was shown to interact specifically with certain membrane lipids, there is no evidence that these lipids act as primary toxin receptors. The delta-endotoxin does not only bind to membrane lipids of susceptible insects but exhibits affinity to a variety of lipids occuring commonly in insensitive invertebrates and vertebrates. Furthermore, its biological activity is not affected by any of these lipids. However, we have to take into account that the performed in vitro binding assays do not necessarily reflect the in vivo conditions encountered by the delta-endotoxin. They do not consider the environment, conformation, and accessability of the lipids in question. These very factors, however, could determine whether the lipids act as physiological receptors or not.

While the studies above were in progress Knowles et al. (1984) got evidence that a lectin-like binding could be the initial step in the action of the delta-endotoxin of B.thuringiensis var. kurstaki, another lepidoperan-specific toxin. Since the authors also discovered that preincubation with various lipids had no effect on toxicity of this delta-endotoxin, they carried out a series of experiments to test the ability of different lectins and monosaccharides to neutral-ize the toxin. They found that the toxin was completely inactivated by prior incubation with N-acetylgalactosamine (GalNAc) and N-acetyl-neuraminic acid as well as by wheat germ agglutinin and soybean agglutinin, which both bind to terminal GalNAc. Based on these results, they suggested that the kurstaki toxin recognizes a specific plasma membrane glycoconjugate receptor with a terminal N-GalNAc residue.

There is one hypothesis on the mode of action of lepidopteran-active delta-endotoxins which would harmonize with both their binding to membrane lipids and with their being inactivated by certain glycoproteins. It is conceivable that the delta-endotoxin first binds to a specific glycoprotein thereby getting into close contact with the membrane lipids. The subsequent interaction with some lipids would then induce destabilization of membrane integrity and finally lead to cell lysis. A similar two component receptor system has been postulated for thyrotropin (Kohn, 1978).

Since so little is known about the composition of insect membranes and the interaction of the delta-endotoxins with these membranes, many speculations on the molecular modes of actions of the toxins are still imaginable. At present it is difficult to conceive the role of the de-monstrated in vitro binding of lepidopteran-specific delta-endotoxins to membrane lipids and glycoproteins in their in vivo mechanism of action. To elucidate their molecular modes of action it would be important to know whether the toxins act on the surface of cell membranes exclusively, or insert into the membrane bilayer, or have to penetrate into the cytoplasm.

Acknowledgement

I would like to express my gratitude to Drs. D.G. Braun, R. Hütter,
P. Lüthy, G. Rosenfelder, and H. Towbin for supporting this work with
valuable advice and helpful discussions. I am also indebted to Dr.
M.G. Wolfersberger for revising the manuscript. This project has been
supported by the Swiss National Science Foundation, grant no.
3.098-081.

REFERENCES

Alouf, J.E., 1977, Cell membranes and cytolytic bacterial toxins, in:
    The Specificity and Action of Animal, Bacterial and Plant Toxins.
    P. Cuatrecasas, ed., Chapman and Hall, London, 219-270.
Angus, T.A., 1968, Similarity of effect of valinomycin and Bacillus
    thuringiensis parasporal protein in larvae of Bombyx mori. J.
    Invertebr. Pathol. 11:145-146.
Brockhaus, M., Magnani, J.L., Blaszezyk, M., Steplewski, Z., Koprowski,
    H., Karlsson, K.-A., Larson, G., and Ginsburg, V., 1981, Mono-
    clonal antibodies directed against the human Le$_b$ blood group
    antigen. J. Biol. Chem. 256:13223-13225.
Bulla, L.A., Bechtel, D.B., Kramer, K.J., Shethna, Y.I., Aronson, A.I.,
    and Fitz-James, P.C., 1980, Ultrastructure, physiology, and bio-
    chemistry of Bacillus thuringiensis. CRC Crit. Rev. Microbiol.
    147-204.
Bulla, L.A., Kramer, K.J., Cox, D.J., Jones, B.L., Davidson, L.I., and
    Lookhard, G.L., 1981, Purification and characterization of the
    entomocidal protoxin of Bacillus thuringiensis. J. Biol. Chem.
    256:3000-3004.
Charles, J.F., 1983, Action de la delta-endotoxine de Bacillus thurin-
    giensis var. israelensis sur cultures de cellules de Aedes aegyp-
    ti 1. Ann. Microbiol. 134:365-381.
Cooksey, K.E., Donninger, C., Norris, J.R., and Shankland, D., 1969,
    Nerve-blocking effect of Bacillus thuringiensis protein toxin.
    J. Invertebr. Pathol. 13:461-462.
Ebersold, H.R., Lüthy, P., and Müller, M., 1977, Changes in the fine
    structure of the gut epithelium of Pieris brassicae induced by
    the delta-endotoxin of Bacillus thuringiensis. Bull. Soc. Ent.
    Suisse 50:269-276.
Ebersold, H.R., Lüthy, P., Geiser, P., and Ettlinger, L., 1978, The
    action of the delta-endotoxin of Bacillus thuringiensis: an
    electron microscopy study. Experientia 34:1672.
Ebersold, H.R., Lüthy, P., and Huber, H., 1980, Membrane damaging
    effect of the delta-endotoxin of Bacillus thuringiensis. Expe-
    rientia 36:495-496.
Fast, P.G., and Angus, T.A., 1970, The delta-endotoxin of Bacillus
    thuringiensis var. sotto: a toxic low molecular weight fragment.
    J. Invertebr. Pathol. 16:465.

Fast, P.G., and Morrison, I.K., 1972, The delta-endotoxin of Bacillus thuringiensis. IV. The effect of delta-endotoxin on ion regulation by midgut tissue of Bombyx mori larvae. J. Invertebr. Pathol. 20: 208-211.

Griego, V.M., Moffett, D., and Spence, K.D., 1979, Inhibition of active K+ transport in the tobacco hornworm (Manduca sexta) midgut after ingestion of Bacillus thuringiensis endotoxin. J. Insect. Physiol. 25:283-288.

Harvey, W.R., and Wolfersberger, M.G., 1979, Mechanism of inhibition of active potassium transport in isolated midgut of Manduca sexta by Bacillus thuringiensis endotoxin. J. exp. Biol. 83:293-304.

Heimpel, A.M., and Angus, T.A., 1959, The site of action of crystalliferous bacteria in Lepidoptera larvae. J. Insect. Pathol. 1:152-170.

Holmes, K.C., and Monro, R.E., 1965, Studies on the structure of parasporal inclusions from Bacillus thuringiensis. J. Mol. Biol. 14: 572-581.

Huber, H.E., and Lüthy, P., 1981, Bacillus thuringiensis delta-endotoxin: composition and activation, in: Pathogenesis of Invertebrate Microbial Diseases, E.W. Davidson, ed., Allanheld, Osmun & Co. Publishers, Totowa, New Jersey, 209-234.

Huber, H.E., Lüthy, P., Ebersold, J.R., and Cordier, J.L., 1981, The subunits of the parasporal crystal of Bacillus thuringiensis: size, linkage and toxicity. Arch. Microbiol. 129:14-18.

Huber-Lukač, M., Lüthy, P., and Braun, D.G., 1983, Specificities of monoclonal antibodies against the activated delta-endotoxin of Bacillus thuringiensis var. thuringiensis. Infect. Immun. 40:608-612.

Klier, A., Fargette, F., Ribier, J., and Rapoport, G., 1982, Cloning and expression of the crystal protein genes form Bacillus thuringiensis strain berliner 1715. EMBO J. 1:791-799.

Knowles, B.H., Thomas, W.E., and Ellar, D.J., 1984, Lectin-like binding of Bacillus thuringiensis var. kurstaki lepidopteran-specific toxin is an initial step in insecticidal action. FEBS Lett. 168: 197-202.

Kohn, L.D., 1978, Relationships in the structure and function of receptors for glycoprotein hormones, bacterial toxins and interferon, in: Receptors and Recognition, P.Cuatrecasas and M.F. Greaves, eds., Chapman and Hall, London, 133-212.

Lüthy, P. and Ebersold, H.R., 1981a, The entomocidal toxins of Bacillus thuringiensis. Pharmac. Ther. 13:257-283.

Lüthy, P. and Ebersold, H.R., 1981b, Bacillus thuringiensis delta-endotoxin: histopathology and molecular mode of action, in: Pathogenesis of Invertebrate Microbial Diseases, E.W. Davidson, ed., Allanheld, Osmun & Co. Publishers, Totowa, New Jersey, 235-267.

Nickerson, K.W., 1980, Structure and function of the Bacillus thuringiensis protein crystal. Biotechnol. Bioeng. 22:1305-1333.

Nishiitsutsuji-Uwo, J. and Endo, Y., 1981, Mode of action of Bacillus thuringiensis delta-endotoxin: changes in hemolymph pH and ions of Pieris, Lymantria and Ephestia larvae. Appl. Ent. Zool. 16: 225-230.

Nishiitsutsuji-Uwo, J., Endo, Y., and Himeno, M., 1980, Effects of Bacillus thuringiensis delta-endotoxin on insect and mammalian cells in vitro. Appl. Ent. Zool. 15:133-139.

Nishiitsutsuji-Uwo, J., Ohsawa, A., and Nishimura, M.S., 1977, Factors affecting the insecticidal activity of delta-endotoxin of Bacillus thuringiensis. J. Invertebr. Pathol. 29:162-169.

Percy, J. and Fast, P.G., 1983, Bacillus thuringiensis crystal toxin: ultrastructural studies of its effects on silkworm midgut cells. J. Invertebr. Pathol. 41:86-98.

Prasad, S.S.S.V. and Shethna, Y.I., 1974, Purification, crystallization and partial characterization of the antitumour and insecticidal protein subunit from the delta-endotoxin of Bacillus thuringiensis var. thuringiensis. Biochem. Biophys. Acta 363:558-566.

Rao, A.S., Amonkar, S.V., and Phondke, G.P., 1979, Cytotoxic activity of the delta-endotoxin of Bacillus thuringiensis var. thuringiensis (Berliner) on fibrosarcoma in Swiss mice. Indian J. exp. Biol. 17:1208-1212.

Thomas, W.E. and Ellar, D.J., 1983a, Bacillus thuringiensis var. israelensis crystal delta-endotoxin: effects on insect and mammalian cells in vitro and in vivo. J. Cell Sci. 60: 181-197.

Thomas, W.E. and Ellar, D.J., 1983b, Mechanism of action of Bacillus thuringiensis var. israelensis insecticidal delta-endotoxin. FEBS Lett. 154:362-368.

Tojo, A. and Aizawa, K., 1983, Dissolution and degradation of Bacillus thuringiensis delta-endotoxin by gut juice protease of the silkworm Bombyx mori. Appl. Environ. Microbiol. 45:576-580.

Zalunin, I.A., Levin, E.D., Timokhina, E.A., Chestukhina, G.G., and Stepanov, V.M., 1983, Difference of the primary structures of delta-endotoxins formed by various serotypes of B.thuringiensis. Biochemistry 47:1327-1332.

MICROENVIRONMENTAL PROPERTIES AND CONFORMATIONAL

CHANGES IN GLYCOGEN PHOSPHORYLASE

A. E. Evangelopoulos

Department of Biochemistry, Center of Biological
Research, The National Hellenic Research Foundation
48 Vassileos Constantinou Avenue, Athens 116 35, Greece

Glycogen is a major source of energy for muscle contraction. Its breakdown and synthesis are regulated by the contractile state of the tissue as well as by the neural and hormonal control of the muscle, through changes in the phosphorylation state of glycogen phosphorylase and glycogen synthase. When muscle is stimulated electrically, calcium ions released from sarcoplasmic reticulum not only initiate muscle contraction but also activate phosphorylase kinase . As a result, the phosphorylation state of glycogen phosphorylase is increased and glycogenolysis is accelerated to provide the ATP required to sustain muscle contraction.

Glycogen phosphorylase catalyses the first step in glycogen degradation and since the fundamental work of the Coris[1] is one of the best understood examples of a regulatory enzyme. Glycogen phosphorylase was the first enzyme whose activity was shown by Krebs and Fischer[2] to be regulated by phosphorylation-dephosphorylation reactions and was one of the first eukaryotic enzymes for which regulation by allosteric effectors was recognized by Cori and Cori[3], Helmreich and Cori[4] and Monod et al.[5] Phosphorylase exists in two interconvertible forms. Phosphorylase b , found in resting muscle , is inactive except in the presence of AMP and IMP and is inhibited by ATP and glucose-6-phosphate. Thus the enzyme's activity is regulated by the intracellular concentration of these metabolites. In response to nervous or hormonal signals, phosphorylase b is coverted to phosphorylase a by the action of phosphorylase kinase[6] which results in the incorporation of a single phosphate group per monomer at Ser 14. Phosphorylase a is always active, although its activity may be increased by AMP and inhibited by glucose. Inactivation of phosphorylase a is catalysed by phosphorylase phosphatase[7].

Since the discovery of this enzyme by Cori and Cori in 1936 the structure, function and regulation of phosphorylase have been extensively studied by many investigators. In the past few years however a great development has taken place in the amino acid sequencing and X-ray diffraction analysis of this enzyme[8,9,10,11]. X-ray crystallographic groups in Edmonton and Oxford revealed the structural features of the phosphorylase dimer and published the a-carbon backbone stereo drawings of phosphorylase a[12] and phosphorylase b[13]. It was found that a phosphorylase subunit consists of two serarate domains, both of them having a b-sheet core surrounded by a coat of a-helices. The N-terminal domain (residues 1-489) can be divided into two structurally different subdomains, one of them contains predominately b-sheets (residues 1-321) and the other is built up from a-helices (residues 322-489)[12,14]. The active site of the enzyme is situated in a hydrophobic pocket near the pyridoxal phosphate binding site (Lys-679) between the two domains of the subunit[10,15]. The activator site is located at a distance of about 30 A from the active site, and represents the first discovered allosteric site of phosphorylase. It is near the monomer interface and the terminal tail peptide fragment containing the phosphoryl acceptor Ser-14[10,11].

Spectroscopic techniques such as NMR, ESR, Fluorescence and Absorption Spectroscopy offer an alternative possibility in place of X-ray diffraction analysis for studying phosphorylase structure, and can give direct and detailed information about the dynamic behaviour and conformational changes of the enzyme molecule[16,17,18,19,20,21].

In the present report fluorescein derivatives and spin labelling are used to evaluate microenvironmental properties and conformational changes in glycogen phosphorylase during its interaction with various ligands.

ESTIMATION OF THE POLARITY OF PHOSPHORYLASE BINDING SITES

A number of fluorescein dyes have been found useful in studying structure-function relationship of proteins; these studies include adsorption of the small molecules on specific binding sites[22,23] or covalent modification of particular reactive groups[24,25,26].

As it has been shown the polarity of a solvent can be expressed by the Z value, an empirical solvent polarity scale suggested by Kosower[27]. Our studies have indicated that a decrease in solvent polarity caused a red shift in the absorption spectrum maximum of the fluorescein dye. In addition we have demonstrated that a linear relationship exists between the λ_{max} of the fluorescein derivative in an organic solvent and the Z value (Fig. 1).The observed correlation of λ_{max} with Z value indicates that the absorption process in fluorescein dyes, like the one in 1-alkylpyridinium iodides, approximates a process in which a very polar state is changed to a neutral one[28]

The λ_{max} of the four fluorescein derivatives bound to a number

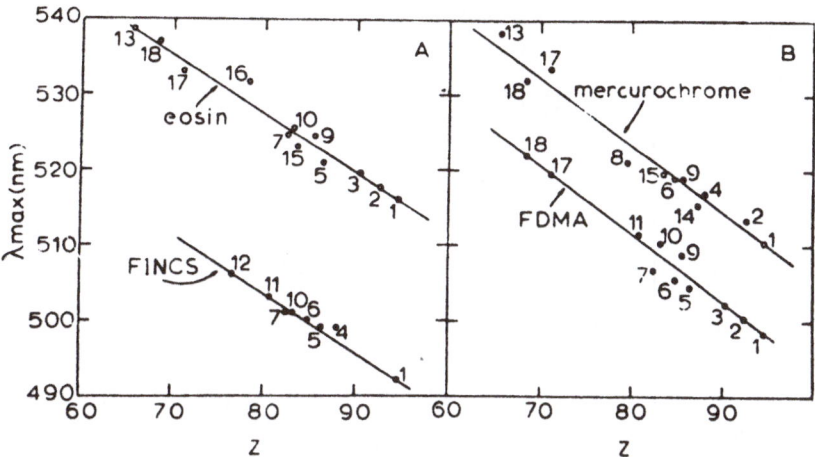

Fig. 1. Relationship between the wavelength of the absorption maxima of fluorescein derivatives and solvent polarity expressed by the Z value. A concentrated solution of fluorescein derivative in a 50 mM NaHCO$_3$- Na$_2$CO$_3$ buffer, pH 8.8 was diluted to a final concentration of 10^3 μM in several solvents before recording the absorption spectrum. (A,$_\circ$) is eosin, (A,$_\bullet$) FlNCS, (B,$_\circ$) mercurochrome and (B,$_\bullet$) FDMA. The numbers on the figure indicate the solvents used: 1 is 50 mM NaHCO$_3$- Na$_2$CO$_3$ buffer , pH 8.8, 2, 3, 4, 5, 6, 7, and 8 are 20, 40, 60, 70, 80, 90 and 100% of ethanol, 9, 10, 11, 12 and 13 contained 60, 70, 80, 90 and 100% of acetone, 14 and 15 are 80 and 100% of methanol, respectively; 16 is n-propanol, 17 is dimethylsulfoxide and 18 is dimethylformamide.

of proteins , obtained from our studies as well as from a few other reports on the subject, are given in Table 1. The data of Fig. 1 and Table 1 , can be used to estimate the Z value of the microenvironment around the conjugated dye. These values (Z_{est}) are also included in Table 1.

The Z_{est} values suggest : (a) that the microenvironment of the -SH groups reacting with fluorescein derivatives of phosphorylase b , apartate aminotransferase and dehydrogenases is basically hydrophobic with a Z_{est} ranged between 79.5 and 86.5, (b) that the amino groups of phosphorylase b and aspartate transaminase which react with fluorescein isothiocyanate hint a microenvironment with low polarity and comparable Z_{est} values, (c) that from the enzymes tested the vitamin B$_6$ enzymes phosphorylase b and aspartate transaminase showed hydrophobic adsorption sites for eosin and (d) that in addition to the above enzymes , albumin was also found to conjugate with the four fluorescein derivatives in specific sites of very low polarity. It is worth noting that the hydrophobicity of the microenvironment of eosin-binding sites of the phosphorylase b molecule can be roughly equated with that of

Table 1. ESTIMATION OF PROTEIN POLARITY SITES FROM THE λ_{max} OF BOUND FLUORESCEIN DERIVATIVES

The fluorescein derivative was titrated with increasing concentrations of protein until no further change in λ_{max} occured. Titrations were carried out in a series of solutions containing 10 μM of dye in 50 mM $NaHCO_3$ - Na_2CO_3 buffer, pH 8.8 and increasing amounts of proteins. Their absorption spectra were recorded 2 hours later at 23°C.

Protein	Eosin		FINCS		Mercurochrome		FDMA	
	λ_{max}(nm)	Z_{est}	λ_{max}(nm)	Z_{est}	λ_{max}(nm)	Z_{est}	λ_{max}(nm)	Z_{est}
Phosphorylase b	519	91	498	87	522	81.5	511	80.5
Aspartate amino-transferase	520	88	497	88	520[18]	83.5	512[16]	79.5
α-Chymotrypsin	516	94.5	492	94.5	513	92	498	94.5
Apyrase	516	94.5	494	92	517	87.5	502	90.5
Urease	516	94.5	492	94.5	514	90.5	498	94.5
Trypsin	516	94.5	492	94.5	512	93	498	94.5
Papain	518	92.5	492	94.5	513	92	498	94.5
Albumin	529	76.5	501	83	522	81.5	508	84
Yeast alcohol de-hydrogenase	-	-	-	-	-	-	508[19]	84
D-Glyceraldehyde-3-P dehydrogenase	-	-	-	-	-	-	506[21]	86.5
Lactate dehydrogenase	-	-	-	-	-	-	512[20]	79.5

a solution composed of 95% propan-1-ol and 5% buffer,

EOSIN AS AN OPTICAL PROBE FOR CALCULATION OF THE DISSOCIATION CONSTANTS OF LIGAND-PHOSPHORYLASE COMPLEXES

Low resolution binding studies with a number of ligands to phosphorylase b have led to identification of the following major binding sites on the enzyme molecule[29] :(a) a site in the N-terminal domain which binds AMP, ATP, glucose-1-P , arsenate and glucose-6-P; (b) an active site in a hydrophobic pocket betweenthe two main domains of the enzyme, near the pyridoxal phosphate binding site,which binds the substrates P_i or glucose-1-P as well as glucose and UDP-glucose; (c) a glycogen storage site and (d) a nucleoside or inhibitor site which binds nucleosides, purine base analogs, dinucleodides and nucleotides.

Eosin has been shown to bind effectively to phosphorylase b[16]. Eosin-phosphorylase b interaction is expressed with a red shift in the spectrum of the dye (the absorption maximum is shifted from 516 to 528 nm). Eosin binding sites (two per subunit of phosphorylase b) have strong affinities for the dye. The K values determined in 100 mM Tris/HCl - 10 mM 2-mercaptoethanol - 1 mM EDTA buffer, pH 7.0 (I=0.091) were 7.7 and 41.7 μM respectively. The influence of substrates and effectors on eosin-enzyme complexes has been used to study ligand-phosphorylase b interactions. IMP completely displaced the dye from the enzyme, indicating that there are two IMP-binding sites per phosphorylase b monomer. A competitive mode of eosin displacement from the enzyme by AMP and IMP exists at least for the nucleoside-binding site (or the low-affinity site for the dye). Glucose-6-P, P_i and glycerol-2-P decrease eosin affinity for both classes of its binding sites, while glucose-1-P and UDP-glucose affect eosin binding to its high-affinity site. In Fig. 2 titrations of a standard eosin-phosphorylase b complex with a number of metabolites are presented. The substantial spectroscopic changes observed during these titrations provided the following results :(a) β-NAD completely displaced the dye from the enzyme, indicating that β-NAD binding sites are related with eosin-binding sites on the enzyme molecule;(b) NADP displaced the dye only from the low affinity eosin-binding site; (c) ADP completely displaced the dye from the enzyme as does IMP. The apparent dissociation constants, K_d , for ligand-enzyme complexes (with respect to eosin binding) calculated from the spectroscopic data of Fig. 2 are shown in Table 2.

CHANGES IN THE STRUCTURE OF PHOSPHORYLASE b INDUCED BY LIGANDS

It has been shown that phosphorylase b was inactivated by fluorescein isothiocyanate and that one $-NH_2$ group per subunit is essential for the catalytic activity of the enzyme[30]. The fluorescein isothiocyanate - phosphorylase complex exhibits a characteristic absorption spectrum with a sharp λ_{max} at 492 nm. an area of the spectrum where neither substrates, cofactors nor effectors absorb. Thus changes in the absorption spectrum of the enzyme-bound fluorescein

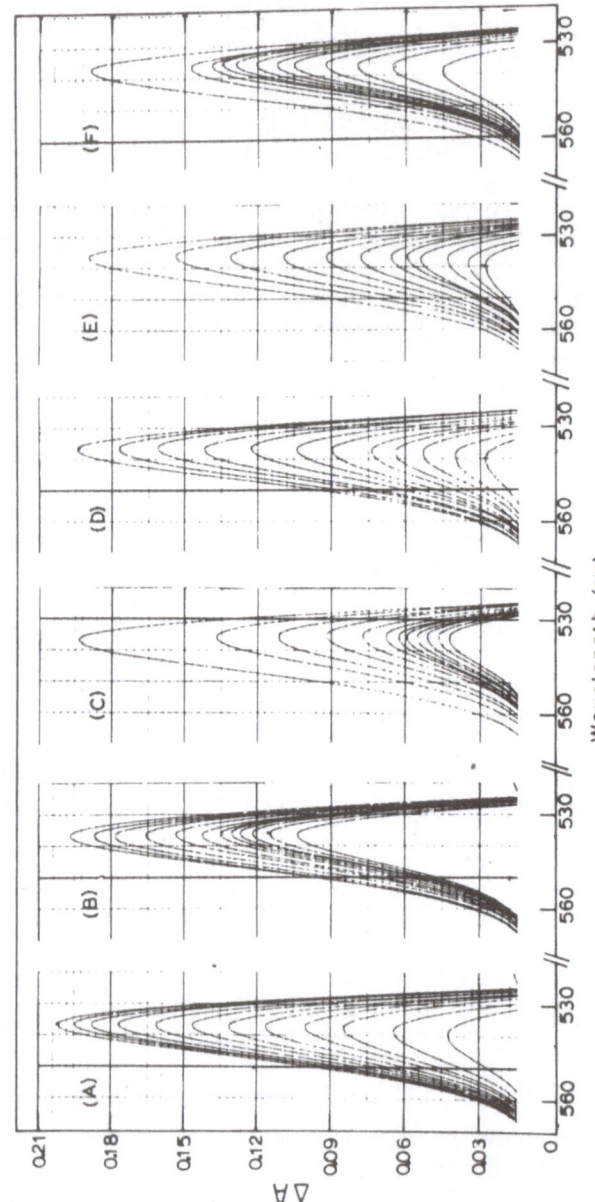

Fig. 2. Effect of ligands on the difference absorption spectrum of eosin-phosphorylase b complex. Titrations of eosin-phosphorylase b complex with a ligand were performed by additions of small volumes from a concentrated solution of the ligand to both a standard eosin-enzyme complex solution (30 μM eosin + 1.67 μM enzyme) in a 100 mM-Tris/HCl/ –.10 mM 2-mercaptoethanol/ 1 mM EDTA buffer, pH 7.0 (23°C) – sample cuvette – and to areference cuvette containing eosin at a concentration equal to that of the sample cuvette. Titrations were carried out with β-NAD (A) and β-NADP (B) at concentrations of 0, 0.16, 0.32, 0.64, 1.11, 1.74, 2.52, 3.44, 4.66, 6.15, 9.06 and 14.54 mM respectively, glucose-6-P (C), IMP (D), ADP (E) and AMP (F) at concentrations of 0,0.20, 0.40, 0.80, 1.39, 2.18, 3.15, 4.30, 5.82, 7.69, 11.32 and 18.18 mM respectively.

154

Table 2. Apparent K_d for Ligand-Phosphorylase b Complexes

Ligand	K_d (mM)	
β-N A D	3.2	10.0
β-N A D P		1.5
Glucose-6-P	0.3	
I M P	1.0	4.1
A D P	0.5	3.5
A M P	0.08	12.9

The data from the spectroscopic titrations after correction for dilution-effect were arranged in the form of Scatchard plots and from the slopes the K_d were calculated.

derivative induced by ligands reflect changes in a region of phosphorylase b critical for its catalytic activity.

Sodium cholate is able to induce activity in rabbit muscle phosphorylase b in absence of AMP. The maximum activation of the enzyme in presence of 7 mM sodium cholate is 24% of that achieved by 1 mM AMP. The spectroscopical properties of a fluorescein isothiocyanate - phosphorylase b conjugate were used to study the effect of sodium cholate on the structure of the enzyme. The difference spectra of the conjugate - sodium cholate system versus the conjugate showed a difference absorption maximum at 507 nm and a minimum at 486 nm. The absorption difference (ΔA) as a function of sodium cholate concentration is shown in Fig. 3. The diagramme is biphasic and is similar to the activation curve of the enzyme by sodium cholate. The concentration of bile salt needed for half maximal absorption or activity change is 3.5 and 4.0 mM respectively.These changes suggest structural alterations of the enzyme molecule in the area of the conjugation of the probe. Of course one cannot exclude the possibility that the spectral changes in the first step of the titration curve of the phosphorylase-dye complex are partly due to a direct interaction of sodium cholate with the bound dye or to an increase of the hydrophobic environment of the dye by direct masking of its binding site by the bile salt.

Phosphorylase b can be also specifically spin-labelled at a site essential for the catalytic action of the enzyme. A paramagnetic analogue of 1-fluoro-2,4-dinitrobenzene was used as a dinitrophenylating agent. Reaction of phosphorylase b with the paramagnetic probe combined with thiolysis , leads to spin labelling of a single -NH_2 group (0.75 groups per subunit) with concomitant loss

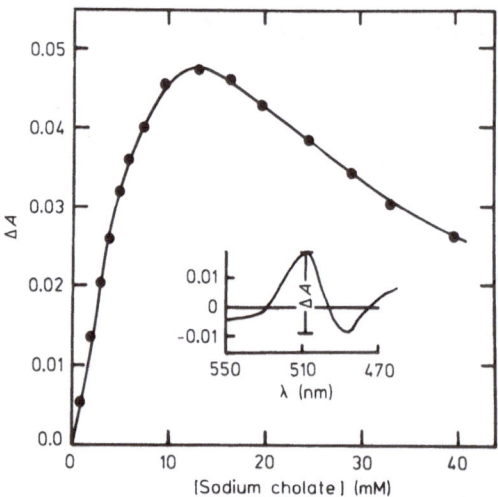

Fig. 3. Titration of the difference absorption spectrum of fluorescein isothiocyanate-phosphorylase b complex versus the complex.
The difference spectra were taken with 45 μM dye-enzyme conjugate
(0.1 moles of dye bound per mole of dimer) in presence of various
concentrations of sodium cholate against 45 μM fluorescein isothiocyanate-enzyme complex in 1 mM EDTA, 20 mM glycerol-2-P buffer, pH
6.8(23°C). Inset shows the difference absorption spectrum in presence of 3.9 mM of sodium cholate. The absorption difference, ΔA is
the absolute difference between the maximum and the minimum of the
difference absorption spectrum.

of 50% of the catalytic activity of the enzyme. The ESR spectrum of
modified phosphorylase b indicates that the attached label has rather
limited segmental mobility and its environment is slightly hydrophobic[17].
    The effect of sodium cholate on the conformational flexibility
of spin-labelled phosphorylase b was studied. The titration of the
paramagnetic 1-fluoro-2,4-dinitrobenzene derivative of phosphorylase b by sodium cholate is shown in Fig. 4. The ESR ratio (low field
to the centre line) decreases gradually with increasing bile salt
concentration and at low steroid concentration the titration curve
is strongly sigmoidal. Hill-plot analysis of the data of Fig. 4,
$\log(R_0-R)/(R_0-R_L)-(R_0-R)$ against log (sodium cholate), where R, R_0
and R_L are the values of the ESR ratio, at a given point in the titration curve, at zero ligand concentration and at saturation,re-
spectively, gave Hill coefficient h=3.6±.0.2 and an apparent dissociation constant, K_d=14.0±0.5 mM. Since the cooperative interaction between ligand binding sites is very strong, we assume[31] that
the slope of the Hill plot approaches the number of sodium cholate
-binding sites which affect the mobility of the spin-label.The ESR
studies show a monophasic strong cooperative behaviour with cholate concentration. A possible explanation of the above observation
could be the existence of two classes of sodium cholate binding si-

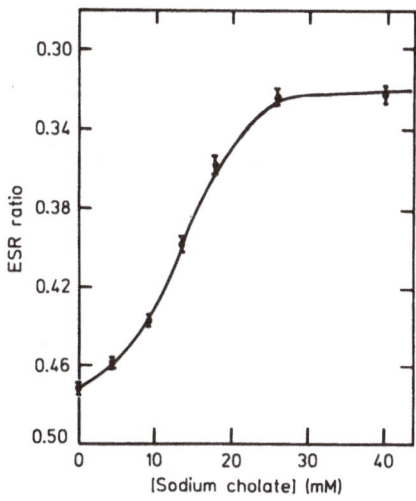

Fig. 4. Changes in the ESR ratio (low field peak height/centre peak height) of spin-labelled phosphorylase b by the 1-fluoro-2,4-dinitrobenzene derivative as a function of sodium cholate concentration. The enzyme (35 μM) was spin-labelled and thiolysed. ESR spectra were recorded immediately after the addition of sodium cholate in the enzyme solution in 1 mM EDTA, 20 mM glycerol-2-P buffer, pH 6.6(23°C). Each point and vertical bar show the mean and standard deviation for six determinations.

tes. The first class of sites which are saturated at lower cholate concentration, are responsible for the enzyme activation and the observed changes in the spectrum of the complex of fluorescein isothiocyanate-phosphorylase b, while the second class of sites which are saturated at higher cholate concentrations are mainly responsible for the conformational changes of the enzyme which lead to the immobilization of the ESR label and for the second step of the activation curve and the titration curve of the dye-labelled enzyme.

In Fig. 5 we present the effect of the physiological ligands of phosphorylase on the spectroscopical properties of the fluorescein isothiocyanate-enzyme complex. It can be seen that glucose-1-P P_i, AMP or glucose-6-P alter the spectrum of the probe during their interaction with the enzyme, causing a decrease in the difference absorption spectrum of the phosphorylase-dye conjugate. These changes reflect structural alterations of the enzyme molecule in a region critical for its catalytic activity.

Spin-labbeled phosphorylase b at a site essential for its catalytic activity can be also used to detect structural changes of the enzyme molecule during the interaction with substrates, effectors or inhibitors. Experiments presented in Table 3, indicate that the ESR spectrum of phosphorylase b, spin-labelled by a paramagnetic analogue of 1-fluoro-2,4-dinitrobenzene at a single amino group per subunit, is sensitive to conformational changes induced by ligands. More specifically, addition of the allosteric activator AMP and the

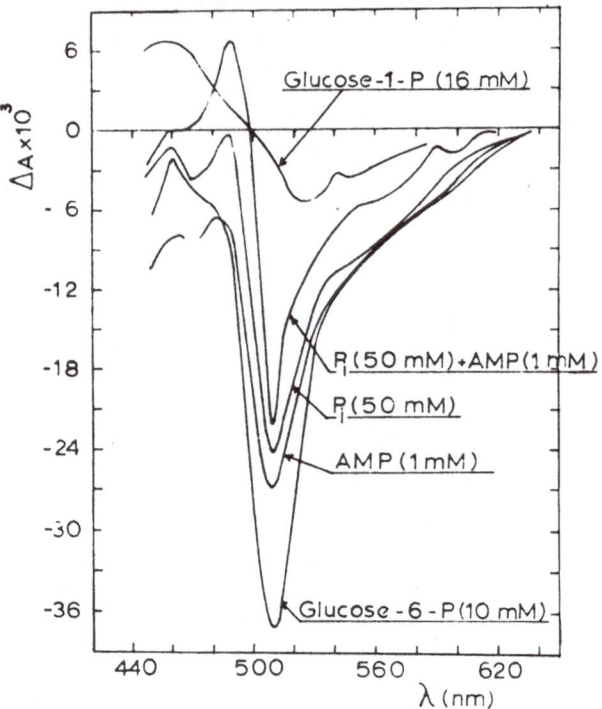

Fig. 5. Effect of various ligands on the difference spectrum of fluorescein isothiocyanate-phosphorylase b complex. The difference spectra were taken with 20 μM fluorescein isothiocyanate-enzyme (1.0 ε-NH$_2$ groups modified per dimer) in the presence of the various ligands, against 20 μM fluorescein isothiocyanate-enzyme complex in a 20 mM sodium β-glycerophosphate, 1.5 mM EDTA buffer pH 6.8(23°C).

allosteric inhibitor glucose-6-P markedly changes the ESR spectrum of the spin-labelled enzyme, causing further immobilization of the label, while the presence of either glucose-1-P or glycogen does not significantly affect the ESR ratio (Table 3). In addition the combined effect of glucose-1-P and AMP is greater than that of either effector alone. The results presented in Table 3 indicate also that the effects of the various ligands on the spin-labelled phosphorylase b are more pronounced at pH 6.8 than at pH 8.0.

Phosphorylase b spin-labelled at -NH$_2$ groups was also used in detecting conformational changes taking place during the modification of the -SH groups of the enzyme. More specifically, the reaction of spin-labelled phosphorylase b with DTNB was monitored by ESR. Profound changes of the resonance spectrum of the spin-labelled enzyme were recorded during -SH group modification. The ratio of the low of the low field to the center peaks of the spectrum has been used to characterize such conformational alterations. While the ESR spectral changes accompanying the reaction of spin-labelled phosphorylase b with DTNB were being recorded with time (Fig. 6) the extent of -SH group modification was determined following the

Table 3. Effect of various Ligands on the ESR Spectrum of Spin-
Labelled Phosporylase b

Ligand	Ratio of the downfield to center bands	
	pH 8.0	pH 6.8
---	0.45	0.43
A M P (3 mM)	0.42	0.38
Glucose-1-P (10 mM)	0.45	0.42
Glycogen (0.5%)	0.45	0.43
Glucose-6-P (20 mM)	0.44	0.40
Glucose-1-P (10mM) + A M P (3 mM)	0.42	0.37
Chloropromazine	---	0.41

The ratio of the low-field line to the centre line has been used to
characterize the spectrum, at pH 8.0 and 6.8. The enzyme (35 µM)
was spin-labelled and thiolysed. Values shown represent.the mean
calculated from five measurements in each case.

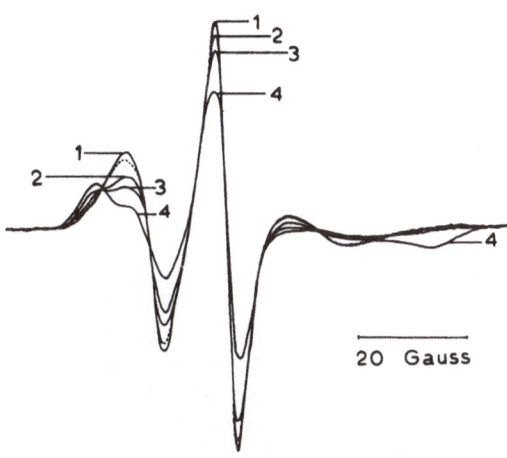

20 Gauss

Fig. 6. ESR spectral changes observed during the reaction of spin-
labelled phosphorylase b with DTNB. ----, ESR spectrum of spin-
labelled enzyme alone; ———, ESR spectra of spin-labelled phosphorylase b in presence of 1 mM DTNB, recorded:(1) 6;(2),47;(3),81 and
(4),600 min (equilibrium) after the addition of DTNB. The reaction
of the lebelled enzyme (45 µM) with DTNB was carried out in 20 mM
glycerol-2-P - 1 mM EDTA buffer, pH 6.8-

reaction rate spectrophotometrically at 412 nM. The results presented in Fig. 7, show that the modification of the highly reactive
groups (about 2) is accompanied by an increase of the motional freedom of the label, while further -SH group modification (up to 6.3
groups per mol of enzyme) results in gradual decrease of the mobility of the spin label. According to Kastenschmidt et al.[32], modification of the fast-reacting -SH groups does not affect either
the activity or the quaternary structure of the enzyme, while modification of the second class of -SH groups of phosphorylase b with
DTNB leads to a mixture of monomers, dimers and polydispersed agre-
gates accompanied by total loss of activity. Although the highly
reactive -SH groups are not directly involved in catalysis,as shown
previously blocking these groups seems to affect the structure of
the enzyme.

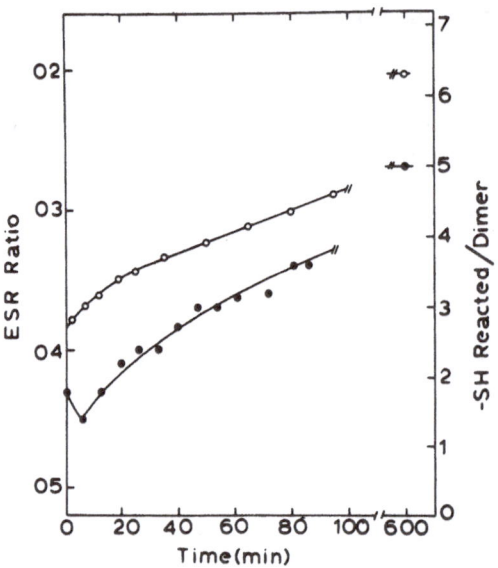

Fig. 7. Effect of -SH group modification (○) on the ESR spectrum
(•) of spin-labelled phosphorylase b. DTNB (1mM) was added at zero
time to 45 μM of the enzyme in 20 mM glycerol-2-P - 1 mM EDTA buffer, pH 6.8. The extent of -SH group modification was determined by
measuring the increase in absorbance at 412 nm using the extinction
coefficient 13600 l mole^{-1} cm^{-1}, to calculate the number of -SH
groups reacting. The changes in the ESR spectrum during the -SH
group modification are expressed as the ratio of the low field to
centre line.

ABBREVIATIONS : FINCS, fluorescein isothiocyanate, isomer 1;FDMA,
fluorescein dimercuric acetate; mercurochrome, dibromohydroxymercu-
rifluorescein; eosin, tetrabromofluorescein; DTNB, 5,5'-dithiobis(2-
itrobenzoic acid).

REFERENCES

1. G. T. Cori and C. F. Cori, The enzymic conversion of phosphorylase a to b. J. Biol. Chem. 158:321(1945).
2. E. G. Krebs and E. H. Fischer, The phosphorylase b to a converting enzyme of rabbit skeletal muscle. Biochim. Biophys. Acta 20:150(1956).
3. C. F. Cori and G. T. Cori, Mechanism of formation of hexosemonophosphate in muscle and isolation of a new phosphate ester. Proc. Soc. Exp. Biol. Med. 34:702(1936).
4. E. Helmreich, M. C. Michaelides and C. F. Cori, Effect of substrates and substrate analogs on the binding of 5'-adenilic acid to muscle phosphorylase. Biochemistry 6:3695(1967).
5. J. Monod, J. Wyman and J. P. Changeux, On the nature of allosteric transitions:a possible model. J. Molec. Biol. 12:88(1965).
6. E. G. Krebs and E. H. Fischer, Molecular properties and transformations of glycogen phosphorylase in animal tissues. Adv. Enzymol. 24:263(1962).
7. T. S. Ingebritsen, J. G. Foulkes and P. Cohen, The protein phosphatases involved in cellular regulation. Eur. J. Biochem. 132:263(1983).
8. K. Titani, A. Koide, J. Hermann, L. H. Ericson, S. Kumar, R. D. Wade, K. A. Walsh, H Neurath and E. H. Fischer, Complete amino acid sequence of rabbit skeletal muscle glycogen phosphorylase. Proc. Natn. Acad. Sci. USA, 74:4762(1977).
9. A. Koide, K. Titani, L. H. Ericson, S. Kumar, H. Neurath and K. A. Walsh, Sequence of the aminoterminal 349 residues of rabbit muscle glycogen phosphorylase including sites of covalent and allosteric control. Biochemistry 17:5657(1978).
10. R. J. Fletterick, J. Sygusch, M. Semple and N. B. Madsen,Structure of glycogen phosphorylase a at 3.0 A° resolution and its ligand binding sites at 6 A°. J. Biol. Chem. 251:6142(1976).
11. I. T. Weber, L. N. Jonson, K. S. Wilson, D. G. R. Yeates, D. L. Wild and J. A. Jenkins, Crystallographic studies on the activity of glycogen phosphorylase b. Nature 273:433(1978).
12. S. Sprang and R. J. Fletterick, The structure of glycogen phosphorylase a at 2.5 A°resolution. J. Mol. Biol.131:523(1979).
13. L. N. Johnson, I. T. Weber, D. L.Wild, K. S. Wilson and D.G.R. Yeates, Crystallographic analysis at low resolution of metabolic binding sites on phosphorylase b. J. Mol. Biol.118:579 (1978).
14. L. N. Johnson, E. A. Stura, K. S. Wilson, M. S. P. Sansom and I. T. Weber, Nucleotide binding to glycogen phosphorylase b in the crystal. J. Mol. Biol. 143:639(1979).
15. E. C. Y. Li, R. J. Fletterick, J. Sygusch and N. B. Madsen, An essential arginine residue in the active-site pocket of glycogen phosphorylase. Can. J. Biochem. 55:465(1977).
16. N. G. Oikonomakos, T. G. Sotiroudis and A. E. Evangelopoulos, Interaction of phosphorylase b with eosin-influence of substrates and effectors on eosin-enzyme complex. Biochem. J. 181:309(1979).

17. C. T. Cazianis, T. G. Sotiroudis and A. E. Evangelopoulos, Spin-labelling of phosphorylase b using a paramagnetic 1-fluoro-2,4-dinitrobenzene derivative. Biochim. Biophys. Acta, 621: 117(1980).

18. T. G. Sotiroudis, C. T. Cazianis, N. G. Oikonomakos and A. E. Evangelopoulos, Effect of sodium cholate on the catalytic and structural properties of phosphorylase b. Eur. J. Biochem. 131:625(1983).

19. M. M. Hoerl, K. Feldmann, K. D. Schnackers and E. J. M. Helmreich, Ionization of pyridoxal-5-P and the interactions of AMP-S and thiophosphoseryl residues in native and succinilated rabbit muscle glycogen phosphorylase b and a inferred from ^{31}P NMR spectra. Biochemistry 18:2457(1979).

20. K. Feldmann and E. J. M. Helmreich , The pyridoxal -5-P site in rabbit skeletal muscle glycogen phosphorylase b - an ultra-violet ^{1}H and ^{31}P nuclear magnetic resonance study. Biochemistry 15:2394(1976).

21. S. J. W. Busby and G. K. Radda, Regulation of the glycogen phosphorylase system - from physical measurements to biological speculations. Curr. Top. Cell Reg. 10:89(1976).

22. L. L. Somerville and F. A. Quiocho , The Interaction of tetra-iodofluorescein with creatine kinase, Biochim. Biophys. Acta 481:493(1977).

23. E. R. Kantrowitz, L.B. Jacobsberg, S. M. Landfear and W. N. Lipscomb, Interaction of tetraiodofluorescein with a modified form of aspartate transcarbamylase. Proc. Natl. Acad. Sci. USA 74:111(1977).

24. T. G. Kalogerakos, N. G. Oikonomakos, C. G. Dimitropoulos, I. Karni-Katsadima and A. E. Evangelopoulos, Interaction of aspartate aminotransferase with mercurochrome. Biochem. J. 167: 53(1977).

25. S. J. D. Karlish, Characterization of conformational changes in (Na,K) ATPase labeled with fluorescein at the active site. J. Bioenerg. Biomembr. 12:111(1980).

26. L. Brand, J. R. Gohlke and D. Sethu Rao, Evidence for binding of rose belgal and anilinonaphthalenesulfonates at the active site regions of liver alcohol dehydrogenase. Biochemistry 6:3510(1967).

27. E. M. Kosower, Effect of Solvent on Spectra - Correlation of spectral absorption data with Z-values. J. Am. Chem. Soc. 80:3261(1958).

28. D. C. Turner and L. Brand, Quantitative estimation of protein binding site polarity. Biochemistry 7:3381(1968).

29. V. Dombradi, Structural aspects of the catalytic and regulatory function of glycogen phosphorylase. Int. J. Biochem. 13:125 (1981).

30. N. Oikonomakos, Chemical modification of an allosteric enzyme. Ph. D. Thesis, University of Athens(1977),Athens,Greece.

31. N. B. Madsen and S. Shechosky, Allosteric properties of phosphorylase b. J. Biol. Chem. 242:3301(1967).

32. L. L. Kastenschmidt, J. Kastenschmidt and E. Helmreich, Subunit interactions and their relationship to the allosteric properties of rabbit skeletal muscle phosphorylase b. Biochemistry 7:3590(1968).

QUANTUM MECHANICAL MODEL FOR PASSIVE TRANSPORT THROUGH

BILAYER LIPID MEMBRANES

Leonor Cruzeiro and Kelo M. C. Da Silva

Lab. Biomecânica
Instituto Gulbenkian de Ciência
Apart. 14, 2781 Oeiras Codex
Portugal

1. INTRODUCTION

Unmodified bilayer lipid membranes (BLMs) are known to be low dielectric media with a correspondingly low ionic conductivity. The mathematical treatment of ion diffusion through them is usually done in classical thermodynamical terms through the so-called Generalized Nernst-Planck equation[1,2]:

$$\Phi = - D \left(\frac{dC}{dx} + zC \frac{d\phi}{dx} + C \frac{dV}{dx} \right) \qquad (1)$$

where Φ is the ion flux per unit area, D is the diffusion coefficient of the ion in the membrane, C is the ion concentration in the membrane, ϕ is the electrical potential, z is the valency of the ion, V is the potential energy of the ion in the membrane and x is the coordinate normal to the membrane surface.

In equation (1) the interaction of the ion with the membrane is taken into account in two ways: (a) implicitly in the diffusion coefficient D and (b) explicitly in the last term which includes V. The low conductivity normally attributed to BLMs is, in equation (1), the mathematical result of the low values attributed to the diffusion coefficient D and of the high values computed for the potential energy V.

In spite of the generalized idea that BLMs are always

very low conductivity media, cases have been found for
which this is not so, viz., (a) the high proton permeabil‐
ity, which is six orders of magnitude, or more, larger
than that of sodium ions[3,4] and (b) the peaks in membrane
permeability observed during phase transitions[5,6]. Cases
like these have not led to the questioning of the solidly
established concept of the BLM's low conductivity, but
instead, they have been interpreted as implying the
existence of local modifications in the lipid bilayer
structure, through which the conductive ions cross. Thus,
situation (a) is explained by postulating lattices of
water molecules, spanning the entire bilayer thickness
and through which the protons would cross the BLMs by
jumping from water molecule to water molecule, in a way
similar to that postulated for their motion in icy
structures[3,4]. Situation (b),on the other hand, is usually
explained off by the consideration of "disordered do
mains" formed at the borders of "clusters", i.e., patches
of lipids whose phase state is different from the remain‐
ing and dominant lipidic matrix[5-7]. In these disordered
domains the packing of the phospholipids is described as
less tight than in any of the phase states[7], so that the
ions find "holes", i.e., emptier regions in the membranes,
through which they can flow more easily. These disordered
domains are considered to exist even below (or indeed
above) the phase transition temperature[7]; at this tempera‐
ture, however, their number is supposed to be larger and,
consequently, the number of "holes" is correspondingly
increased. The inclusion of these hypotheses in a mathe‐
matical treatment by equation (1) might be possible if
adequate values for the diffusion coefficient D and for
the potential energy of the ion in the corresponding
structures were provided. However, the calculations would
also have to involve a statistical parameter describing
the number of modified regions (either water lattices or
disordered domains) present per unit area, and this need
would probably be serious enough to perclude the use of
equation (1) altogether.

In this article we use a non-classical approach to
develop a model for ion diffusion through BLMs. As it will
be seen, our model sheds new light on the nature of this
phenomenon and eliminates difficulties encountered in the
classical treatments such as those mentioned above.

2. QUANTUM MECHANICAL MODEL

Our description of ion the diffusion through unmodi‐
fied BLMs is a quantum mechanical one and its mathematical

basis is the unidimensional stationary Schrödinger equation:

$$-\frac{\hbar^2}{2m}\frac{d^2\Psi}{dx^2} + V(x)\,\Psi = E\,\Psi \qquad (2)$$

where $\hbar = h/2\pi$ and h is Planck's constant, m is the mass of the ion, Ψ is the ionic wave function, E is the total energy of the ion, V(x) is, as before, the potential energy of the ion in the BLM, and x is the coordinate perpendicular to the membrane surface.

The only unknown parameter in equation (2) is the potential energy profile V(x), for which we assume a shape as is shown in figure 1. This potential energy profile is very different from the one which is inserted in equation (1), the reason being that we no longer treat the membrane as an homogeneous medium solely characterized by a low dielectric constant, as authors like Neumcke and Läuger have done[1,2]. We think that, given the sizes of the ions, these particles are going to "feel" the molecular structure of the membrane and hence, will have a different interaction with each group which constitutes the phospholipid molecule. The potential energy profile was thus postulated according to what is known of the BLM's structure[8,9]: the higher energy barriers at the edges, designated by VP in figure 1, correspond to the position of the polar groups wherein we assume that the ion will face the highest repulsive field; the sequence of smaller barriers corresponds to the sequence of CH_2 groups in the hydrocarbon chains, the one designated by VCH in figure 1 corresponding to the interaction with each CH_2 group and the one designated by VL corresponding to the region in between the CH_2 groups; the barrier designated by VM represents the discontinuity associated with the region of interaction of the two monolayers. The distance between successive VCH barriers is taken to be 1.25 Å, in agreement with the crystallographic data[8,9]. The values taken for the barriers's lengths were, typically 0.8 Å for the VP barriers, 0.75 Å for the VLs, 0.5 Å for the VCHs and 0.5 Å for VM. The constitution of V(x) as a series of rectangular elements was adopted in order to solve equation (2) by the method developed by Hosur[10].

When the potential energy function of figure 1 is inserted in equation (2) one can find its solution Ψ and from it compute the quantity $|\Psi|^2$ which is the BLM's transmission coefficient for the ion of mass m[10]. This quantity, which is the probability of the ion getting through the BLM, is proportional to the membrane perme-

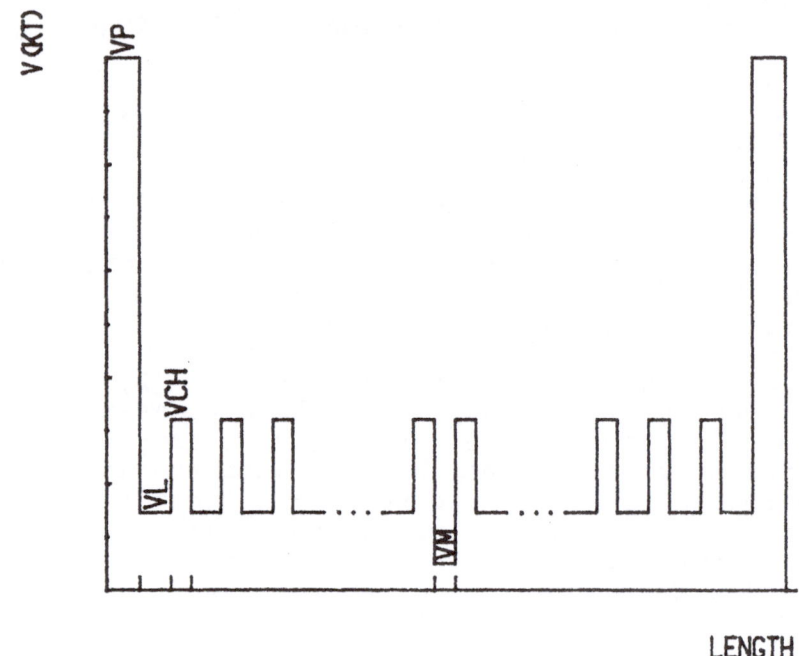

Fig. 1. Potential energy function of an ion in a BLM.

ability to the given ion and here we shall designate it
by the letter P and call it the permeability of the BLM.

The variation, predicted by our model, of the permeability with the mass of the ion and with the potential
energy of the ion at headgroup level is illustrated in
figures 2 and 3, respectively. Both curves are essentially of an exponential type, the permeability decreasing
as either VP or the mass increases. The permeability
versus mass curve also shows a modulation superimposed
on the exponential curve, but the peaks are too narrow
and their location depends on the values of other parameters like VCH, VL and VP; an interpretation of these
peaks in terms of ion selectivity is therefore difficult
and we shall ignore them.

If the value of the energy barrier VP was the same
for protons and for sodium ions, the ratio of their respective permeabilities would be, according to figure 2,
of the order of 10^{12}, a much greater value than the one
measured experimentally[3,4]. However, ions do not only
differ in the value of their masses, but they may also

differ in their charge, or, as is the case for protons and sodium ions, they may have different ionic radius. Indeed, the fact that the ionic radius of protons (~0.5Å) is smaller than that of sodium ions (~0.95Å) will result in a value of VP greater for protons than for sodium ions. In Table 1 we show the values of the membrane permeability to protons and to sodium ions, calculated for different values of VP. Thus, if VP has a value of 2.0 kT for protons and a value of 1.7 kT for sodium ions, then their corresponding permeability ratio will be of the order of 10^6, in agreement with the experimental measurement.

The high ratio of proton to sodium permeabilities can thus be interpreted by our model as resulting only from the interaction of the ions with the undisturbed BLM structure, with no need to resort to artificial devices such as water bridges through the hydrophobic portion of the membrane which, incidently, seem to be very unlikely from an energetic point of view[11].

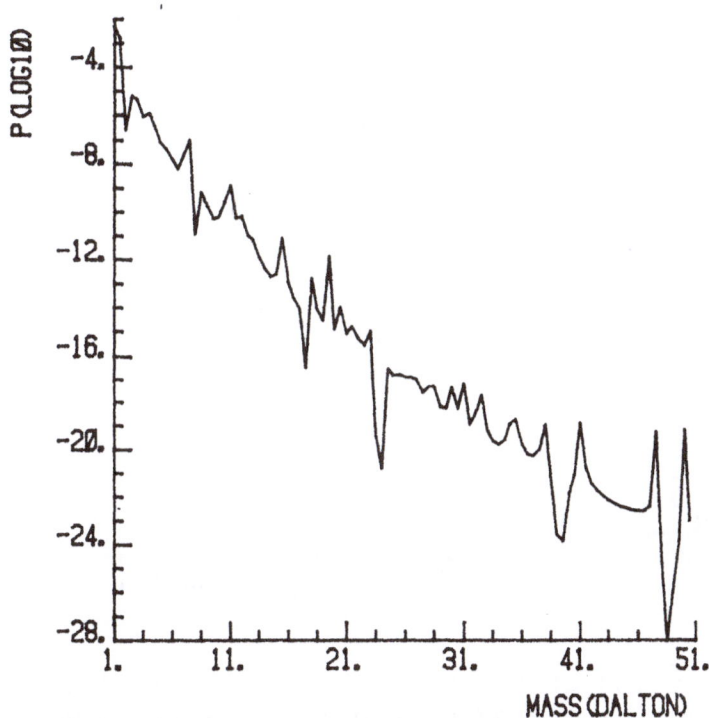

Fig. 2. Variation of the permeability of the BLM with the mass of the particle, keeping all other parameters constant.

Figure 4 shows, for the potassium ion (m–39.102 daltons), the variation of the membrane permeability with the heights of the energy barriers VCH and VL, which are related to potential energy of the ion in the hydrocarbon region (see figure 1). Figure 4 is thus a three dimensional plot where the third dimension is given by the degree of grey colour: the lighter a point is the higher will the permeability be at the corresponding values VCH, VL. Figure 5, which shows a profile along the fixed value of VCH indicated by the arrow in figure 4, allows a better understanding of this dark/light representation. According to figure 4, the permeability variation with VCH and VL is the following: superimposed on a very rough planar surface there are deep wells, i.e., regions where the permeability lowers abruptly, such as regions A and B. We think that such regions may be associated with the physical states of the membrane, while the neighbouring

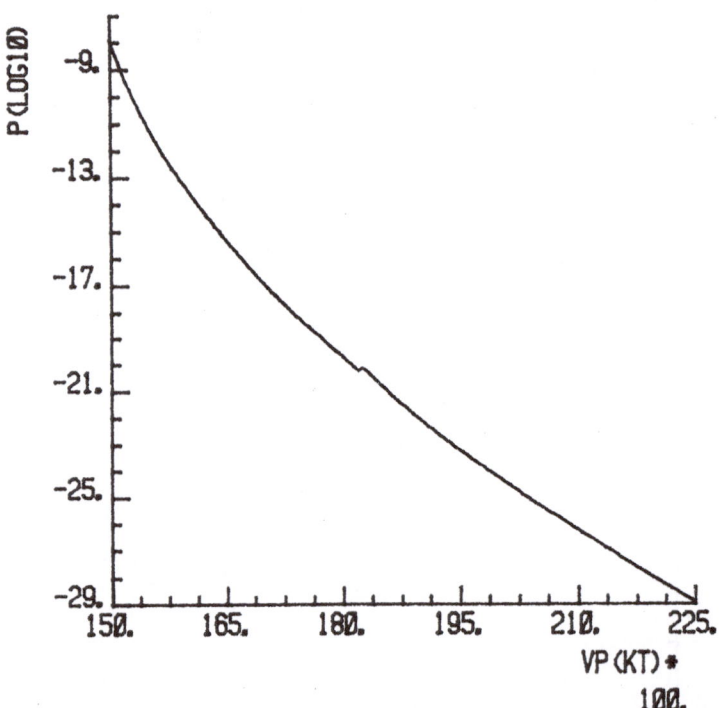

Fig. 3. Variation of the permeability of the BLM with the potential energy of the ion at headgroup level (see figure 1), when all other parameters are kept constant.

Table 1. BLM permeability to protons and sodium
         ions for different values of VP (see
         figure 1):

VP(kT)	2.0	1.7
Na$^+$	1.0×10^{-15}	2.5×10^{-9}
H$^+$	5.0×10^{-3}	2.5×10^{-1}

regions may correspond to the phase transition. According
to this interpretation, the peaks in the permeability
which occur during the phase transition can be explained
in the following way: let us consider the membrane in a
gel-like sate; the phospholipid molecules are packed tight
and the values of the potential energy of the ion in the
membrane are large, being situated, for instance, in
region A of figure 4 where the permeability is on average very low as expected for a gel state. As the temperature increases, the hydrocarbon chain mobility increases
and the values of the energy barriers VCH and VL start
fluctuating towards the borders of region A and the permeability, on average, increases. At the phase transition
the chain mobility is as high as possible for a gel-like
packing, and the values of VCH and VL, on average, move
out of region B to points where the permeability has very
large values. When the phase transition takes place and
the whole lattice changes abruptly, phospholipid molecules
acquire a different conformation and a different packing
and the whole potential energy profile suffers a sudden
change. In particular, the values of VCH and VL undergo
a discontinuous variation and jump to another region of
figure 4, to region B for instance, where the permeability,on average,is again low.

    Again, and as had happened with regard to the high
proton permeability, our model accounts for the peaks
displayed by the membrane permeability at the phase
transition in a straightforward manner and without any
need to resort to extraneous phenomena. According to our
interpretation, the increased permeability results from
an increased hydrocarbon chain mobility which occurs
without necessarily involving a changing in the lattice
packing; indeed, the hydrocarbon chains have been experimentally shown by ESR and NMR techniques to be the most
mobile portions of the phospholipid molecules[12,13], this

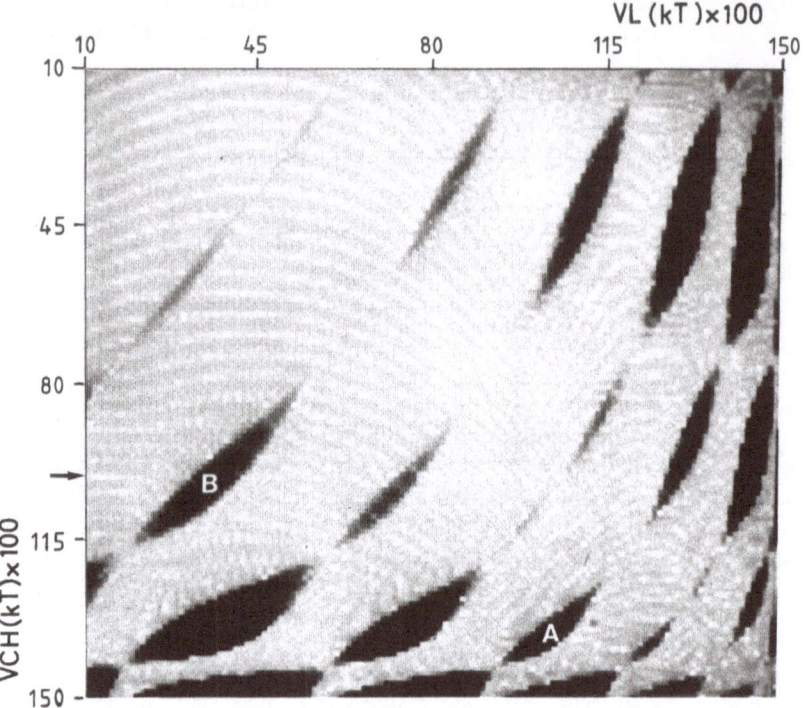

Fig. 4. Variation of the permeability of the BLM with
the energies VCH and VL (see figure 1), when all
other parameters are kept constant. The third
dimension, i.e., the permeability, is given by
the degree of photographic grey: the darker a
point is, the lower is the permeability for
the corresponding values VCH and VL.

mobility increasing from the polar region to the membrane
middle.

3. DISCUSSION

The present quantum mechanical model with its main
postulate regarding the potential energy profile cannot,
at this stage, claim any capacity for quantitative predictions and the main objective of this article is to
bring out its potentialities in explaining what has
proven to be difficult phenomena in classical approaches.
In this context it is however important to stress that

the results we have shown in figures 2-5 are not dependent
on the specific values the energy barriers shown in
figure 1 prove to have for the different ions in the
different membrane states. In fact, our results are only
dependent on the validity of the shape of the potential
energy profile shown in figure 1, i.e., on the existence
of higher barriers at the edges, where the energy of the
ion is assumed to be lower than VP, and of a quasi-
periodic sequence of smaller energy barriers in the
hydrophobic region. It can also be mathematically proven
that the use of rectangular potential energy barriers
instead of the smooth rounded ones which naturally occur
does not alter very much the solution of equation (2),
provided the fundamental shape is not greatly modified in
the process. It is the validity of this shape which has
to be ultimately checked by quantum mechanical calcula-

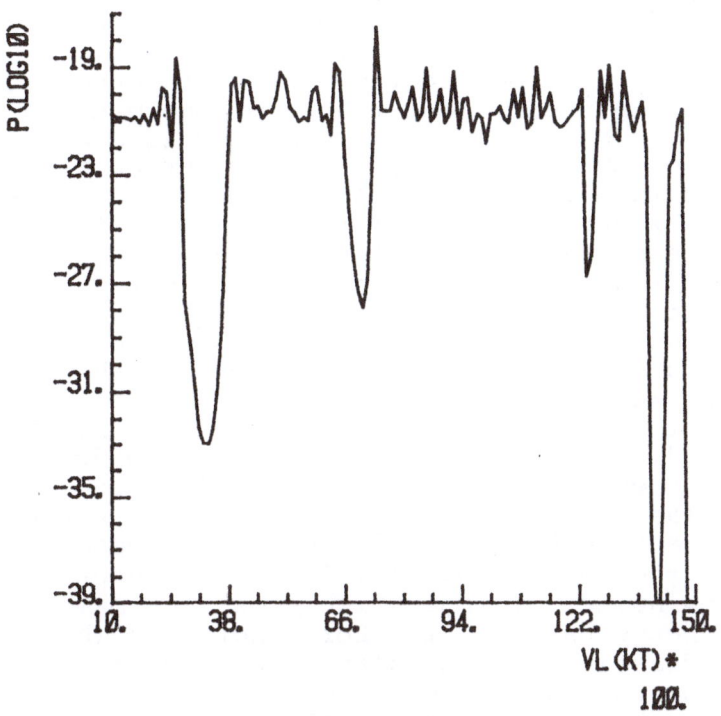

Fig. 5. Variation of the permeability of a BLM with the
energy VL, when the other parameters are kept
constant, the energy VCH being fixed in the
value indicated by the arrow shown in figure 4.

tions of the energy of an ion in a BLM.

Let us now point out another very fundamental difference between the quantum mechanical approach and the classical one regarding the movement of the ion accross the BLM. In the classical view the time taken by an ion to cross the membrane can be estimated through the Einstein formula for the one dimensional random walk:

$$<r^2> = 2 D t \qquad (3)$$

where r represents the distance a particle has moved from a starting point during a time t and D is, as before, the diffusion coefficient of the particle in the medium.

Diffusion coefficients for ions in BLMs vary with the ion species, the phospholipid composition, the temperature and other factors. Taking advantage of the fact that the diffusion coefficient D and the permeability satisfy the relationship D=P.l, where l is the membrane thickness, and considering the bilayer thickness to be 50 Å, diffusion coefficient values of the order of 10^{-22} cm^2/sec at T=4°C and 10^{-18} cm^2/sec at T=23°C may be computed from experimentally measured permeability values for egg lecithin liposomes[14] and for phosphatidylserine liposomes[3], respectively. For the purposes of our reasoning we choose the greatest of these two values and find from formula (3), with an r=50 Å, that the time an ion takes to cross the BLM is around 10^5 sec. Quantum Mechanics, on the other hand, invalidates the use of formula (3) and considers that the velocity with which the ion crosses the BLM is the same with which it moves in the adjacent aqueous solution, i.e., around 10^4-10^5 cm/sec, leading to a crossing time of the order of 10^{-11}-10^{-12} sec, respectively 10^{16}-10^{17} times shorter than the classical one. This disparity is a striking one and illustrates clearly the difference in mechanisms assumed by the two approaches. Classical theories attribute the low number of ions which cross the membrane to the high energy barrier, the Born energy[1,2] which slows down in a drastic way the speed with which the ions enter the BLM from the aqueous solution. Thus, in order to explain greater fluxes, classical theories seek ways of increasing the ionic speed of crossing and, therefore, they have to postulate conductive devices alien to the very structure of the BLM. On the other hand, Quantum Mechanics explains the reduced number of ions crossing the bilayer not as a result of a very slow motion in

that media, but as a consequence of the very low probability the ions have of getting through it. But still,
when the very improbable event of an ion crossing the
BLM occurs, it does so very quickly, i.e., in $10^{-11}-10^{-12}$
sec. From our point of view, the way to increase the ion
flux is to increase the ionic probability of crossing and
this does not necessarily imply a totally different
structure. We have indeed seen how a different dynamical
state in the same BLM structure is able to accomplish
that result. Experimental evidence of ion channels in
unmodified BLMs has been reported and shown to be
responsible for ionic currents which are much larger than
the ones predicted by classical theories[15,16]; their
random burst-like character[15] is in agreement both with a
high crossing speed and with our results of variation of
permeability with VCH and VL (see figure 4). Although the
measured transient currents have also been interpreted in
terms of disordered domains by the authors[15], we think
that they reflect thermal fluctuations in the hydrocarbon
layers and consider this experimental fact to give
further support to our quantum mechanical model.

We have seen how the peaks in permeability at the
phase transition could be explained without the need to
postulate extra "pores" specifically confined to that
membrane condition. We should point out that our
interpretation of the phase transition as resulting from
increased motion of the hydrocarbon chains is strongly
supported by the quantitative theories of phase transitions[17-21]. There are, however, other authors[7], who
contest this point of view and prefer to think of it as
resulting from a progressive increase in size of a large
number of small clusters dispersed throughout the basic
BLM matrix. According to this view, the relevant parameter in our model related to the phase transition would
be the potential energy barriers VP, whose magnitude
would first decrease with the spreading of the clusters
and then increase again to a different value when the
new phase spread to the whole of the BLM; these changes
in VP would be accompanied by a corresponding change in
the permeability and could assumedly account for the
permeability peaks during the phase transition. However,
this view of the phase transition stems directly from
the need classical theories have to create a device to
explain the events which occur at the phase transition.
Our model cannot decide of the validity of this view but
it does diminish its importance in what BLM transport
properties are concerned in so far as it provides an
alternative mechanism to explain those properties, which,
as has already been pointed out, is legitimated by a

solid theoretical rationale[17-21]. Furthermore, the mechanism we propose bears close resemblances to other permeability theories which attribute the peaks at the phase transition to enhanced fluctuations of the phospholipid molecules[22,23]. The difference is that those authors[22,23] are also thinking of larger spaces at headgroup level whereas, according to our model, enhanced fluctuations mainly limited to the hydrocarbon chains can explain the permeability increase in a very straightforward way.

The interpretations our model gives for high proton permeability and the peaks of permeability during phase transitions create an unified picture of ion diffusion through the BLM, as they respectively attribute the BLM selectivity property only to the ion characteristics and its interaction with the BLM, and link the permeability properties in the different BLM states and in the phase transition between these states to the dynamical properties of its molecules rather than to drastic structural modifications.

As a broader conclusion, our quantum mechanical model definitely questions the well established idea of the BLM necessarily low conductivity by explaining the "exceptions" to this idea not as such but as phenomena intrinsically related to the BLM's structure. We may therefore presume that the roles played by the lipid bilayers in biological membranes can go a long way beyond the commonly accepted and more trivial one of providing a passive resistive barrier for diffusion. For instance, we think, along with other authors[15], that the lipids may supply the transport medium of many processes, namely those occuring in excitable membranes.

REFERENCES

1. B.Neumcke and P. Läuger, Non electrical effects in lipid bilayer membranes, II. Integration of the Generalized Nernst-Planck equations, Biophys. J. 9: 1160 (1969)
2. P. Läuger, Ion transport through lipid bilayer membranes, in: "Membranes and Intercellular communication", R. Balian, M. Chabre, P. F. Devaux, eds., North-Holland, Amsterdam, New York, Oxford (1981)
3. J. W. Nichols and D. W. Deamer, Net proton-hydroxyl permeability of large unilamellar liposomes measured by an acid-base titration technique, Proc. Natl. Acad. Sci. USA 77: 2038 (1980)

4. M. Rossignol, P. Thomas and C. Grignon, Proton permeability of liposomes from natural phospholipid mixtures, Biochim. Biophys. Acta 684:195 (1982)

5. D. Papahadjopoulos, K. Jacobson, S. Nir and T. Isac, Phase transitions in phospholipid vesicles. Fluorescence polarization and permeability measurements concerning the effect of temperature and cholesterol, Biochim. Biophys. Acta 311:330 (1973)

6. M. C. Blok, E. C. Van Der Neut-kok, L. L. M. Van Deenen and J. De Gier, The effect of chain length and lipid phase transitions on the selective permeability properties of liposomes, Biochim. Biophys. Acta 406:187 (1975)

7. A.G. Lee, Lipid phase transitions and phase diagrams, I. Lipid phase transitions, Biochim. Biophys. Acta 472:237 (1977)

8. P. B. Hitchcock, R. Mason, K. M. Thomas and G. G. Shipley, Structural Chemistry of 1,2 Dilauroyl-DL-phosphatidylethanolamine: Molecular Conformation and Intermolecular Packing of Phospholipids, Proc. Natl. Acad. Sci. USA 71:3036 (1974)

9. H. Hauser, I. Pascher, R. H. Pearson and S. Sundell, Preferred Conformation and Molecular Packing of Phosphatidylethanolamine and Phosphatidylcholine, Biochim. Biophys. Acta 650:21 (1981)

10. R. V. Hosur, Quantum Mechanical Theory for transport Across Biological Membranes, Intern. J. Quant. Chem. 13:411 (1978)

11. K. A. Dill, D. E. Koppel, R. S. Cantor, J. D. Dill, D. Bendedouch and S.-H. Chen, Molecular Conformations in surfactant micelles, Nature 309:42 (1984)

12. J. Seelig and A. Seelig, Lipid Conformation in Model Membranes and Biological Membranes, Quart. Revs. Biophys. 13:19 (1980)

13. J. N. Israelachvili, S. Marcelja and R. G. Horn, Physical principles of membrane organization, Quart. Revs. Biophys. 13:121 (1980)

14. H. Hauser, M. C. Phillips and M. Stubbs, Ion permeability of Phospholipid Bilayers, Nature 239:342 (1972)

15. V. F. Antonov, V. V. Petrov, A. A. Molnar, D.A. Predvoditelev and A. S. Ivanov, The appearance of single-ion channels in unmodified lipid bilayer membranes at the phase transition temperature, Nature 283:585 (1980)

16. G. Boheim, W. Hanke and Eibl, Lipid phase transition in planar bilayer membrane and its effect on carrier- and pore-mediated ion transport, Proc. Natl. Acad. Sci. USA 77:3403 (1980)

17. J. F. Nagle, Theory of Lipid Monolayer and Bilayer Phase Transitions: Effect of Headgroup Interactions,

J. Membrane Biol. 27:233 (1976)

18. S. Marcelja, Chain Ordering in Liquid Crystals, II. Structure of Bilayer Membranes, Biochim. Biophys. Acta 367:165 (1974)

19. D. W. R. Gruen, A Statistical Mechanical Model of the Lipid Bilayer above its phase transition, Biochim. Biophys. Acta 595:161 (1980)

20. H. L. Scott, Jr., Phosphatidylcholine Bilayers. A Theoretical Model which describes the main and the lower transitions, Biochim. Biophys. Acta 643:161 (1981)

21. S. Mitaku, T. Jippo and R. Kataoka, Termodynamic properties of the lipid bilayer transition. Pseudo-critical phenomena, Biophys. J. 42:137 (1983)

22. J. F. Nagle and H.L. Scott, Jr., Lateral compressibility of the lipid Mono- and bilayers. Theory of Membrane Permeability, Biochim. Biophys. Acta 513:236 (1978)

23. S. Marcelja and J. Wolfe, Properties of bilayer membranes in the phase transition or phase separation region, Biochim. Biophys. Acta 557:24 (1979).

ORIGIN OF THE PLASTID ENVELOPE MEMBRANES

Roland Douce, Jacques Joyard, Albert-Jean Dorne
and Maryse A. Block

Physiologie Cellulaire Végétale (UA CNRS n° 576)
DRF/BV, CENG and USMG, 85 X
38041 Grenoble cedex (France)

INTRODUCTION

The endosymbiotic theory, first proposed by Schimper (1885) and
Mereschkovsky (1905), postulates that the plastids in eukaryotic
cells originated as free-living prokaryotes which found shelter within primitive cells (protoeukaryotes) and then became permanent
symbiotic elements within them, probably by providing more efficient
energy-transducing systems. This suggestion is supported by numerous
biochemical studies which favor the evolutionnary relationship between plastids and blue-green algae (see for instance Fredrick, 1981;
Schiff, 1982, Whatley, 1983). However, most of the studies done in
order to prove the endosymbiotic theory generally focus on the structure and function of DNA and RNA, consider also the major functions
(photosynthesis, respiration), but generally pay little attention
to an important structural feature of the cell organelles and free-
living prokaryotes, i.e. their limiting membranes. Cell organelles,
such as all kind of plastids, and some prokaryotes (see Stanier and
Cohen-Bazire, 1977), such as gram-negative bacteria and especially
blue green algae (cyanobacteria) are limited by a double membrane
system. However, it is generally assumed, without experimental evidences, that the inner envelope membrane from chloroplasts could
derive from the plasma membrane of the prokaryotic ancestor whereas
the outer envelope membrane of chloroplasts could derive from the
endomembrane system of the protoeukaryote that engulfed the free
living prokaryote.

The development of methods to isolate the envelope membranes
paved the way for studies on the structure, the chemical composition and the functions of the plastid envelope (for reviews, see
Douce and Joyard, 1979 ; Chua and Schmidt, 1979 ; Heber and Walker,

1979 ; Heber and Heldt, 1981 ; Douce et al., 1984). It is now possible to compare the plastid envelope membranes with the other cell
membranes and those of some prokaryotes (Douce and Joyard, 1979,1981 ;
Mudd, 1982 ; Keegstra et al., 1984). The purpose of this article is
to focus on some of the prokaryotic and/or eukaryotic features of
the plastid envelope membranes, especially those dealing with their
polar lipid composition and their role in plastid biogenesis, and
therefore to challenge the idea that the outer envelope membrane
from plastid could derive from the endomembrane system of the cell,
i.e. from the hypothetical protoeukaryote ancestor.

ISOLATION OF PLASTID ENVELOPE MEMBRANES

   A review entirely devoted to the preparation and characterization of the plastid envelope membranes has been published (Douce and
Joyard, 1982). Therefore, only the outlines and recent developments
will be summarized here.

   The technique used to prepare envelope fractions containing
both the outer and inner envelope membranes is based on the gentle
swelling of intact and purified chloroplasts in the presence of Mg^{2+}
(Fig. 1) and further purification of the fractions on a discontinuous sucrose gradient (Douce et al., 1973). The buoyant density of
the whole envelope membranes (d = 1.12) proved sufficiently different from that of thylakoids (d = 1.17) to allow clean separation
of the two membrane systems. This method, which produce high yields
(4-5 mg envelope protein/kg spinach leaves) is very rapid and re-
liable. Unfortunately, irreversible changes take place in the membranes (fusions ?) during the osmotic shock (Douce and Joyard, 1982)
thus making the separation of the outer envelope membrane from the
inner one impossible. Such a separation can only be achieved by
strikingly different procedures. Cline et al. (1981) have described
a method for purifying two membrane fractions from the pea chloroplast envelope. The procedure used includes freeze-thaw lysis of
intact.chloroplasts kept in a hypertonic medium, followed by flota-
tion centrifugation through a discontinuous sucrose gradient to separate the envelope membranes and a further fractionation of the
vesicles obtained by isopycnic density-gradient centrifugation.
Block et al. (1983a) have developed a different and faster procedure
for the separation of two membrane fractions respectively enriched
in outer and inner envelope membranes from purified intact spinach
chloroplasts (Fig. 2). This separation was achieved by osmotically
shrinking the inner envelope membrane (by incubation in 0.6M mannitol),
thus widening the intermembrane space, and then subsequently removing the "loosened" outer envelope by applying low pressure to the
shrunken chloroplasts and slowly extruding them through the small
aperture of a Yeda press under controlled conditions. By centrifugation of the mixture obtained through a discontinuous sucrose gradient, Block et al. (1983a) were able to separate two membrane fractions having different densities (light fraction, d = 1.08 ; heavy

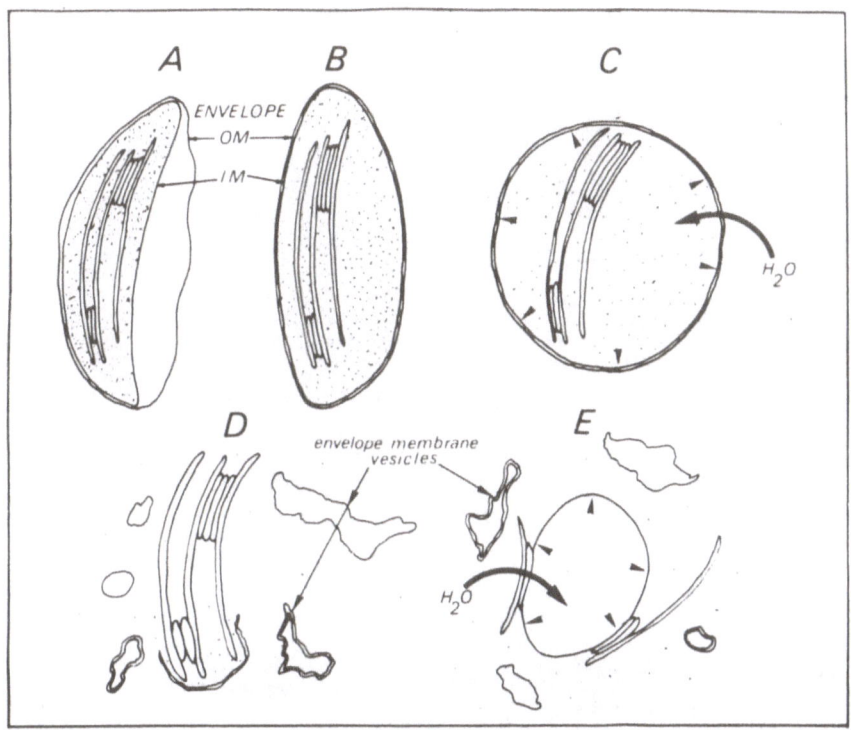

Fig. 1.   Schematic mechanism for the bursting of intact chloroplasts
in a low osmolarity medium. (A) Intact chloroplasts in 0.6 M
mannitol (hypertonic medium). The outer membrane of the
chloroplast envelope appears to be loosely attached to the
inner envelope membrane with large empty spaces between
them ; such chloroplasts are used for the separation of the
outer membrane from the inner one (see Fig. 2).(B) Intact
chloroplasts in 0.25 M mannitol ; (C) Intact chloroplast
in 0.05 M mannitol : water enters very quickly into the
chloroplast, the envelope is unable to support the high
pressure and burst in a few seconds (D). Under these conditions, the bulk of the stroma enzymes is released into
the medium. Very often, both the inner and the outer membranes can fuse along their breaking edges. Consequently,
the large vesicles thus formed, which are very unstable,
derive at the same time from the inner and the outer envelope membranes. (E) The ruptured envelope is present as
vesicles in the medium ; then water enters slowly into the
internal space of the thylakoids. A swollen thylakoid
looks like a balloon.

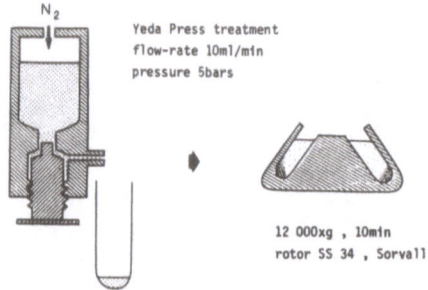

Intact and purified chloroplasts suspended in hypertonic medium : 0.6M mannitol, 4mM MgCl$_2$, 10mM tricine-NaOH pH 7.8 for 10 min

N$_2$

Yeda Press treatment
flow-rate 10ml/min
pressure 5bars

12 000xg , 10min
rotor SS 34 , Sorvall

Supernatant diluted to 80ml with 10mM tricine-NaOH pH 7.8, 4mM MgCl$_2$. 13ml of the mixture layered on top of a discontinuous sucrose gradient made of 3 layers (8ml each) containing 1M ; 0.65M ; 0.4M sucrose , 10mM tricine-NaOH pH 7.8 and 4mM MgCl$_2$.

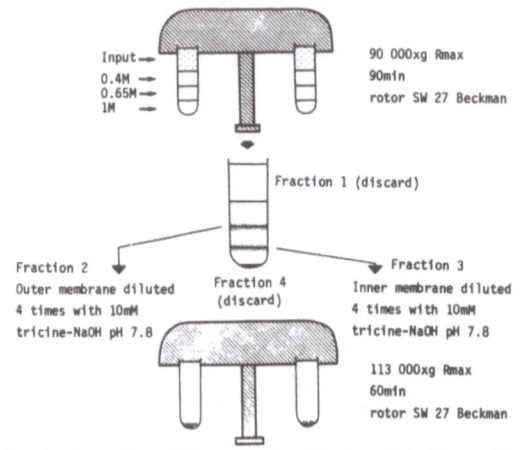

Input→
0.4M →
0.65M→
1M →

90 000xg Rmax
90min
rotor SW 27 Beckman

Fraction 1 (discard)

Fraction 2
Outer membrane diluted
4 times with 10mM
tricine-NaOH pH 7.8

Fraction 4
(discard)

Fraction 3
Inner membrane diluted
4 times with 10mM
tricine-NaOH pH 7.8

113 000xg Rmax
60min
rotor SW 27 Beckman

Pellets of outer membrane (∼500μg protein) and inner membrane (∼1mg protein)

Fig. 2. Preparation of membrane fractions enriched in outer (fraction 2) or inner (fraction 3) envelope membranes. Reproduced from Block et al. (1983a).

fraction, d = 1.13). The characterization of polypeptides localized on the outer envelope membrane from spinach chloroplasts (E10 and E24) by Joyard et al. (1983) made possible the characterization of the two membrane fractions. Analyses of the polypeptides by SDS-polyacrylamide gel electrophoresis and immunoblotting have shown that the light fraction was strongly enriched in outer envelope membrane proteins (more than 90 %) whereas the heavy fraction contained about 80 % of inner membrane proteins. Further characterization was provided by biochemical studies (polar lipid, pigments, enzymatic activities...) (Block et al., 1983b).

The envelope membrane fractions prepared either by the procedure of Douce et al. (1973) or by that of Block et al. (1983a) differ strikingly from "microsomal" membranes, mitochondria, peroxisomes in that they exhibit no oxidoreductase activities, viz., NAD(P)H:cytochrome c, NAD(P)H:ferricyanide, cytochrome $c:O_2$ and glycollate:O_2 oxidoreductases. In addition, unlike the outer membrane of plant mitochondria (Douce et al., 1972), envelope membranes are also devoid of b-type cytochrome. Indeed, when compared with the other cell membranes, the two plastid envelope membranes exhibit a most unusual chemical composition. This is particularily true at the level of their constituent polar lipids.

POLAR LIPID COMPOSITION OF PLASTID MEMBRANES

Plastid membranes are characterized by a high content of glycolipids (galactolipids, sulfolipid) and a low content of phospholipids (Douce et al. 1973) (Table 1) : envelope membranes as well as thylakoids are devoid of phosphatidylethanolamine (PE), the major polar lipid in the other cell membranes, and of cardiolipin, a specific lipid of the inner membrane of mitochondria (Douce and Joyard, 1979, 1980). In addition, the results obtained with mature chloroplasts are also valid for etioplasts, chromoplasts or amyloplasts (Douce and Joyard, 1979, 1980). Interestingly, except for phosphatidylcholine (PC), the polar lipid composition of chloroplasts from higher plants or algae are very similar to that of cyanobacteria, and also to *Prochloron* -which is an endosymbiotic prokaryote, from a special algal division, the Prochlorophyta (Lewin, 1981).

The polar lipids of both types of envelope membranes are identical, but the proportions in which they are present is different (Cline et al., 1981 ; Dorne et al., 1982a ; Block et al., 1983b). For instance, as shown in Table 1, monogalactosyldiacylglycerol (MGDG) and digalactosyldiacylglycerol (DGDG) are present in the outer envelope membrane in the ratio 0.6:1 and the main phospholipids, PC and phosphatidylglycerol (PG), in the ratio 3:1. This situation is totally reversed in the inner envelope membrane, which is highly comparable to thylakoids (Cline et al., 1981 ; Block et al., 1983b). In this case, MGDG and DGDG are present in a ratio of 1.6:1 and PC and PG in a ratio of 0.5-0.8:1; respectively (Table 1). Interestingly, in the cyanobacteria *Anacystis nidulans*, thylakoids and the cytoplasmic membrane (inner membrane) have also a similar polar lipid composition (Omata and Murata, 1983). The higher level of DGDG in the cell envelope (outer membrane) of *Anacystis nidulans* (Murata et al., 1981), when compared with thylakoids, is similar to the difference observed between outer envelope membrane and thylakoids from spinach chloroplasts (Cline et al., 1981 ; Block et al., 1983b).

However, the major difference between chloroplasts and cyano-

TABLE 1   POLAR LIPID COMPOSITION OF SOME PROKARYOTES AND OF PLANT
CELL ORGANELLES (CHLOROPLAST AND MITOCHONDRIA)

	MGluDG	MGDG	DGDG	SL	PG	PC	PI	PE	DPG
PROKARYOTES									
Prochloron sp. (a)	3.2	55.3	10.4	25.7	5.4	0	0	0	0
Anabaena variabilis (b)	1	54	17	11	17	0	0	0	0
Anacystis nidulans (c)	–	50.8	19	15.9	14.3	0	0	0	0
CHLOROPLAST (Spinach) (d)									
Thylakoids	–	52	26	6.5	9.5	4.5	1.5	0	0
Inner envelope membrane	–	49	30	5	8	6	1	0	0
Outer envelope membrane	–	17	29	6	10	32	5	0	0
MITOCHONDRIA (Mung Bean)(e)									
Inner membrane	–	0	0	0	1	29	2	50	17
Outer membrane	–	0	0	0	2	68	5	24	0

Abreviation used : MGluDG, monoglucosyldiacylglycerol ; PI, Phosphatidylinositol ; DPG, diphosphatidylglycerol or cardiolipin ; for the
others, see text. Monoglucosyldiacylglycerol has been found in some
prokaryotes. This glycolipid is in fact a precursor in MGDG synthesis in these prokaryotes and therefore just accumulates (when it
does accumulate) to only a minute extent. Note that the cyanobacteria analyzed apparently do not contain cardiolipin although they
usually contain cytochrome oxidase (Peschek, 1984). The results are
expressed as dry weight percent of total lipids. References, (a)
Murata and Sato (1983) ; (b) Murata and Sato (1982) ; (c) Allen et
al. (1966), in their publication, these authors have presented their
results relative to SL, we have recalculated their data for this
table ; (d) Block et al. (1983b) ; (e) Bligny and Douce (1981).

bacteria is the high level of PC (a typical eukaryotic polar lipid)
in the outer envelope of chloroplasts (Cline et al., 1981 ; Block
et al., 1983b). This situation, together with the presence of sterols in envelope membranes (Hartmann-Bouillon, 1980 ; Moeller and
Mudd, 1982 ; Douce et al., 1984), could reflect a possible dual origin of the envelope membrane. However, if the outer envelope membrane is shown as homologous with the endomembrane system of the
host cell, it is difficult to understand why the outer envelope membrane contains large amounts of typical cyanobacteria lipids, such
as galactolipids and sulfolipid, and does not contains PE, a characteristic endomembrane polar lipid. Since PC, in the outer chloroplast envelope, is solely localized in the outer leaflet (Dorne et
al., 1984), it could be possible that at least only the leaflet facing the cytosol could derive from the endomembrane system. We be-

lieve that this hypothesis is unlikely for the following reasons :
(a) cytochemical analyses have shown that both envelope membranes,
together with the internal thylakoids membrane system, share characteristic chemical constituents which are responsible for the
specific staining of plastid membranes within the plant cell
(Hurkmann et al., 1979 ; Carde et al., 1982) ; (b) galactolipids
and sulfolipid are accessible to specific antibodies from the cytosolic side of isolated intact chloroplasts (Billecocq et al., 1972 ;
Billecocq, 1974, 1975). Therefore, the major conclusion for these
observations is that both the outer and the inner envelope membrane
have a typical plastid origin. However, the problem of the origin
of PC in the outer envelope membrane remains to be elucidated. Since
envelope membranes are probably unable to synthesize PC (Joyard and
Douce, 1976), it is possible that PC accumulation in the outer leaflet of the outer envelope could be due to PC transfer from endoplasmic reticulum to the envelope, for instance via a phospholipid
transfer protein (Wirtz, 1974 ; Mazliak and Kader, 1980). Therefore,
one can suggest that during the course of the integration of the
prokaryote within the protoeukaryotic ancestor, one of the possible
adaptation was addition of PC into the outer leaflet of the outer
membrane. Indeed, gram-negative bacteria (DiRienzo et al., 1979),
including cyanobacteria (Peschek, 1984), have a highly asymmetric
outer membrane which contains much less lipids in the outer leaflet
than in the inner one. Such a situation would highly favor integration of additional lipid molecules in the outer leaflet. Interestingly, the outer membrane of other cellular organelles, such as
mitochondria (Bligny and Douce, 1980), is also strongly enriched in
PC.

Another characteristic feature of polar lipids is their fatty
acid content. Indeed, the galactolipids, sulfolipid, phospholipids
from envelope and thylakoid membranes of spinach chloroplasts have
been analyzed with respect to proportions, positional distribution,
and pairing of fatty acids (Siebertz et al., 1979). All the specific characteristics of the diacylglycerol moiety of lipids known
from previous analyses of thylakoid lipids or whole leaves were also found in envelope lipids (Fig. 3 and 4). For example, C16 fatty
acids, whenever analyzed in typical chloroplast lipids such as MGDG,
DGDG, sulfolipid (SL) and PG are mainly concentrated at the sn-2
position of the glycerol backbone. In contrast, PC is separated
from the aforementionned lipids by the fact that it excludes C16
fatty acids (in this case 16:0) from its sn-2 position and therefore may be considered to belong to a different group. In this context, it is interesting to note that prokaryotic cyanobacteria direct C16 fatty acids to sn-2 position and C18 fatty acids to sn-1
position of the glycerol backbone (Heinz, 1977 ; Murata and Sato,
1982). Analyses of the diacylglycerol part of glycolipids led to
the conclusion that these structures can be separated into two
groups : the first one has C16 or C18 fatty acids at sn-1 position
of the glycerol backbone and C16 fatty acids at the sn-2 position

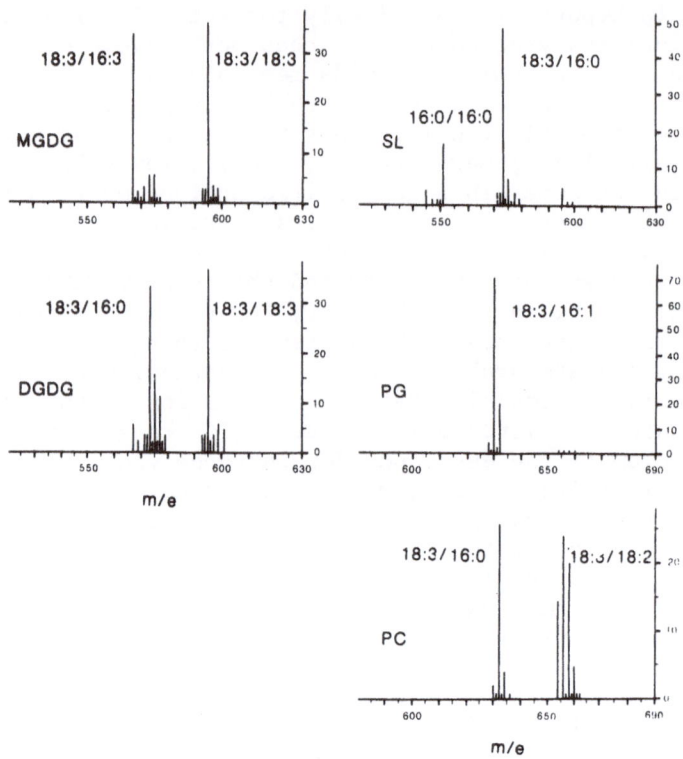

Fig. 3. Pairing of fatty acids in lipids from chloroplast envelopes as analyzed by mass spectrometry. Portions of spectra cover molecular ions from acetylated diacylglycerol (derived from phosphatidylglycerol (PG) and phosphatidylcholine (PC) after phospholipase C hydrolysis to release diacylglycerol from parent lipid) or fragment ions of diacylglycerol structure from glycolipid derivatives. Ordinates: relative abundances as percentages. Abscissas: m/e values. Major species are identified by constituent fatty acids. Lipid abbreviations are defined in the text. Adapted from Siebertz *et al.* (1979).

whereas the second one has C16 or C18 fatty acids at sn-1 position of the glycerol backbone and C18 fatty acids at sn-2 position. Since the first structure is characteristic for cyanobacteria therefore it is called "prokaryotic" structure, whereas the second one is called "eukaryotic" structure and is present in all eukaryotic cells (Heinz, 1977). Galactolipids, and especially MGDG, can have both prokaryotic and eukaryotic structure : in plants such as pea (18:3 plant), only the eukaryotic structure is present in MGDG whereas in plants such as *Anthriscus* (16:3 plant) only the prokaryotic structure is present in MGDG, spinach (another 16:3 plant) contains both structure

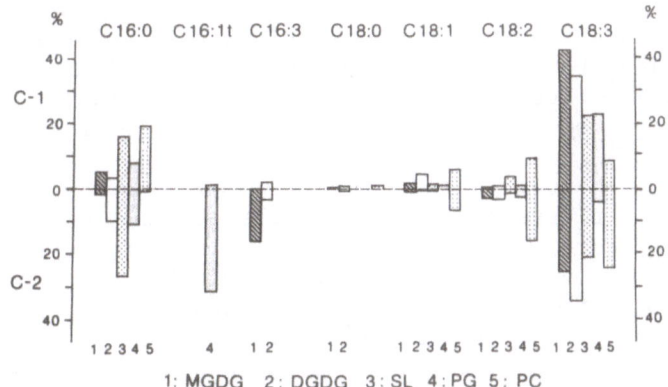

Fig. 4. Positional distribution of fatty acids in envelope lipids.
Each fatty acid is represented by a bar indicating its proportion (in moles/100 moles of fatty acids present in this
particular lipid) and distribution between C-1 and C-2 positions at the sn-glycerol backbone. (C16:1t) *Trans* isomer.
Lipid abbreviations are defined in the text. Adapted from
Siebertz *et al.* (1979).

(Heinz, 1977). This structure could reflect their biosynthetic route
(Heinz and Roughan, 1982 ; Williams et al., 1982) as we shall see
below.

Finally, the major fatty acids present in galactolipids are polyunsaturated fatty acids (16:3 and/or 18:3) (Douce and Joyard, 1980).
This is in contrast with the situation in most cyanobacteria, such as
Anacystis nidulans (Nichols et al., 1965), or in *Prochloron* (Murata and
Sato, 1983) which contain saturated and monounsaturated fatty acids
(see Peschek, 1984). However, this is not a general situation since
filamentous cyanobacteria, such as *Anabaena variabilis*, contain polyunsaturated fatty acids in appreciable amounts (Nichols et al.,
1965). Therefore, the degree of unsaturation cannot be entirely considered as a characteristic feature of eukaryotic structures.

THE ENVELOPE AND THE ORIGIN OF PLASTID COMPONENTS

Analyses of envelope membranes from different plastids (proplastids, amyloplasts, etioplasts, chromoplasts, chloroplasts) suggest
that they all contain the same characteristic components (polar lipids, pigments, prenylquinones...) (see Douce and Joyard, 1979 ;
Douce et al., 1984). Since the envelope is a permanent structure of
all the plastid types which preceeds, in chloroplasts, the existence
of thylakoids, it was reasonable to suggest that the envelope membranes are directly or indirectly involved in the synthesis of the

typical constituents that are found in plastid membranes. This was indeed demonstrated for the synthesis of plastid polar lipids, such as galactolipids, but also for prenylquinones or pigments (see Douce and Joyard, 1979 ; Douce et al., 1984). In this article, we shall examine the importance of the chloroplast envelope in lipid biosynthesis only.

Polar lipid biosynthesis requires the assembly of three parts : fatty acids, glycerol, and a polar head (galactose, in the case of galactolipids). In plant cells, plastids are the sole source of fatty acids (16:0 , 18:0 and 18:1) for membranes of developing organelles and serve as precursors of glycolipids and phospholipids (see Harwood, 1979 ; Stumpf, 1980, Roughan and Slack, 1982). Fatty acid synthetase is localized in the plastid stroma and consists of several different enzymes which can be separated from each other. Therefore, this complex has a typical prokaryotic structure, analogous to that described as a non-associative structure of fatty acid synthetase in *Escherichia coli* (for a discussion, see Stumpf et al., 1982 ; Ohlrogge, 1982). During their *de novo* biosynthesis, their elongation and desaturation to 18:1, the plastid fatty acids remain attached to ACP (acyl carrier protein), therefore, the major products of fatty acid biosynthesis in the plastid stroma are 16:0-ACP and 18:1-ACP (see Stumpf et al., 1982). These molecules are destined for further metabolism outside the chloroplast (for biosynthesis of phospholipids such as PC and PE) or within the organelle (for galactolipid, SL and PG biosynthesis) (see Douce and Joyard, 1980 ; Roughan and Slack, 1982).

The first step in glycerolipid biosynthesis is the acylation of a water soluble precursor, *sn*-glycerol 3-phosphate, which is provided by the cytosol. Douce and Guillot-Salomon (1970) demonstrated that isolated chloroplasts are able to catalyze the acylation of *sn*-glycerol 3-phosphate to form phosphatidic acid, and then by dephosphorylation, to form diacylglycerol, the precursor of galactolipids. All the enzymes involved in this synthesis, known as the Kornberg-Pricer pathway, have been localized on envelope membranes (Joyard and Douce, 1977) and more precisely on the inner envelope membrane (Block et al., 1983b). The whole pathway is summarized in Fig. 5. The first acyltransferase is "soluble", but probably functions in close association with the envelope. It catalyzes the acylation of *sn*-glycerol 3-phosphate to yield lysophosphatidic acid. This acyltransferase possesses a high positional specificity for the acylation of *sn*-1 position of the glycerol (Bertrams and Heinz, 1979 ; Joyard et al., 1979 ; Joyard, 1979). In addition, the same authors have clearly demonstrated that during the *in vitro* experiments, lysophosphatidic acid contains almost exclusively C18 fatty acids at this *sn*-1 position. The true physiological acyl donor for this enzyme is probably 18:1-ACP, as shown by Frentzen et al. (1982). The second acyltransferase is firmly membrane bound and directs C16 fat-

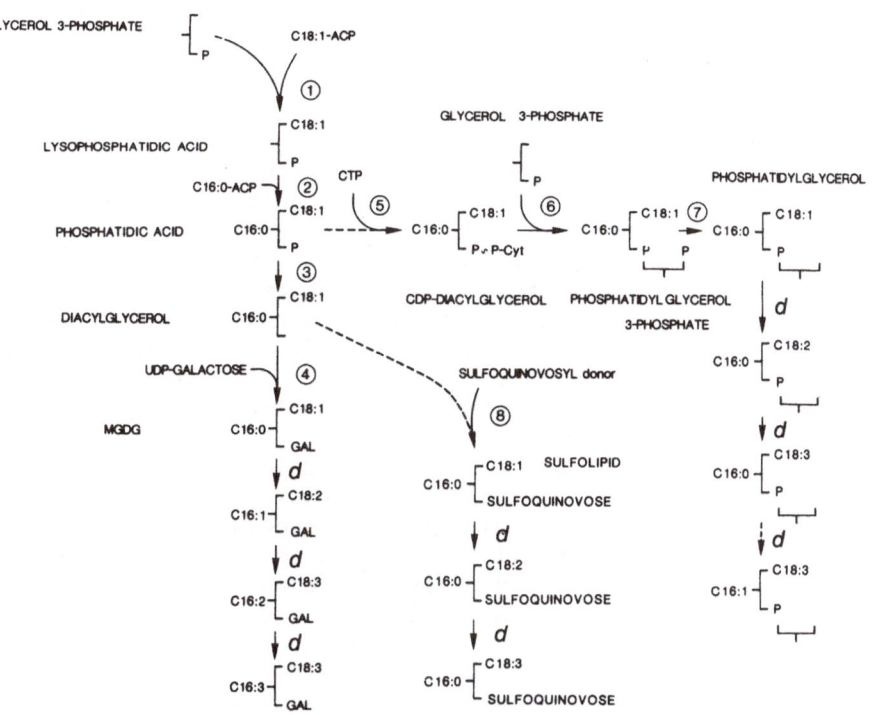

Fig. 5. Pathways of lipid metabolism in chloroplasts. Isolated intact chloroplasts are able to synthesize most of their polar
lipids : galactolipids, SL and PG. The backbone for these polar lipids probably comes from phosphatidic acid and diacylglycerol, which are synthesized on the chloroplast envelope,
owing to an acyl-ACP:sn-glycerol 3-phosphate acyltransferase
(1), which is responsible for the synthesis of lysophosphatidic acid ; to an acyl-ACP:monoacylglycerol 3-phosphate acyltransferase (2), which is responsible for the formation of
phosphatidic acid ; and to a specific phosphatidic acid phosphatase (3), which yields diacylglycerol. MGDG is then synthesized by galactosylation of diacylglycerol on the inner
envelope membrane (4). PG synthesis (7) probably involves
the formation of intermediates such as CDP-diacylglycerol(5)
and phosphatidylglycerolphosphate (6). All these steps have
been clearly demonstrated on isolated envelope membranes.
The other reactions shown in this figure occur in isolated
chloroplasts and probably take place on envelope membranes :
SL synthesis (8) and desaturation (d), which occurs after
polar lipid synthesis, on MGDG, SL and PG. The molecular
species synthesized all have the "prokaryotic" structure, i.e.
with C16 fatty acids at sn-2 position. However, origin of the
"eukaryotic" structure, i.e. with C18 fatty acids at sn-2
position, is still controversial.

ty acids at sn-2 position of the glycerol (Joyard et al., 1979 ; Joyard, 1979 ; Frentzen et al., 1982). In fact, the two envelope acyltransferase play a decisive role in establishing a specific fatty acid distribution at the glycerol backbone : the acyltransferase responsible for the acylation of sn-2 position seems to exclude C18 fatty acids from this position and therefore led to the formation of phosphatidic acid and then of diacylglycerol molecules having a typical "prokaryotic" structure, i.e. with C18 fatty acids at sn-1 position and only C16 fatty acids at sn-2 position (see above). Diacylglycerol molecules play a key role in galactolipid biosynthesis and therefore envelope membrane contains a characteristic phosphatidic acid phosphatase (Joyard and Douce, 1977 ; 1979) localized on the inner membrane (Block et al., 1983c) which is able to provide diacylglycerol molecules for further galactosylation (Fig. 5).

Douce (1974) has demonstrated that the chloroplast envelope is the site of UDP-galactose (UDP-gal) incorporation into galactolipids in leaf cells and that the activity is not associated with endoplasmic reticulum (which was actually considered to be *the* site of polar lipid biosynthesis). The ability of the isolated envelope membranes to catalyze the last steps of galactolipid was confirmed on other plastid types (see Douce and Joyard, 1980). At least two distinct enzymes are responsible for the synthesis of MGDG and DGDG. One is specific for the synthesis of MGDG using UDP-gal as a galactosyl donor (Fig. 5). The second enzyme is responsible for the formation of DGDG and is either a UDP-gal:MGDG galactosyltransferase (Ongun and Mudd, 1977) or a galactolipid:galactolipid galactosyltransferase (Van Besouw and Wintermans, 1978). Galactolipid formation has also been studied in some cyanobacteria, such as *Anabaena variabilis* (Sato and Murata, 1982a,b,c), and in *Prochloron* (Sato and Murata, personal communication). Most of the pathway is very similar to that found in chloroplasts, however, the last steps involve the transfer of a glucosyl group, instead of a galactosyl group, from UDP-glucose (instead of UDP-gal) to diacylglycerol. The product formed, monoglucosyldiacylglycerol, is found only in minute amounts in the membranes (see Table 1) since it is rapidly transformed into MGDG by epimerisation of the glucosyl group (Sato and Murata, 1982c). Chloroplasts apparently lack this epimerase activity, another type of epimerase is present in the cytosol which transform UDP-glucose into UDP-gal therefore the source of UDP-gal in the cell is the cytosol. It is possible that during the course of evolution, the prokaryotic epimerase activity disappeared on behalf of the epimerase activity present in the protoeukaryote, which in turn resulted into a slight modification of the pathway for galactolipid biosynthesis.

All together, these results demonstrate that owing to their envelope membranes, chloroplasts are able to synthesize their major lipid constituents, galactolipids, and that the pathway for this synthesis as well as the diacylglycerol structure was highly similar to that which is actually found in free-living or symbiotic prokaryotes.

Indeed, the same situation can be found for two other specific plastid polar lipids, SL (Haas et al., 1980) and PG (Mudd and de Zacks, 1981 ; Sparace and Mudd, 1982 ; Andrews and Mudd, 1984) which are also synthesized in *Anabaena variabilis* (Murata and Sato, 1982).

However, most of the galactolipid molecules which are found in chloroplasts do not have the same diacylglycerol pattern as the molecules formed through the pathway described above : up to now only molecules having a "prokaryotic" structure can be formed through the envelope Kornberg-Pricer pathway. How plastids acquire the typical eukaryotic structure of MGDG having C18 fatty acids at sn-2 position ? Frentzen et al. (1982) and Gardiner and Roughan (1983) have demonstrated that chloroplasts from 18:3-plants have a much lower capacity to accumulate diacylglycerol, owing to a less active phosphatidic acid phosphatase in envelope membranes from 18:3-plants. For instance, the turn-over rate of plastid phosphatidic acid was 20-30 fold faster in spinach (16:3-plant) than in pea (18:3-plant) (Gardiner and Roughan, 1983). This difference may explain why 18:3-plants, which are in fact able to synthesize phosphatidic acid having 16:0/18:1 pairing (probably for SL or PG synthesis), do not contain the corresponding diacylglycerol in MGDG. However, this observation does not explain the origin of C18/C18 backbone in MGDG. Since this structure is readily formed in whole leaves, it has been proposed that the diacylglycerol part of MGDG having two C18 fatty acids could have been synthesized outside the chloroplast, and could originate from PC (see Roughan and Slack, 1982). This pathway is often called the "eukaryotic pathway" (Heinz and Roughan, 1982 ; Roughan and Slack, 1982). Such a hypothesis is very attractive but, in our opinion, the postulated role of PC in providing the C18/C18 backbone to the chloroplast is not yet clear since, (a) the outer envelope membrane accumulates PC in pea (Cline et al., 1981) as well as in spinach (Block et al., 1983b), and especially in the outer leaflet of the membrane (Dorne et al., 1984), an observation which does not fit with a precursor-product relationship between PC and MGDG ; (b) isolated envelope membranes from spinach chloroplasts are devoid of the phospholipase C-type activity which would yield diacylglycerol from PC (Douce and Joyard, 1979) ; (c) MGDG synthesis occurs on the inner envelope membrane in spinach (Block et al., 1983b), therefore PC or diacylglycerol should be transported from the outer to the inner membrane, which has not been demonstrated. It is obvious that cooperation between plastids and endoplasmic reticulum occurs in the cell (for instance plastids provide fatty acids to endoplasmic reticulum for PC and PE synthesis) but, despite extensive studies (Heinz and Roughan, 1982 ; Roughan and Slack, 1982 ; Williams et al., 1982 ; Dubacq et al., 1983), the mechanism which could supply diacylglycerol having C18/C18 fatty acid pairing from endoplasmic reticulum to envelope membranes for further galactosylation into MGDG are completely unknown. On the contrary, the "prokaryotic" pathway of galactolipid, SL or PG synthesis (Fig. 5) is strongly supported by numerous experimental evidences.

Finally, the polar lipid molecules having either the prokaryotic or the eukaryotic structure contain, and this is particularily
true in chloroplasts, almost exclusively polyunsaturated fatty acids.
After very strong discussions (see Douce and Joyard, 1980), it is
now generally accepted that desaturation occurs on polar lipids containing monounsaturated or saturated fatty acids, i.e. after their
formation (see Douce and Joyard, 1980 ; Roughan and Slack, 1982).
It is interesting to note that the same observation was made in cyanobacteria such as *Anabaena variabilis*:desaturation of fatty acids
also occurs after esterification of the glycerol moiety, on galactolipid, SL or PG molecules having saturated or monounsaturated
(Murata and Sato, 1982).

ORIGIN OF THE PLASTID ENVELOPE PROTEINS

Analyses of envelope membrane polypeptides by sodium dodecyl
sulfate-polyacrylamide gel electrophoresis (SDS-PAGE) have shown
that this membrane system contains at least 75 polypeptides ranging
in molecular weight from 140,000 to less than 12,000 (Joyard et al.,
1982, see Douce et al., 1984). On the basis of their staining intensity, the envelope polypeptides fall roughly into two groups (Fig. 6).
There are several major polypeptides (E54, E37, E30, E14, E12) and
a multitude of faint bands in the high molecular weight region of
the gel but also in the other parts. The polypeptide pattern of envelope membranes is very different from that of the stroma or the
thylakoids : very few polypeptides from the three fractions have the
same electrophoretic mobility in the different gel systems used
(Joyard et al., 1982). Two of these, E54 and E14, were identified by
crossed immunoelectrophoresis as the large and the small subunit of
ribulose bisphosphate carboxylase (RuBPcase), respectively. Flügge
and Heldt (1976) have demonstrated that polypeptide E30, one of the
major envelope polypeptide, plays a key role in the phosphate translocation across the inner envelope membrane. The outer envelope membrane has a polypeptide pattern distinct from that of the inner membrane (Cline et al., 1981 ; Block et al., 1983a). Surprisingly, the
outer membrane fraction, although containing a high weight ratio of
lipids to proteins (Dorne et al., 1982a), contained more than 30 different polypeptides with 5 major components (Fig. 7). None of them
have been identified yet, although several enzymatic activities have
been found in this membrane (Douce et al., 1984), but two polypeptides (E10 and E24) have been localized precisely on the cytoplasmic
face of the outer envelope membrane (Joyard et al., 1983). We believe that the polypeptide distribution in the inner envelope membrane
reflects the dual function of this membrane. It is possible that the
major polypeptides are involved in the regulation of the inflow of
raw material for photosynthesis and the outflow of photosynthetic
products (this is at least the case for E30, which corresponds to the

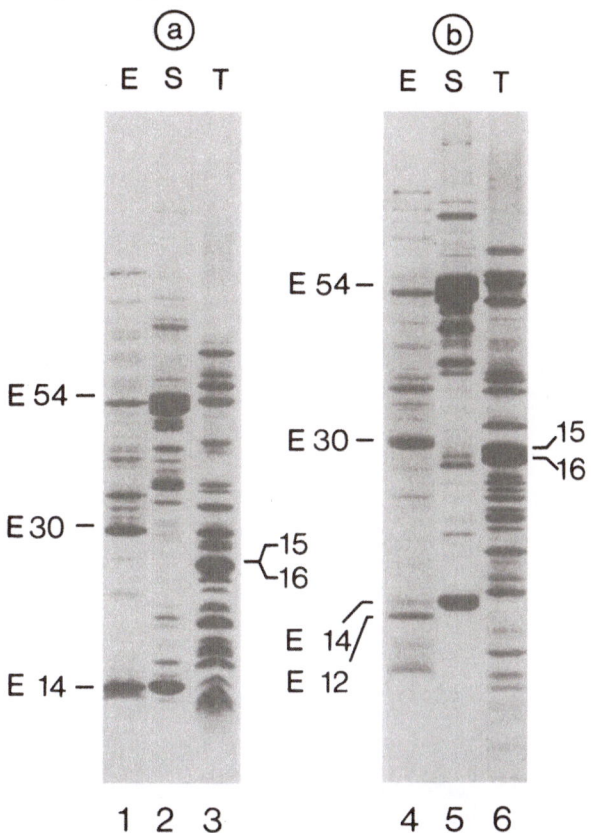

Fig. 6. Electrophoretic analyses of envelope (E), stroma (S), and
thylakoid (T) polypeptides from purified and intact spinach
chloroplasts. (a) LDS-PAGE at 4°C, with a 7.5-15% acrylamide gradient ; (b) SDS-PAGE at room temperature, with a
12-18% acrylamide gradient in the presence of 8 M urea.
This gradient system gives better resolution for envelope
and thylakoid polypeptides of molecular weights below
15,000. Experimental conditions and polypeptide nomencla-
ture are as described by Joyard *et al.* (1982).

phosphate translocator, but the dicarboxylate or the adenine nucleotide translocators have not been yet identified). On the other hand,
the minor polypeptides could be involved in the synthesis of chloroplast constituents such as polar lipids, prenylquinones, carotenoids,
and others (see Douce and Joyard, 1979 ; Douce et al., 1984). The
enzymatic activities so far identified in envelope membranes are listed in Table 2.

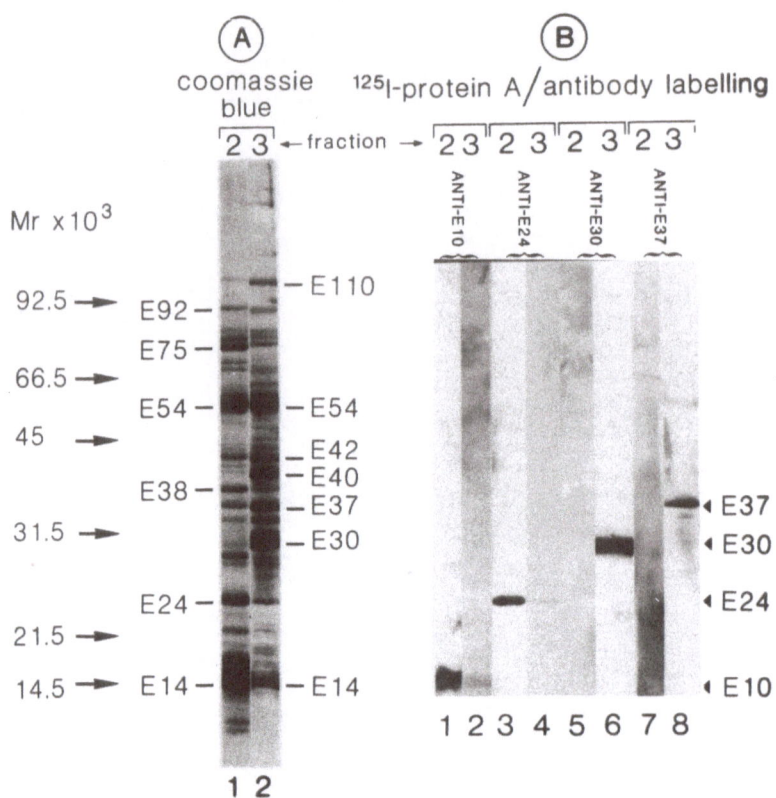

Fig. 7. Electrophoretic and immunochemical characterization of membrane fractions enriched in outer and inner envelope membranes from spinach chloroplasts. A:Coomassie blue staining
of polypeptides from fractions 2 and 3 separated by SDS-
polyacrylamide gel electrophoresis at room temperature with
a 7.5 to 15% acrylamide gradient. *Lane 1*:fraction 2 (outer
membrane fraction), *lane 2*:fraction 3 (inner membrane fraction).B:autoradiography of antigen-antibody-^{125}I-protein A
complexes after electrophoretic transfer of polypeptides
from polyacrylamide gels to nitrocellulose sheets. Antibodies to envelope polypeptides E10 (*lanes 1* and *2*), E24
(*lanes 3* and *4*), E30 (*lanes 5* and *6*), and E37 (*lanes 7* and
8) were used. *Lanes 1, 3, 5*, and *7*:fraction 2; *lanes 2, 4,
6*, and *8*:fraction 3. Fractions *2* and *3* were prepared as
described in Fig. 3 (reproduced from Block et al., 1983a).

TABLE 2 ENZYMATIC ACTIVITIES ASSOCIATED WITH THE PLASTID ENVELOPE MEMBRANES

ENZYMES	LOCALIZATION		
ENZYMES INVOLVED IN LIPID BIOSYNTHESIS	nd	OM	IM
acyl-ACP:sn-glycerol 3-phosphate acyltransferase			x(a)
acyl-ACP:monoacylglycerol 3-phosphate acyltransferase		x	x
acyl-CoA synthetase		x	
acyl-CoA thioesterase			x
phosphatidic acid phosphatase			x
UDP-galactose:diacylglycerol galactosyltransferase			x
UDP-galactose:MGDG galactosyltransferase	x		
galactolipid:galactolipid galactosyltransferase		x	
acylgalactosyldiacylglycerol-forming enzyme	x		
acylsterylglucosyl-forming enzyme	x		
ENZYMES INVOLVED IN PIGMENT, PRENYLQUINONES, or FLAVONOID BIOSYNTHESIS			
4-hydroxyphenylpyruvate dioxygenase	x		
geranylgeraniol derivatives synthesis			x
SAM(b):methyl-6-solanesylquinol methyltransferase			x
SAM:methyl-6-geranylgeranylquinol methyltransferase			x
SAM:methyl-6-phytylquinol methyltransferase			x
SAM:γ-tocopherol methyltransferase			x
SAM:γ-tocotrienol methyltransferase			x
phytoene synthase	x		
phytoene dehydrogenase	x		
zeaxanthin epoxidase	x		
SAM:flavonol O-methyltransferase	x		
monoterpene biosynthesis	x		
ENZYMES INVOLVED IN METABOLITE TRANSPORT			
adenine nucleotide translocator			x
dicarboxylate translocator			x
glucose translocator			x
sulfate translocator			x
pore-forming enzyme		x	
OTHERS			
Mg^{2+}-dependent ATPase (DCCD insensitive)(c)			x
protein receptor		x	

(a) these enzymes are found in the soluble phase but are associated with the envelope membranes ; (b) SAM, S-adenosylmethionine ; (c) DCCD, N,N'-dicyclohexylcarbodiimide. nd: not determined. OM:outer membrane ; IM:inner membrane. The references relative to these enzymes are detailed by Douce et al. (1984).

The origin of chloroplast envelope proteins responsible for all these enzymatic activities is difficult to understand since we know little on their molecular nature. All the attempts made in order to identify the site of synthesis of envelope proteins have led to the conclusion that most of them are coded for by nuclear DNA. This has been demonstrated for the phosphate translocator (Flügge, 1982), and very strong evidences have been presented for the enzymes involved in polar lipid synthesis (Dorne et al., 1982b). This is surprising since chloroplasts contain all the components necessary for a biological system to be autonomous, i.e. DNA, DNA polymerase, RNA polymerase, 70S ribosomes, and additional factors required for protein synthesis (see Hoober, 1976 ; Bogorad, 1981). Chloroplast DNA has a size that corresponds roughly to a coding potential of about 200 polypeptides of average molecular weight 30,000 (Hermann et al., 1980 ; Edelman, 1981) and therefore do not code for all the chloroplast proteins (Ellis, 1976, 1981 ; Chua and Gillham, 1977). Indeed, this is probably the case for all the envelope membrane proteins. It is likely that during the course of its evolution, the initial prokaryotic ancestor became adapted to a symbiotic existence and lost various functions which have been partly taken over by the host nucleus. The shifting of genes from one genome to another within a eukaryotic cell by a gene transfer mechanism seems a reasonable possibility (Bogorad, 1975). This is probably the case for envelope proteins that are involved in plastid biogenesis (see Table 2) since these enzymatic activities are present in living prokaryotes. On the other hand, during the proliferation of the cyanobacteria which had to become synchronized with the cell division of the host, it is likely that the prokaryotic plasma membrane acquired the appropriate carriers (phosphate translocator, dicarboxylate translocator...) necessary for its metabolism to be integrated with that of its host. Therefore, it is possible that the specific carriers are the only entirely new components necessary in the evolutionary transition from the plasma membrane of cyanobacteria to the inner membrane of the chloroplast envelope. This is also probably the case for several outer envelope membrane proteins. For instance, cytoplasmically synthesized proteins must, somehow, cross the chloroplast envelope to reach their final sites (envelope, stroma, thylakoids) during plastid growth and division (see Chua and Schmidt, 1979). The protein receptor involved in this transport should have been acquired during the course of evolution. The same hypothesis can be made for the acyl-CoA synthetase. This enzyme, localized on the outer envelope membrane (Block et al., 1983d), is probably involved in the transport of fatty acids from the plastids to the cytosol for PC and PE synthesis. On the contrary, the porine activity, which allows all molecules either charged or uncharged (up to a molecular weight of 10,000) to gain free access to the inner envelope membrane from the external compartment (Flügge and Benz, 1984), is similar to that described in the outer membrane of gram-negative bacteria (see DiRienzo et al., 1978).

Consequently, even if all the proteins of the envelope membranes are coded for by nuclear DNA, it is possible to trace back to those, either localized on the inner or the outer membrane, which originated from the prokaryotic ancestor or which appeared during the course of the evolution within the protoeukaryotic ancestor.

CONCLUSION

We believe that all the structural and functional characteristics of the cell organelles should be considered if we want to understand the origin and evolution of these eukaryotic intracellular organelles. The chemical composition of the inner envelope membrane (the nature and structure of its polar lipids with their specific polar head groups and fatty acid pairing, but also the occurence of prenylquinones -compare for instance the chapters on quinones in Douce et al, 1984 and Peschek, 1984), the major role of this membrane in plastid biogenesis (site of galactolipid, PG and probably SL synthesis, site of plastoquinone-9 and α-tocopherol synthesis,...) provide strong evidences in favor of the hypothesis that the inner envelope membrane from higher plant plastids could derive from the plasma membrane of a cyanobacteria-type or *Prochloron*-type of prokaryotic ancestor. On the contrary, all the structural and biochemical analyses of the outer envelope membrane do not provide any evidence to support the hypothesis that this membrane derives from an endocytotic membrane of a protoeukaryotic ancestor. The outer envelope membrane does not resemble in its chemical composition to any kind of endomembrane system found in eukaryotic cells. The presence of galactolipids and sulfolipid, especially in the outer leaflet of the outer envelope membrane, their structure, the absence of PE, a characteristic endomembrane phospholipid, the presence of pigments (Block et al., 1983b) and of prenylquinones (Soll et al., 1984), which are all characteristic constituents of plastid structures but are absent from extrachloroplastic membranes, are strong evidences for the hypothesis that the outer envelope membrane is analogous to the outer membrane of cyanobacteria and therefore could derive from a similar sturcture present in the prokaryotic ancestor. The presence, on the outer envelope membrane (Flügge and Benz, 1984), of a pore-forming activity analogous to that found in gram-negative bacteria strongly support this hypothesis.

A great deal of research is needed if we want to understand the origin of the plastid envelope membranes. We believe that research done on primitive eukaryotic algae (which have envelope membranes made of two, three or even four membranes -see for instance, Fredrick, 1981),on prokaryotes that are strongly related to plastids (i.e. *Prochloron*...), and on symbiotic prokaryotes or eukaryotes (Fredrick, 1981 ; Schiff, 1982) may provide numerous informations for this understanding.

REFERENCES

Allen, C.F., Hirayama, O., and Good, P., 1966, Lipid composition of
    photosynthetic systems, in:"Biochemistry of Chloroplasts,
    volume 1", T.W. Goodwin, ed., pp. 195-200, Academic Press,
    London.
Andrews, J., and Mudd, J.B., 1984, Phosphatidylglycerol synthesis
    in pea chloroplasts:characterization and localization, in:
    "Structure, Function and Metabolism of Plant Lipids", P.A.
    Siegenthaler, ed., in press, Elsevier, Amsterdam.
Bertrams,M., and Heinz, E., 1979, Soluble, isomeric forms of glyce-
    rophosphate acyltransferase in chloroplasts, in:"Advances in
    the Biochemistry and Physiology of Plant Lipids"(L.A.
    Appelqvist and C. Liljenberg, eds.), pp. 139-144,Elsevier/
    North-Holland, Amsterdam.
Billecocq, A., 1974, Structure des membranes biologiques:Localisa-
    tion des galactosyldiglycérides dans les chloroplastes au
    moyen des anticorps spécifiques. II.Etude en microscopie élec-
    tronique à l'aide d'un marquage à la peroxydase, Biochim. Bio-
    phys. Acta 352:245-251.
Billecocq, A., 1975, Structure des membranes biologiques:Localisa-
    tion du sulfoquinovosyldiglycéride dans les diverses membranes
    des chloroplastes au moyen des anticorps spécifiques, Ann.
    Immunol. (Inst. Pasteur) 126c:337-352.
Billecocq, A., Douce, R., and Faure, M., 1972, Structure des mem-
    branes biologiques:Localisation des galactosyldiglycérides
    dans les chloroplastes au moyen des anticorps spécifiques,
    C.R. Acad. Sci.,275:1135-1137.
Bligny, R., and Douce, R., 1980, A precise localization of cardio-
    lipin in plant cells, Biochim. Biophys. Acta, 617:254-263.
Block, M.A., Dorne,A.-J.,and Douce, R., 1983a, Preparation and
    characterization of membrane fractions enriched in outer and
    inner envelope membranes from spinach chloroplasts.
    I. Electrophoretic and immunochemical analyses, J. Biol. Chem.
    258:13273-13280.
Block, M.A., Dorne, A.-J., and Douce, R., 1983b, Preparation and
    characterization of membrane fractions enriched in outer and
    inner envelope membranes from spinach chloroplasts.
    II. Biochemical characterization, J. Biol. Chem. 258:13281-13286.
Block, M.A., Dorne, A.-J., and Douce, R., 1983c, The phosphatidic
    acid phosphatase of the chloroplast envelope is located on the
    inner envelope membrane, FEBS Lett., 169:111-115.
Block, M.A., Dorne, A.-J., and Douce, R., 1983d, The acyl-CoA syn-
    thetase and the acyl-CoA thioesterase are located respectively
    on the outer and on the inner membrane of the chloroplast enve-
    lope, FEBS Lett. 153:377-381.
Bogorad, L., 1981, Chloroplasts, J. Cell Biol., 91:256-270.
Bogorad, L., 1975, Evolution of organelles and eukaryotic genomes,
    Science, 188:891-898.

Carde, J.P., Joyard, J., and Douce, R., 1982, Electron microscopic studies of envelope membranes from spinach plastids, Biol. Cell., 44:315-324.

Chua, N.-H., and Gillham, N.W., 1977, The sites of synthesis of the principal thylakoid membrane polypeptides in *Chlamydomonas reinhardii*, J. Cell Biol., 74:441-452.

Chua, N.-H., and Schmidt, G.W., 1979, Transport of proteins into chloroplasts and mitochondria, J. Cell Biol., 81:461-483.

Cline, K., Andrews, J., Mersey, B., Newcomb, E.H. and Keegstra, K., 1981, Separation and characterization of inner and outer envelope membranes of pea chloroplasts, Proc. Natl. Acad. Sci. U.S.A., 78:3595-3599.

DiRienzo, J.M., Nakamura, K., and Inouye, M., 1978, The outer membrane proteins of gram-negative bacteria:Biosynthesis, assembly, and functions, Ann. Rev. Biochem., 47:481-532.

Dorne, A.-J., Block, M.A., Joyard, J., and Douce, R., 1982a, Studies on the localization of enzymes involved in galactolipid metabolism in chloroplast envelope membranes, in:"Biochemistry and Metabolism of Plant Lipids"(J.F.G.M. Wintermans and P.J.C. Kuiper, eds.), pp. 153-164, Elsevier, Amsterdam.

Dorne, A.-J., Block, M.A., Joyard, J., and Douce, R., 1984, Localization of phosphatidylcholine in chloroplast envelope membranes from spinach, in:"Structure, Function and Metabolism of Plant Lipids", P.A. Siegenthaler, ed., in press, Elsevier, Amsterdam.

Dorne, A.-J., Carde, J.P., Joyard, J., Börner, T., and Douce, R., 1982b, Polar lipid composition of a plastid ribosome-deficient barley mutant, Plant Physiol., 65:1467-1470.

Douce, R., 1974, Site of biosynthesis of galactolipids in spinach chloroplasts, Science, 183:852-853.

Douce, R., and Guillot-Salomon, T., 1970, Sur l'incorporation de la radioactivité du sn-glycerol 3-phosphate-^{14}C dans le monogalactosyldiglycéride des plastes isolés, FEBS Lett., 11:121-126.

Douce, R., and Joyard, J., 1979, Structure and function of the plastid envelope, Adv. Bot. Res., 7:1-116.

Douce, R., and Joyard, J., 1980, Plant galactolipids, in:The Biochemistry of Plants, Vol. 4, Lipids:Structure and Function"(P.K. Stumpf, ed.), pp. 321-362, Academic Press, New-York.

Douce, R., and Joyard, J., 1981, Does the plastid envelope derive from the endoplasmic reticulum ?, Trends Biochem. Sci. 6:237-239.

Douce, R., and Joyard, J., 1982, Purification of the chloroplast envelope, in:Methods in Chloroplast Molecular Biology"(M. Edelman, R. Hallick, and N.-H. Chua, eds.), pp. 239-256, Elsevier/North-Holland, Amsterdam.

Douce, R., Christensen, E.L., and Bonner, W.D., 1972, Preparation of intact plant mitochondria, Biochim. Biophys. Acta, 275:148-159.

Douce, R., Holtz, R.B., and Benson, A.A., 1973, Isolation and properties of the envelope of spinach chloroplasts, J. Biol. Chem., 248:7215-7222.

Douce, R., Block, M.A., Dorne, A.J., and Joyard, J., 1984, The plastid envelope membranes:Their structure, composition, and role in chloroplast biogenesis, in:"Subcellular Biochemistry, Volume 10", D.B. Roodyn, ed., pp. 1-84, Plenum, New-York.

Dubacq, J.-P., Drapier, D., and Trémolières, A., 1983, Polyunsaturated fatty acid synthesis by a mixture of chloroplasts and microsomes from spinach leaves:Evidence for two distinct pathways of the synthesis of trienoic acids, Plant Cell Physiol. 24:1-9.

Edelman, M., 1981, Nucleic acids of chloroplasts and mitochondria, in:The Biochemistry of Plants, Vol. 6, Proteins and Nucleic Acids"(A. Marcus, ed.), pp.249-301, Academic Press, New-York.

Ellis, R.J., 1976, Protein and nucleic acid synthesis by chloroplasts, in:The Intact Chloroplast"(J. Barber, ed.), pp. 335-364, Elsevier, Amsterdam.

Ellis, R.J., 1981, Chloroplast proteins:Synthesis, transport, and assembly, Annu. Rev. Plant Physiol., 32:111-137.

Flügge, U.I., 1982, Biogenesis of the chloroplast phosphate translocator, FEBS Lett., 140:273-276.

Flügge, U.I., and Benz, R., 1984, Pore-forming activity in the outer membrane of the chloroplast envelope, FEBS Lett., 69:85-89.

Flügge, U.I., and Heldt, H.W., 1976, Identification of a protein involved in phosphate transport of chloroplasts, FEBS Lett., 68:259-262

Fredrick, J.F., 1981, Origins and evolution of eukaryotic intracellular organelles, Ann. N.Y. Acad. Sci.,361.

Frentzen, M., Heinz, E., McKeon, T., and Stumpf, P.K., 1982, De novo-biosynthesis of galactolipid molecular species by reconstituted enzyme systems from chloroplasts, in:Biochemistry and Metabolism of Plant Lipids"(J.F.G.M. Wintermans and P.J.C. Kuiper, eds.), pp. 141-152, Elsevier, Amsterdam

Gardiner, S.E., and Roughan, P.G., 1983, Relationship between fatty-acyl composition of diacylgalactosylglycerol and turnover of chloroplast phosphatidate, Biochem. J., 210:949-952.

Haas, R., Siebertz, H.P., Wrage, K., and Heinz, E., 1980, Localization of sulfolipid labelling within cells and chloroplasts, Planta, 148:238-244.

Hartmann-Bouillon, M.-A., 1980, Les stérols des membranes des plantes:Distribution et métabolisme, Thèse de Doctorat d'Etat, Université de Strasbourg, France.

Harwood, J.L., 1979, The synthesis of acyl lipids in plant tissues, Prog. Lipid Res., 18:55-86.

Heber, U., and Heldt, H.W., 1981, The chloroplast envelope:Structure, function and role in leaf metabolism, Annu. Rev. Plant Physiol., 32:139-168.

Heber, U., and Walker, D.A., 1979, The chloroplast envelope-Barrier or bridge?, Trends Biochem. Sci., 4:393-421.

Heinz, E., 1977, Enzymatic reactions in galactolipid biosynthesis, in:Lipids and Lipid Polymers in Higher Plants"(M. Tevini and H.K. Lichtenthaler, eds.), pp. 102-120, Springer-Verlag, Berlin.

Heinz, E., and Roughan, P.G., 1982, De novo synthesis, desaturation
     and acquisition of monogalactosyldiacylglycerol by chloroplasts
     from "16:3"- and "18:3" -plants, in:"Biochemistry and Metabolism
     of Plant Lipids"(J.F.G.M. Wintermans and P.J.C. Kuiper, eds.),
     pp. 169-182, Elsevier, Amsterdam.
Hermann, R.G., Seyer, P., Schedel, R., Gordon, K., Bisanz, C.,
     Winter, P., Hildebrandt, J.W., Wlaschek, M., Alt,J., Driesel,
     A.J., and Sears, B.B., 1980, The plastid chromosomes of seve-
     ral dicotyledons, in:Biological Chemistry of Organelle Forma-
     tion"(T. Bücher, W. Sebald, and H. Weiss, eds.), pp. 97-112,
     Springer-Verlag, Berlin.
Hoober, J.K., 1976, Protein synthesis in chloroplasts, in:Protein
     Synthesis, Vol. 2"(E.H. McConkey, ed.), pp. 169-248, Marcel
     Dekker, New-York and Basel.
Hurkmann, W.J., Morré, D.J., Bracker, C.E., and Mollenhauer, H.H.,
     1979, Identification of etioplast membranes in fraction from
     soybean hypocotyls, Plant Physiol., 257:1095-1101.
Joyard, J., 1979, L'enveloppe des chloroplastes, Thèse de Doctorat
     d'Etat, Université de Grenoble, France.
Joyard, J., and Douce, R., 1976, L'enveloppe des chloroplastes est-
     elle capable de synthétiser la phosphatidylcholine?, C.R. Acad.
     Sci. 282:1515-1518.
Joyard, J., and Douce, R., 1977, Site of synthesis of phosphatidic
     acid and diacylglycerol in spinach chloroplasts, Biochim. Bio-
     phys. Acta 486:273-285.
Joyard, J., and Douce, R., 1979, Characterization of a phosphatidate
     phosphatase activity associated with chloroplast envelope mem-
     branes, FEBS Lett., 102:147-150.
Joyard, J., Chuzel, M., and Douce, R., 1979, Is the chloroplast en-
     velope a site of galactolipid synthesis? Yes!, in:Advances in
     the Biochemistry and Physiology of Plant Lipids"(L.A. Appelqvist
     and C. Liljenberg, eds.), pp. 181-186, Elsevier/North-Holland,
     Amsterdam.
Joyard, J., Grossman, A.R., Bartlett, S.G., Douce, R., and Chua, N.H.,
     1982, Characterization of envelope membrane polypeptides from
     spinach chloroplasts, J. Biol. Chem., 257:1095-1101.
Joyard, J., Billecocq, A., Bartlett, S.G., Block, M.A., Chua, N.H.,
     and Douce, R., 1983, Localization of polypeptides to the cyto-
     solic side of the outer envelope membrane of spinach chloro-
     plasts, J. Biol. Chem., 258:10000-10006.
Keegstra, K., Werner-Washburne, M., Cline, K., and Andrews, J.,
     1984, The chloroplast envelope:is it homologous with the double
     membranes of mitochondria and gram-negative bacteria ? J. Cell.
     Biochem., 24:55-68.
Lewin, R.A., 1981, Prochloron and the theory of symbiogenesis, in:
     "Origins and evolution of eukaryotic intracellular organelles",
     J.F. Fredrick, ed., pp. 325-329, Ann. N.Y. Acad. Sci., New-York.
Mereschkowsky, C., 1905, Ober natur and ursprung der chromatophoren
     im Pflanzenreiche, Biol. Zentralbl., 25:593-604.

Moeller, C.H., and Mudd, J.B., 1982, Localization of filipin-sterol complexes in the membranes of *Beta vulgaris* roots and *Spinacia oleracea* chloroplasts, Plant Physiol., 70:1554-1561.

Mudd, J.B., and de Zacks, R., 1981, Synthesis of phosphatidylglycerol by chloroplasts from leaves of *Spinacia oleracea* L. (spinach), Arch. Biochem. Biophys., 209:584-591.

Mudd, J.B., 1982, Lipid Metabolism, in:"On the origins of chloroplasts", J.A. Schiff, ed., pp. 131-148, Plenum, New-York.

Murata, N., and Sato, N., 1982, In vivo synthesis of lipids in the blue-green alga, *Anabaena variabilis*, in:"Biochemistry and Metabolism of Plant Lipids"(J.F.G.M. Wintermans and P.J.C. Kuiper, eds.), pp. 165-168, Elsevier, Amsterdam.

Murata, N., Sato, N., Omata, T., and Kuwabara, T., 1981, Separation and characterization of thylakoid and cell envelope of the blue green alga (cyanobacterium) *Anacystis nidulans*. Plant Cell Physiol., 22:855-866.

Nichols, B.W., Harris, R.V., and James, A.T., 1965, The lipid metabolism of blue-green algae, Biochem. Biophys. Res. Commun., 20:256-262.

Ohlrogge, J.B., 1982, Fatty acid synthetase:Plant and bacteria have similar organization. Trends Biochem. Sci., 7:386-387.

Omata, T., and Murata, N., 1983, Isolation and characterization of the cytoplasmic membranes from the blue-green alga (cyanobacterium) *Anacystis nidulans*, Plant Cell Physiol., 24:1101-1112.

Ongun, A., and Mudd, J.B., 1968, Biosynthesis of galactolipids in plants, J. Biol. Chem. 243:263-275.

Peschek, G.A., 1984, Structure and function of respiratory membranes in cyanobacteria (blue-green algae), in: "Subcellular Biochemistry, Volume 10", D.B. Roodyn, ed., pp. 85-191, Plenum, New-York.

Roughan, P.G., and Slack, C.R., 1982, Cellular organization of glycerolipid metabolism, Annu. Rev. Plant Physiol., 33:97-132.

Sato, N., and Murata, N., 1982a, Lipid biosynthesis in the blue-green alga, *Anabaena variabilis*.
I. Lipid classes, Biochim. Biophys. Acta, 710:271-278.

Sato, N., and Murata, N., 1982b, Lipid biosynthesis in the blue-green alga, *Anabaena variabilis*.
II. Fatty acids and molecular species, Biochim. Biophys. Acta, 710:279-289.

Sato, N., and Murata, N., 1982c, Lipid biosynthesis in the blue-green alga, *Anabaena variabilis*.
III. UDP-glucose:diacylglycerol glucosyltransferase in vitro. Plant Cell Physiol. , 23:115-1120.

Schimper, A.F.W., 1885, Untersuchungen über die chlorophyllkörner und die ihnen homologen gebilde, Jahrb. Wiss. Botan., 16:1-247.

Schiff, J.A., 1982, "On the origins of chloroplasts", Elsevier/North-Holland, New-York.

Siebertz, H.P., Heinz, E., Linscheid, M., Joyard, J., and Douce, R., 1979, Characterization of lipids from chloroplast envelopes, Eur. J. Biochem., 101:429-438.

Soll, J., Schultz, G., Joyard, J., Douce, R., and Block, M.A., 1984, Localization and synthesis of prenylquinones in isolated outer and inner envelope membranes from spinach chloroplasts, in: "Structure, Function and Metabolism of Plant Lipids", P.A. Siegenthaler, ed., in press. Elsevier, Amsterdam.

Sparace, S.A., and Mudd, J.B., 1982, Phosphatidylglycerol synthesis in spinach chloroplasts:Characterization of the newly synthesized molecule, Plant Physiol., 70:1260-1264.

Stanier, R.Y., and Cohen-Bazire, G., 1977, Phototrophic prokaryotes-The cyanobacteria, Annu. Rev. Microbiol., 31:225-274.

Stumpf, P.K., 1980, Biosynthesis of saturated and unsaturated fatty acids, in: "The Biochemistry of Plants, Vol. 4, Lipids:Structure and Function" (P.K. Stumpf, ed.), pp. 177-204, Academic Press, New-York.

Stumpf, P.K., Shimakata, T., Eastwell, K., Murphy, D.J., Liedvogel, B., Ohlrogge, J.B., and Kuhn, D.N., 1982, Biosynthesis of fatty acids in a leaf cell, in: "Biochemistry and Metabolism of Plant Lipids" (J.F.G.M. Wintermans and P.J.C. Kuiper, eds.), pp. 3-11, Elsevier, Amsterdam.

Van Besouw, A., and Wintermans, J.F.G.M., 1978, Galactolipid formation in chloroplast envelopes. I. Evidence for two mechanisms in galactosylation, Biochim. Biophys. Acta, 529:44-53.

Whatley, J.M., 1983, Plastids. Past, present and future, Int. Rev. Cyt., 514:329-373.

Williams, J.P., Khan, M., and Mitchell, K., 1982, Galactolipid biosynthesis in *Brassica napus* and *Vicia faba*:A comparison of lipid biosynthesis in 16:3- and 18:3-plants, in: "Biochemistry and Metabolism of Plant Lipids" (J.F.G.M. Wintermans and P.J.C. Kuiper, eds.), pp : 183-186, Elsevier, Amsterdam.

Wirtz, K.W.A., 1974, Transfer of phospholipids between membranes, Biochim. Biophys. Acta, 344:95-117.

PIGMENT-PROTEIN COMPLEXES OF THYLAKOID MEMBRANES:

ASSEMBLY, SUPRAMOLECULAR ORGANIZATION

George Akoyunoglou and Joan Argyroudi-Akoyunoglou

Biology Department
Nuclear Research Center "Demokritos"
153 41 Aghia Paraskevi, Attiki, Greece

INTRODUCTION

The biosynthesis and assembly of the pigment-protein complexes reflects in fact the biogenesis of the photosynthetic membrane, i.e., the etioplast to chloroplast differentiation. As it is known, in dark-grown plants the proplastids are transformed into etioplasts. Initially the etioplasts contain a number of perforated prothylakoids; the prolamellar bodies appear later, attached to the prothylakoids. The prolamellar bodies increase in size as darkness is prolonged[1]. It seems that in the early stages of etioplast development (in young etiolated leaves) all thylakoid components are synthesized at a relatively low rate; however, only some of these components accumulate, the rest being digested. The accumulation rate for each component follows a sigmoidal curve: it is initially low, then it increases and finally stops after reaching a steady level[2]. As the age of the etiolated tissue increases some of these components stop from being synthesized. Depending, therefore, on the age of the etiolated tissue, i.e., on the developmental stage of the etioplast, one may or may not observe the expression of some of these components in the dark-grown plants[2]. Etioplasts do not contain chlorophyll (Chl) but its precursor, protochlorophyllide (PChlide). The PChlide accumulates in the etioplast in the form of a complex with the enzyme PChlide-oxidoreductase, and it is located in the prothylakoids and the prolamellar bodies[3]. A number of other polypeptides can also be found in prothylakoids, while a variety of lipids have been extracted from prolamellar bodies[4,5].

In angiosperms light is required for the differentiation of etioplasts into chloroplasts. Chloroplasts contain a system of lamellae (thylakoids) embedded in the stroma matrix. The internal membrane arrangement varies widely between different organisms. It ranges from the single unappressed lamellae of red algae, the bundle sheath chloroplasts of C_4-plants, the chloroplasts of C_3-mutants and of plastids in the early stages of greening, to the appressed in groups of two or three lamellae (*Euglena gracilis*), and the highly granal structures of the mesophyll chloroplasts of fully greened plants. The enzymes of the Calvin cycle, which carry out the reduction of CO_2, are in the stroma, while the photoreduction of $NADP^+$ by H_2O, and the photophosphorylation of ADP to ATP takes place in the thylakoids. The electron flow from H_2O to $NADP^+$ is driven by light absorbed by two pigment assemblies, the photosystem (PS) I and PSII, according to the generally accepted Z scheme. These two pigment assemblies, called "photosynthetic units", consist of proteins associated with Chl, carotenoids (car), lipids and electron carriers. Each photosynthetic unit is supposed to be consisted of (a) the "core" of the unit which contains the reaction center, i.e., the minimum components required to carry out the primary photochemical event, a small number of Chla and b-car which act as antennae, and the enzymes and cofactors that stabilize the primary charge se-

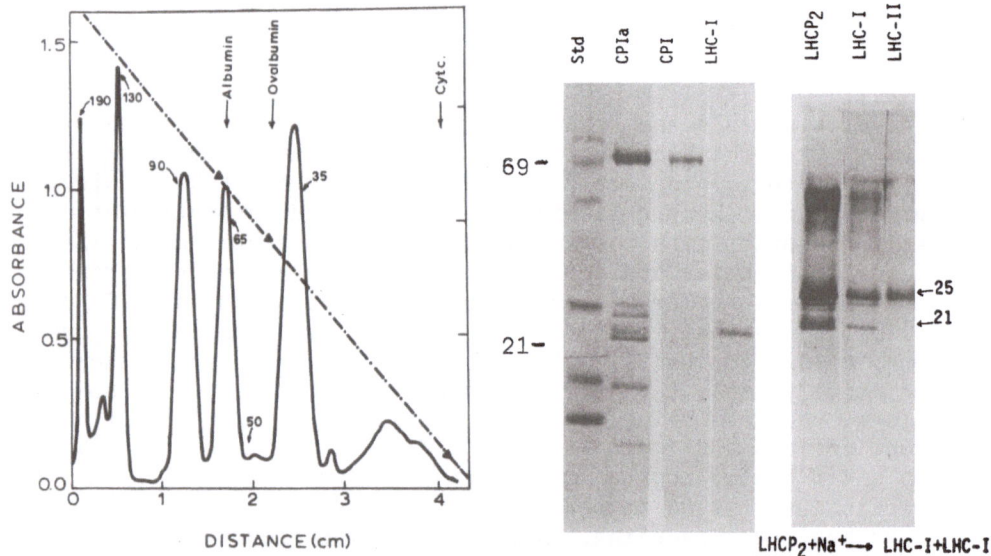

Fig. 1. Electrophoretic profile of the pigment-protein complexes of SDS-solubilized bean thylakoids, with apparent Mr. SDS-PAGE as in (10). Main complexes from left to right: CPIa, CPI, LHCP$_1$, LHCP$_2$, CPa and LHCP$_3$.

Fig. 2. Polypeptide analysis by dissociating SDS-PAGE of the pigment-protein complexes bands excised from the gel, eluted in H_2O and lyophilized. CPI and LHC-I were separated from CPIa, and LHC-I and LHC-II from LHCP$_2$ by rerunning them on a gel containing 100 mM Na^+ in the stacking gel. Left: coomassie staining; Right: silver nitrate staining.

paration; (b) the light-harvesting pigments (Chla, Chlb and car) bound along with lipids to proteins. The reaction center communicates with the antennae and the light-harvesting pigments by utilizing the transfer of excitation energy. The chloroplast lamellae consist of lipids and proteins in about equal proportions; of the lipids 23% by weight are pigments. There is an uneven distribution of the thylakoid components in grana and stroma lamellae. Grana are enriched in PSII units, Chlb and xanthophylls (xanth), while stroma lamellae are enriched in PSI units, Chla and b-car[6],[7]. All the pigments together with lipids and proteins are assembled into pigment-protein complexes which compose the two photosynthetic units, PSI and PSII.

SUPRAMOLECULAR ORGANIZATION OF THE PIGMENT-PROTEIN COMPLEXES

Isolation of the Pigment-Protein Complexes

Mild SDS-polyacrylamide gel electrophoresis procedures have allowed the separation from SDS-solubilized thylakoids of six major pigment-protein complexes, and three unidentified minor ones. These complexes are real entities existing *in vivo* as shown by (^{14}C)-labelling experiments[8], and they contain more than 90% of the thylakoid Chl[9],[10]. The complexes in order of increasing mobility are the CPIa, CPI, LHCP$_1$, LHCP$_2$ CPa and LHCP$_3$ (fig. 1). SDS-sucrose gradient centrifugation has been also used for the separation of the complexes[11]; in this case four distinct bands are resolved, while the free pigment, removed by the action of the detergent, remains on the top of the gradient. Band I contains LHCP$_3$, band II is a mixture of LHCP$_1$ and LHCP$_2$, band III contains CPI and band IV CPIa.

Isolation of the complexes from SDS-solubilized subchloroplast fractions enriched in PSI or PSII units revealed that CPIa and CPI originate in PSI, and LHCP$_1$, LHCP$_2$, LHCP$_3$ and CPa originate in PSII. For example, grana lamellae which are enriched in PSII units are also enriched in LHCPs and CPa; stroma lamellae, on the contrary, which are enriched in PSI units are also enriched in CPIa and CPI[10],[12]. Similarly, the PSII particle DTS-III, obtained after sucrose density gradient centrifugation of the digitonin-sonication-Triton X-100 treated thylakoids[13], is enriched in LHCPs[12]; SDS-PAGE of the SDS-solubilized PSI-110 particle, prepared by treatment of the chloroplast membrane with Triton X-100 in a low ionic strength medium[14], revealed the presence of only CPIa, CPI and LHC-I (the light-harvesting complex of PSI)[15]. In addition, isolation of the complexes from SDS-solubilized thylakoids of plants greened in intermittent light, which contain only the core of the PSI and PSII units[10], or from barley mutants[9], revealed that CPI and CPa originate in the core of the PSI and PSII unit respectively. CPI has an apparent Mr of 130 kDa[16]. It contains Chla, P700, b-car and a polypeptide of 110 kDa when undissociated (68 kDa when

Table 1. Pigment composition of the pigment-protein complexes isolated from thylakoids of green bean leaves by sucrose gradient centrifugation

Complex	Chl$\underline{a}$/Chl$\underline{b}$	b-car	xanth	lut	nx	vx	P700
			(moles/100 moles Chl)				
CPIa	4.5	18.5	3.0	3.0	-	-	1.0
CPI	13.0	12.0	-	-	-	-	3.3
LHCP$_3$	1.4	2.1	21.5	14.4	5.0	2.1	-
LHCP$_{1+2}$	1.2	1.2	27.0	16.5	7.8	2.7	-
CPa[a]	13.0	10.2	-	-	-	-	-
LHC-I[b]	3.4	-	11.5	10.2	-	1.3	-

[a]CPa was isolated by SDS-PAGE; [b]LHC-I was separated from CPIa.

dissociated)[16]; its fluorescence emission peak at 77°K is mainly at 725 nm[10,17,18] (see Table 1 and figs 2 and 3). CPIa is a highly organized complex with an apparent Mr of 190 kDa[16]; its polypeptide

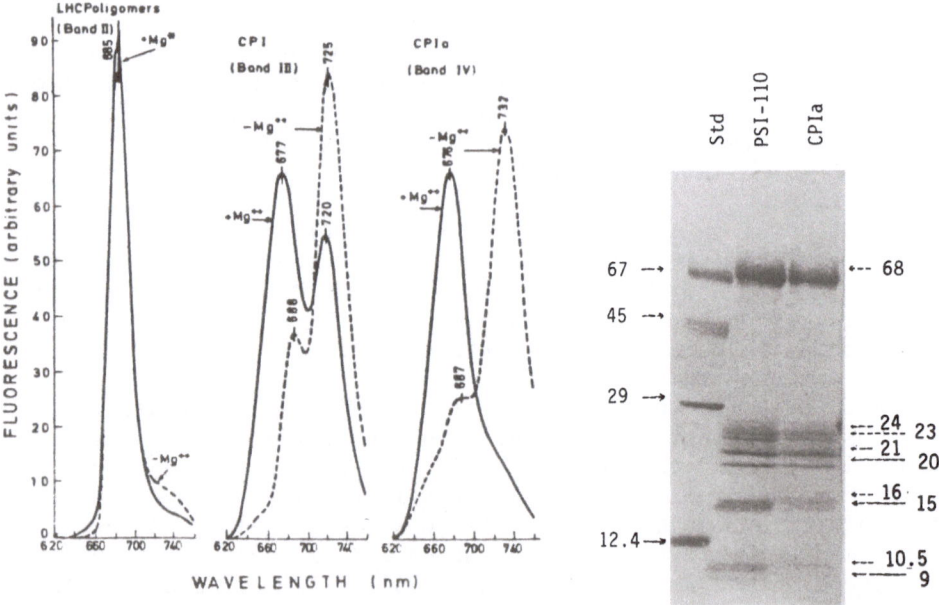

Fig. 3. Fluorescence spectra at 77°K of *Pisum sativum* pigment-protein complexes separated according to (11). Spectra recorded in 0.2% deoxycholate-0.2% Triton-0.1% SDS±10mM Mg2+
Fig. 4. Polypeptide analysis by dissociating SDS-PAGE of PSI-110 particle[14], and of CPIa complex separated as in (11).

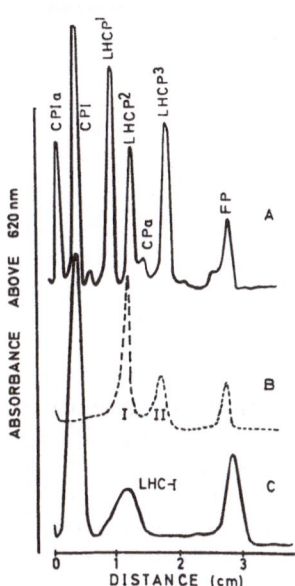

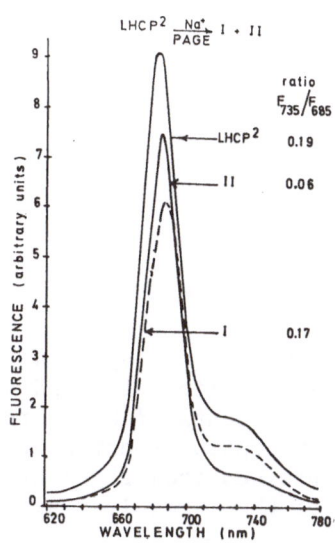

Fig. 5. Electrophoretic profiles of (A) the pigment-protein complexes separated from SDS-solubilized thylakoids after mild
SDS-PAGE; (B) LHCP$_2$ and (C) CPIa bands excised from the gel
(A) and reelectrophoresed on a gel containing 100 mM Na$^+$
in the stacking gel.

Fig. 6. Fluorescence spectra at 77°K (*in situ*) of LHCP$_2$ excised
from the gel after mild SDS-PAGE of thylakoids, and of its
components LHC-I (I) and LHC-II (II) resolved after partial
dissociation of LHCP$_2$.

composition is identical to that of the Triton X-100 extracted PSI-
110 particle, i.e., it contains CPI, the 24, 23, 21 and 20 kDa polypeptides, the low molecular weight, Fe-S ones, of 16, 15, and 10.5
and 9 kDa[15,19] (also shown in fig. 4). CPIa contains Chl$\underline{a}$, Chl$\underline{b}$,
P700, lutein (lut) and b-car (Table 1); its emission band at 77°K
is at 732 nm (fig. 3)[17,18]. CPIa can be dissociated into CPI and
LHC-I if the CPIa band, after excision from the gel, is reelectrophoresed on a gel containing 100 mM Na$^+$ in the stacking gel[19]. Fig.
5 shows such a dissociation. The polypeptide composition of the separated complexes, and their 77°K emission spectra are shown in
figs. 2 and 7. It should be mentioned that the 77°K emission band
of the complexes isolated by SDS-PAGE is located at shorter wavelength than that of complexes isolated by sucrose density gradient
centrifugation (compare figs. 3 and 7). LHC-I has an apparent Mr of
75 kDa, and its electrophoretic mobility is identical to that of
LHCP$_2$[15,19]. It contains Chl$\underline{a}$, Chl$\underline{b}$, lutein, violoxanthin (vx) and
a 21 kDa polypeptide (Table 1 and fig. 2). Its emission band at 77°K
ranges from 717 to 730 nm, depending on the isolation procedure;

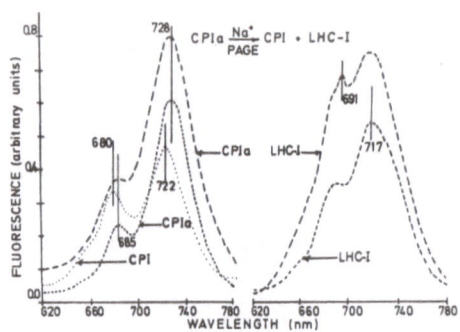
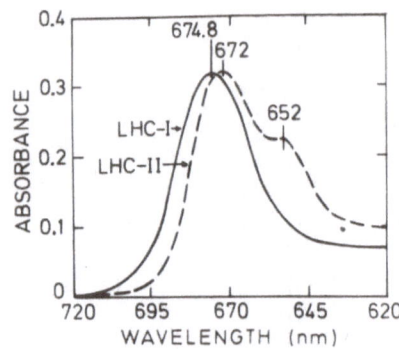

Fig. 7. Fluorescence spectra at 77°K (*in situ*) of CPIa excised from
a gel after mild SDS-PAGE of thylakoids, and of CPI and LHC-I,
components of CPIa, resolved after its partial dissociation.

Fig. 8. Absorption spectra at room temperature (*in situ*) of LHC-II
excised from a gel after mild SDS-PAGE of thylakoids, and
of LHC-I resolved after partial dissociation of CPIa.

its absorption peak at room temperature in the red part of the spectrum is at 675 nm (figs. 6, 7 and 8)[14-18]. CPa contains Chl$\underline{a}$, b-car, the 47 and 42 kDa polypeptides and a number of low Mr ones. Its emission band at 77°K is at 695 nm[20]. LHCP$_1$, LHCP$_2$ and LHCP$_3$ act as light-harvesting antennae of the PSII unit[21]. LHCP$_1$ and LHCP$_2$ are oligomers of LHCP$_3$, all containing the 25-23 kDa polypeptide doublet, Chl$\underline{a}$, Chl$\underline{b}$ and xanthophylls (Table 1 and fig. 2)[11,17]. Their 77°K emission spectra have a peak at 685 nm and a shoulder at 730 nm (fig. 3)[18]. The F_{685}/F_{730} ratio is higher in LHCP$_1$ and LHCP$_3$ than in LHCP$_2$ (see fig. 9). This is due to the fact that the LHCP$_2$ band separated by SDS-PAGE from SDS-solubilized thylakoids, and the LHCP$_{1+2}$ band separated by SDS-sucrose density gradient centrifugation, are contaminated with LHC-I[19]. The presence of LHC-I in these bands is more pronounced whenever the thylakoid starting material is a fraction derived from stroma lamellae, which are enriched in PSI particles. For example, the composition of the LHCP$_2$, or the LHCP$_{1+2}$ bands isolated from stroma lamellae is different from that of the complexes isolated from grana[7]. The LHCP$_{1+2}$ band of grana is enriched in Chl$\underline{b}$ and xanthophylls, contains mainly the 25-23 kDa polypeptide doublet, and fluoresces at 77°K exclusively at 685 nm (F_{685}/F_{730} = 14). On the contrary, the LHCP$_{1+2}$ band of stroma lamellae contains less Chl$\underline{b}$ (Chl$\underline{a}$/Chl$\underline{b}$ = 2.0), is poor in xanthophylls, contains mainly the 21 kDa polypeptide, heavily stained with silver nitrate (fig. 10), and fluoresces at 77°K at both 685 and 730 nm (F_{685}/F_{730} = 4.0)[7,18]. In addition, the LHCP$_{1+2}$ band of stroma lamellae has less pigment/protein than the corresponding band of grana. A partial separation of the LHCP$_2$ band into LHC-I and LHC-II is

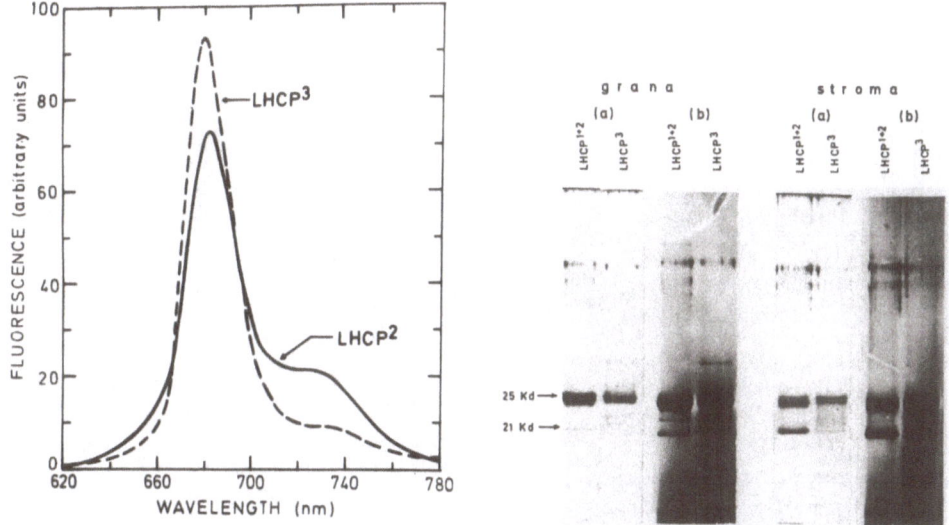

Fig. 9. Fluorescence spectra at 77°K (*in situ*) of LHCP$_2$ and LHCP$_3$ excised from a gel after mild SDS-PAGE of thylakoids.

Fig. 10. Electrophoretic analysis by dissociating SDS-PAGE of the pigment-protein complexes LHCP$_{1+2}$ and LHCP$_3$ separated from SDS-solubilized grana and stroma lamellae according to (11). (a) Coomassie staining; (b) silver nitrate staining.

achieved by reelectrophoresis of the excised from the gel LHCP$_2$ band on a gel containing 100 mM Na$^+$ in the stacking gel (fig. 5)[19]. The polypeptide composition of the separated complexes and their 77°K emission spectra are shown in figures 2 and 6.

Effect of Cations on the Supramolecular Organization of the Pigment-Protein Complexes.

As it was already mentioned, Na$^+$ induces the dissociation of CPIa into CPI and LHC-I; other cations such as Mg^{2+} and H$^+$ have also the same effect. Cations (Mg^{2+}, Na$^+$, H$^+$) induce also the dissociation of the oligomeric forms of the LHC-II (LHCP$_1$ and LHCP$_2$) into their monomeric form[16-18,22]. The separation of LHC-I and LHC-II from the LHCP$_2$ band described previously, is based on this cation effect; i.e., in the presence of Na$^+$ the oligomeric form of LHC-II, present in the LHCP$_2$ band, dissociates into its monomeric form, LHCP$_3$, which runs on the gel faster than the LHC-I. The degree of dissociation of the complexes depends on the cation concentration present in the solubilization buffer, or present in the acrylamide gel. At low cation concentration, or high pH, the more organized complexes, CPIa, LHCP$_1$ and LHCP$_2$ predominate. As the cation concentration is increased, or the pH lowered, the organized structures disappear, giving rise to their dissociation products CPI, LHC-I

211

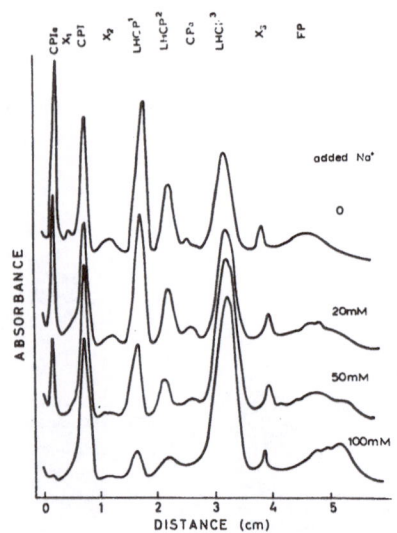
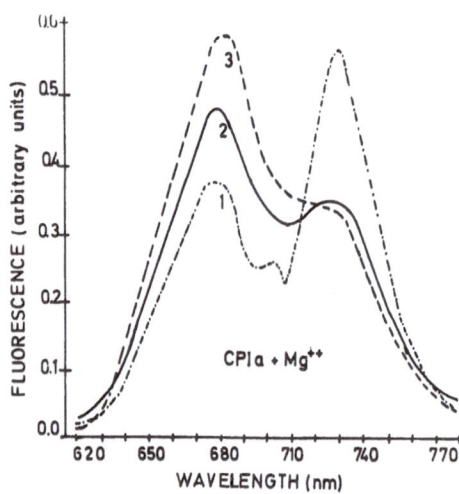

Fig. 11. Electrophoretic profiles of the pigment-protein complexes of bean thylakoids as affected by Na⁺ addition to the SDS-buffer. SDS/Chl = 10; 520 ug Chl/ml.

Fig. 12. The effect of $MgCl_2$ concentration on the 77⁰K fluorescence spectra of CPIa separated from SDS-solubilized thylakoids as in (11). (1):0.09mM Mg^{2+}; (2):1mM Mg^{2+}; (3):1.6mM Mg^{2+}.

and LHCP$_3$[16,22]. This is shown clearly in fig. 11 (see also fig. 1 of ref. 22). The CPIa, LHCP$_1$ and LHCP$_2$ are prominent in the sample having no Na⁺ in the solubilization buffer; addition of Na⁺ induces a decrease in the concentration of the organized forms, and an increase in CPI and LHCP$_3$; finally, in the sample which is dissolved in SDS containing 100 mM Na⁺ practically all Chl ends up in CPI and LHCP$_3$. The effect is more pronounced (on a concentration basis) when the gels are supplemented with cations; in this case for full effect 1-2 mM Mg^{2+} or 50-100 mM Na⁺ are required[16]. The organization, therefore, of the isolated pigment-protein complexes of mature chloroplasts depends greatly on ionic environment.

The changes in the supramolecular organization of the pigment-protein complexes, induced by cations, are closely followed by a drastic change in the 77⁰K fluorescence emission spectra of the detergent solubilized thylakoids, and of the isolated complexes; this change is especially evident in the 77⁰K emission spectra of the complexes originating in the PSI unit[15-19,22,23]. As shown in fig. 3, the isolated CPIa has a 77⁰K emission peak at 732 nm, which shifts to 676 nm upon addition of 10 mM Mg^{2+}. CPI emits at 725 and 688 nm; the emission at 725 nm is much higher than that at 688 nm, and the F_{688}/F_{725} ratio is low; in the presence of 10 mM Mg^{2+} a

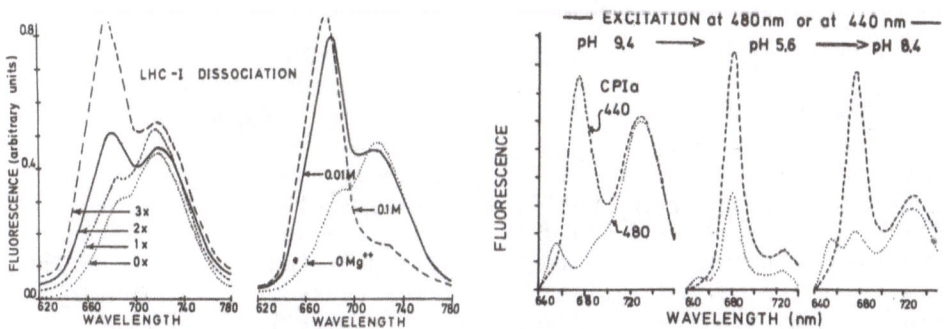

Fig. 13. Fluorescence spectra at 77°K (*in situ*) of LHC-I as affected by freeze-thawing (left), or MgCl$_2$ (right). Left: the
gel slice containing LHC-I was frozen in liquid N$_2$ and the
spectra recorded (ox), then thawed and refrozen succesively
1-3 more times (1x-3x). Right: the gel slice was immersed
for 1 min at 25°C in solution containing 0.01 or 0.1 M
MgCl$_2$, and then the spectra were recorded.

Fig. 14. The reversal of the pH effect on the 77°K emission spectra of CPIa complex isolated as in (11). Excitation at
440 and 480 nm.

small shift in both emission peaks and an increase in the F_{677}/F_{720}
ratio, which reaches value greater than one, takes place. The presence or absence of Mg^{2+} has no effect on the emission spectrum of
the monomeric form of LHC-II, which ahows a peak at 685 nm. However,
the oligomeric forms of LHC-II, LHCP$_1$+LHCP$_2$, show in addition to the
685 nm peak, a shoulder at 730 nm, which disappears upon the addition of 10 mM Mg^{2+}; the shoulder is due to the contamination of the
LHCP$_1$+LHCP$_2$ band with LHC-I. Similar changes in the 77°K emission
spectra of the complexes are also induced by pH changes. For example, by lowering the pH from 9.4 to 5.6 the emission peak of the
CPIa shifts to shorter wavelength, and the long wavelength peak disappears (see fig. 14). The effect of Mg^{2+} addition on the 77°K spectra of LHC-I, separated from CPIa, is shown in figure 13. It is
clear that upon Mg^{2+} addition the long wavelength peak, which ranges
from 717 to 730 nm depending on the isolation procedure, is blue-
shifted and the F_{730}/F_{685} ratio is drastically reduced. The degree
of change in the emission spectra of the complexes depends on cation
concentration. Figures 12 and 13 show the effect of increasing concentrations of Mg^{2+} added on the gradual changes in the emission
spectra of CPIa and LHC-I. In figure 13 the effect of freeze-thawing
on the 77°K emission spectra of LHC-I is also shown. It is clear
that the effect of freeze-thawing is similar to the effect of cation
addition. Prolonged action of SDS on the complexes has also the same
effect[12]. All these results indicate that the changes in the 77°K
emission spectra of the pigment-protein complexes, and especially

those of the complexes originating in the PSI unit, reflect structural changes in their organization towards dissociation to less-organized structures.

The dissociation of the CPIa by cations, and the oligomer to monomer transformation of LHC-II, is a reversible process. The reversibility has been observed in a number of cases, e.g., washing off Mg^{2+} from thylakoids reverses the dissociation process, and the highly organized pigment-protein complexes reassociate[16]. LHC-II precipitated by Mg^{2+} from SDS-solubilized thylakoids is in the monomeric state; removal of Mg^{2+} by Tricine washing results in stabilization of the olifomers[26]. Treatment of thylakoids with phospholipase transforms the oligomeric forms of LHC-II into the monomer; addition of liposomes reconstitutes the oligomeric structures[24]. The reversibility has also been observed in the isolated CPIa complex; the F_{730}/F_{685} ratio decreases by lowering the pH, and it is restored by increasing the pH to 8.4 (fig. 14)[23,25].

It is known that the room temperature fluorescence yield of isolated chloroplasts is highly dependent on the ionic conditions[26]. Cations added to low-salt chloroplasts induce an increase in the variable fluorescence yield, attributed to structural changes induced by the cations[26]. The 77^oK fluorescence emission spectrum of chloroplasts is similarly affected by cations[27]; upon Mg^{2+} addition the 685 nm emission peak (due to the antennae of PSII) is increased, while that at 730 nm (due to PSI) is depressed (see fig. 15). Since under these conditions the quantum yield of the PSI reaction is depressed, while that of the PSII reaction is enhanced, it has been proposed that Mg^{2+} controls the distribution of excitation energy between the two photosystems (spill-over), i.e., in the presence of Mg^{2+} the excitation energy transfer from PSII to PSI is blocked, and thus the Chla fluorescence emitted from PSII is increased[27]. Preillumination of chloroplasts with PSII light has been found to suppress the low-temperature F_{685} emission, while increasing the F_{730} one[28]; this has been attributed to the increase of PSII to PSI excitation energy transfer. Furthermore, it has been shown that in algae two types of fluorescence states can be attained by manipulation of the light quality[29]. State I (high fluorescence-hight enhancement) is induced by illumination with PSI light, while State II (low fluorescence-low enhancement) is induced by PSII light. These results have been interpreted in terms of changes in the initial distribution of light energy between the two photosystems[29]. Later, it has been shown that State I in algae corresponds to the state with Mg^{2+} in chloroplasts (i.e., quanta absorbed by PSII remain in PSII and they can not be transferred to PSI when PSI light is in excess or Mg^{2+} is present), and State II corresponds to the state without Mg^{2+} (i.e., excitation energy from PSII is transferred to PSI when PSII light is in excess or no Mg^{2+} is present)[30]. It is obvious that there is a very good correlation between the cation induced changes in the 77^oK fluorescence emission spectra of the

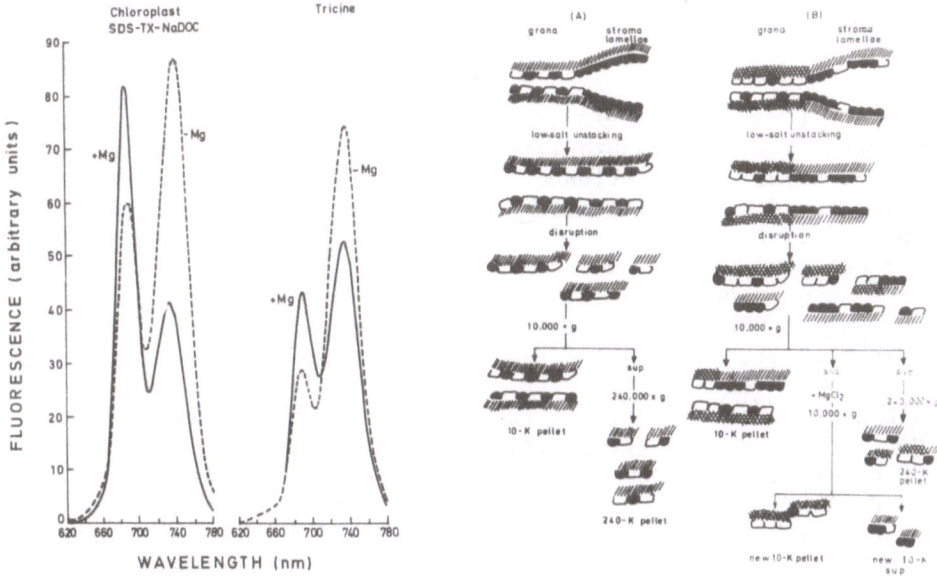

Fig. 15. Fluorescence spectra at 77ºK of low-salt chloroplasts in
Tricine ± 10 mM Mg^{2+} (right), or in 0.2% deoxycholate -
0.2% Triton X-100 - 0.1% SDS ± 10 mM Mg^{2+} (left).

Fig. 16. Simplified models of grana unstacking in chloroplasts suspended in low-salt buffer. (A): model of lateral protein
diffusion proposed by Barber[34]. (B): model explaining the
results of our studies. ⊔: the LHC-II complex;   : the
CPI complex; ⫿⫿⫿⫿: the lipid bilayer of the thylakoid.
Heavily shaded areas of lipid identify the thylakoid part
coming from grana.

chloroplasts, and those of the isolated pigment-protein complexes
of the PSI unit, which correspond to the cation-induced dissociation
of the CPIa. Similar changes are also observed with the detergent
solubilized thylakoids. This is shown clearly in figure 15. Addition of 10 mM Mg^{2+} to low-salt chloroplasts, or to low-salt thylakoids solubilized in detergent, increase the 77ºK F_{685}/F_{730} ratio
in both samples in a similar way. It is tempting, therefore, to
suggest that the dissociation-reassociation of CPIa, and the changes
in the supramolecular organization of the CPI and LHC-I, controlled
by cations, may regulate the distribution of the absorbed energy between the two photosystems by changing the absorption properties of
the PSI unit. In other words, during photosynthetic electron transport the light induced uptake of H^+ (and the concomitant extrusion
of Mg^{2+}) may cause changes in the absorption properties, i.e., the
cross-section, of the PSI unit, and accordingly, changes in the distribution of excitation energy between the two photosystems. It
should be mentioned that Zn^{2+} or Cd^{2+}, which have no effect on the

215

fluorescence yield of chloroplasts at room temperature, nor on their 77°K emission spectra[31,32], have neither an effect on the dissociation of the isolated CPIa complex and its 77°K emission spectrum[22].

Low-salt conditions, in addition to inducing lowering of the fluorescence yield in chloroplasts, induce also unstacking of their grana[6,33]. As already mentioned, PSII is localized mainly in grana, and PSI in stroma lamellae. It is, therefore, difficult to explain how spill-over may occur from PSII to PSI, under conditions of no contact between PSI and PSII units. The explanation was offered by Barber[34], who proposed that low salt unstacking induces also lateral movement of CPI from stroma thylakoids to grana, and of LHC-II from grana to stroma lemallae, so that a random distribution of complexes is achieved. According to Barber, therefore, the light subchloroplast fraction (10,000xg sup; stroma thylakoid fraction) obtained from low-salt chloroplasts after their disruption by French press, should contain one type of thylakoids depleted in CPI and enriched in LHC-II (see model A of fig. 16). We have found, however, that the enrichment of this fraction in LHC-II is due to contamination of stroma lamellae with grana fragments, due to extensive disruption of the unstacked thylakoids. This fraction was found to be a mixture of thylakoid fragments coming from grana (rich in LHC-II) and from stroma lamellae (rich in CPI), as judged by the selective precipitation of the LHC-II rich fragment by Mg^{2+} addition to the light subchloroplast fraction (low-salt 10,000xg sup)(see model B of fig. 16). This finding, therefore, suggests that upon low-salt unstacking there is no intermixing of complexes to allow for spill-over of excitation energy from PSII to PSI to occur.This further supports our proposal that the cation-induced fluorescence changes cannot be attributed to the control of spill-over from PSII to PSI, but rather to the dissociation of the photosystem I unit components, and the decrease in the effective cross section of the PSI unit.

ASSEMBLY OF THE PIGMENT-PROTEIN COMPLEXES

The biosynthesis and assembly of the pigment-protein complexes is a very complicated and multistep process. It involves: Biosynthesis of pigments, lipids and proteins; transport of these components from the point of synthesis to the target membrane; recognition of the target membrane; assembly to multimeric complexes; reorganization of components in the pigment-protein complexes; organization into supramolecular structures.

Biosynthesis of pigments and proteins

Chloroplasts, like mitochondria, contain the entire genetic system, i.e., DNA, DNA polymerase, RNA polymerase and the protein synthesizing machinery[35]. The chloroplast genome, however, has a potential coding capacity of approximately $6x10^6$ dalton polypeptides, and it codes only for a small, but essential, number of chloroplast polypeptides. The majority of the polypeptides are encoded in nuclear genes and are synthesized on cytoplasmic ribosomes. This is based on *in vivo* and *in vitro* experiments such as: (1) protein synthesis by isolated subcellular systems, i.e., proplastids and etioplasts[36], chloroplasts[37,38], free and bound chloroplast ribosomes[37], chloroplast or cytoplasmic mRNA[39,40], etc.; (2) use of selective protein synthesis inhibitors (e.g., chloramphenicol or cycloheximide), specific for inhibiting protein synthesis either on chloroplast or cytoplasmic ribosomes[37,41]; (3) use of heat treated plants[42,43], the chloroplasts of which are deficient in ribosomes, but contain normal plastid DNA. Chloroplast proteins, therefore, can be synthesized in the cytoplasm but not in the chloroplast; (4) use of mutants deficient in chloroplast components[44].

A number of experiments with cell-free systems have shown[35,45], that the nuclear encoded polypeptides are first released from the cytoplasmic ribosomes and then they are transported through the envelope by a post-translational transport mechanism. These polypeptides are synthesized as higher molecular weight precursors, and they are processed to their mature size by a soluble enzyme after being transported to the stromal compartment. It is probable that the additional sequence contains the information necessary for specific binding to a receptor located in the chloroplast envelope. The transport of the polypeptides is stimulated by light, which can be substituted by ATP, and it does not require cytoplasmic or chloroplast protein synthesis[46]. The chloroplast encoded polypeptides of the thylakoid are synthesized on thylakoid-bound ribosomes[47], and they are inserted into the thylakoid by a cotranslational transport mechanism; in other words, their insertion depends on the concomitant protein synthesis by ribosomes bound on the membrane to be traversed.

Etioplasts do not contain Chl but its precursor PChlide. The conversion of Pchlide to Chl in angiosperm leaves is light-dependent. Darkness acts as an inhibitor of the conversion, leading to the accumulation of Pchlide. The amount of PChlide present in the etiolated tissue is age dependent, and reaches a plateau (in bean leaves 30-35 ug/g fresh weight) at about 10 days of age[3]. Brief illumination of the etiolated leaves with visible light of sufficient intensity converts 90-95% of PChlide to Chl. The dark resynthesis of PChlide following a single irradiation shows a similar time course[3]. There is no lag phase in the resynthesis through the 5th day. Beginning with the 6th day there is a lag period which increases in length with time. Under continuous illumination (CL) the biosynthesis of Chl$\underline{a}$ and Chl$\underline{b}$ proceeds normaly, i.e., after a lag phase, which depends on the age of the etiolated tissue, both pigments are synthesized in a parallel way, and they are integrated into the developing thylakoid, so that they soon reach the relative concentration of the mature chloroplast[48]. The lag phase in Chl formation is abolished when

prior to continuous light irradiation the etiolated plants are exposed for a short time to white or red light, followed by 3-6 hrs
dark period. Far-red light reverses the effect of red light.

When etiolated plants are exposed to periodic light-dark cycles
(LDC; 2 min light - 98 min dark), the main Chl molecules accumulated
are those of Chla. There is no lag-phase in Chla synthesis, and the
final concentration reached is almost 1/6th of that of mature green
leaves. A small amount of Chlb can be detected in young etiolated
tissue, and this only after 40 LDC; therefore, the Chla/Chlb ratio
is high. Transfer of the LDC-leaves to CL induces the synthesis of
Chlb, more Chla and a decrease in the Chla/Chlb ratio. In young etiolated leaves the amount accumulated, and the value of the Chla/Chlb
ratio reached depend on preexposure to LDC, decreasing as the time
in LDC increases; in old etiolated leaves the amount of Chl accumulated during the continuous illumination is independent of preexposure to LDC, and equal to that of etiolated leaves of the same age
exposed directly to CL (see fig. 2 of ref. 48).

Carotenoids are present in the etiolated leaves in considerable
amounts. These are mainly xanthophylls, and so the xanth/b-car ratio
is high. Under continuous illumination all carotenoids are synthesized at a high rate; at the same time the xanth/b-car and the Chl/
car ratios change and soon reach the composition of the mature green
leaves[48]. Recent experiments with plastoglobuli-free thylakoids have
shown that the relative composition of the isolated thylakoids of
etiolated plants exposed directly to continuous light remains constant throughout the greening process, and similar to that of the
mature chloroplasts[49]. In other words, thylakoids of the same pigment
composition are synthesized from the beginning of greening in CL.
The difference observed between whole leaves and isolated thylakoids
is due to the presence of plastoglobuli, which exist in considerable
amounts in the etiolated leaves. With time in CL the number of plastoglobuli decreases, and the relative pigment composition of mature
leaves approaches that of thylakoids. In LDC the concentration of
carotenoids present in the etiolated tissue increases slightly; their
relative composition, however, in the plastoglobuli-free thylakoids
changes considerably, and the xanth/b-car ratio reaches values much
lower than those of green controls. Transfer of the LDC-leaves to
continuous light induces the synthesis of all carotenoids, and a
change in the xanth/b-car and Chl/car ratios, which reach either
values of the green controls (short preexposure to LDC), or lower
ones (long preexposure to LDC)[49].

Organization of Pigments and Proteins into Pigment-protein Complexes

The assembly of pigment-protein complexes takes place in a step-
wise manner. This became evident from experiments with etiolated
plants greened in LDC and then transferred to CL. SDS-PAGE of SDS-
solubilized thylakoids have shown that early in LDC only CPI and CPa

(the core of the PSI and PSII units respectively) are formed. As time in LDC is prolonged one can detect the LHC-II, mostly in the monomeric form. Later, during greening in CL, the rest of the complexes are also synthesized. It seems that the monomeric form of LHC-II is formed first, and then it is gradually organized to give rise to multimeric structures. At the same time the LHC-I is synthesized and organized together with CPI into CPIa[25]. A similar trend is also observed when etiolated leaves are exposed directly to CL. An increase in the concentration of $LHCP_3$ is noticed on exposure to CL, with a maximum at about 30 h; then its concentration declines, while the concentration of the multimeric forms, $LHCP_1$ and $LHCP_2$, increases steadily and reaches a plateau after 100 h. Similar changes are observed in CPI and CPIa during greening (see fig. 1 of ref. 50). The formation mainly of CPI and CPa (i.e., small in size PSI and PSII units) in LDC is also supported from fluorescence and photosynthetic activity measurements. For example, plastids ("protochloroplasts") obtained from etiolated leaves greened in LDC show high PSI/Chl, PSII/Chl and P700/Chl; they require high light intensity for saturation of the PSI and PSII activity, and the half-rise time of the fluorescence induction ($t\frac{1}{2}$) in the presence of DCMU is severalfold higher than the corresponding value of the mature chloroplast. Moreover, the absence of Chlb and grana in "protochloroplasts", which are characterized by the presence of single "primary" thylakoids, indicate also the absence of LHC-II. After transfer of the LDC-leaves to CL the PSI/Chl, PSII/Chl, P700/Chl, the light intensity requirement for saturation of the PSI and PSII activities, and the $t\frac{1}{2}$ reach the values of the mature chloroplast, indicating the gradual growth of the PSI and PSII units, i.e., the formation of the light-harvesting complexes LHC-I and LHC-II[3,48]. Since the amount of CPI and CPa formed at the very beginning of greening is very little, it is difficult to distinguish which one of the two complexes is formed first. However, studies on the appearance of the PSI and PSII activities during greening have shown that PSI appears soon after the etiolated leaves are exposed to LDC, while the time of appearance of the PSII activity varies, and depends on the age of the etiolated tissue[48]. In the beginning of greening the PSI/Chl is very high, then decreases and finally levels off. The PSII/Chl, on the contrary, is initially low, then increases steadily and finally reaches a plateau. The values at the plateau for both activities are many times higher than those of the mature chloroplast (see fig. 3 of ref. 48). These results indicate that the formation of CPI precedes that of CPa.

Parallel to the assembly of the pigment-protein complexes, and their organization into supramolecular structures, changes in the room temperature Chla fluorescence yield (Fmax/Chl), and the 77° K fluorescence emission spectra of the developing chloroplasts take place. It has been observed[23,25], that there is a close correlation between the changes in the Fmax/Chl, and the changes in CPI/CPIa and $LHCP_3/LHCP_{1+2}$ ratios. The close correlation observed, i.e., high

Fmax/Chl when monomers predominate, and low when they become organized into multimers (see fig. 1 of ref. 50), suggests that the relative concentration of monomeric to multimeric forms, which reflects the organizational state of the photosynthetic units, may be responsible for the changes in the Fmax/Chl. Similar changes in the absolute fluorescence yield at 77^OK, which is accompanied by a pronounced red-shift of the long wavelength peak, takes also place during greening[23,25]. Figure 17 shows the 77^OK emission spectra of whole leaves obtained from etiolated bean plants exposed to 23 LDC, and then transferred to CL. In the LDC-leaves the main emission peak is at 690nm. The maximum at 735nm (due to the antennae of PSI[51]) is not present, but a peak is noticed at 725nm, which is probably due to the core of the PSI unit. The peak at 655nm is probably due to the presence of a small amount of PChlide. After transfer of the LDC-plants to CL the 725nm peak increases and reaches values higher than the value at 690nm; at the same time its maximum is red-shifted and reaches the wavelength of 735nm. The changes in the absolute fluorescence yield at 77^OK of plastids obtained from etiolated leaves exposed to LDC and then transferred to CL are shown in figure 18. In this case the spectra were taken in the presence of the internal fluorescence standard phycocyanin. The emission maximum at 653nm is mainly due to phycocyanin, but partly also to PChlide. In the LDC-plastids an increase in both F720 and F690 is clearly noticed; however, after estimation of the PChlide contribution to F653, one can calculate that during exposure to LDC the F690 remains constant, while the F720 increases; thus, the F690/F720 ratio diminishes from about 2.2 to 1.5 (after 100 LDC). During this time, the emission at 717nm shifts to 725nm. Upon transfer of the LDC-plants to CL a sharp increase in both F690 and F730 emissions occurs, followed by a gradual decline as illumination is prolonged. Parallel to this decrease a farther red-shift to 735nm is clearly noticed. The increase and the subsequent decrease in both F690 and F730 emissions follow a more or less similar trend as that observed in the Fmax/Chl. However, in all cases the increase in the F730 emission is greater than that at 690nm, so that as greening proceeds the F690/F730 ratio continuously declines and reaches a value of 1.2, found also in green plant chloroplasts. Taking into account the 77^OK fluorescence emission spectra of the isolated pigment-protein complexes, and the changes observed in the 77^OK fluorescence spectra of plastids during development, we can correlate the fluorescence changes with the biosynthesis and organization of components within each complex, the subsequent gradual organization of CPI and LHC-I into CPIa, and of the monomeric form of LHC-II into its multimeric forms, resulting in photosynthetic units more efficient for capturing the light energy.

Studying the changes in the pigment content of the pigment-protein complexex during thylakoid development, it has been found that the relative concentration of the pigments in each complex is constant from the beginning of their formation; the pigment/protein ratio, however, increases as the duration in continuous light in-

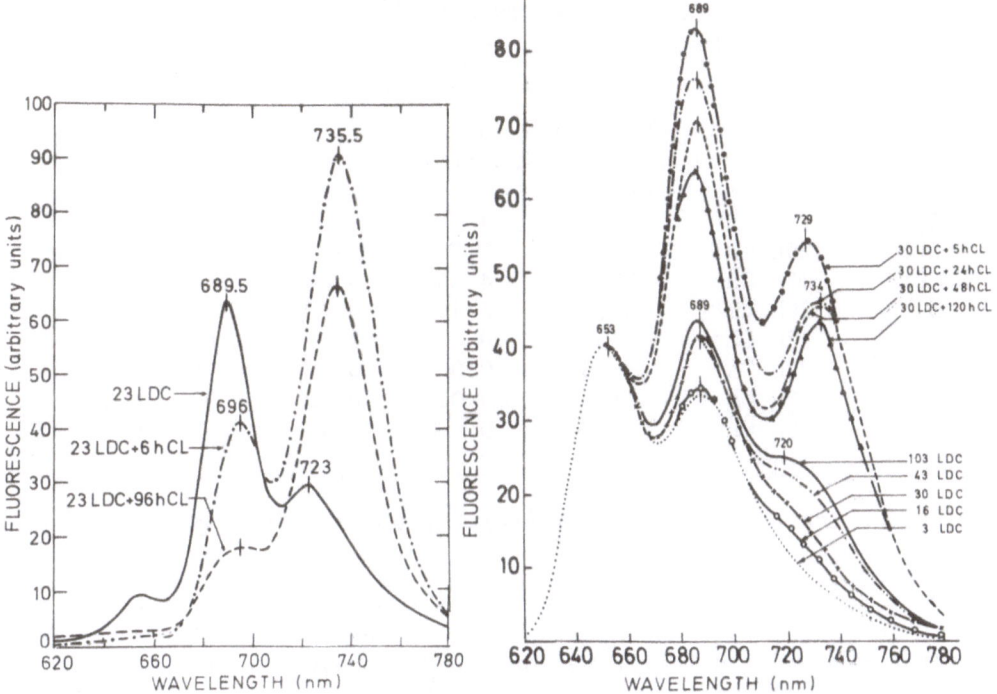

Fig. 17. Low temperature fluorescence emission spectra of 6-day etiolated bean leaves exposed to intermittent light (LDC), and
then transferred to continuous light (CL).

Fig. 18. Low temperature fluorescence emission spectra, in the presence of the internal fluorescence standard phycocyanin, of
bean plastids obtained from etiolated leaves esposed to intermittent light (LDC), or to 30 LDC plus continuous light
(CL). Phycocyanin (PC) was present at a PC/Chl ratio of 0.8
(w/w). The normalization of the fluorescence intensity at
653nm was done by changing the intensity of the exciting
light.

creases (see Table 2). This indicates that the assembly of a complex
is a step-wise process, and that there are many pigment-binding sites
on the apoprotein of each complex, which are gradually filled with
pigments, but at a constant ratio. The step-wise assembly, therefore,
of the complexes involves the organization of relatively ·simple components into organized supramolecular structures.

The changes in the rate of accumulation of the pigment-protein
complexes during chloroplast development under different environmental conditions, i.e., the way the thylakoid develops, has also an
effect on the final stage of chloroplast development. This can be
observed in etiolated plants exposed to LDC, and then transferred to

Table 2. Changes in pigment content of the pigment-protein
complexes isolated from thylakoids of etiolated
bean leaves exposed to continuous light (CL)

Sample		Chl($\underline{a}$+$\underline{b}$)	b-car	lut	nx	vx
			(mole/Mr of complex)			
CPI	18h CL	9.4	1.10	-	-	-
CPI	120h CL	19.8	2.70	-	-	-
CPIa	18h CL	22.2	4.30	0.60	-	-
CPIa	120h CL	33.5	6.10	0.40	-	-
LHCP$_3$	18h CL	1.8	0.04	0.26	0.05	0.04
LHCP$_3$	120h CL	4.3	0.07	0.58	0.30	0.08
LHCP$_{1+2}$	18h CL	1.9	0.02	0.29	0.14	0.05
LHCP$_{1+2}$	120h CL	6.7	0.06	1.10	0.50	0.20

(From P. Antonopoulou and G. Akoyunoglou, in preparation)

CL, i.e., during etioplast to "protochloroplast" and "protochloroplast" to chloroplast dufferentiation. It has been found[2,21,48],
that the degree of differentiation of "protochloroplast" to chloroplast depends on the time of preexposure to LDC. The preexposure
time affects the final amount of Chla, Chlb and LHC-II accumulated,
the size of the PSI and PSII units, the PSII/PSI ratio, and the
structure of the chloroplast. This can be seen clearly in Table 3.
In plants transferred to CL after short preexposure to LDC the Chl
accumulated, and the amount of LHC-II formed, are equal to those of
the green control; in plants transferred to CL after long preexposure to LDC the Chl accumulated, and the amount of LHC-II formed,
are much less. The final concentration of the PSII units (PSII/g fr
wt), after transfer to CL reaches always the concentration of the
green control. The final concentration of the PSI units (PSI/g fr wt)
or (P700/g fr wt), however, decreases with time of preexposure to
LDC. Thus the PSII/PSI ratio drops from 2.4 (in LDC) to the value
of the green control (1) in plants preexposed to LDC for a short
time, or to a higher value (between 1 and 2) in plants preexposed to
LDC for a long time. The size of the PSI unit increases after transfer to continuous light, and it reaches always the size of the green
control. The size of the PSII unit increases also in continuous
light, but it reaches either the size of the green control (short
preexposure to LDC), or a smaller size (long preexposure to LDC). In
other words, prolonged preexposure to LDC affects especially the
growth of the PSII unit, which reaches an intermediate stage of development, and the concentration of the PSI units.

The final concentration of the cytochromes (cyt$\underline{f}$, cytb$_{563}$ and
cytb559_{LP}), after the LDC-plants are transferred to $\overline{C}$L, reaches in

Table 3. PSI, PSII, P700 and cyt$\underline{f}$ content of plastids obtained from 6-day etiolated bean leaves exposed first to intermittent light (LDC), and then transferred to continuous light (CL)

Sample	Chl(a+b) (ug/g fr wt)	P700	Chla Chlb	Activity g fr wt.h PSI	PSII	PSII PSI	cytf nmoles/ g fr wt
28 LDC	210	1.45		102	200	2.4	1.68
57 LDC	393	2.83	26.0	188	373	2.4	3.18
8b LDC	610	4.26	14.0	289	543	2.3	4.72
28 LDC+49h CL	2765	6.96	3.0	553	470	1.0	4.70
57 LDC+49h CL	1529	3.84	3.9	344	463	1.7	4.67
8b LDC+49h CL	1214	3.10	4.7	279	466	200	4.73
Green Control	2800	7.00	3.0	560	476	1.0	4.70

PSII activity: umoles DCIP reduced; PSI activity: umoles oxygen consumed; The ratio PSII/PSI was estimated on the assumption that in chloroplasts of green leaves it is equal to 1.

all cases the concentration of the green control, i.e., it parallels the concentration of the PSII units (see Table 3). Thus the number of PSII units, and the number of electron transport chains per chloroplast remain constant in all steady states studied, while the number of PSI units per chloroplast changes. This indicates that the number of PSII units per electron transport chain remains constant, but the number of PSI units per chain varies.

Reorganization of the Pigment-protein Complexes

The question that arises is whether the pigment-protein complexes after their assembly, and organization into supramolecular structures, are stable, or a reorganization can take place. It has been found[21,52-55]. that during development, and under certain environmental conditions, a reorganization of the thylakoid components may take place. During this reorganization the following may happen: (a) dissociation of one or more pigment-protein complexes leading to liberation of Chla, which binds to newly synthesized polypeptides forming other pigment-protein complexes; (b) insertion of newly synthesized polypeptides into the thylakoid, competition for the Chla already bound to another complex, removal of the Chla to form new complexes, and degradation of the old one. The first case is observed in young etiolated leaves exposed to LDC for six days (86LDC),

Table 4. Chlorophyll, P700 and cytochrome content of chloroplasts obtained from 6-day etiolated bean leaves exposed first to continuous light (CL), and then transferred to dark (D)

Sample	Chl$\underline{a}$	Chl$\underline{b}$	P700	cyt$\underline{f}$	Cytb$_{559LP}$	cytb$_{559HP}$	cytb$_{563}$
	ug/leaf			nmoles/leaf			
14h CL	14.4	4.8	0.05	0.04	0.10	0.03	0.11
14h CL+48h D	14.2	1.0	0.11	0.22	0.54	0.08	0.56
40h CL	63.7	25.5	0.24	0.17	0.40	0.21	0.38
40h CL+48h D	62.8	12.3	0.33	0.42	0.97	0.33	0.98
65h CL	172.9	72.1	0.64	0.42	1.03	0.69	1.10
65h CL+48h D	172.7	72.0	0.64	0.44	1.10	0.71	1.17

and then transferred to CL in the presence of the protein synthesis inhibitor chloramphenicol. Under these conditions no net Chl synthesis occurs during the continuous illumination, but the biosynthesis of the LHC-II apoprotein is not inhibited. As soon as the LDC-plants are transferred to CL a number of their PSII units, i.e., the CPa complex, are disorganized. Since the 25kDa polypeptide is synthesized in the presence of chloramphenicol, the Chl$\underline{a}$ liberated from CPa binds to it; at the same time a number of Chl$\underline{a}$ molecules are transformed into Chl$\underline{b}$, and finally the LHC-II is assembled, and organized around the core of the remaining PSII units, increasing their size. This is demonstrated by the decrease in PSII/Chl, PSII/leaf, the light intensity required for saturation of the PSII activity, the half-rise time of the fluorescence induction kinetics (t$\frac{1}{2}$) in the presence of DCMU, and the Chl$\underline{a}$/Chl$\underline{b}$ ratio during the CL, and the increase in LHC-II and the 25kD$\underline{a}$ polypeptide. No change is observed in the Chl($\underline{a+b}$), PSI/Chl, PSI/leaf and the P700/Chl, indicating that the PSI units are unaffected during this process. If the LDC-plants are transferred to CL in the presence of cycloheximide, which inhibits the biosynthesis of the LHC-II apoprotein, then during the CL a number of PSII units are also disorganized, but the LHC-II complex is not formed. In this case, therefore, the CPa is disorganized first, and then its Chl$\underline{a}$ is reassembled into LHC-II[21,25,52].

The second case is observed in young etiolated leaves exposed to CL for various durations, and then transferred to darkness[52-55]. In this case the reorganization involves: disorganization of the LHC-I and LHC-II, degradation of Chl$\underline{b}$, digestion of their apoproteins, and reuse of their Chl$\underline{a}$ for the formation of more PSI and PSII units, mainly CPI and CP$\underline{a}$. This is demonstrated by the following: In continuous light the etiolated leaves accumulate both Chl$\underline{a}$ and Chl$\underline{b}$ with a high rate, and all the pigment-protein complexes are

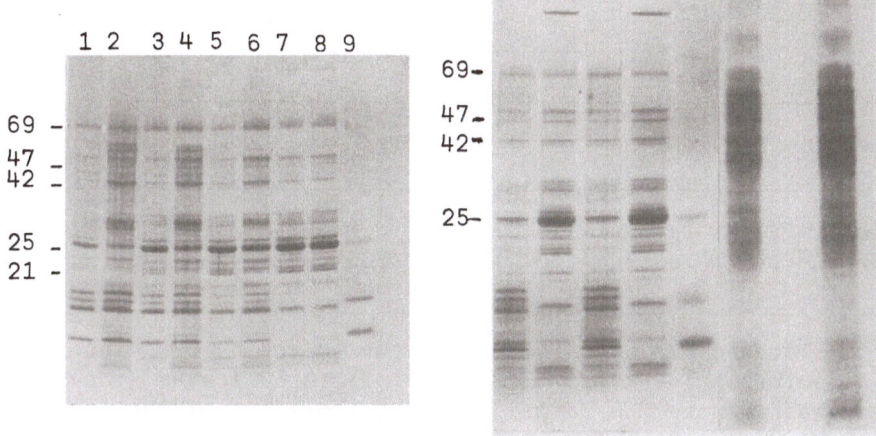

Fig. 19. SDS-PAGE of SDS-solubilized thylakoids obtained from 6-day
         etiolated bean leaves exposed first to CL for 14h (slot 1),
         24h (slot 3), 40h (slot 5) or 65h (slot 7), and then trans-
         ferred to darkness for 48h (slots 2, 4, 6 and 8 respective-
         ly). Slot 9: standard proteins.
Fig. 20. (A): SDS-PAGE of SDS-solubilized thylakoids obtained from
         6-day etiolated bean leaves exposed first to CL for 14h
         (slots 1 and 3), or 65h (slots 2 and 4) followed by trans-
         fer to darkness for 24h in the presence of (^{14}C)-leucine.
         100 ug protein was loaded in slots 1 and 2; 120 ug protein
         was loaded in slots 3 and 4. Slot 5: standard proteins.
         (B): Fluorogram of the same slab gel.

formed. After transfer to darkness the Chla/leaf remains constant,
while the Chlb/leaf decreases; thus, the Chla/Chlb ratio increases.
The decrease in Chlb is more pronounced in leaves transferred to
darkness after a short preexposure to CL. As preexposure to CL increases, the decrease in Chlb gradually becomes smaller, and finally
stops completely (see Table 4). The decrease in Chlb is accompanied
by a decrease in the relative concentration of LHC-II and CPIa, and
an increase in CPa and CPI; Thus, the LHC-II/CPa, and CPIa/CPI ratios
decrease drastically. The decrease in the concentration of CPIa during the dark period is due to the destruction of the LHC-I. This is
clearly reflected by the polypeptide composition shown in figure 19.
In each pair of samples the same amount of Chla was loaded on the
gel. Since the Chla content remains constant during the dark period,
the thylakoid polypeptide pattern shows the real changes in the concentration of the polypeptides in each case. As it is evident, a decrease in the 25kDa polypeptide, component of LHC-II, and in the 24,
23 and 21 kDa polypeptides, components of LHC-I, is observed in the
dark; on the other hand the 47 and 42 kDa polypeptides, components

Table 5. Thylakoid protein synthesis in 6-day etiolated
bean leaves exposed first to continuous light
(CL), and then transferred to darkness (D) in
the presence of (^{14}C)-leucine

Sample	Chl$\underline{a}$/Chl$\underline{b}$	cpm/mg protein
14h CL+((^{14}C)-leucine+24h D)	4.6	606,000
65h CL+((^{14}C)-leucine+24h D)	2.6	31,000

(^{14}C)-leucine was administered to the leaves by brushing
3 ml solution (0.08 mCi, 342 uCi/umol) to 60 leaves; they
were washed with H_2O 2h prior to the isolation of the
thylakoids.

of CPa,and the 69 kDa polypeptide, component of CPI, are increased.
In addition, an increase in the concentration of a great number of
other polypeptides, and a decrease in the 27 kDa polypeptide takes
also place in the dark.

The synthesis of new polypeptides in the dark is further supported from *in vivo* labelling experiments, in which young etiolated
bean leaves were exposed to CL for various periods of time and then
transferred to darkness in the presence of (^{14}C)-leucine. It was
found that in leaves exposed to CL for 14h and then transferred to
darkness for 24h the thylakoids incorporated (^{14}C)-leucine; on the
contrary, in leaves exposed to CL for 65h and then transferred to
darkness for 24h there was no incorporation of (^{14}C)-leucine in thylakoids (see Table 5). SDS-PAGE of the thylakoids and detection of
the (^{14}C)-labelled polypeptides by fluorography showed that, in the
case of the 14h CL+24h D sample the 69, 47 and 42 kDa polypeptides,
and a great number of others, are (^{14}C)-labelled; on the contrary,
in the case of the 65h CL+24h D sample no polypeptide is (^{14}C)-la-
belled (see fig. 20). In addition, the fluorogram shows that even
the 25, 24, 23 and 21 kDa polypeptides are also (^{14}C)-labelled, indicating that the LHC-II and LHC-I apoproteins continue to be synthesized during the dark period. This is in agreement with the results of Bennett[56] who found that, after initiation by a 24h illumination period, the synthesis of the LHC-II apoprotein continues
in darkness at readily detectable rate. Our results show that not
only the LHC-II apoprotein, but also the LHC-I apoproteins continue
to be synthesized in darkness.

The decrease in the concentration of the LHC-I in the dark is
also supported from fluorescence measurements. As it was mentioned

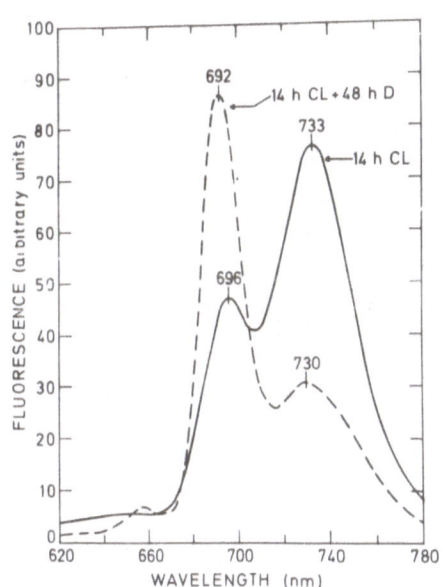

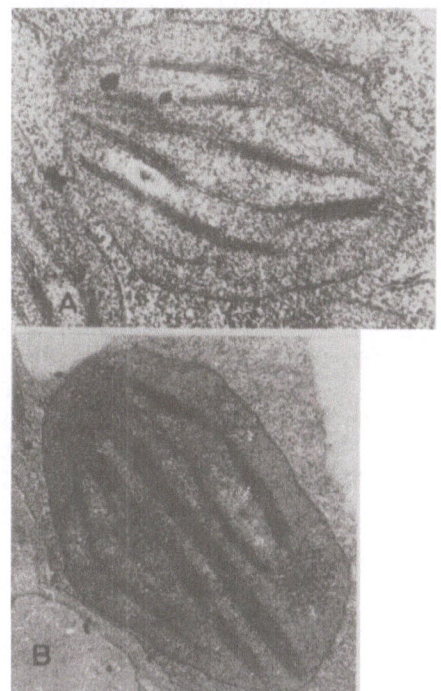

Fig. 21. Low temperature fluorescence emission spectra of chloroplasts obtained from 6-day etiolated bean leaves exposed
first to CL for 14h (————), and then transferred to darkness (D) for 48h (-----).

Fig. 22. Electron micrographs of chloroplasts present in 6-day etiolated bean leaves exposed to CL for 13h (A), and then transferred to darkness for 48h (B).

earlier, the presence of the 730-735 nm band in the $77^{o}K$ fluorescence emission spectra of leaves or isolated chloroplasts depends
on the presence of LHC-I, and its incorporation along with CPI into
CPIa. Accordingly, one should expect a high 730nm emission band in
the spectra of leaves exposed to CL for 14h, where all the complexes
are formed, and a low emission at 730nm in the same leaves transferred to darkness, where dissociation of CPIa and destruction of
LHC-I takes place. Figure 21 shows that this is the case. The 730nm
emission band of the leaves greened in CL for 24h is high compared
to the 690nm band, reflecting the synthesis of the LHC-I, and its
association with CPI into CPIa. Upon transfer to darkness for 48h,
the 730nm emission band declines, indicating the dissociation of
CPIa and the destruction of LHC-I.

Considering that the LHC-II acts as the adhesive force holding
the grana thylakoids into appressed stacks, one should expect un-

Table 6. PSI, PSII, cytochromes and P700 content of plastids obtained from 6-day etiolated bean leaves exposed first to continuous light (CL), and then transferred to darkness (D).

Sample	Activity (Vmax)				PSI P700	PSII cytf	PSI cytf	PSII PSI
	PSII	PSI	PSII	PSI				
	umol/mg Chl.h		umol/leaf.h					
14h CL	513	218	9.8	4.2	1.1	1.0	0.9	1.1
14h CL+48h D	2857	526	49.1	9.0	1.1	1.0	0.3	2.7
40h CL	500	213	44.6	19.0	1.0	1.1	1.0	1.1
40h CL+48h D	1333	357	99.8	26.8	1.1	1.0	0.6	1.8
65h CL	400	197	98.00	48.3	1.0	1.0	1.0	1.0
65h CL+48h D	395	202	96.7	49.4	1.0	0.9	1.0	1.0

PSII activity: umol DCIP reduced; PSI activity: umol O_2 consumed. Vmax was determined from the intercept of the plot $1/V$ versus $1/I$ at $1/I = 0$. PSI/P700, PSI/cytf, PSII/cytf and PSII/PSI were estimated on the assumption that in green leaves (65h CL) they are equal to one.

stacking of some of the preexisting grana, and the appearance of single long thylakoids to occur in plants exposed to CL and then transferred to darkness. Figure 22 shows that unstacking of grana really takes place when etiolated leaves are exposed to CL for 13h, and then transferred to darkness for 48h. The unstacking of grana takes place as long as the transfer from light to darkness occurs early in development. Thus, one can find conditions where no grana stacks are present at all (13h CL+48h D), or conditions where no morphological changes in the grana structure can be noticed (60h CL+60h D), (see fig. 5 of ref. 53)

During the reorganization of the pigment-protein complexes taking place in darkness, a severalfold increase in the PSI and PSII activities, and the P700/Chl ratio, can be observed, (Table 6). The increase, however, is lower in PSI than in PSII, and thus, the PSII/PSI ratio, which is constant throughout the CL exposure, increases. This indicates that the rate of formation of the PSI and PSII units is different in darkness than in light; and that darkness affects especially the rate of formation of the PSI unit. This effect is similar to the one observed with etiolated bean leaves greened in LDC, where the PSII/PSI ratio is 2.5 times higher than that of the mature chloroplast. Since this occurs without any net Chl synthesis per leaf, the results indicate that the Chla in the dark transferred plants is organized into new PSI and PSII units. These units are small in size, as it is evident from the high light intensity required for saturation of the PSI and PSII activities, and the increase in the half-rise time of the fluorescence induction (t½) in the presence of DCMU. Parallel to the formation of the new PSI and PSII units, an increase in the concentration of the thylakoid cytochromes is also observed (see Table 5). Considering the changes in the concentration of cytochromes (e.g., cytf/leaf), and those of PSI and PSII units (see Table 6), we can conclude that, in leaves transferred to darkness after a short preexposure to CL, the rate of cytochrome formation (i.e., electron transport chains/leaf) follows closely the rate of formation of the PSII units (PSII/cytf = constant), while the rate of formation of the PSI units is much lower (PSI/cytf decreases). It has been proposed that a number of PSII and PSI units operate on the same electron transport chain[48]. It seems, therefore, that the number of the PSII units per chain remains constant during development in the dark, and equal to that of the green control (65h CL), while the number of the PSI units per chain is smaller. The changes taking place in darkness diminish as preexposure to light increases, and finally stop. It seems that both PSI and PSII units, as well as the cytochrome content stop from changing at the same time.

Concerning the mechanism of the reorganization process taking place in the dark there are two possibilities: (a) the apoproteins of the PSI and PSII reaction centers are synthesized first, inserted into the developing thylakoid, compete for the Chla already bound to LHC-I and LHC-II, remove it to form CPI and CPa, and then the LHC-I and LHC-II are disorganized and their apoproteins digested; or (b) the LHC-I and LHC-II are disorganized first, and then their Chla becomes available to bind to the apoproteins of the PSI and PSII reaction centers. In the first case the destruction of the LHC-I and LHC-II will depend on the rate of synthesis of the reaction center polypeptides, while in the second case it will not. Experiments with etiolated leaves exposed first to LDC, then to CL for a short time and then to darkness[54], showed that the first hypothesis is correct. As it is known, in LDC small PSI and PSII units containing only the core of the unit are formed; their number increases as the time in LDC increases. Transfer of the LDC-plants to CL induces the accumulation of LHC-I and LHC-II, and the units increase in size. After transfer to darkness the reorganization process starts. It was found that the reorganization of the thylakoid components during the dark period depends on the time of preexposure to LDC, and not on the time of preexposure to CL. For example, in leaves exposed to CL for 24h after a preexposure to 14 LDC, a reorganization occurs after transfer to darkness. In leaves, however, exposed to CL for 24h after a preexposure to 42 LDC, no reorganization occurs. In both cases, however, LHC-I and LHC-II accumulate during the 24h continuous illumination, but only in the first case, where the thylakoids have reduced amount of PSI and PSII units, and the reaction center polypeptides continue to be synthesized dur-

229

ing the dark period, the reorganization is noticed. In other words, it seems that first the Chl$\underline{a}$ of the LHC-I and LHC-II is removed and used for the formation of new PSI and PSII units; then the LHC-I and LHC-II are disorganized and their components degraded and digested. So, the amount of LHC-I and LHC-II disorganized would depend on the concentration of the PSI and PSII units present in the thylakoid, i.e., on the preexposure time to light. It is not known whether the Chl$\underline{a}$ of LHC-II is used specifically for the formation of the new PSII units, and the Chl$\underline{a}$ of LHC-I for the PSI units, or the Chl$\underline{a}$ of both LHC complexes is used unspecifically for the formation of both PS units. The first possibility seems to be more probable, since, as it has been shown[21], the growth of the PSII unit is independent of that of the PSI unit. A modulation of PSI and PSII organization was also observed in a temperature sensitive mutant of *Chlorella pyrenoidosa*[58]. It was found that the Chl$\underline{a}$ organized in the CPI could be used for the formation of additional PSI reaction centers, i.e., dissociation of CPI takes place, and a reassembly of its Chl$\underline{a}$ into new PSI-reaction center-antenna-protein complexes.

Mechanism of Complex Assembly

The experiments desrcibed suggest the following mechanism operating on the assembly of the pigment-protein complexes, and the development and growth of the thylakoid. According to this mechanism, in the early stages of etioplast development all thylakoid components are synthesized, but with a relatively low rate, except Chl$\underline{a}$ which accumulates as PChlide $\underline{a}$. From all thylakoid components synthesized in darkness during etioplast growth only a number of them accumulate, the rest are digested, since they can not be stabilized in the absence of Chl$\underline{a}$. As the age of the etiolated tissue increases a number of the thylakoid components stop to be formed. Depending, therefore, on the developmental stage of the etioplast, one may or may not observe the expression of some of these components in the dark-grown plants. Illumination of young etiolated leaves transforms the PChlide $\underline{a}$ into Chlide $\underline{a}$, and stimulates also the synthesis of the other thylakoid components. In old etiolated leaves light re-induces also the synthesis of those components, which have been stopped to be synthesized in darkness. The development of thylakoid depends on the rate by which the different components are synthesized. If one, or more of the components are synthesized at a low rate, then their synthesis controls the way the thylakoid develops. The rate of Chl$\underline{a}$ formation is a determining factor in thylakoid development, since most of the polypeptides are stabilized by forming with Chl$\underline{a}$ pigment-protein complexes. If Chl$\underline{a}$ is formed at a low rate, then its synthesis regulates the assembly of the pigment-protein complexes, and accordingly the development of the thylakoid.

According to this mechanism we can explain the formation of the pigment-protein complexes in plants greened in LDC, and the reorganization of the thylakoid components in plants transferred to dark-

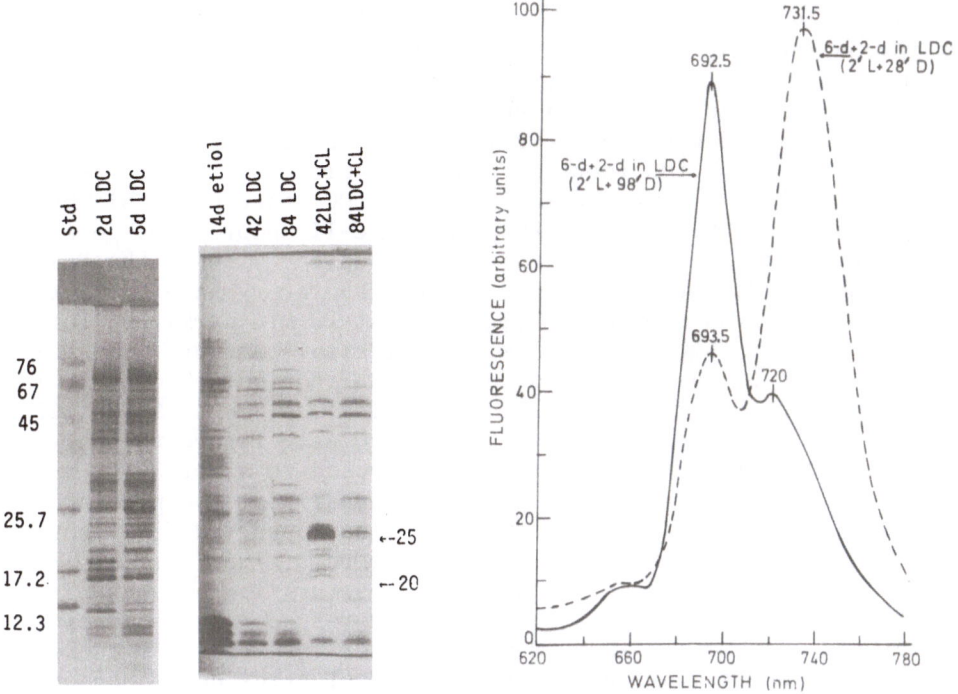

Fig. 23. Slab gel electrophoresis of SDS-solubilized thylakoids obtained from 6-day etiolated bean leaves exposed to (2 min light+28 min dark) LDC (Left); or to (2 min light+98 min dark) LDC (Right).

Fig. 24. Low temperature fluorescence emission spectra of 6-day etiolated bean leaves exposed to (2 min light+28 min dark LDC; or to (2 min light+98 min dark) LDC.

ness after being exposed to CL, as follows: (a) Complex assembly in plants exposed to intermittent light. In the case of plants greened in LDC the Chla formed during the 2 min illumination of every cycle is very small, and so the rate of Chla accumulation becomes the limiting factor. There is, therefore, a competition for the small amount of Chla; it seems that the apoproteins of the reaction centers have a higher affinity for Chla than the apoproteins of LHC-I and LHC-II, so they form CPI and CPa, i.e., the core of the PSI and PSII units, while the apoproteins of LHC-I and LHC-II, not being able to be stabilized, are digested. It is not known whether, during the intermittent illumination, the LHC-I and LHC-II are assembled first, and then in the dark they are disorganized, since their Chla is taken up to be used for the formation of CPI and CPa, and their apoproteins digested. In other words, whether in this case also there is initially formation and then reorganization of the thylakoid components, similar to the one observed in plants exposed to CL for a short time, and then transferred to darkness. Both hypotheses seem probable. This explains the absence of LHC-II in the LDC-plants[10], even though the mRNA of its apoprotein is present and it is translated[57]. If during the intermittent illumination the relative rate of Chla formation increases, either by shortening the dark period in the LDC (e.g., 2 min light every 30 min), or by decreasing the rate of synthesis of the reaction center apoproteins, as in the case of blue light LDC, or after a long time in LDC, then some Chla becomes available to bind on the LHC-I and LHC-II apoproteins forming the respective complexes, and increasing thus the size of the PSI and PSII units. Recent experiments in our laboratory have shown that thylakoids of leaves exposed to (2 min light+28 min dark) LDC for 2 or 5-days, contain LHC-II and CPIa in appreciable amounts. The polypeptide pattern of these thylakoids is shown in figure 23. As it is obvious, the apoproteins of LHC-I and LHC-II (i.e., the 25, 24, 23 and 21 kDa polypeptides) are present. In the same figure the polypeptide pattern of thylakoids of leaves exposed to 42 or 84 (2 min light+98 min dark) LDC is also shown for comparison. The presence of LHC-I and LHC-II in these plastids is also supported from photosynthetic activity and fluorescence measurements. The PSII/Chl, the P700/Chl, the light intensity required for saturation of the PSII activity, and the half-rise time of the fluorescence induction in the presence of DCMU are about 2-times higher than those of the green control, but many times lower than those of plastids developed under the (2 min light+98 min dark) LDC regime. The LHC-I present in these thylakoids is organized into CPIa as the 77°K fluorescence emission spectra shows (i.e., high 730nm) (see fig. 24). Similarly, leaves exposed to LDC for 6-days (86LDC), where the rate of synthesis of the reaction center polypeptides decreases, contain LHC-II in appreciable amounts (see fig. 1 of ref. 52). (b) Reorganization of the complexes in plants transferred to darkness. During the continuous illumination the rate of Chla synthesis is high, and all polypeptides are stabilized by forming complexes. After transfer to darkness, and as long as the thylakoid is still in the process of development, the reaction center polypeptides, as well as a number of other polypeptides, continue to be synthesized. Since no Chla is formed in the dark, the reaction center polypeptides compete for the Chla already bound to LHC-I and LHC-II, remove it, and bind to it to form CPI and CPa. The removal of Chla from LHC-I and LHC-II results in the disorganization of the complexes, degradation of Chlb and digestion of their polypeptides.

The disorganization of the pigment-protein complexes, and the reassembly of Chla into new and different pigment-protein complexes, during chloroplast development, seems to be a general mechanism in thylakoid biosynthesis. Moreover, the digestion of the newly inserted, but still unstable, polypeptides, as well as those which become unstable during the reorganization process, may be one of the mechanisms which control the growth and differentiation of the photosynthetic membrane, as well as the adaptation of plants to their environment. It is not known whether this control mechanism operates in all biological membranes; however, this possibility can not be excluded.

REFERENCES

1. D. Von Wettstein, The formation of plastid structure, Brookhaven Symposia in Biology, 11:138 (1958).
2. G. Akoyunoglou, Biosynthesis of the pigment-protein complexes, in: "Protochlorophyllide Reduction and Greening", C. Sironval and M. Brouers, eds., Martinus Nijhoff/Dr. Junk Publishers, The Hague (1984).
3. G. Akoyunoglou and H. W. Siegelman, Protochlorophyllide resynthesis in dark-grown bean leaves, Plant Physiol. 43:66 (1968).
4. C. Lütz, U. Röper, N. S. Beer and W. T. Griffiths, Sub-etioplast localization of the enzyme NADPH:Protochlorophyllide oxidoreductase, Eur. J. Biochem. 118:347 (1981)
5. S. Murakami and M. Ikeuchi, Biochemical characterization and localization of the 36,000 dalton NADPH-Protochlorophyllide oxidoreductase in squash etioplasts, Progr. Clin. Biol. Res. 102B:13 (1982).
6. J. H. Argyroudi-Akoyunoglou, Effect of cations on the reconstitution of heavy subchloroplast fractions (grana) in disorganized low-salt agranal chloroplasts, Arch. Biochem. Biophys. 176:267 (1976).
7. P. Antonopoulou, J.H. Argyroudi-Akoyunoglou and G. Akoyunoglou, The composition of stroma and grana thylakoid pigment-protein complexes, in: "Protochlorophyllide Reduction and Greening", C. Sironval and M. Brouers, eds., Martinus Nijhoff/Dr. Junk Publishers, The Hague (1984).
8. J. H. Argyroudi-Akoyunoglou and A. Castorinis, Specificity of the chlorophyll to protein binding in the chlorophyll-protein complexes of the thylakoid, Arch. Biochem. Biophys. 200:326 (1980).
9. J. M. Anderson, J. C. Waldron and S. W. Thorne, Chlorophyll-protein complexes of spinach and barley thylakoids, FEBS Lett. 99:227 (1978).
10. J. H. Argyroudi-Akoyunoglou and G. Akoyunoglou, The chlorophyll-protein complexes of the thylakoids in greening plastids of Phaseolus vulgaris, FEBS Lett. 104:78 (1979).
11. J. H. Argyroudi-Akoyunoglou and H. Thomou, Separation of the pigment-protein complexes by SDS-sucrose density gradient centrifugation, FEBS Lett. 135:177 (1981).
12. J. H. Argyroudi-Akoyunoglou and G. Akoyunoglou, On the formation of photosynthetic membranes in bean plants, Photochem. Photobiol. 18:219 (1973).
13. D. I. Arnon, R.K. Chain, B. D. McSwain, H. Y. Tsujimoto and D. B. Knaff, Evidence from chloroplast fragments for three photosynthetic light reactions, Proc. Natl. Acad. Sci. 67:1404 (1970).

14. J. E. Mullet, J. J. Burke and C. J. Arntzen, Chlorophyll-proteins of photosystem I, Plant Physiol. 65:814 (1980).

15. T. Y. Kuang, J. H. Argyroudi-Akoyunoglou, H. Y. Nakatani, J. Watson and C. Arntzen, On the origin of the long-wavelength fluorescence emission band (77°K) from photosystem I, Arch. Biochem. Biophys. 235:618 (1984).

16. J. H. Argyroudi-Akoyunoglou, Cation-induced transformation of the oligomeric to monomeric forms in the pigment-protein complexes of the thylakoid, Photobiochem. Photobiophys. 1:279 (1980).

17. J. H. Argyroudi-Akoyunoglou, Polypeptide composition of the pigment-protein complexes of *Phaseolus vulgaris* thylakoids. Cation-induced disaggregation of oligomeric to monomeric forms correlates with the increase in the ratio F685/F730 in their fluorescence spectra, Progr. Clin. Biol. Res. 102B:277 (1982).

18. J. H. Argyroudi-Akoyunoglou, A. Castorinis and G. Akoyunoglou, Cation-induced increase in the low-temperature fluorescence F685/F730 ratio in detergent solubilized pigment-protein complexes separated by sucrose gradient centrifugation, Photobiochem. Photobiophys. 4:201 (1982).

19. J. H. Argyroudi-Akoyunoglou, The 77K fluorescence spectrum of the photosystem I pigment-protein complex CPIa, FEBS Lett. 171:47 (1984).

20 K. Satoh, P-695 emission form from the purified photosystem II chlorophyll-protein complex, FEBS Lett. 110:53 (1980).

21. G. Akoyunoglou, S. Tsakiris and J. H. Argyroudi-Akoyunoglou, Independent growth of the photosystem I and II units. The role of the light-harvesting pigment-protein complexes, in: "Photosynthesis, Vol. V, Chloroplast Development", G. Akoyunoglou, ed., Balaban Internat. Sci. Services, Philadelphia, Pa (1981).

22. J. H. Argyroudi-Akoyunoglou and G. Akoyunoglou, Supramolecular structure of chlorophyll-protein complexes in relation to the Chla fluorescence of chloroplasts at room or liquid nitrogen temperature, Arch. Biochem. Biophys. 227:469 (1983).

23. J. H. Argyroudi-Akoyunoglou, A. Castorinis and G. Akoyunoglou, Biogenesis and organization of the pigment-protein complexes: Relation to the low-temperature fluorescence characteristics of developing thylakoids, Israel J. Bot. 33:65 (1984).

24. R. Remy, A. Tremolieres, J. C. Duval, F. A. Ambart-Breteville and J. R. Dubecq, Study of the supramolecular organization of the LHCP Chl-protein, FEBS Lett. 137:271 (1982)

25. G. Akoyunoglou, Thylakoid biogenesis in higher plants: Assembly and reorganization, in: "Advances in Photosynthesis Research, Vol. IV", C. Sybesma, ed., Martinus Nijhoff/Dr. W. Junk Publishers, The Hague (1984).

26. P. Homann, Cation effects on the fluorescence of isolated chloroplasts, Plant Physiol. 44:932 (1969).

27. N. Murata, Control of excitation transfer in Photosynthesis. II. Magnesium ion-dependent distribution of excitation energy

between the two pigment systems in spinach chloroplasts, Biochem. Biophys. Acta 189:171 (1969).

28. N. Murata, Control of excitation transfer in Photosynthesis: I. Light-induced change of Chla fluorescence in *Porphyridium cruentum*, Biochim. Biophys. Acta 172:242 (1969).

29. C. Bonaventura and J. Myers, Fluorescence and oxygen evolution from *Chlorella pyrenoidosa*, Biochim. Biophys. Acta 189:366 (1969).

30. P. Bennoun, Correlation between states I and II in algae and the effect of magnesium on chloroplasts, Biochim. Biophys. Acta 368:141 (1974).

31. Y. S. Li, Salts and chloroplasts fluorescence, Biochim. Biophys. Acta 376:180 (1975).

32. N. Murata, H. Tashiro and A. Takamiya, Effects of divalent metal ions on Chla fluorescence in isolated spinach chloroplasts, Biochim. Biophys. Acta 197:250 (1970)

33. S. Izawa and N. E. Good, Effects of salts and electron transport on the conformation of isolated chloroplasts:II. Electron microscopy, Plant Physiol. 41:544 (1976).

34. J. Barber, An explanation for the relationship between salt-induced thylakoid stacking and the Chl fluorescence changes associated with changes in spillover of energy from photosystem II to photosystem I, FEBS Lett. 118:1 (1980)

35. R. J. Ellis, Chloroplast proteins: Synthesis, transport and assembly, Ann. Rev. Plant Physiol. 32:111 (1981)

36. S. G. Siddell and R. J. Ellis, Protein synthesis in chloroplasts. VI. Characteristics and products of protein synthesis *in vitro* in etioplasts and developing chloroplasts from pea leaves, Biochem. J. 146:675 (1975)

37. R. J. Ellis, Protein synthesis by isolated chloroplasts, Biochim. Biophys. Acta 463:185 (1977).

38. R. E. Zielinski and C. A. Price, Synthesis of thylakoid membrane proteins by chloroplasts isolated from spinach, J. Cell Biol. 85:435 (1980).

39. W. Bottomley and P. R. Whitfeld, Cell-free transcription and translation of total spinach chloroplast DNA, Eur. J. Biochem. 93:31 (1979).

40. J. R. Bedbrook, G. Link, D. M. Cohen, L. Bogorad and A. Rich, Maize plastid gene expressed during photoregulated development, Proc. Natl. Acad. Sci. USA 75:3060 (1978).

41. J. K. Hoober, Protein synthesis in chloroplasts, in: "Protein Synthesis", E. H. McConkey, ed., Dekker, New York (1976).

42. J. Feierabend, Cooperation of cytoplasmic and plastidic protein synthesis in rye leaves, in: "Chloroplast Development", G. Akoyunoglou and J. H. Argyroudi-Akoyunoglou, eds., Elsevier/ North Holland, Amsterdam (1978).

43. R. G. Hermann and J. Feierabend, The presence of DNA in ribosome-deficient plastids of heat-bleached rye leaves, Eur. J. Biochem. 104:603 (1980).

44. N. W. Gillham, J. E. Boynton and N.-H. Chua, Genetic control

of chloroplast proteins, Curr. Adv. Bioenerg. 9:211 (1978).

45. N.-H. Chua and G. W. Schmidt, Transport of proteins into mito-chondria and chloroplasts, J. Cell Biol. 81:461 (1979).

46. A. Grossman, S. Bartlett and N.-H. Chua, Energy-dependent up-take of cytoplasmically synthesized polypeptides by chloro-plasts, Nature 285:625 (1980).

47. E.-I. Minami and A. Watanabe, Thylakoid membranes: The trans-lational site of chloroplast DNA-regulated thylakoid poly-peptides, Arch. Biochem. Biophys. 235:562 (1984).

48. G. Akoyunoglou, Assembly of functional components in chloro-plast photosynthetic membranes, in: "Photosynthesis, Vol. V, Chloroplast Development", G. Akoyunoglou, ed., Balaban In-ternational Science Services, Philadelphia, Pa (1981).

49. P. Antonopoulou and G. Akoyunoglou, Changes in the pigment com-position of the thylakoids of *Paseolus vulgaris* during chlo-roplast development, in: "Advances in Photosynthesis Re-search, Vol. IV", C. Sybesma, ed., Martinus Nijhoff/Dr. W. Junk Publishers, The Hague (1984).

50. A. Castorinis, G. Akoyunoglou and J. H. Argyroudi-Akoyunoglou, Correlation between the organization of the pigment-protein complexes and the Chla fluorescence yield of chloroplasts during development in *Phaseolus vulgaris*, Photobiochem. Pho-tobiophys. 4:283 (1982).

51. W. L. Butler and M. Kitajima, Energy transfer between photosys-tem II and photosystem I in chloroplasts, Biochim. Biophys. Acta 396:72 (1975).

52. G. Akoyunoglou, Reorganization of thylakoid components during chloroplast development in higher plants, Progr. Clin. Biol. Res. 102B:171 (1982).

53. J. H. Argyroudi-Akoyunoglou, A. Akoyunoglou, K. Kalosakas and G. Akoyunoglou, Reorganization of the PSII unit in develop-ing thylakoids of higher plants after transfer to darkness, Plant Physiol. 70:1242 (1982).

54. A. Akoyunoglou and G. Akoyunoglou, Mechanism of thylakoid re-organization during chloroplast development in higher plants, Israel J. Bot. 33:   (1984).

55. A. Akoyunoglou and G. Akoyunoglou, Reorganization of thylakoid components during chloroplast development in higher plants. Changes in PSI unit components and in cytochromes, Plant Physiol. (in press).

56. J. Bennett, Biosynthesis of the light-harvesting Chla/b pro-tein. Polypeptide turnover in darkness, Eur. J. Biochem. 118:61 (1981).

57. M. Viro and K. Kloppstech, Expression of genes for plastid mem-brame proteins in barley under intermittent light conditions, Planta 154:18 (1982).

58. L. Lavintman, G. Galling and I. Ohad, Modulation of PSI and PS II organization during loss and repair of photosynthetic ac-tivity in a temperature sensitive mutant of *Chlorella pyre-noidosa*, Plant Physiol. 68:1264 (1981).

SPECIFIC PROPERTIES OF HIGHER PLANT MITOCHONDRIA

Roland Douce, Richard Bligny, Etienne-Pascal Journet,
and Michel Neuberger

Physiologie Cellulaire Végétale, DRF/BV
(U.A au CNRS n° 576) CEN-G and USM-G - 85 X
F 38041 Grenoble, Cedex, France

INTRODUCTION

One of the major functions of mitochondria from all organisms is to provide ATP as the principal energy source for the cell. This is true also of plant mitochondria and it is therefore no surprise that many basic features of mitochondrial membranes have been conserved between animals and plants despite a billion years of divergent evolution. Thus the morphology of plant mitochondria closely resembles that of their animal counterparts, as do their cyt chain, ATPase complex, energy conservation (H^+ ejection) mechanisms and membrane phospholipid composition (Douce, 1985). Presumably these basic features of mitochondrial membranes are essential for their functioning in energy transduction.

Nonetheless, there are many unique features of plant mitochondria which one assumes reflect their functioning in autotrophic metabolism. The unique features of plant mitochondria include the followings : a) cyanide-insensitive respiration ; b) respiratory-linked oxidation of external NADH and rotenone-insensitive oxidation of internal NADH ; c) matrix-located malic enzyme ; d) oxaloacetate transport ; e) carrier-mediated net uptake of nucleotides, NAD, coenzyme A and thiamin pyrophosphate ; f) the size and complexity of their DNA. In addition, the rate of O_2 consumption on a protein basis is much higher than that in animal mitochondria, while fatty acid oxidation is not detectable in plant mitochondria (the bulk of fatty acid oxidation in the plant cell being confined to microbodies) (Douce, 1985).

In plant cells, the active cytoplasm occupies a peripheral shell between the central vacule and a rigid cell wall. The vacuole is a

237

large watery compartment surrounded by a thick membrane (tonoplast) and containing a wide variety of substances harmful to other orga- nelles (flavonoids, various colorless phenolic compounds, etc..) and hydrolases (lipases, proteinases, etc..). When a plant tissue is homogenized in order to isolate mitochondria, cellular compartmen- talization is destroyed and the "vacuolar compounds" are released into the medium with effects ranging from undesirable to devasting depending on the tissue and on the isolation technique. Thus well- tried methods for animal tissues generally cannot be applied direc- tly to plants. Consequently, considerable expertise is required in order to prepare plant mitochondria displaying the same biochemical and morphological characteristics as mitochondria *in vivo*. It is difficult, if not impossible, at present, to prepare 100 % intact and fully functional mitochondria ; it is a simple matter to prepare a "mitochondrial fraction" (i.e. a subcellular fraction containing mi- tochondria). Although many methods that are currently employed yield preparations containing an appreciable proportion of intact mitochondria, they also result in substantial contamination by extra- mitochondrial components (carotenoids, galactolipids, harmful hy- drolases, etc..) and may lead, in consequence, to ambiguous and pos- sibly misleading results.

With the exception of the DNA, most of the above mentioned dif- ferences between animal and plant mitochondria have been known for many years but were often attributed to artefacts generated by da- mage or contamination of mitochondria during their isolation from plant tissues. On the contrary, plant mitochondria purified by iso- pycnic centrifugation in density gradients of modified silica (Per- coll) have practically no contamination by extramitochondrial enzymes and plastid membranes containing galactolipids and carotenoids. Hence, after many years of rigorous testing (it is now possible to obtain preparations of plant mitochondria which display better than 97 % intactness and which are virtually completely free of enzymes and membranes originating from other cell compartments ; Neuburger et al, 1982), we can now be sure that the observed differences in plant mitochondria are representative of real (and often unique) proper- ties of fully functional organelles.

This chapter considers the general principles of the isola- tion of purified mitochondria from plant tissues and set out the unique features of plant mitochondria which one assumes reflect their functioning in autotrophic metabolism.

PREPARATION OF PLANT MITOCHONDRIA

If the investigations to be undertaken do not demand a particu- lar species, difficulties can usually be minimized by careful choice of plant material (cauliflower buds, white and sweet potato tubers, castor bean endosperm, spinach and pea leaves, etc..). Fresh mate-

rial (0.5 Kg) is cut into 3 liters of chilled medium containing 0.3 M
mannitol, 4 mM cysteine, 1 mM EDTA, 30 mM   3-(N-morpholino)-propane
sulfonic acid (Mops) buffer (pH 7.5), 0.6 % (w/v) insoluble polyvi-
nyl pyrrolidone (PVP) and 0.1 % (w/v) defatted bovine serum albumin
(BSA). Acidity, phenolic compounds, tannin formation and oxidation
products lead to the rapid inactivation of mitochondria. These difficulties are usually overcome by adding alkaline buffers, insoluble PVP (Loomis and Battaile, 1966) and cysteine (Bonner, 1967) to
the isolating medium. The uncoupling of oxidative phosphorylation
by free fatty acids formed by the action of acylhydrolase enzymes
during extraction is usually overcome by adding relatively high
concentrations of defatted BSA. Other strategies to minimize the
effect of vacuoles on plant mitochondria are to keep a low ratio of
tissue to breakage medium (Bonner, 1967) and to separate the organelles from soluble elements of the cellular brei as rapidly as
possible (Bonner, 1967). A rapid and convenient method for breaking
open tissue cells is to disrupt the tissue at low speed for 2 sec
in a 1-gallon Waring Blender. Invariably, the more violent the ho-
mogenization, the lower the quality of the isolated mitochondria.
The shear forces must be great enough to break the cells without
stripping the outer membrane from the mitochondria. In other words
tissue disruption should be carried out very gently. The brei is
rapidly squeezed through 8 layers of muslin and 50-μm nylon net.
After an initial low-speed centrifugation the mitochondria are sedi-
mented at 10,000 x g, and, after washing, resedimented at 10,000 x g.
This method, when followed carefully, will produce mitochondria with
a high degree of intactness (fig. 1). The crucial point in the
whole procedure is the grinding of the tissue. In order to obtain
intact mitochondria it is absolutely necessary to restrict the grinding procedure to a minimum. Longer blending improves the yield of
recovered mitochondria but considerably increases the percentage of
envelope-free mitochondria and of mitochondria that have resealed
following rupture and the loss of matrix content. It is clear, therefore, that the final yield of intact mitochondria capable of rapid substrate-dependent O_2 uptake with pronounced respiratory control and ADP/O ratios approaching "theoretical" limits is very low.
A number of recent investigations have been encouraging, however,
in demonstrating the potential usefulness of plant protoplasts for
physiological research (Nishimura et al, 1976). Cell wall materials
are enzymatically degraded during the preparation of protoplasts,
and subsequent gentle disruption of the protoplast plasmalemma is
ideal for isolating intact mitochondria with a high yield
(Nishimura et al, 1982).

    Eventhough the final preparation often contains more than 95 %
intact mitochondria, the material is not necessarily devoid of
various extramitochondrial membranes and soluble enzymes (vacuolar
enzymes, glycolytic enzymes, catalase, lipoxygenase, etc..). In order
to avoid the danger of long and fruitless controversy, the intact
mitochondria obtained by differential centrifugation must be puri-

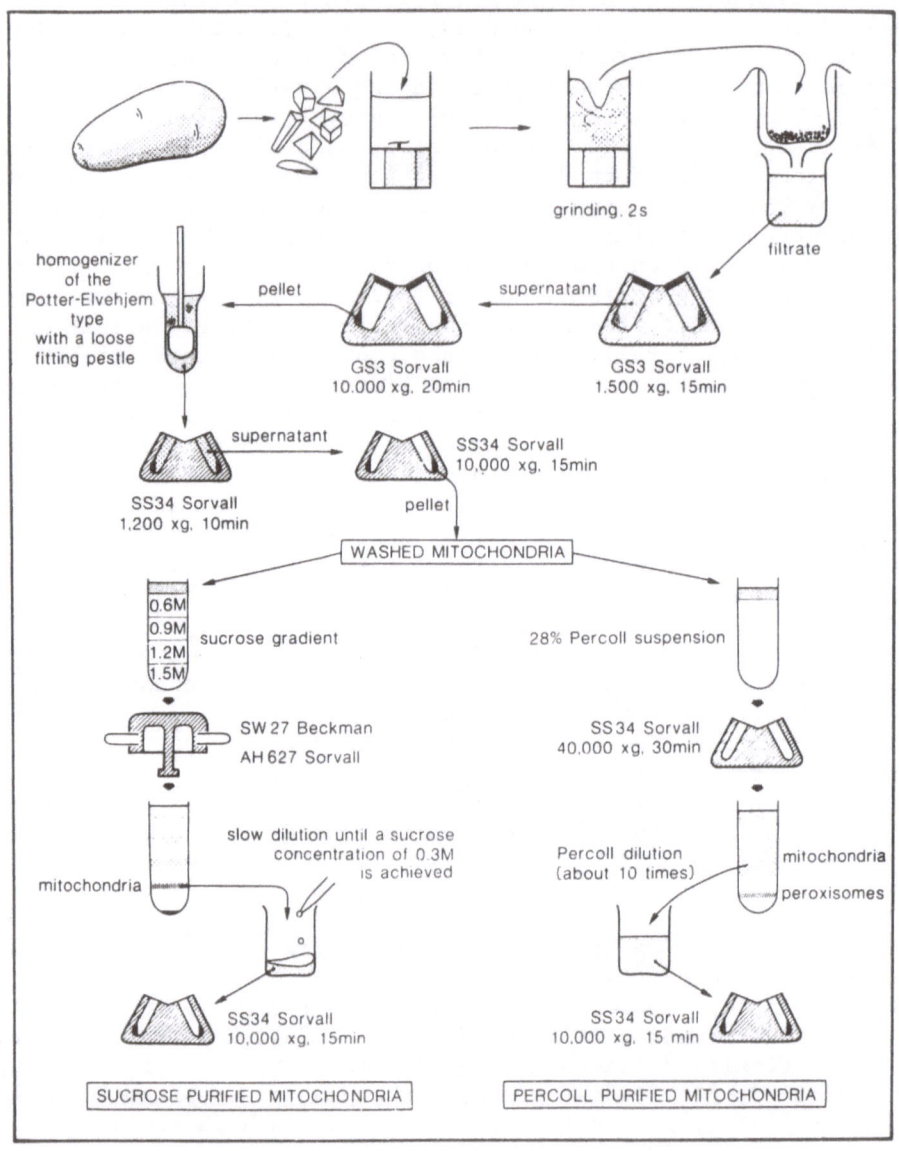

Fig. 1.

Diagram summarizing the preparation of mitochondria
from potato (*Solanum tuberosum*) tubers.

fied by isopycnic centrifugation in a sucrose (Douce et al, 1972)
or silica-sol (Jackson et al, 1979 ; Bergman et al, 1980 ; Neuburger
et al, 1982) gradient). The recent introduction of a nontoxic silica-
sol gradient material, Percoll ^{TM} [by Parmacia Fine Chemicals Ltd,
Uppsala, Sweden (Pertoft and Laurent, 1977)] , has permitted the
development of a rapid purification procedure utilizing iso-osmotic
and low-viscosity conditions. Consequently, Percoll, which consists
of colloidal silica particles of 15-30 nm in diameter
coated with PVP, is nearly an ideal gradients material for the separation of mitochondria from higher plants. Unlike sucrose, Percoll
does not penetrate biological membranes and does not exert large
osmotic effects. Subcellular organelles, therefore, band at densities
lower than on sucrose gradients and densities from 1.04 to 1.10 g/ml
have been recorded for different types of mitochondria (Neuburger
et al , 1982). The general method is to layer a suspension of washed
mitochondria over a density gradient of the silica sol and centrifuge to equilibrium. Mitochondria from material such as potato tubers can be separated on a continuous self-generated gradient. A
centrifuge tube is filled with Percoll medium 28 % (v/v) Percoll ;
300 mM sucrose ; 10 mM phosphate buffer (pH 7.2) ; 1 mM EDTA and
0.1 % (w/v) BSA (the final concentration of Percoll and sucrose
should be carefully determined according to the origin of the mitochondria and the nature of contaminations present in the preparation).Mitochondria are then carefully layered onto the gradient
and fractionation is undertaken for 30 min at 40,000 ^{x} g (the auto-
matic rate controller must be used). Most of the mitochondria appeared as a broad band located beneath a yellow layer at the top of
the gradient. The passage of mitochondria from etiolated tissues
through a Percoll density gradient is an advantageous step because
it not only removes extramitochondrial membranes such as microbodies
and amyloplast membranes containing carotenoids (Douce and Joyard,
1979) but also removes contaminating hydrolases such as lipolytic
acyl hydrolase. In addition, plant mitochondria separated from
contaminating organelles and membrane vesicles on Percoll gradients
showed markedly higher rates of tricarboxylic acid cycle substrate
oxidations than did unpurified mitochondria (Neuburger et al, 1982).
Under these conditions the physiological integrity of isolated mitochondria can be maintained at least for 3 days (Neuburger and Douce,
1983). It seems likely therefore that colloidal silica particles
coated with PVP may act beneficially by binding these hydrolases. In
support of this suggestion, the purification of potato mitochondria
by centrifugation through a discontinuous sucrose density gradient
does not completely remove these contaminating hydrolases (Neuburger
et al, 1982).

UNIQUE FEATURES OF PLANT MITOCHONDRIA

Electron transports

    For many years, the organization of the respiratory chain of

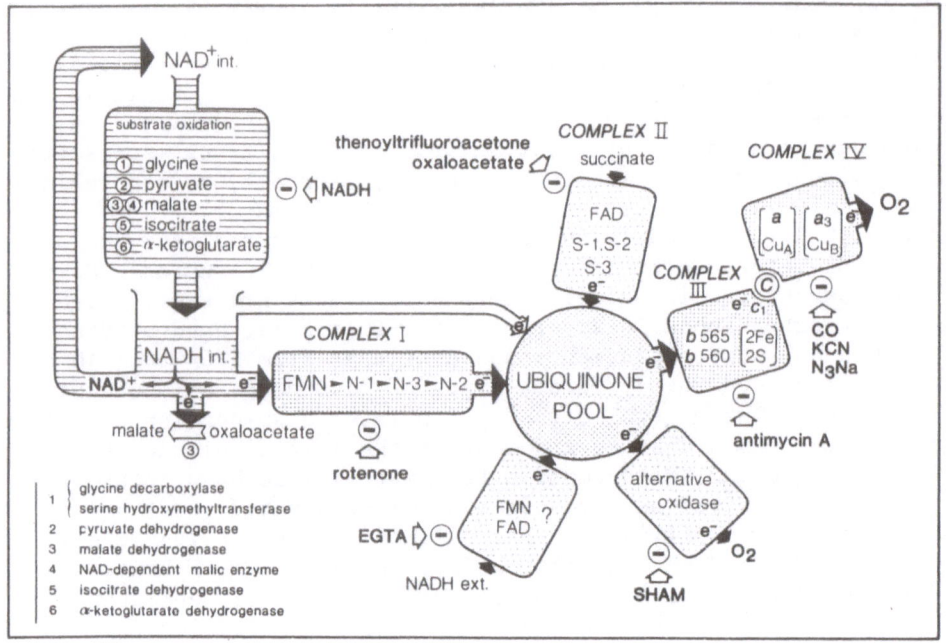

Fig. 2.

Schematic diagram of the components of the higher plant respiratory chain arranged as a continuous sequence from NADH (low potential) generated by various internal NAD^+-linked dehydrogenases to oxygen (high potential), via either complex IV (cytochrome oxidase) or the alternative oxidase.

The ability to oxidize glycine via the respiratory chain is specific for mitochondria from leaf tissues having the C_3 pathway of photosynthesis (C_3 plants). During the course of glycine oxidation, 2 moles of glycine are converted to 1 mole each of serine, NH_3 and CO_2 in a complex reaction catalysed by glycine decarboxylase and serine hydroxymethyltransferase.

plant mitochondria was thought to be very similar to mitochondria from more extensively studied animal sources such as rat liver or beef heart. In fact, the plant mitochondrial cyanide-sensitive electron pathway (i.e. the sequence of electron carriers which mediate the flow of electrons from respiratory substrates to O_2 via cytochrome oxidase) and phosphorylation system appear similar to those found in mitochondria from animal tissues (Douce, 1985). It is clear that this basic system, developed at an early stage of evolution, has been highly conserved throughout the development and divergence of animal and plant species (fig. 2). However, it is now recognized that there are a number of distinct differences between plant and animal mitochondria. These include the cyanide-resistant electron pathway, which is also encountered in the mitochondria of microorganisms (Lloyd, 1974), and the rotenone-resistant electron pathway (Palmer, 1976) (fig. 2). At one time these differences were not felt to be real but rather to be artifacts due to difficulties associated with isolating mitochondria from plant tissues. Fortunately this view is no longer widely held.

Cyanide resistant electron pathway :

Practically all the plant tissues examined so far show a residual respiration in the presence of CN^-, CO or N_3^- (Henry and Nyns, 1975). Cyanide-insensitive tissues generally yield mitochondria that are also cyanide-insensitive. Bendall and Bonner (1971) established that cyanide and antimycin A-insensitive respiration was mediated by an additional electron transport pathway consisting of the same set of dehydrogenases as the normal respiratory chain, but entirely by-passing the cytochromes via a second oxidase (alternative oxidase). In fact, the cyanide-resistant electron transport system consists of a branch from the conventional electron transport system beginning with ubiquinone and terminating with an "alternative oxidase" (Storey, 1976) (fig. 2). In isolated mitochondria the K_m for O_2 of the alternative pathway is found to be somewhat higher than that of cytochrome oxidase but nonetheless quite low (1-2 μM) (Bendall and Bonner, 1971). Substituted benzohydroxamic acids (Schonbaum et al, 1971) inhibited the cyanide-insensitive respiratory pathway in isolated plant mitochondria.

Since the alternative oxidase is indistinct both in its electron paramagnetic resonance and in its spectrophotometric parameters the nature of the alternative oxidase has remained elusive. A further problem encountered in attempts at characterization of the alternative oxidase is the lack of an artificial donor which may be used for the direct assay of the alternative oxidase and the apparent lability of the oxidase. According to Huq and Palmer (1978) and Rich (1978) the alternative oxidase might be a quinol oxidase because plant mitochondria oxidize p-quinols in a cyanide-insensitive hydroxamate-sensitive manner.

The physiological role of the alternative oxidase is still uncertain (Day et al, 1980). Several workers have concluded that energy conservation is not associated with the alternative oxidase and therefore ascribe a thermogenic role to this independent "non-cytochrome" oxidase. In a few cases such as in *Arum* lilies, the inhibitor resistant electron transport is apparently directly related to thermogenic metabolism (Meeuse, 1975). On the day of flowering the alternative pathway becomes operational and starch which has been accumulated up to this stage is burned up in a few hours due to a prodigiously high rate of glycolysis and respiration (Meeuse, 1975). The site of this crisis is generally the sterile part of the inflorescence and the excess of energy that is not used for ATP synthesis is liberated as heat. At climax, this phenomenon can lead to an increase in temperature 15°C above ambient in *Arum maculatum* and even more in *Alocasia pubera*, and the heat releases odoriferous amines that attract insects involved in the pollinisation process. However, in many tissues there is no such correlation and the existence of a cyanide-insensitive pathway appears to be a wasteful energy process. Obviously much yet remains to be done in order to understand how the control of the alternative pathway operates under physiological situations. The mechanism whereby the alternative path is engaged and the extent to which it operates are of overriding importance in the physiological role of the alternative path.

Rotenone-resistant electron pathway :

Lehninger (1964) found that NADH added to liver mitochondria was not oxidized. However, if the mitochondria were gently disrupted by hypotonic swelling, oxidation of NADH was considerably enhanced via the universally distributed rotenone-sensitive respiratory chain-linked NADH dehydrogenase (Complex I). Indeed isotopic studies clearly showed that NADH readily penetrated the outer mitochondrial membrane but exchanged at an insignificant rate with NADH in the matrix compartment (Von Jagow and Klingenberg, 1970). It is now generally accepted that in animal mitochondria the inner membrane is totally impermeable to NADH.

Oxidation of exogenous NAD(P)H by plant mitochondria : in contrast to animal mitochondria the mitochondria of higher plants catalyze a rapid oxidation of exogenous NADH in the absence of added cytochrome c (Bonner, 1967). Exogenous NADH is oxidized by an external NADH dehydrogenase located on the outer surface of the inner membrane (fig. 3). This dehydrogenase is specific for the β-hydrogen of NADH and feeds electrons directly to complex III (Douce et al, 1973) by-passing complex I and the first site of H^+ translocation (see fig. 2). This pathway, which is inhibited by antimycin A, does not seem to be connected with the alternative oxidase (except in the case of mitochondria from *Arum* lilies). Consequently NADH oxidation by this external NADH dehydrogenase, which does not require NADH translocase, is insensitive to rotenone and has an ADP/O ratio similar to that of succinate. Parenthetically plant mitochondria also oxidize exogenous NADPH, apparently via a Ca^{2+}-dependent dehydrogenase located on the outer surface of the inner membrane (Koeppe and Miller, 1972). The metabolic significance of this external dehydrogenase capable of oxidizing cytosolic NADH very rapidly, and present in all the plant mitochondria isolated so far, is unknown. It

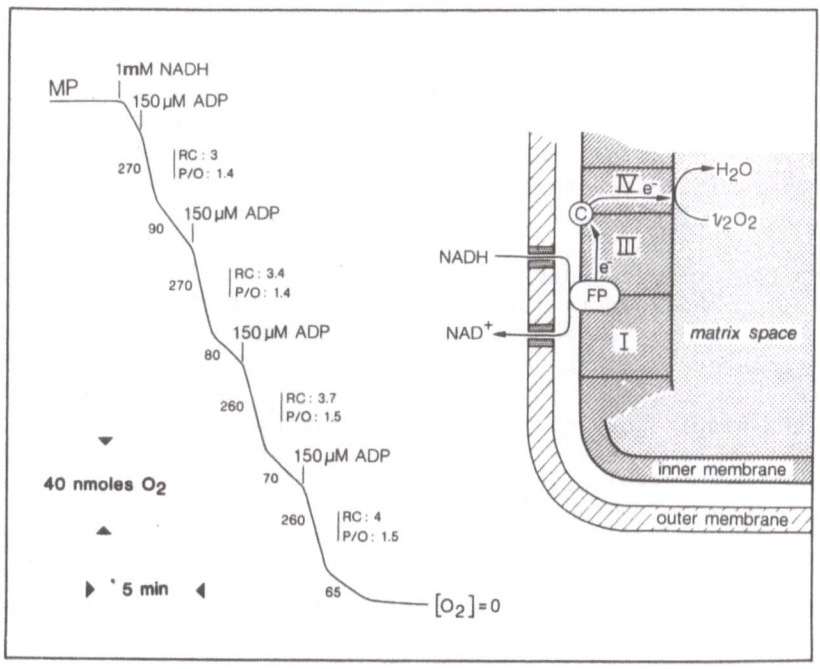

Fig. 3.

Rates and mechanism of NADH oxidation by mitochondria isolated from higher plants.
Note that external NADH is directly oxidized by a specific
dehydrogenase located on the outer surface of the inner
membrane and which does not require NADH translocase. This
dehydrogenase which is not inhibited by rotenone feeds
electron directly to complex III

is clear that this dehydrogenase will favour the conversion of gly-
ceraldehyde 3-P to glycerate 3-P and therefore the forward direction
of glycolysis.

Oxidation of endogenous NADH : the oxidation of endogenous
NADH in plant mitochondria appears to be more complex than its counterpart in mammalian mitochondria. The most obvious indication of
this is that inhibitors such as piericidin A or rotenone, which inhibit the oxidation of endogenous NADH in animal mitochondria by interacting with the iron-sulphur centres associated with complex I,
only cause a partial and sometimes an imperceptible inhibition in
the plant mitochondrial system. By measuring the ATP formation, it

is clear that the rotenone-resistant pathway is not coupled to the first site of ATP synthesis, whereas the rotenone-sensitive pathway is (Palmer, 1976). It seems that plant mitochondria, in contrast with animal mitochondria, possess two internal NADH dehydrogenases on the inner surface of the inner membrane. One of these internal dehydrogenases oxidizes endogenous NADH readily in a rotenone-sensitive manner. This dehydrogenase is therefore coupled to the synthesis of three molecules of ATP and is probably similar to complex I characterized in mammalian mitochondria (fig. 2). Current evidence suggests that this dehydrogenase has an apparent Km for NADH of 8 μM (Møller and Palmer, 1982). We also believe that complex I which operates in close relationship with all the NAD^+-linked TCA cycle dehydrogenases utilizes a common pool of NAD^+ present in the matrix space (Neuburger and Douce, 1983). The second dehydrogenase connected to the respiratory chain via the ubiquinone pool is insensitive to inhibition by rotenone and is coupled to the synthesis of only two moles of ATP (Palmer, 1976). This dehydrogenase in contrast with complex I exhibits a low affinity for internal NADH (Møller and Palmer, 1982) and differs from the rotenone-resistant NADH dehydrogenase associated with the outer face of the inner membrane inasmuch as it is not sensitive to EGTA or Ca^{++} (Møller and Palmer, 1982) (fig. 2). The physiological significance of the rotenone-resistant internal NADH dehydrogenase is not understood. It has been suggested that complex I may be associated with the cyanide-sensitive oxidase, whereas the internal non-phosphorylating dehydrogenase is associated with the cyanide-resistant oxidase providing a totally non-phosphorylating pathway for the oxidation of endogenous NADH when the energy charge is high. Thus the rotenone-resistant dehydrogenase may play a role when the phosphate potential restricts electron flow through the normal respiratory chain. Such an idea is very attractive especially in the case of mitochondria from thermogenic tissues. Again it is clear that the mechanism whereby the rotenone-insensitive pathway is engaged and the extent to which it operates are of the utmost importance in the physiological role of this pathway. The concentration of NADH in the matrix space seems to play an important role in the regulation of the pathways responsible for endogenous NADH oxidation because the affinity of the rotenone-sensitive NADH dehydrogenase for NADH is greater than the affinity of the internal rotenone-resistant NADH dehydrogenase.

Anion transports

In addition to the basic metabolite transport systems found in mammalian mitochondria (phosphate, adenine nucleotide, pyruvate, dicarboxylate and tricarboxylate transports) specific carriers have been found in mitochondria isolated from plant tissues. In particular plant mitochondria possess a specific carrier with high affinity and activity for oxaloacetate and several carriers for the uptake of large molecular weight cofactors such as NAD^+, which have not been described in mitochondria from other organisms.

Oxaloacetate transport

Oxaloacetate is an intermediate in several physiologically important metabolic cycles and sequences, both catabolic and anabolic in nature. Furthermore oxaloacetate may perform a catalytic function in the exchange of reducing equivalents between the mitochondrial and cytosolic compartments of the cell. In mammalian cells it has been assumed that the inner of the two membranes enclosing the mitochondrion may be considered to be impermeable to oxaloacetate under normal physiological conditions, although reports have shown that it can be transported very slowly via either the α-ketoglutarate translocator or the dicarboxylate translocator (for review see Day and Wiskich, 1984). In marked contrast, oxaloacetate has been found to traverse rapidly the membranes of all the plant mitochondria isolated so far (Douce and Bonner, 1972). At an extramitochondrial oxaloacetate concentration of 100 to 500 μM, the influx of oxaloacetate was so severe that NAD^+-linked TCA cycle substrate-dependent O_2 consumption stopped because of the competition for NADH by malate dehydrogenase (fig. 4). Alleviation of respiratory inhibition subsequently occured as the oxaloacetate became reduced. Phthalonate severely restricts oxaloacetate transport (Day and Wiskich, 1984).

For the present, the details of oxaloacetate transport in plant mitochondria remain a mystery but we do believe that the carrier may be specific for oxaloacetate. The extraordinary low half saturation of oxaloacetate transport (~ 20 μM) would make it possible for a very active malate oxaloacetate shuttle to occur between the mitochondrial and the cytosolic compartments of a plant cell under physiological conditions (Douce and Bonner, 1972). Recent measurements in our laboratory using pea leaf mitochondria suggest an uptake rate of close to 1000 nmol/min per mg of protein.

NAD^+ transport

The mitochondrial inner membrane is generally considered to be impermeable to nicotinamide ("pyridine") nucleotides (Von Jagow and Klingenberg, 1970). However, we have recently shown that isolated intact plant mitochondria actively accumulate NAD^+ from the external medium leading to a substantial increase in the matrix concentration of the cofactor and stimulating dehydrogenase and electron-transport activity (Neuburger and Douce, 1983). Plant mitochondria apparently possess a specific NAD^+ carrier, since NAD^+ uptake is concentration dependent and exhibits Michaelis-Menten kinetics (Neuburger and Douce, 1983). Furthermore, the rate of NAD^+ transport is strongly temperature-dependent and the analogue N-4-azido-2-nitrophenyl-4-aminobutyryl NAD^+ (NAP_4-NAD) inhibits (almost completely) net NAD^+ import (Neuburger and Douce, 1983)(fig. 5). It should be emphasized that this transport does not reflect membrane damage during organelle isolation ; the mitochondria in question have been

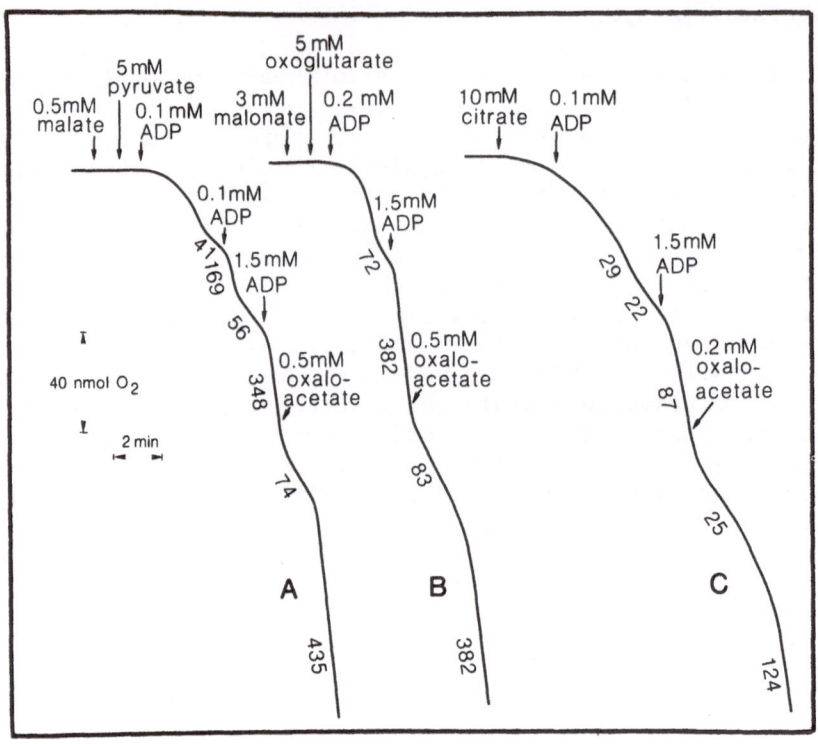

Fig. 4.

Effect of limiting amounts of oxaloacetate on pyruvate
α-ketoglutarate (oxoglutarate) and citrate oxidations
by isolated mitochondria from potato tubers.
Numbers on traces refer to nmol O_2 consumed/min per mg of
protein. Note that when oxaloacetate is added to state
3, a clear inhibition of the respiration rates occurs
which is gradually reversed. Enzymic analysis showed
that, during the time of inhibition, the oxaloacetate
is converted to malate, a conversion that is dependent
on the intramitochondrial NADH generated during the
course of substrate oxidation.

rigorously tested and found to be greater than 90% intact.

The physiological role of this NAD^+ carrier still remains uncertain, but we can speculate that, *in vivo*, the matrix NAD pool is
maintained against a concentration gradient by virtue of a one-way
movement of nicotinamide nucleotides. This net uptake can also ex-

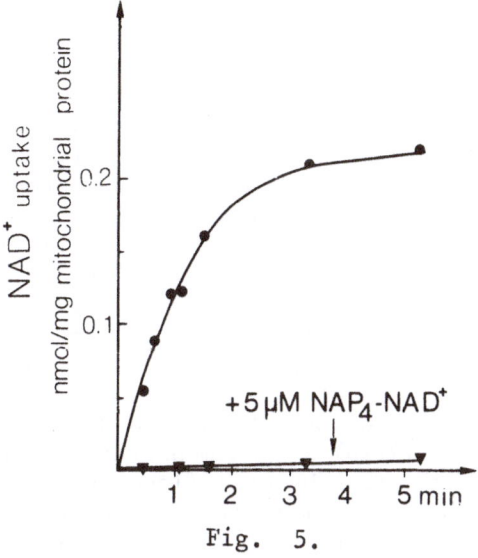

Fig. 5.

Effect of the analogue N-4-azido-2-nitrophenyl-4-aminobutyryl NAD$^+$ (NAP$_4$ - NAD$^+$) on [^{14}C] - NAD$^+$ uptake by potato mitochondria.

[^{14}C] - NAD$^+$ uptake was measured at 25°C by silicone oil filtration. Labelled NAD$^+$ was added at 10 μM and α-ketoglutarate was added as substrate to drive NAD$^+$ uptake.

plain how the total nicotinamide-nucleotide content is established and maintained during mitochondrial proliferation. In addition, we have shown the need for a continued energy supply to maintain NAD$^+$ in the matrix, and perhaps the same conditions are necessary to maintain nucleotides, coenzyme A and thiamin pyrophosphate at high levels. Consequently, the NAD$^+$ carrier could play its part in mitochondrial biogenesis. It is also possible that this carrier has an important regulatory function *in vivo* by allowing manipulation of matrix NAD$^+$ concentration and thus regulating the internal rotenone-insensitive pathway. In support of this suggestion we have found that the NAD carrier also functioned in the efflux of NAD$^+$ from isolated mitochondria. We believe that the NAD$^+$-transporter may play a general role in the coarse regulation of plant respiration during transition from dormancy to a stage of active plant growth and may be particularly important during processes such as seed germination.

The mechanism of malate oxidation

Plant mitochondria in contrast with mammalian mitochondria, readily oxidize malate without the necessity of removing oxaloacetate because they possess a specific NAD^+-linked malic enzyme and a specific oxaloacetate carrier. In the absence of thiamin pyrophosphate, O_2 uptake with malate as substrate is attributed solely to malate dehydrogenase and/or NAD^+-linked malic enzyme. Pyruvate and/or oxaloacetate are therefore the major products formed during the course of malate oxidation (Palmer, 1976). NAD^+-linked malic enzyme was discovered in plant mitochondria by Macrae and Moorhouse (1970) and is not found in most animal mitochondria. This enzyme is specific for L-malate, has an absolute requirement for Mn^{2+} and is characterized by low substrate affinity. It is inhibited by bicarbonate and this inhibition is relieved by NAD^+ and CoASH (Neuburger and Douce, 1980). Macrae (1971b) also demonstrated that the rates and products of malate oxidation by cauliflower bud mitochondria alter in response to relatively small changes in the pH of the incubation medium. The effect of pH on malate oxidation is interpreted in terms of the pH-activity profiles of the NAD^+ requiring malic enzyme and malate dehydrogenase. Thus during the course of malate oxidation, addition of a known amount of NaOH which causes a rapid increase of the pH from 6.5 to 7.5 immediately triggers oxaloacetate formation and stops pyruvate production (fig. 6). Conversely, addition of a known amount of HCl which causes a rapid fall of the pH from 7.5 to 6.5 immediately triggers pyruvate formation (Neuburger and Douce, 1980) and stops oxaloacetate production. The ratio of the products, oxaloacetate and pyruvate, during the course of malate oxidation reflects the balance of the two malate-oxidizing enzymes (Neuburger and Douce, 1980). When the activity of the NAD^+-linked malic enzyme is weakened (high bicarbonate concentration, low CoA concentration, alkaline pH), oxaloacetate is preferentially excreted and there is a decrease in the rate of malate oxidation as the reaction proceeds owing to the accumulation of oxaloacetate and the unfavorable equilibrium of the reaction catalysed by malate dehydrogenase. One of the reasons why potato mitochondria are able to consume O_2 at significant rates, at high pH with malate as substrate in the absence of a system to remove oxaloacetate, is because they excrete oxaloacetate in the external medium. Phthalonate is a potent inhibitor of oxaloacetate uptake and efflux in plant mitochondria ; consequently adding phthalonate to plant mitochondria respiring malate at alkaline pH induces a marked inhibition of O_2 uptake. On the other hand, when the activity of the NAD^+-linked malic enzyme is powerful (low bicarbonate concentration, high CoA concentration, acidic pH) oxaloacetate concentration is maintained at a low level and pyruvate is rapidly excreted. In other words, with all the plant mitochondria isolated so far, whenever the NAD^+-linked malic enzyme activity is weakened, the rate of oxaloacetate production is higher than that of pyruvate. Consequently it is clear that malate dehydrogenase and NAD^+-linked malic enzyme are competing at the level of

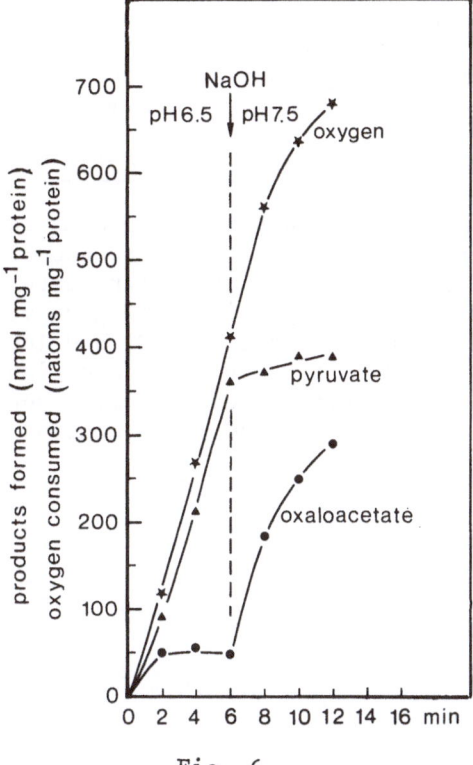

Fig. 6.

Production of pyruvate and oxaloacetate
and O_2 consumption during malate oxidation by mitochondria from potato tubers
as a function of time.

the pyridine nucleotide pool and that the regulation *in vivo* of malate dehydrogenase can be readily accounted for by equilibrium effects alone (Neuburger and Douce, 1980).

The physiological significance of mitochondrial NAD[+]-linked
malic enzyme is unclear. This enzyme is present in all the plant
mitochondria isolated so far and may play a key role in the organic
acid metabolism. When stored reserves of malate within the vacuole
(the available evidence suggests that malate accumulated in vacuoles
is synthesized by a sequence involving phosphoenolpyruvate carboxylase and malate dehydrogenase and rapidly transferred to the vacuole
after synthesis) are mobilized, NAD[+]-linked malic enzyme allows
their complete oxidation via conversion of the malate to pyruvate.
Pyruvate is then converted to acetyl-CoA, which in turn can be com-

pletely oxidized in the normal reactions of the TCA cycle (fig. 7). In other words, malic enzyme allows the conversion of C_4 acids into acetyl-CoA, the normal respiratory substrate without the necessity of supplying pyruvate from glycolysis (Palmer, 1976).

The size and complexity of plant mitochondrial DNA

The size of the plant mitochondrial genome is large and variable and it is generally agreed that the mitochondrial genome of higher plants is the largest one identified to date. Kinetic-complexity measurements of a range of plant mt-DNAs give values of 240 Mdal for pea, 320 Mdal for corn and between 220 and 1660 Mdal in the cucurbitaceae family (Ward et al, 1981). In addition electron microscopy has been used to identify a heterogeneous population of circular mt-DNA ranging from 0.5 μm to 30 μm in contour length and linear

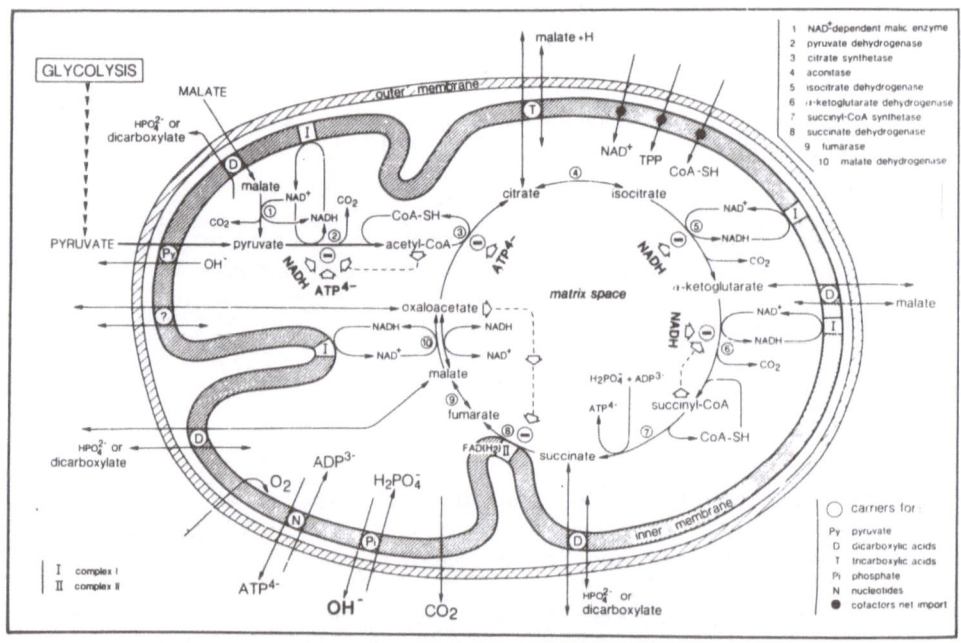

Fig. 7.

Metabolic regulation of the tricarboxylic acid (TCA) cycle in higher plant mitochondria.
This scheme indicates that there exists a concerted action of malate dehydrogenase, NAD^+-linked malic enzyme and cytosolic glycolysis to provide pyruvate and oxaloacetate in the anaplerotic function of the TCA cycle.

molecules of exceptional length. Consequently electron microscopical studies have failed to demonstrate existence of a unique DNA molecule which could represent the complete mitochondrial genome. Furthermore, restriction endonuclease patterns of plant mt-DNA$_s$ are unexpectedly complex, being characterized by a large number of fragments of varying stoichiometries (Quetier and Vedel, 1977). Restriction analysis reveals between 40-68 fragments whose additive molecular weight is appreciably higher than the estimated physical molecular weights. These properties distinguish plant mt-DNA from that of all eukaryotes examined to date. For example, with mammalian or yeast mt-DNA, restriction endonuclease patterns are readily interpretable and the fragment molecular weights sum to a value consistent in each case with the sequence complexity (determined by DNA-DNA reassociation kinetics) and physical molecular weight (estimated by electron microscopy) (Leaver and Gray, 1982). Parenthetically, chloroplast genomes are highly conserved in size (120-180 Kb) and in sequence arrangement (Palmer and Thompson, 1982). To date, no correlation has been demonstrated between mitochondrial genome size and the number of polypeptide products made by isolated mitochondria (Leaver and Gray, 1982). It seems likely, therefore, that higher plant mt-DNA, consists largely of noncoding sequences. In support of this suggestion Stern and Palmer (1984) have demonstrated the widespread presence of chloroplast DNA in the mitochondrial genomes of diverse families of angiosperms.

Based on the first complete restriction map for a higher plant mitochondrial genome, that of chinese cabbage (*Brassica campestris*) (Palmer and Shields, 1984), a general structure for higher plant mt-DNA can be predicted which can account for all the observed phenomena. This genome is organized as three physically distinct circular molecules. The largest circle, 218 Kb in size, bears the entire sequence complexity of the genome (including two copies of an approximately 2 Kb element present as direct repeats separated by 135 Kb and 83 Kb) which constitutes a "master" chromosome while two smaller circles contain distinct 135 Kb and 83 Kb subsets of the master molecule and one copy each of the 2 Kb repeat element. These three chromosomes appear to interconvert via reciprocal recombination across the major repeat element (2 Kb) present on all three molecules. Likewise in maize the structure of the mitochondrial genome has been predicted from restriction mapping data to exist in one of two states, either as a single large circular molecule of approximately 1200 Kb or three circular molecules, two of which arise by intramolecular recombination between paired direct repeats from the larger of the three circles (Lonsdale, 1984). In other words, an active recombinational system will result in sequence "flip-flop" between inverted repeats, or sequence "loop-out" from the interaction of direct repeats (fig. 8). The remarkable similarity of these two unrelated mitochondrial genomes explains many of the puzzling features of plant mitochondrial DNA. Under these circumstances, these very large circular molecules would be extremely unlikely to

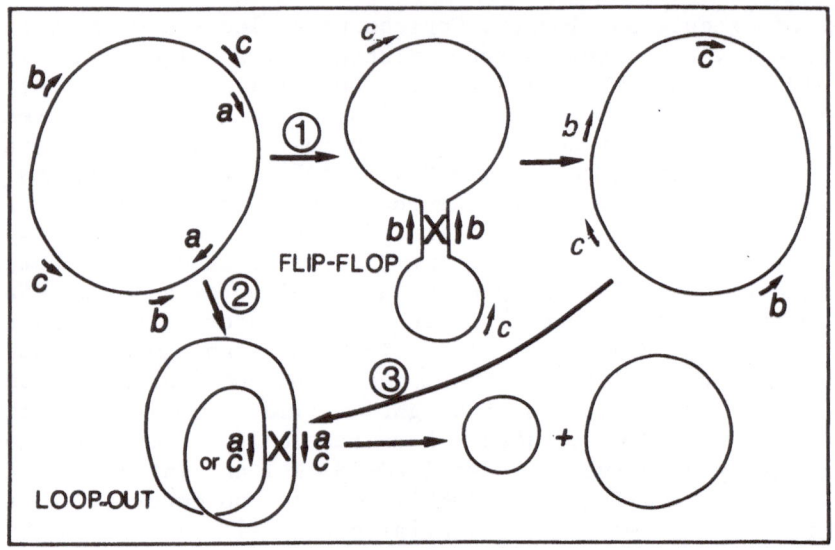

Fig. 8.

Possible arrangements for repeated sequences in the mitochondrial genome.
1) single pair of inverted repeats, recombination leads
to sequence inversion, "flip-flop". 2) and 3) single
pair of direct repeats, recombination leads to the generation of two circular products, "loop-out". The larger
the number of direct and inverted repeats, the more complex the genome organisation becomes.

survive mt-DNA isolation procedures. Therefore it is not unexpected
that isolated mt-DNA consists of a series of fragmented linear molecules, with a low proportion of circular DNA species (Lonsdale, 1984).

CONCLUSION

In vitro, plant mitochondria prove to be biochemically very
flexible organelles. One example of this is substrate metabolism in
mitochondria, where the reoxidation of NADH produced in the matrix
upon operation of various dehydrogenases can be reoxidized equally
well by either the respiratory chain (in which case substrate oxidation will be linked to the energy status of the mitochondria) or
substrate shuttles (in which case the NADH is transferred out of the
mitochondria and substrate oxidation will not be directly linked to
the energy status of the mitochondria) (see fig. 2). One is faced
with another choice at the level of the respiratory chain since
electron transport to O_2 can be either phosphorylating (via cyt

oxidase) or non-phosphorylating (via the cyanide-insensitive alternative oxidase). Another example of the flexibility and complexity of plant mitochondrial metabolism can be found in the active uptake of NAD^+ via a specific transporter (Neuburger and Douce, 1983). Recent evidence show that both influx and efflux of NAD^+ from mitochondria can occur via this carrier and that the NAD^+ transporter operates *in vivo* to alter the concentration of NAD^+ in the mitochondrial matrix. Since the concentration of NAD^+ has such a profound influence on matrix enzyme activity and O_2 uptake by isolated mitochondria, it is potentially a very powerful regulator of respiration *in vivo* and could play an important role in the coarse control of metabolism, particularly during transition from a dormant stage to a stage of active growth and *vice versa*.

Both linear and circular mt-DNA molecules in varying proportions and length distribution have been observed in different plant species. While some of these discrepancies undoubtedly have their origin in preparative artifacts and differences in the analytical techniques, the possibility remains that real differences in the organization of the mitochondrial genome may exist in different plants. This raises the important question of how such heterogeneity is transmitted from generation to generation with regard to biological aspects of plant mt-DNA replication. It is possible, however, as suggested by Lonsdale (1984) that in the specialized tissues where cell division is occurring, mt DNA recombination is suppressed, such that the only replicating DNA species would be the master chromosome.

REFERENCES

Bendall, D.S., and Bonner, W.D., Jr., 1971, Cyanide-insensitive respiration in plant mitochondria, Plant Physiol., 47:236.
Bergman, A., Gardeström, P., and Ericson, I., 1980, Method to obtain a chlorophyll-free preparation of intact mitochondria from spinach leaves, Plant Physiol., 66:442.
Bonner, W.D., 1967, A general method for the preparation of plant mitochondria, Methods Enzymol., 10:126.
Day, D.A., and Wiskich, J.T., 1984, Transport processes of isolated plant mitochondria, Physiol. Vég., 22:241.
Day, D.A., Arron, G.P., and Laties, G.G., 1980, "Nature and control of respiratory pathways in plants : the interaction of cyanide-resistant respiration with the cyanide-sensitive pathway" in: "The Biochemistry of Plants, Vol. 2, Metabolism and Respiration," D.D. Davies, ed. Academic Press, New York.
Douce, R., 1985, Mitochondria in higher plant cells : Structure. function and Biogenesis, Academic Press, New-York.
Douce, R., and Bonner, W.D., Jr., 1972, Oxaloacetate control of Krebs cycle oxidations in purified plant mitochondria, Biochem. Biophys. Res. Commun., 47:619.

Douce, R., and Joyard, J., 1979, Structure and function of the plastid envelope, Adv. Bot. Res., 7:1.

Douce, R., Christensen, E.L., and Bonner, W.D., 1972, Preparation of intact plant mitochondria, Biochim. Biophys. Acta, 275:148.

Douce, R., Mannella, C.A., and Bonner, W.D., Jr., 1973, The external NADH dehydrogenases of intact plant mitochondria, Biochim. Biophys. Acta, 292:105.

Jackson, C., Dench, D.E., Hall, D.O., and Moore, A.L., 1979, Separation of mitochondria from contaminating subcellular structures utilizing silica sol gradient centrifugation, Plant Physiol., 64:150.

Henry, M.F., and Nyns, E.J., 1975, Cyanide-insensitive respiration. An alternative mitochondrial pathway, Sub. Cell. Biochem., 4:1.

Huq, S., and Palmer, J.M., 1978, Oxidation of durohydroquinone via the cyanide-insensitive respiratory pathway in higher plant mitochondria, FEBS Lett., 92:317.

Koeppe, D.E., and Miller, R.J., 1972, Oxidation of reduced nicotinamide adenine dinucleotide phosphate by isolated corn mitochondria, Plant Physiol., 49:353.

Leaver, C.J., and Gray, M.W., 1982, Mitochondrial genome organization and expression in higher plant, Ann. Rev. Plant Physiol., 33:373.

Lehninger, A.L., 1964, "The mitochondrion", Benjamin, New-York.

Lloyd, D., 1974, "The mitochondria of microorganisms", Academic Press, London.

Lonsdale, D.M., 1984, A review of the structure and organization of the mitochondrial genome of higher plants, Plant molecular Biology, in press.

Loomis, W.D., and Battaile, J., 1966, Plant phenolic compounds and the isolation of plant enzymes, Phytochemistry, 5:423.

Macrae, A.R., 1971, Effect of pH on the oxidation of malate by isolated cauliflower bud mitochondria, Phytochemistry, 10:1453.

Macrae, A.R., and Moorhouse, R., 1970, The oxidation of malate by mitochondria isolated from cauliflower buds, Eur. J. Biochem., 16:96.

Meeuse, B.J.D., 1975, Thermogenic respiration in aroids, Annu. Rev. Plant Physiol., 26:117.

Møller, I.M., and Palmer, J.M., 1982, Direct evidence for the presence of a rotenone resistant NADH dehydrogenase on the inner surface of the inner membrane of plant mitochondria, Physiol Plant., 54:267.

Neuburger, M., and Douce, R., 1980, Effect of bicarbonate and oxaloacetate on malate oxidation by spinach leaf mitochondria, Biochim. Biophys. Acta, 589:176.

Neuburger, M., and Douce, R., 1983, Slow passive diffusion of NAD^+ batween intact isolated plant mitochondria and suspending medium, Biochem. J., 216:443.

Neuburger, M., Journet, E.P., Bligny, R., Carde, J.P., and Douce, R., 1982, Purification of plant mitochondria by isopycnic centrifugation in density gradients of Percoll, Arch. Biochem. Biophys., 217:312.

Nishimura, M., Graham, T., and Akazawa, T., 1976, Isolation of intact chloroplasts and other cell organelles from spinach protoplasts, Plant Physiol., 58:309.

Nishimura, M., Douce, R., and Akazawa, T., 1982, Isolation and characterization of metabolically competent mitochondria from spinach leaf protoplasts, Plant Physiol., 69:916.

Palmer, J.M., 1976, The organization and regulation of electron transport in plant mitochondria, Ann. Rev. Plant Physiol., 27:133.

Palmer, J.D., and Thompson, W.F., 1982, Chloroplast DNA rearrangements are more frequent when a large inverted repeat sequence is lost, Cell, 29:537.

Palmer, J.D., and Shields, C.R., 1984, Triparticle structure of the Brassica campestris mitochondrial genome, Nature, 307:437.

Pertoft, H., and Laurent, T.C., 1977 "Isopycnic separation of cells and cell organelles by centrifugation in modified colloidal silica gradients", in: "Methods of cell separation, N. Catsimpoolas, ed., Plenum, New-York.

Quetier, F., and Vedel, F., 1977, Heterogeneous population of mitochondrial DNA molecules in higher plants, Nature, 268:365.

Rich, P.R., 1978, Quinol oxidation in Arum maculatum mitochondria and its application to the assay, solubilisation and partial purification of the alternative oxidase, FEBS Lett., 96:252.

Schonbaum, G.R., Bonner, W.D., Jr., Storey, B.T., and Bahr, J.T., 1971, Specific inhibition of the cyanide-insensitive respiratory pathway in plant mitochondria by hydroxamic acids, Plant Physiol., 47:124.

Stern, D.B., and Palmer, J.D., 1984, Extensive and widespread homologies between mitochondrial DNA and chloroplast DNA in plants, Proc. Natl. Acad. Sci. USA, 81:1946.

Storey, B.T., 1976, Respiratory chain of plant mitochondria XVIII. Point of interaction of the alternate oxidase with the respiratory chain, Plant Physiol., 58:521.

Von Jagow, G., and Klingenberg, M., 1970, Pathways of hydrogen in mitochondria of Saccharomyces carlsbergensis, Eur. J. Biochem., 12:583.

Ward, B.L., Anderson, R.S., and Bendich, A.J., 1981, The size of the mitochondrial genome is large and variable in a family of plants (Cucurbitaceae), Cell, 25:793.

FAST MEMBRANE TRANSFORMATION IN A FLASHING ENDOPLASMIC RETICULUM

Jean-Marie Bassot

Laboratoire de Bioluminescence
C.N.R.S. 91190 Gif sur Yvette, France

Action potentials are known to propagate along plasma membranes, in axons as well as conducting epithelia[1] . Synaptic transmission or electrical coupling via gap junctions allows cell to cell communication[2,3] . Nevertheless, in the muscle system, the existence of dyad junctions coupling the transverse tubules of the plasma membrane with terminal saccules of the sarcoplasmic reticulum indicates that intracellular membrane pathways can also carry a fast signal.

We present here a new example of conducting endoplasmic reticulum. It responds to electrical stimulation with flashes of bioluminescence. This remarkable property, as well as a related fluorescence, opens up the possibility of accurate studies in vivo. The response is easy to measure photometrically and also reveals its own microsources[4] . We provide evidence that dyad junctional complexes are an absolute requirement for microsources to flash, that they are optional, and that they differentiate and dedifferentiate in milliseconds.

The system in question is in the elytra of scale-worms. Among the extreme diversity of bioluminescent reactions and light organs indicative of multiple evolutive origins[5,6,7,8], the scale-worm system exhibits a particularly suitable set of favorable features [4] .

Scale-worms.

Scale-worms are Polychaete Annelids belonging to the family Aphroditidae and the sub-family Polynoïnae. Numerous species are common on European and American sea shores. Some live under

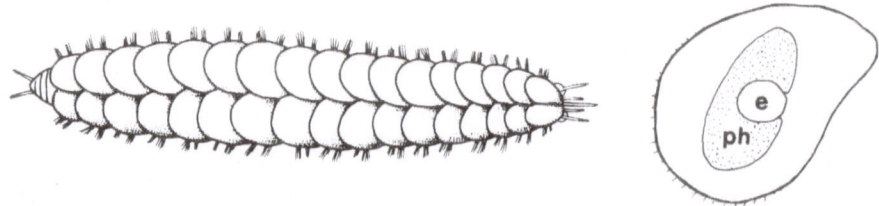

Figure 1. A scale-worm, Harmothoe lunulata, with its double row of
scales or elytra. On each elytrum (right) the photogenic area ph
extends around the disc of insertion of the elytrophore e.

stones, other are commensals of various invertebrates such as sea
urchins, starfish or other annelids. The major morphological
characteristic of scale-worms is the presence on their back of a
double row of scales or elytra, which are parapodial cirri that
have become flat, often transparent discs (Fig.1). Some species
raise their larvae under the shelter of the elytra.

In certain species, the elytra are bioluminescent [9]. When the
animal is disturbed, waves of light illuminate its back,
flickering on each elytrum. Some elytra may detach by autotomy.
These isolated scales emit rhythmically long series of flashes.
The striking fact about these autonomous emissions is the
extremely gradual but nevertheless very large variation in the
intensity of the successive flashes[10]. Intensities first increase
during a period of facilitation and then decrease throughout a
longer period of decay (Fig.4).

The behavioral significance of this modulated rhythmic
flashing, be it predator avoidance or display in mating, is at
present only conjectural[6]. Nevertheless, the question of the
physiological control of the flash and of its modulation during
facilitation is in itself of extreme interest in this isolated and
relatively simple system. The bioluminescent elytrum is a model
for the study of fast cellular activities. It will be shown that
the microsources as well as the priming mechanism of their
activation are indeed intracellular, and that the problematics
concerned are completely dependent on membrane conduction,
membrane coupling and membrane behavior, in highly dynamic
processes.

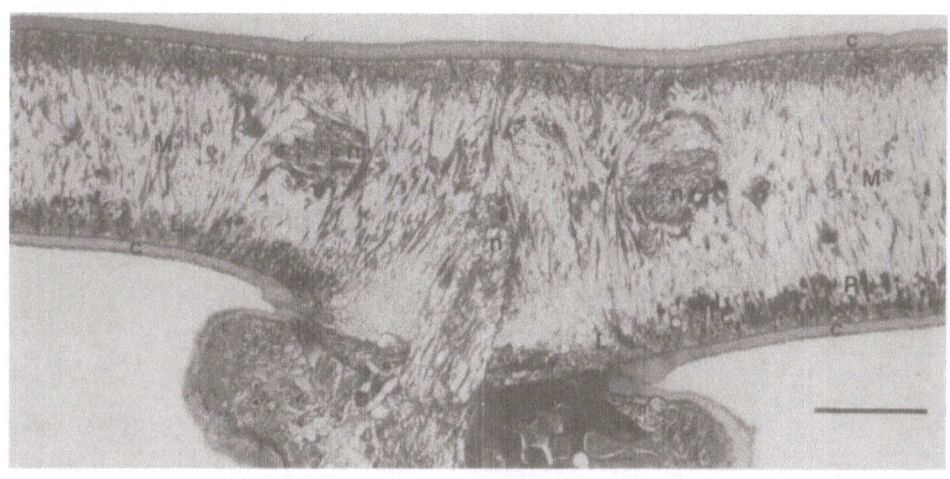

Figure 2. Acholoe astericola, transversal section of an elytrum; epon, toluidine blue. e, elytrophore; c, cuticle; n, nerve; S, epithelium of the upper face; L, epithelium of the lower face, with photogenic cells P characterized by dark photosomes. In the wide extracellular middle compartment M, pillars arising from both epithelial faces intercross. Scale : 100 µ. Document courtesy A. Bilbaut.

Structure

Elytra consist of a single layer epithelium, covered with a continuous cuticle. (Fig.2). Adjacent epithelial cells are linked by desmosomes, septate junctions and numerous gap junctions. The cell bodies are prolonged by several long expansions filled with fibrils, the pillars, which cross the wide extracellular space constituting the middle compartment and are anchored by large desmosomes in the invaginations of the epithelial cells on the opposite face of the elytrum. A nervous ganglion, or more precisely a plexus, is situated in the elytrophoral disc; its horizontal branches pass through the framework of crisscrossing pillars arising from each epithelial surface and connect sensili at the periphery of the elytrum. There are no synapses or direct connections between neurons and epithelial cells, except at the border of the elytrophoral disc, where neuronal processes have been observed intermingled with the nearest epithelial cells[11]. This could account for the transmission from neuronal to epithelial conduction.

The photogenic area is an elliptical or reniform territory of epithelial cells belonging to the lower face of the elytrum. In

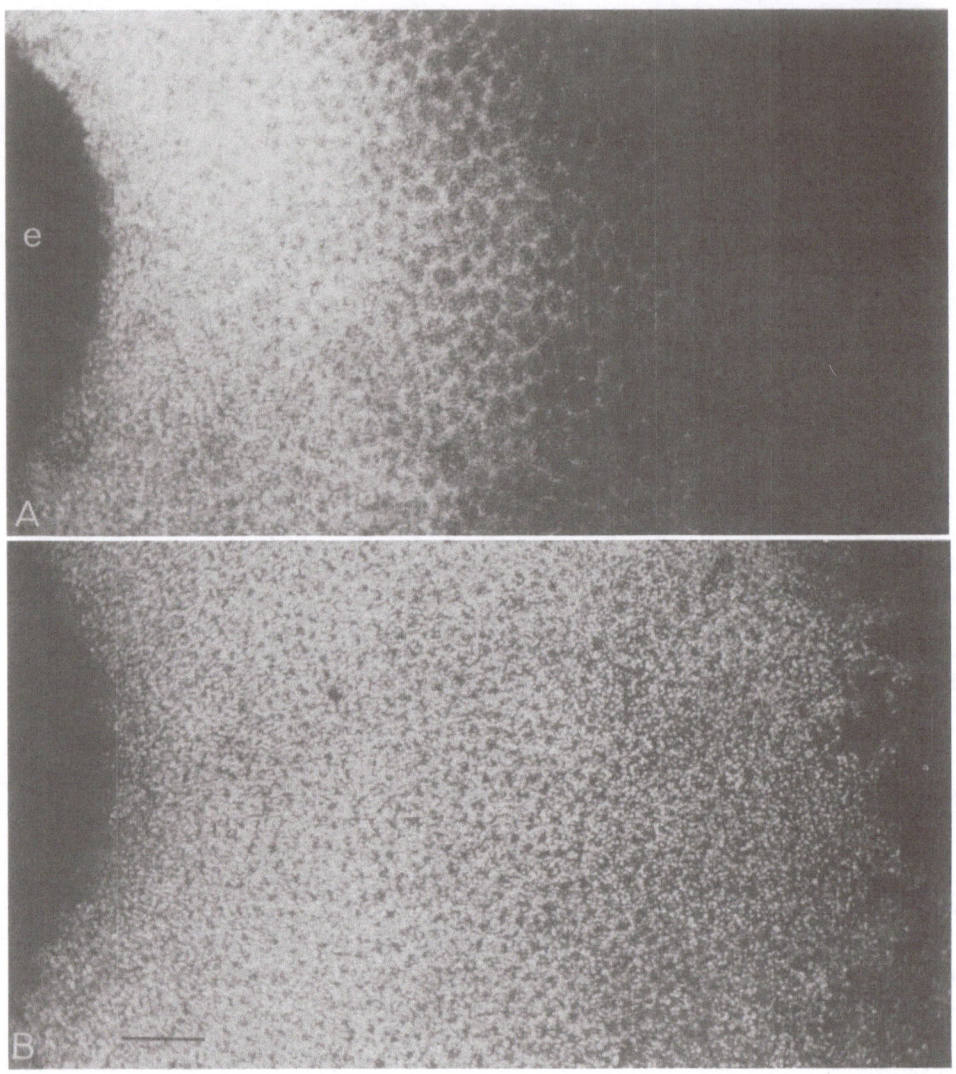

Figure 3. Harmothoe lunalata. Part of the photogenic area, in whole and living isolated elytra. e, elytrophoral disc. The photosomes are sharply and specifically revealed by their yellow-green fluorescence under UV illumination. Fluorescence intensity increases with each new flash of bioluminescence (Fig.5). Picture A was taken after the first ten flashes of a typical emission, and shows the centrifugal gradient of activity. At the periphery of the photogenic area, the outer rings of photosomes, in each cell, form a mosaic pattern. Picture B was taken in a similar elytrum at the end of a typical emission. All the photosomes have then their maximal fluorescence. Scale : 100 μ. Exposure time : 20 sec.

most species of scale-worms, it extends excentrically around the elytrophoral disc. Depending on the species, it comprises a few hundred to several thousand cells. The outstanding characteristic of the photocytes is that each one contains 20 to 30 photosomes, 1 to 8 μ, in diameter, which are spherical or polyhedral. These specific organelles are histochemically rich in protein and sulphydril groups. The best way to observe them is on whole and living isolated elytra, under deep blue or UV illumination. The photosomes are then sharply and individually defined by their intrinsic fluorescence, in the practically bidimensional preparation of the single layered epithelium[13] . In each photocyte, the photosomes are arranged along two concentric rings around the nuclear region. In some preparations, outer rings of adjacent cells form a hexagonal pattern. By all standards, the living elytrum is one of the most beautiful and sharply defined preparations observable in fluorescence microscopy (Fig.3).

Paracrystals

With electron microscopy, photosomes reveal an extraordinary organization[14] . They are made up of membrane tubules, 200 Å in diameter, regulary and repeatedly curved along a straight axis. The repetition of this pattern in parallel axes, forming planes which themselves are either parallel or intersect at 60° results in a highly complex three dimensional lattice, which develops a considerable surface of membrane in relation to the volume of the photosome (Fig.13).

Photosome tubules are in continuity with cisternae of smooth or rough endoplasmic reticulum, and sometimes with the outer membrane of the nuclear envelope. Thus, they are undoubtedly part of the endoplasmic reticulum (ER). In the paracrystalline array of the ER in the photosome (PER), the membrane separates two compartments of comparable volume, one intracisternal and the other cytosolic. On freeze fractures, the tubules of PER exhibit numerous particles.

PER with a similar ultrastructure have occasionally been observed in a variety of plants invertebrate cells (often in the eyes), in vertebrate and even in human pathological tissues such as liver (see 15 for references) but their significance is always enigmatic. Nevertheless, two of these scattered examples are particularly interesting. One is precisely found in the eyes of scale-worms of both luminous and non luminous species, and more generally in all the Aphroditidae. The lens of their eyes consists of apposed distal pigment cell elements filled with PER[16]. The presence of similar paracrystals in the photoemitter and photoreceptor organs of the same animal remains puzzling. The other example of such a PER has been described in an ascidian (prochordates) muscle[17,18] . The sarcoplasmic reticulum (SR) may

263

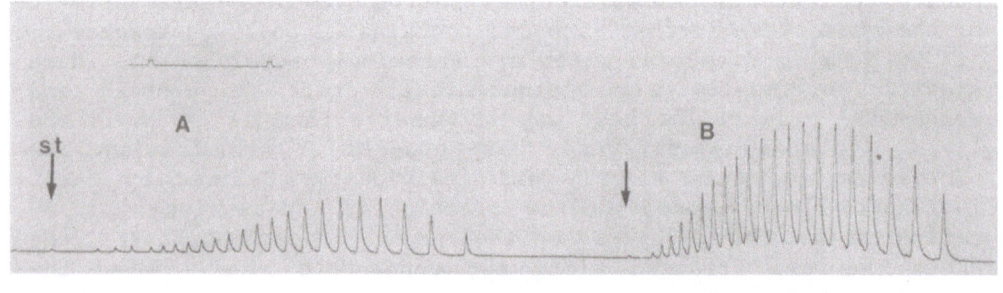

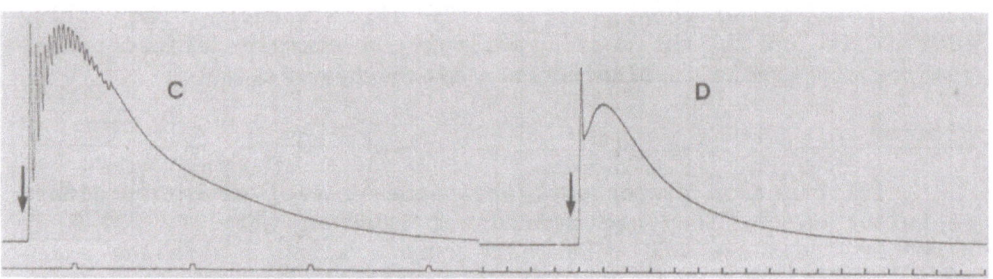

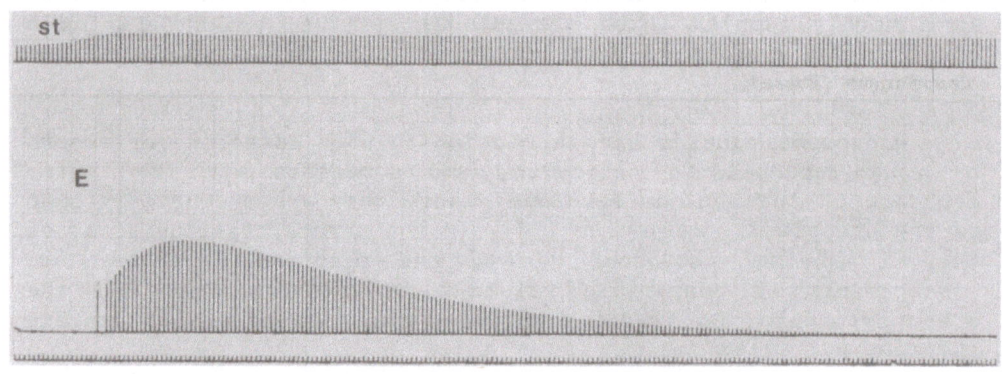

Figure 4. Different modes of luminous responses on the elytra.
A and B : autonomous rythmic flashing induced by a strong (15 v)
electrical shock (st, arrow). Scale : 1 sec.
C and D : explosive responses to a strong stimulation, electrical
(15v) in C, chemical (.1M KCl) in D, with superimposed rythmic
flashing in C.
E : Successive flashes elicited by a juxtathreshold electrical
stimulation st (here 1.3 v) repeated every second. The series of
flashes constitute a typical emission.
Note that all types of responses begin by a period of
facilitation.

form PER which in that case are called multilamellar aggregates. These are connected to conventional SR arrays, themselves connected to subsarcolemnal cisternae running parallel to the plasma membrane in peripheral dyad junctions.

Flashes

Elytra are easy to isolate from anesthetized animals (they regenerate in 10-20 days[19]. Isolated elytra are a convenient preparation, easy to manipulate, which remains alive for several days in sea water. When stimulated by a strong electric shock, an isolated elytrum emits a strong, explosive response which completely exhausts its photogenic capacities within 2 or 3 sec. Such strong stimulation sometimes triggers the pacemaker activity which results in an autonomous rhythmic flashing similar to the one displayed by autotomized elytra (Fig.4).

When weak electrical stimulation is used and progressively increased to reach the threshold value, the response is a single flash. When repeated, such stimulation can elicit hundred of flashes as unitary responses until the complete exhaustion of the system, which does not recover in isolated elytra.

The striking feature of such series of unitary responses, called <u>typical emissions</u>[20], is again the progressive but large variation in the amplitude of the successive flashes. During the period of facilitation, the duration of the flashes also increases progressively, while the response lag time and the ascending slope remain constant. On the contrary, during the period of decay, the lag time increases and ascending slopes decrease exponentially (Fig.6). The general shape of the flashes, as well as the initial lag time and flash duration, differ greatly and unpredictably depending on the elytra, even in the same individual. Lag times range from 2 to 20 msec, and flash durations from 40 to 200 msec. Although peaks can be single, double or irregular in shape, the successive flashes in a given typical emission are all of the same particular shape.

The rate of facilitation, and therefore the relative duration of the facilitation period, is an exponential function of the frequency of the repetitive stimulation. Facilitation is negligible at .1 Hz, but maximal at 1 or 2 Hz. At higher frequencies, up to 15-18 Hz, summation occurs in addition to facilitation, and the emission looks like the explosive responses to strong electrical shocks.

When an emission which starts with repetitive stimulation at, for example 1 Hz, is interrupted for a while and then restarted with the same repetitive stimulation, the new sequence always begins with a period of facilitation. This refacilitation occurs whenever the period of interuption takes place during the typical

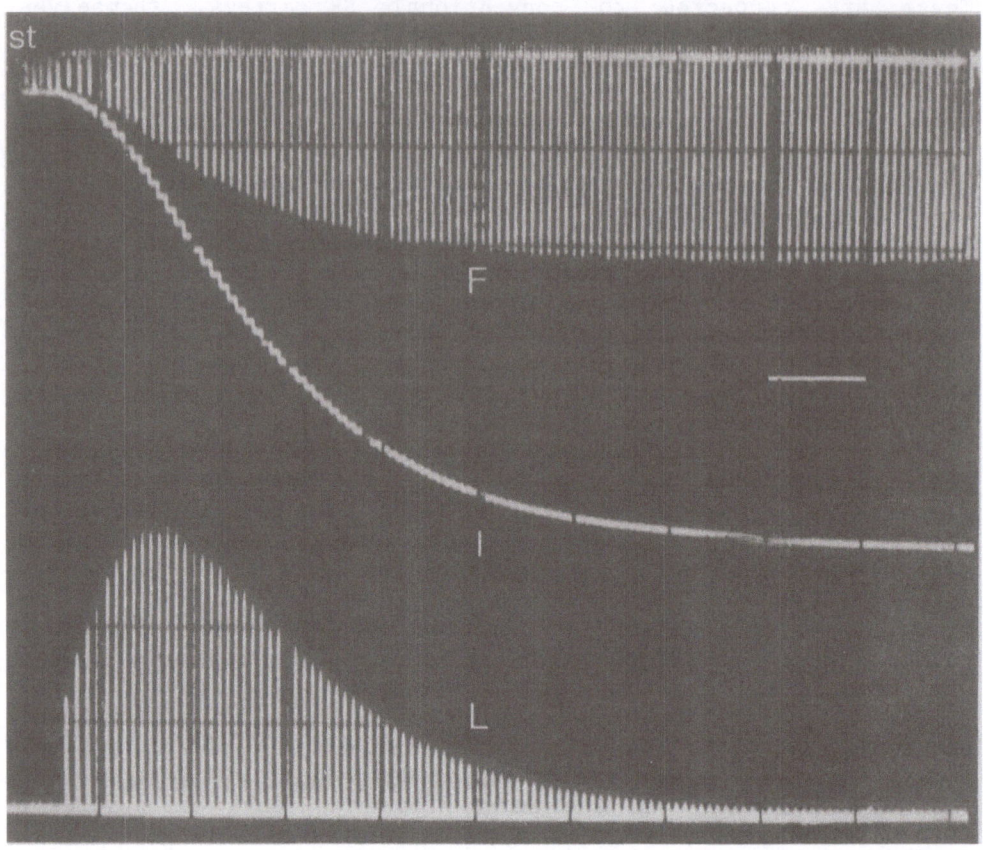

Figure 5. Fluorescence intensity F, measured between each flash L of a typical emission stimulated every second st, increases like the integration I of the luminous activity. Scale : 12 sec.

emission, even during the period of decay. Here again, its rate is exponentially related to the duration of the interruption period[21][22].

Fluorescence

The specific fluoresence of photosomes (max : 520 nm) has the same green color as bioluminescence[23][24] . Nevertheless, unstimulated elytra are completely devoid of green fluoresence, and are referred to as "virgin" elytra. Fluorescence appears with the first bioluminescent response and increases with each flash. Fluorescence intensity is a function of the amount of luminescence previously emitted[13] (Fig.5). Since it is stable at rest (except for the photochemical decay consequent to UV excitation) it measures the past luminescence activity and allows estimation of the amount of photogenic reserves still available[21][22] .

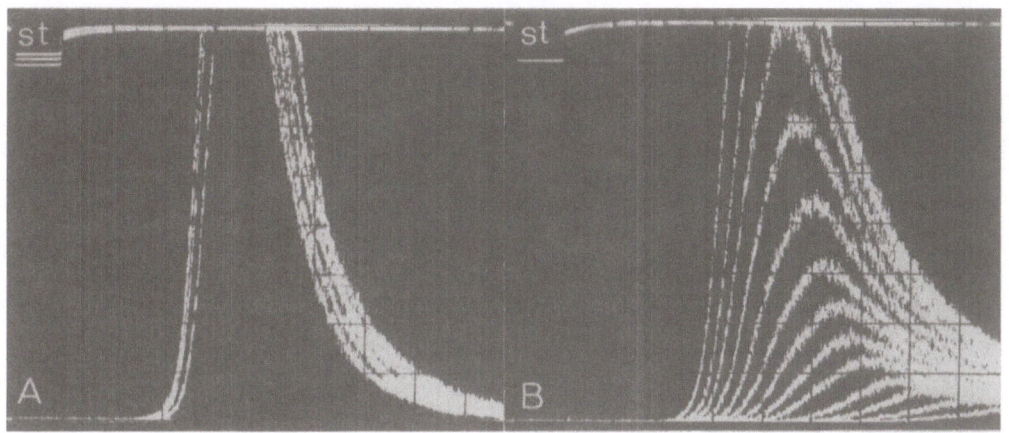

<u>Figure 6</u> A : the first 5 flashes of a typical emission stimulated at 1 Hz. <u>st</u>, 10 msec electrical stimulation.
B : same typical emission. Flashes of the period of decay. Note the progressive modification of the ascending slopes and of the delays.

Spatial and temporal correlation

To understand what determines flash modulation, the events occurring in the photogenic area during bioluminescence activity will now be observed under the microscope, either by luminescence or fluorescence. In practice, an image intensifier is indispensable. Observations, sequential pictures and video recording show that :

1) Photosomes are indeed the microsources of both bioluminescence and fluorescence. They can flash repeatedly and their fluorescence gradually increases. Photosomes can therefore be considered as the units of activity of the elytral system. In a virgin elytrum, they are all loaded with a full <u>photogenic charge</u> which decreases, probably exponentially, flash after flash.

2) During a flash, the photogenic area does not fire synchronously, but is traversed by a <u>sweep of ignition</u>. The speed and trajectory of this wave determine the duration and shape of the flash. Ignition starts at the border of the elytrophoral disc; then, depending on the elytra, it propagates either radially, or in a circular motion which lengthens the flash and makes "bumps" in it, or doubles its if the trajectory is sinuous. The trajectory is repeated flash after flash, whatever its shape.

3) During the period of facilitation, the <u>number</u> of photosomes which fire during a flash increases with each response. This crucial observation, which correlates facilitation with recruitment, was first made on species with a particularly large photogenic area, such as <u>Acholoe astericola</u> or <u>Polynoe scolopendrina</u>[21,25]

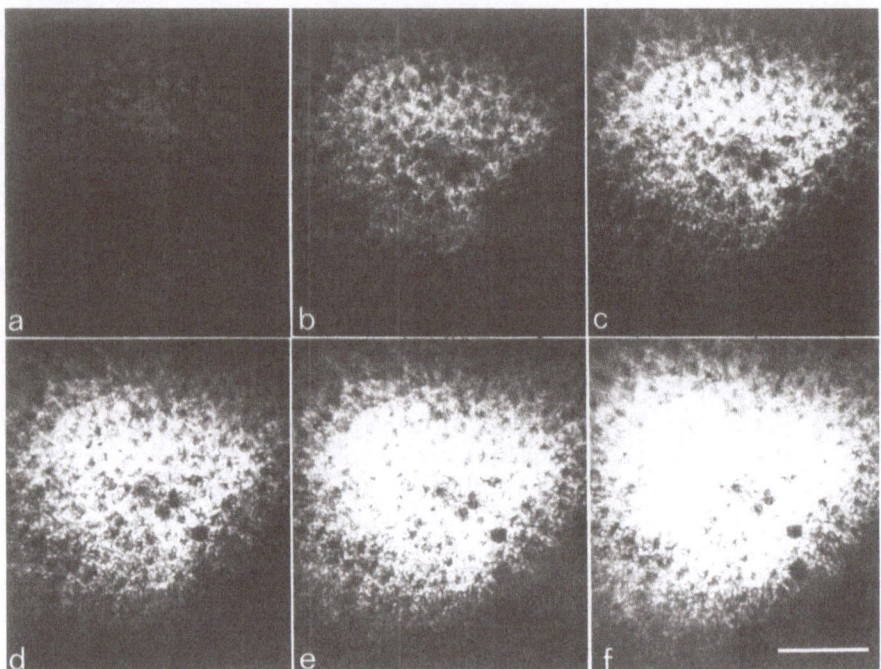

Figure 7. Fluorescence pictures of a field of photogenic cells. Pictures were taken between the first seven flashes of a typical emission stimulated at 1 Hz, through microscope and image intensifier, which allowed a 10 msec exposure time. From a to f, in correlation with facilitating responses, note the progressive extension of the active territory. The mosaic pattern made by the outer rings of active photosomes, in each photocyte, is also progressively filled with photosomes of inner rings which are set into activity. Scale : 100 μ.

Even at low magnification, one can see that only a small territory of the photogenic area lights up during the first flash. Then, step by step, this active territory expands centrifugally with each flash until it reaches the limits of the photogenic area at the end of the period of facilitation. At that time, all the photogenic cells and all their photosomes are active, but some are at their 30th or 40th response, while others have only just started to respond. During the period of decay, it is thus normal

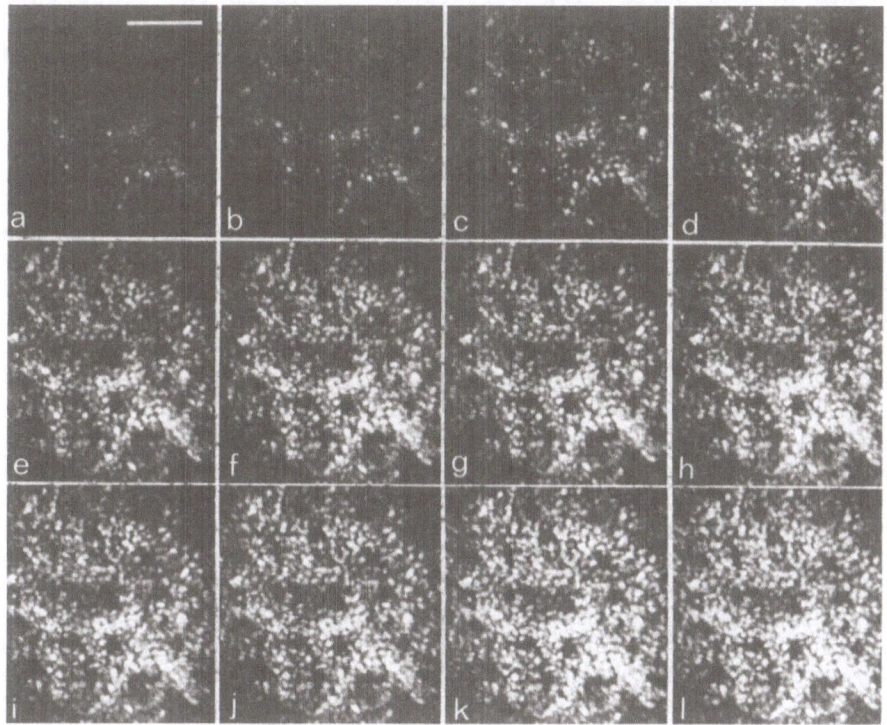

Figure 8. As in Fig.7, but at higher magnification (scale : 20 μ). Fluorescence pictures of the same photogenic field taken between the facilitating responses of a typical emission. Facilitation is correlated with an intracellular recruitment of active photosomes.

to observe similar progressive enlargement of a dark, exhausted territory. Naturally, this enlargement is correlated with the increase in the delay and with the decrease of the ascending slopes already noted.

Recently[22], observation of the same species and several others (particularly Harmothoe lunulata) at higher magnification has shown that the territorial extension just described is in fact a peculiar case of the most fundamental mechanism of recruitment, which is basically intracellular. In a given photogenic cell within the active territory, the first flash is due to the activity of a few photosomes of the outer ring. During the next falsh, the same photosomes fire again, plus neighbouring ones. Response after response the recruitment proceeds until all the photosomes of the outer ring participate; this is the stage when the fluorescence photosomes form beautiful hexagonal patterns. Later, the photosomes of the inner ring are progressively involved in this activity (Figs.7,8).

Thus, three parameters combine to modulate the basic response to stimulation. The amount of light emitted depends on the number of photosomes responding and the sum of their individual photogenic charges. For a given amount of light, the ratio of the intensity of the response to its duration, as well as the shape of the response are directly related to the speed and trajectory of the ignition wave.

Observation of recruitment implies that, in a given photocyte at a given time, photosomes can be either coupled and respond to stimulation, or uncoupled and unresponsive. The experiments on stimulation frequency and interrupted emissions indicate that the coupled state has a limited lifetime, which is basically of the order of 1 sec; that this lifetime can be lengthened up to about 10 sec by reinforcement; and that coupling, the action by which uncoupled photosomes become coupled in the propagated process of recruitment, occurs in a matter of milliseconds, as illustrated by the whole process of facilitation in explosive responses or in elytra stimulated at high frequency. Some photosomes must be precoupled to respond to the very first suprathreshold stimulation.

A mathematical modelization of the elytral system has just been achieved with P. Courrège and M. Girard. On the basis of four parameters only – the photogenic charge, the rate of coupling, the lifetime of the coupled state and its initial amount – it leads to an equation which justifies the variations in the intensity of the successive responses and fits remarkably well with the experimental records of typical emissions [22,26].

By this stage, three questions have been clearly defined and point to different strategies. The first is the meaning of the term "photogenic charge"; here biochemistry is the right tool for indentifying the components of the luminous reaction. The second question deals with the physiological trigger of the flash, or excitation-bioluminescence coupling; this calls for an electrophysiological study, which has also to consider the wave of ignition. Last but not least is the question of the nature of the coupling, understood here in the sense of the coupling mechanism which results in facilitation; owing to the short lifetime of the coupled state, fast fixation is required to freeze the mechanism's morphological supports in situ.

270

Biochemistry

In vitro studies of the bioluminescent reaction have proven difficult in the scale-worm system. First, it was shown that the reaction is not of the luciferine-luciferase type, and that it could be triggered by sodium dithionite[27]. Then photosomes were isolated on sucrose gradient[28]. The fraction containing them consisted only of the characteristic PER and membrane vesicles. As expected, this fraction also contained the maximal luminous activity. Triton X was needed to solubilize the active principle, which was then isolated by chromatography and characterized as a new photoprotein, polynoidin[29, 30].

Activated forms of oxygen, probably superoxide radicals, trigger the luminous reaction of polynoidin. Such radicals can be produced by Fenton reagent, sodium dithionite, the xanthine-xanthine oxidase system or flavin oxidation, and they have been studied together with a variety of competitors and scavengers. Calcium ions have no triggering effect on polynoidin, which is thus clearly different from aequorin. The luminous reaction of polynoidin is not a flash, but lasts for seconds or minutes depending on the reagent. Its λ max. (510 nm) is slightly lower than living bioluminescence (520 nm). Polynoidin is not fluorescent, either before or after its luminous emission. Its further characterization depends mainly on overcoming the difficulty of collecting a sufficient amount of material.

The fluorescent substance of photosomes, easily soluble, is located in the supernatant of the first fractionation. Spectral analysis characterizes it as riboflavin. Fluorescence has been considered as a product of the luminous reaction[31]. On the contrary, we proposed the conception according to which the flavin oxidation shown by the fluorescence increase at each flash produces the superoxide radicals needed to trigger polynoidin luminescence[30].

As a membrane protein, polynoidin could well correspond to the particles observed in freeze fracture on the PER tubules. The fluorescent flavin might be stored in one or other of the photosome compartments, and, among other possibilies, it is conceivable that the reaction could be triggered by depolarization of the PER membrane allowing a rapid ionic exchange.

Action potentials

In the scale-worm the first control of bioluminescent activity is nervous. Bursts of action potentials associated with flashes can be recorded in the nerve ganglia of each metamere as well as in the elytrophore[32]. The next event occurs in the elytrum. The nerve plexus is histochemically rich in

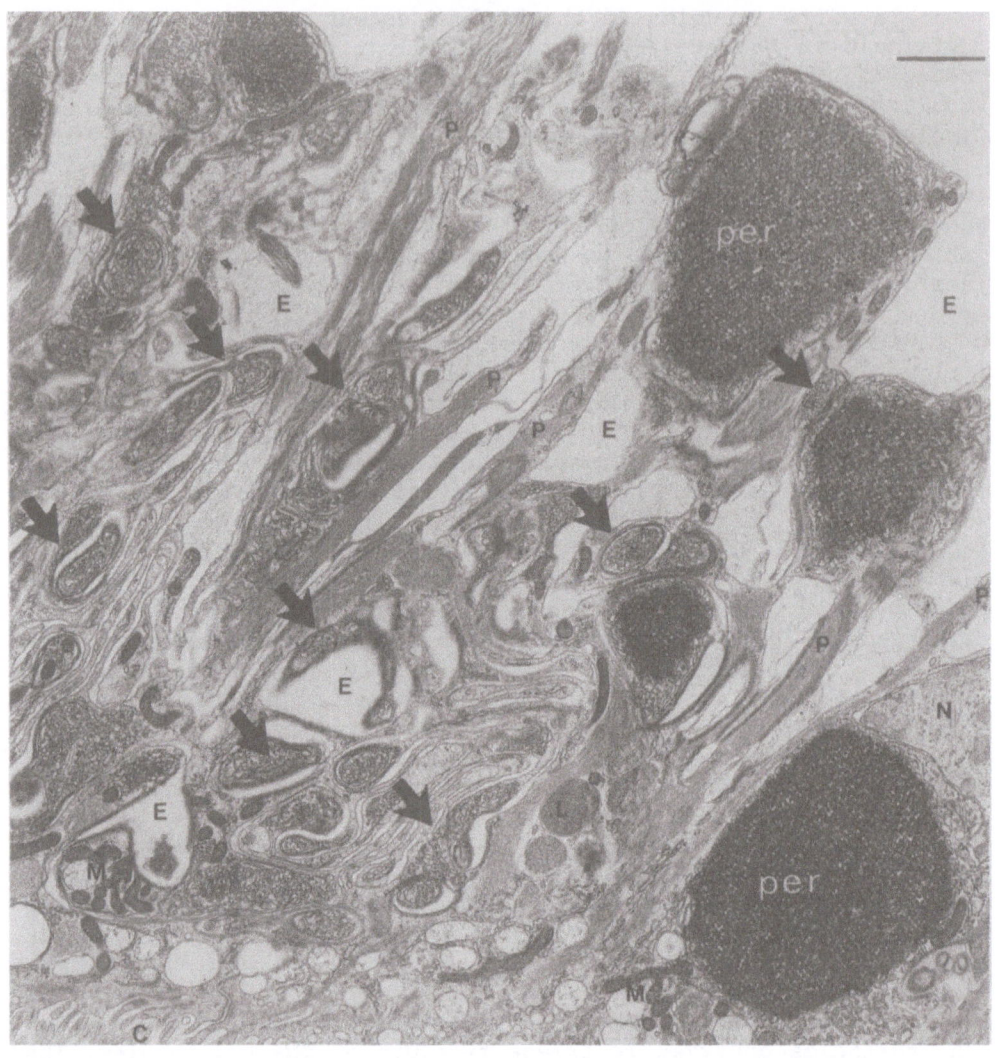

Figure 9. Harmothoe lunulata. Transversal section of the photogenic epithelium. Fast freeze fixation during repetitive stimulation, substitution in acetone and OsO_4, epon.
C, cuticle; E, extracellular space; L, lysosomes; M, mitochondria; N, nucleus; PER, photosomes, P, pillars.
The photosomes of the inner ring (lower right) are situated deeper in the cell than those of the outer ring, which are enclosed in pouches (upper right). Between the straight pillars, numerous tentacles twin together and thus create narrow clefts of extracellular space. Their enlarged heads are filled with intermediate reticulum ending to terminal dyadic saccules (arrows). Scale : 1 μ.

acetylcholinesterase activity[11]. In vivo, flashes can be elicited as unitary responses to minute microinjections of acetylcholine in the elytral ganglion[33]. The responses are reversibly blocked by antimuscarinic drugs such as atropin.

The second part of the excitation path is the elytral epithelium, with the relay probably occurring at the boundaries of the elytrophoral disc. Herrera[27,34] and Bilbaut[35,36] have shown independently that all the epithelial cells of the elytrum, whether photogenic or banal, are electrically coupled. Stimulation evokes an action potential which spreads over all the conducting epithelium and can be recorded in any cell. Nevertheless, this action potential differs markedly in banal cells, where it consists of an Na^+ influx only and in the photocytes, where it is composed of both a Na^+ and Ca^{2+}. The amplitude of the Ca^{2+} peak is related to the external Ca^{2+} concentration in accordance with the Nernst equation, and no luminous response can be obtained in calcium-free sea water. It has thus been concluded that calcium triggers the luminous reaction.

The electrical conduction of the elytral epithelium is in good agreement with the morphological observation of numerous gap junctions between adjacent epithelial cells[35]. Nevertheless, calcium does not trigger bioluminescence in vitro. Furthermore the electrophysiological studies have not taken into consideration the existence of coupled and uncoupled photosomes, or the possibility that an internal excitation route supported by ER membranes could reach them.

Dyad junctions

Previous electron microscope observations have occasionally and unexpectedly shown[11,37] that photosomes can be connected to long, flat ER saccules running parallel to the plasma membrane. Such associations were called dyad junctions in reference to the similar arrangement of the sarcoplasmic reticulum in the muscle. To demonstrate that such junctions might well be the morphological supports of coupled photosomes, the ultrastructure of non stimulated (virgin) and stimulated elytra were compared after fast freeze fixation. This recently developped technic consists of freezing a biological sample by its fast application over a metal plate cooled to about -270°C with liquid helium[38,39,40,41,42]. Freezing is achieved in 1 msec. The ice of the sample is substituted in acetone, in the presence of OSO_4, at -90°C. After thawing, embedding and sectioning are done as usual in electron microscopy. This procedure avoids the osmotic artefacts induced by liquid fixatives and indeed affords remarkably good preservation of a layer 10 to 15 μ thick devoid of ice crystals; this is sufficient to observe the flat photogenic epithelium.

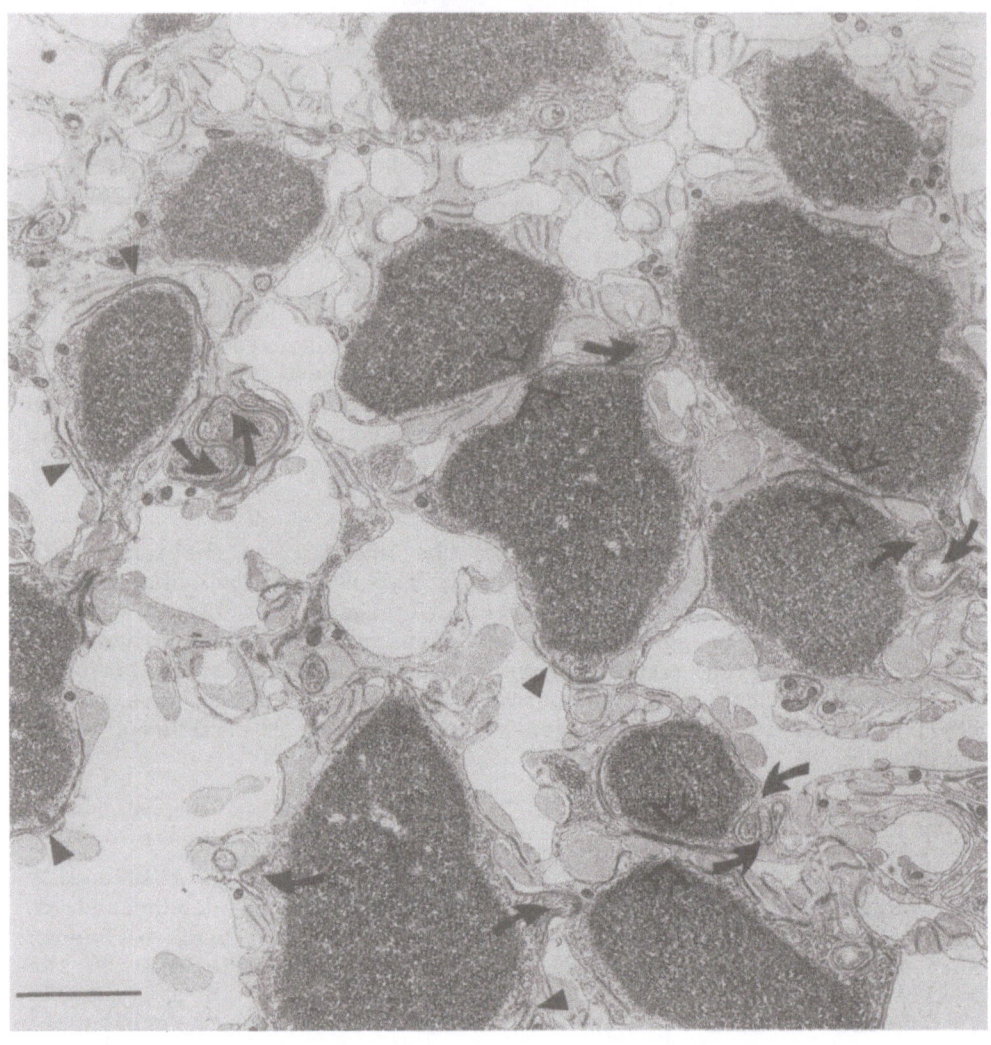

Figure 10. Harmothoe lunulata; frontal section of the photogenic epithelium. Fast freeze fixation during facilitation, substitution in acetone and OSO_4, epon.

The frontal section is slightly oblique. The photosomes of the inner ring, in the upper part of the picture, are surrounded by numerous invaginations bearing desmosomes which anchor the pillars. The photosomes of the outer ring, in the lower part of the picture, are enclosed in pouches. Pouches contact each other either by apposition (hollow arrows) or by lateral tentacles which wrap around the pouches (arrowheads) or twin together (curved arrows). Dyad junctions differentiate along the narrow extracellular spaces thus created. Scale : 2 μ.

One of the first results of the fast freeze-substitution technic was to show the existence of an irregular ER network around the photosomes. This network arises from PER tubules, but does not possess their regular diameter or organization. It is called intermediate ER (IER) because it eventually connect terminal dyadic saccules (DER)[43] . IER is poorly preserved with the usual liquid fixatives such as glutaraldehyde or osmium, which transform it into irregular vesicles and myelin-like bodies, suggesting that its intracisternal osmotic pressure could differ markedly from that of the average cytosol.

Second, fast freeze substitution visualized the development of the IER which was actually seen to increase considerably during repetitive stimulation. In virgin elytra, the IER is thinly distributed as short strands around some facets of the photosomes. These strands do not reach the plasma membrane, except for those of certain photosomes close to the lateral cell boundaries, which connect short terminal saccules. These few dyad junctions might characterize the precoupled photosomes ready to fire at the first stimulation.

Stimulated elytra, on the contrary, exhibit considerable development of the IER around the photosomes; it forms thick strands obviously growing and evaginating the plasma membrane. As already mentioned, the internal part of the epithelial cells bathes directly, without any basal lamina, in the extracellular medium of the middle compartment. Between the pillars, which probably constitute a relatively rigid skeleton, the cell surface turns out to be amazingly flexible and deformable. Its topography is extremely complex in stimulated elytra. The photosomes of the outer ring are enclosed in pouches, sometimes with a narrow neck. Long bract-like pseudopods envelop the pouches. Other long, slender digitations, the tentacles, join up together and their enlarged heads are entwined in Yin-Yang figures. There are numerous gap junctions between adjacent pouches, between the pouches and the bract-like pseudopods, and between the coiled tentacles (Figs.9,12).

The pouches, pseudopods and tentacles contain continuous strands of IER, which connect very large terminal DER saccules measuring up to 10 μ. These saccules lie strictly parallel to the plasma membrane. Their junctional gaps and the cisternal spaces have about the same thickness (160 nm). No foot processes have been observed, even after treatment with tannic acid.

DER only differentiate near areas including narrow extracellular clefts, and most often symmetrically, in triads. These narrow clefts, reminiscent of the T system, are obtained by different topographical arrangements resulting from the apposition of pouches, pseudopods and tentacles. Dyads or triads often occur

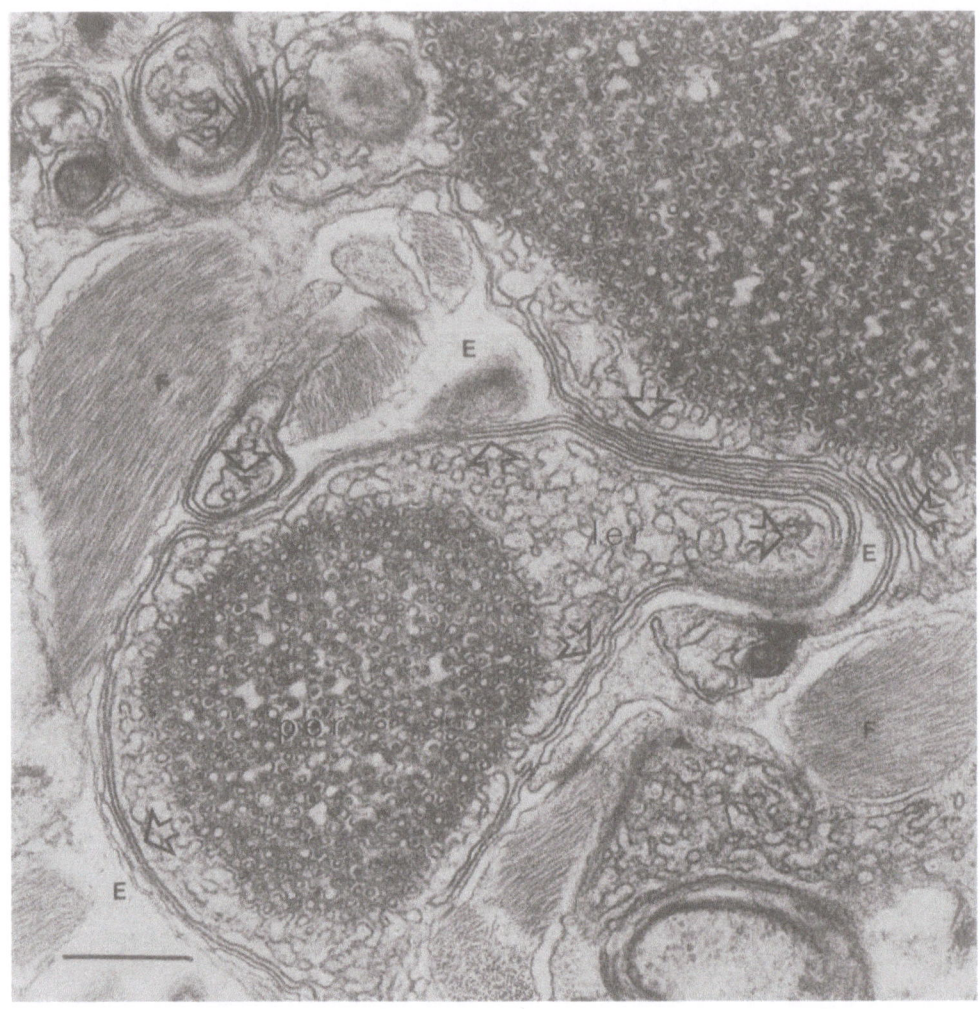

Figure 11. _Harmothoe_ _lunulata_; fast freeze fixation during facilitation, substitution in acetone and OSO_4, epon.
The intermediate reticulum _IER_ surrounds the photosome _PER_ with both membranous and intracisternal continuity. _IER_ connects long terminal saccules which constitute a nearly continuous dyad junction (arrows) with the plasma membrane of the pouch. Growing tentacles filled with _IER_ penetrate in other pouches or contact together. Symmetrical junctional complexes (triads) differentiate along such areas of apposition. They comprise occasionally additional terminal saccules. Scale : .5 μ.

over gap junctions, but never at the level of desmosomes of septate junctions. As regards the photosomes of the inner ring, they establish contact with a different type of cleft, formed by invaginations deeply built into the cell. It is in these funnel shaped invaginations that the pillars from the opposite surface of the elytrum are anchored by desmosomes.

The coupled photosomes are therefore connected to a continuous network of IER, itself connected to terminal DER. This junctional complex is either missing or incomplete in virgin elytra. Its development involves extremely important phenomena of growth, association and transformation, by different means, depending on the local topography, but all tending to the same result.

DISCUSSION

Our exploration of the elytra system started with the simple question of how to explain why the successive flashes of the typical emissions evolve strongly and progressively, even though the repetitive stimulation remains constant. Photosomes, the microsources of the system, could be observed during this activity, because they bear natural probes of luminescence and fluorescence and are arranged in a monolayer. The correlation between the spatial and temporal aspects showed that the amount of light emitted during a flash is a function, not only of the photogenic charge remaining in the system, but also of the number of photosomes firing, which increases gradually. This recruitment explains the facilitation; it is basically intracellular. To account for the degree of freedom enabling photosomes of the same cell to be either coupled to the excitation pathway and thus respond to the stimulation, or uncoupled and remain still, and because the coupled state has a basically short lifetime (1 sec), we used fast freeze fixation to capture the morphological supports of the coupling mechanism. In view of the evidence thus obtained that coupled photosomes are indeed characterized by a continuous reticular network coupling them to the plasma membrane by DER plus the necessary length of IER, it is clear that the problematics of the elytral system are entirely based on membrane conduction and membrane dynamics.

Since photosomes flash, but only if coupled, the excitation-bioluminescence pathway must follow the internal route of the ER. The notion of conducting ER, directly observed here, is in many respects similar in the muscle system. Both systems involve Ca^{2+}. In both, the junctional ER or SR have the same associative features and the same successive differentiations. The existence of paracrystalline SR in a prochordate muscle strengthens the grounds for the comparison[17][18].

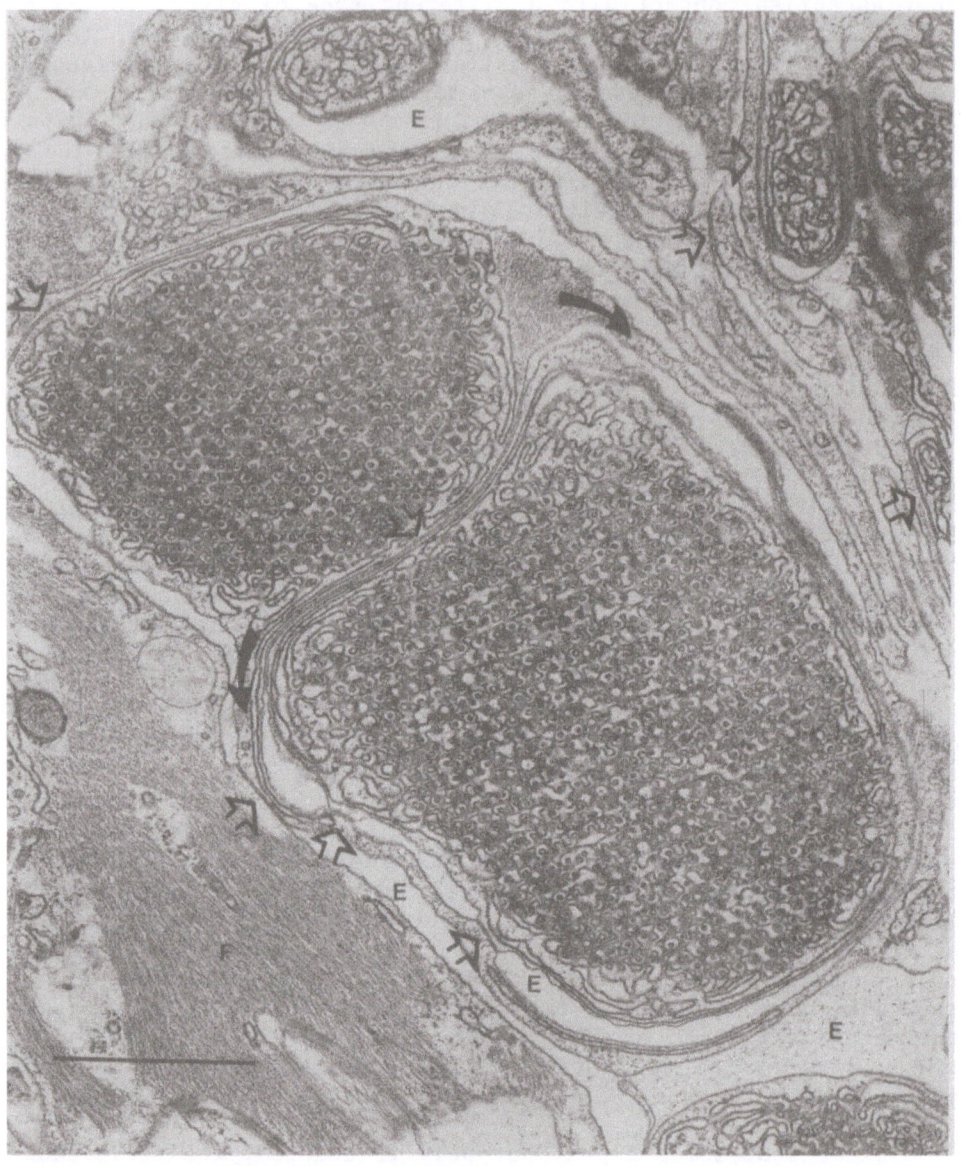

Figure 12. _Harmothoe lunulata_; fast freeze fixation during facilitation, substitution, epon.
Two photosomial pouches are coupled first by a long gap junction along which triad junctions extend. In addition, a bract-like pseudopod arising from the upper pouch (curved arrows) completely envelops the lower pouch. Gap junctions are indicated by hollow arrows. Scale : 1 μ.

In the elytral system, further experimentation and a reinterpretation of the electrophysiological data will be needed to understand the ionic and electrical events which the conducting ER might vehicle. At present, one of the first difficulties lies in the apparent absence of any bridge or foot processes which are assumed to constitute an electromechanical coupling of SR dyads [44] [45] [46]. Another question concerns the missing link between the biochemical studies[29] [30] which resulted in the isolation of a membrane photoprotein triggered by O_2^-, and the electrophysiological studies [34] [35] [36] which led to the conclusion that Ca^{2+} triggers the flash in vivo.

Nevertheless, the notion of conducting ER is clear enough in the luminous system to be strongly emphasized. In the conducting epithelium, the internal route must offer a preferential way to the conduction of the signal which reaches the photosomes and triggers the light reaction. We may further assume that the signal continues on its ER route, reaches other photosomes and even passes from cell to cell. The triad junctions, particularly when they include a gap junction as in the tentacles, might constitute the most likely sites of direct communication between the internal compartments of adjacent cells. In that case, the internal ER route would progressively extend its limits and reinforce its network during the period of facilitation. During a flash, the sweep of ignition would thus be a visible expression of the wave of excitation. This conception explains the exponential increase of the flash latencies as well as the decrease in the ascending slopes observed during the decay of the typical emission, because the conduction path comprises an increasing length of exhausted PER.

The ER of the photocytes exhibits an impressive and continuous series of differentiations. These consist of the DER, IER and PER, all involved in rapid conduction phenomena, and also of the rough and smooth membranes and the nuclear envelope, not to mention the related but not connected Golgi apparatuses and lysosomes involved in much slower elaboration processes. It would be interesting to know whether these functions can interact, and whether the fast one is of a general character. The conducting ER undoubtedly constitutes the fastest of the cell strategies based on secondary messengers for travel through cytoplasm.

Although continuous, the ER membranes must have strikingly different mechanical properties, particularly of fluidity, to be firmly stabilized in the curved tubules of PER, to be so obviously elastic and flexible in IER, and to be maintained parallel to the plasma membrane in the DER. However, while the PER remains relatively stable and apparently unchanged, the remarkable property of the IER and DER is their extremely fast differentiation during repetitive stimulation, as well as their extremely

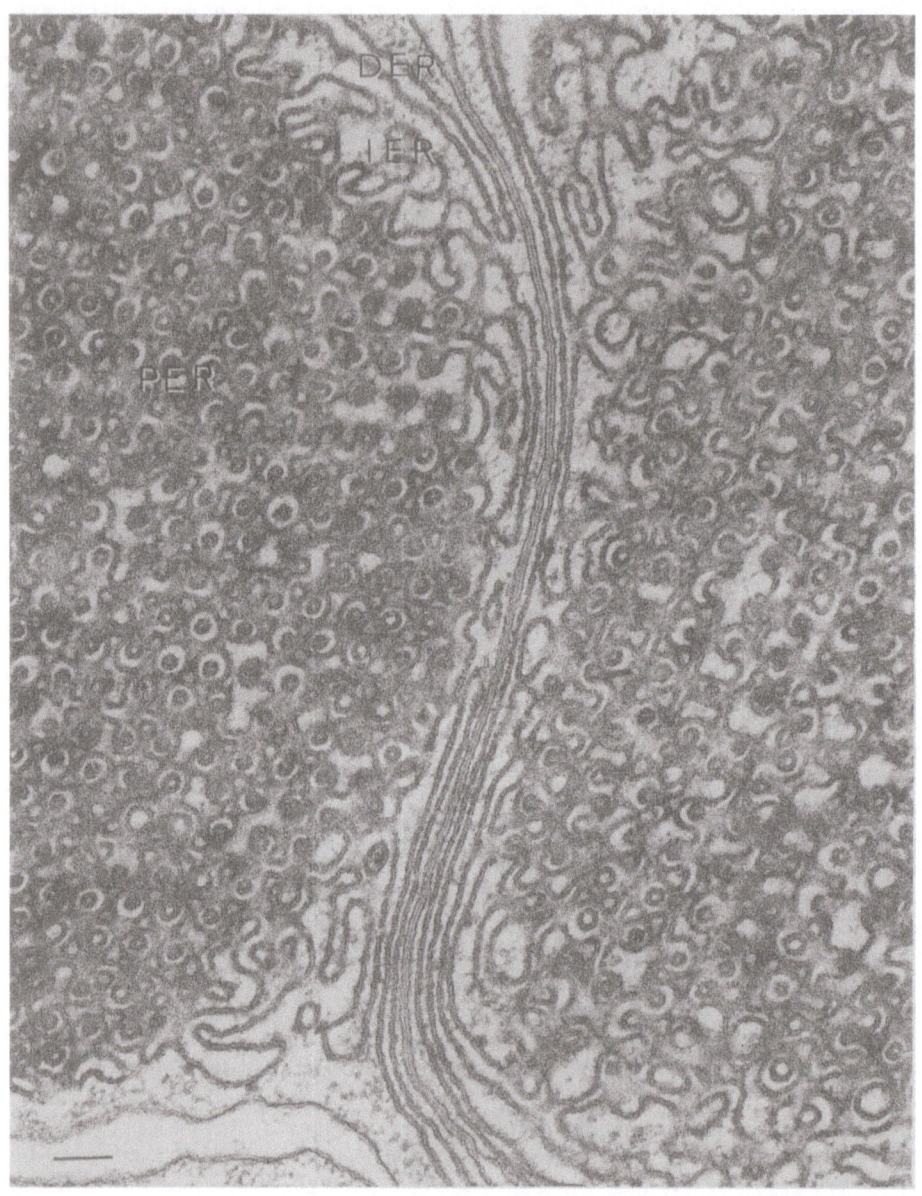

Figure 13. Enlarged part of Fig.12, showing the continuity between the tubules of the photosomes <u>PER</u>, the intermediate reticulum <u>IER</u>, here nearly virtual, and the terminal dyadic saccules <u>DER</u>. Note the unusual, collapsed aspect of the gap junction and the clear definition of the unit membranes. Scale : .1 μ.

fast dedifferentiation at rest. The glimpses of these processes provided by fast freezing suggest the occurrence of highly dynamic transformations involving elongating pseudopods, growing strands of membranes, twining tentacles and differentiating junctions. These processes in turn involve membrane growth, recognition and transformation on a surprisingly large scale and within a surprizingly short time.

The growth of the IER might in part be the passive result of the action of osmotic forces. However, its orientation towards the plasma membrane, the symmetry of the triad arrangement in relation to the narrow extracellular spaces, and the considerable elongation of the tentacles suggest a more dynamic mechanism, such as an interaction between the parts of the plasma membrane bordering narrow clefts on the one hand, and certain receptive portions of the ER on the other. Such a mechanism might depend on a "coupling inducing factor" which in turn would be directly dependent on the electrical stimulation or on any of its consequences. In other words, the local concentration of a presulmably diffusible coupling inducing factor might determine the extent of the junctional ER and thus the lifetime of the coupled state.

CONCLUSION

In the scale-worm system, fast freeze fixation reveals the existence of an important and rapid membrane <u>behavior</u> resembling a dance, for the purpose of creating <u>communications</u>. This behavior occurs at the integration level at which the membranes give shape and individuality to the organelles.

The optional coupling mechanism and ER conduction described here might well prove to be a cell strategy more widespread than in the system which revealed it with luminous evidences. Functionally, this mechanism translates the stimulation frequency into intensity of response. It manages the photogenic reserves economically. It enables the responses to improve with repetition, at least at the beginning, and prepares the next response in the event of further stimulation.

This facilitation therefore displays characteristics of elementary memory and anticipation. As with every biological memory, it is only revealed as such when expressed by an activity, be it movement, thought or light. In the system examined here, its inscription is clearly not static, but is represented by the <u>potential</u> capacity of membranes to recognize each other and couple.

References

1. P.A.V. Anderson, Epithelial conduction : its properties and functions, Progress in Neurobiol., 15:161-203 (1980).
2. L.A. Staehelin, Structure and function of intercellular junctions, Int. Rev. Cytol., 39:191-283 (1974).
3. M. Pavans de Ceccatty, "Communications et interactions cellulaires", Presses Universit. France, Paris (1983).
4. B.M. Sweeney, Intracellular sources of bioluminescence, Int. Rev. Cytol. 68:173-195 (1980).
5. E.N. Harvey, "Bioluminescence", Acad. Press, New York (1952).
6. J.M. Bassot, On the comparative morphology of some luminous organs, in :"Bioluminescence in Progress", F.H. Johnson and Y. Haneda ed. Princeton Univ. Press, Princeton (1966).
7. J. Buck, Functions and evolutions of bioluminescence, in : "Bioluminescence in action", P. Herring ed. Acad. Press (1977).
8. J.W. Hastings, Biological diversity, chemical mechanisms, and the evolutionary origins of bioluminescent systems, J. Mol. Evol. 19:309-321 (1983).
9. J.A.C. Nicol, Luminescence in Polynoid worms, J. Mar. Biol. Ass U.K., 32:65-84 (1953).
10. J.A.C. Nicol, Luminescence in Polynoids : II. Different mode of response in the elytra, J. Mar. Biol. Ass. U.K., 36:261-269 (1957).
11. M. Pavans de Ceccatty, J.M. Bassot, A. Bilbaut and M.T. Nicolas, Bioluminescence des élytres d'Acholoe. I. Morphologie des supports structuraux, Biol. cellulaire, 28:57-64 (1977).
12. J.M. Bassot, Données histochimiques et histologiques sur l'organe lumineux des élytres d'Annelides Polynoinae, Cah. Biol. Mar., 7:39-52 (1966).
13. J.M. Bassot and A. Bilbaut, Bioluminescence des élytres d'Acholoe, IV. Luminescence et fluorescence des photosomes, Biol. Cellulaire, 28:163-168 (1977).
14. J.M. Bassot, Une forme microtubulaire et paracristalline de reticulum endoplasmique dans les photocytes des Annelides Polynoiae, J. Cell. Biol., 31:135-158 (1966).
15. M.T. Nicolas, Etude des structures et des mécanismes de la bioluminescence des Annélides Polynoinae, Thèse, Univ. Paris Sud (1982).
16. J.M. Bassot and M.T. Nicolas, Similar paracrystals of endoplasmic reticulum in the photoemitters and the photoreceptors of scale-worms, Experientia 34:726:728 (1978).
17. M.J. Cavey, Multilaminar aggregates of sarcoplasmic reticulum in caudal muscle cells of an ascidian larva, Canad. J. Zool., 58:538-542 (1980).

18. M.J. Cavey, Coextensive sarcoplasmic reticulum in larval muscle cells of a didemnid ascidian, Diplosoma macdonaldi. Canad. J. Zool., 61:732-736 (1983).

19. M.T. Nicolas, Bioluminescence des élytres d'Acholoe, V. Les principales étapes de la régénération, Arch. Zool. exp. gen., 118:103-120 (1977).

20. A. Bilbaut and J.M. Bassot, Bioluminescence des élytres d'Acholoe, II. Données photométriques, Biol. Cell., 28:154-154 (1977).

21. J.M. Bassot, sites actifs et facilitation dans trois systèmes bioluminescents, Arch. Zool. exp. gen., 120, 5-24 (1979).

22. J.M. Bassot, Intracellular recruitment during facilitation of bioluminescent emission, in the scale-worm system. Its characteristics of elementary memory and anticipation. J. Cell Biol., in preparation.

23. J.A.C. Nicol, Spectral composition of the light of polynoid worms, J. mar. biol. Ass. U.K., 36:529-538 (1957).

24. B. Lecuyer and B. Arrio, Some spectral characteristics of the light emitting system of the polynoid worms, Photochem. Photobiol., 22:213-215 (1975).

25. J.M. Bassot and A. Bilbaut, Bioluminescence des élytres d'Acholoe, III. Déplacement des sites d'origine au cours des émissions. Biol. Cell., 28:155-162 (1977).

26. J.M. Bassot, G. Nicolas and J. Escaig, Rapid structural coupling of endoplasmic reticulum with plasma membrane in the bioluminescent system of scale-worms, as revealed by quick freeze fixation. J. gen. Physiol., in preparation.

27. A.A. Herrera, The physiology of bioluminescence in polynoid polychaete worms., Thesis, U.C.L.A. (1977).

28. M.T. Nicolas, Présence de photosomes dans les fractions lumineuses du système élytral des Polynoinae (Annelides polychètes). C.R. Acad. Sci. Paris, 289:177-180 (1979).

29. M.T. Nicolas, J.M. Bassot and O. Shimomura, Caractérisation d'une photoprotéine nouvelle dans le système bioluminescent des Annélides Polynoinae, C.R. Acad. Sci. Paris, 293:777-780 (1981).

30 M.T. Nicolas, J.M. Bassot and O. Shimomura, Polynoidin : a membrane photoprotein isolated from the bioluminescent system of scale-worms, Photochem. Photobiol. 35:201-207 (1982).

31. C. Fresneau, B. Arrio, B. Lecuyer, A. Dupaix, N. Lescure and P. Volfin, The fluorescent product of the scale-worm bioluminescent reaction : an in vitro study, Photochem. Photobiol. 22:213-215 (1984).

32. M. Pavans de Ceccatty, A. Bilbaut and J.M. Bassot, Correlation entre les signaux électriques spontanés, la motricité et la luminescence chez Acholoe astericola Delle Ch. C.R. Acad. Sci. Paris, 275:2523-2526 (1972).

33. M.T. Nicolas, M. Moreau and P. Guerrier, Indirect nervous control of luminescence in the polynoid worm Harmothoe lunulata, J. exp. Zool., 206:427-433 (1978).

34. A.A. Herrera, Electrophysiology of bioluminescent excitable epithelial cells in a polynoid polychaete worm. J. comp. Physiol. 129:67-78 (1979).

35. A. Bilbaut, Cell junctions in the excitable epithelium of bioluminescent scales on a polynoid worm : a freeze fracture and electrophysiological study, J. Cell. Sci. 41:341-368 (1980).

36. A. Bilbaut, Excitable epithelial cells in the bioluminescent scales of a polynoid worm; effects of various ions on the action potentials and on the excitation-luminescence coupling. J. exp. Biol., 88:219-238 (1980).

37. M. Pavans de Ceccaty, J.M. Bassot and A. Bilbaut and M.T. Nicolas, Genèse des paracristaux photogènes et leurs structures d'excitation dans les cellules de l'élytre d'Acholoe astericola Delle Ch. C.R. Acad. Sci. Paris, 275:2363-2366 (1972).

38. A. Van Harveld and J. Crowell, Electron microscopy after rapid freezing on a metal surface and substitution fixation., Anat. Record, 149:381-386 (1964).

39. J.E. Heuser, T.S. Reese and D.M.D. Landis, Preservation of synaptic structure by rapid freezing, Cold Spring Harb. Symp. Quant. Biol., 40:17-24 (1976).

40. J.E. Heuser, T.S. Reese, M.J. Dennis, Y. Jean, L. Jean and L. Evans, Synaptic vesicle excocytosis captured by quick freezing and correlated with quantal transmitter release, J. Cell Biol., 81:275-300 (1979).

41. J. Escaig, G. Geraud and G. Nicolas, Congélation rapide des tissus biologiques. Mesure des températures et des vitesses de congélation par thermocouple en couche mince. C.R. Acad. Sci. Paris, 284:2289-2292 (1977).

42. J. Escaig, New instruments which facilitate rapid freezing at 83 K and 6 K. Microscopie, 126:221-229 (1982).

43. J.R. Sommer, N.R. Wallace and J. Junker, The intermediate cisterna of the sarcoplasmic reticulum of skeletal muscle. J. Ultrastr. Res. 74:126-142 (1980).

44. A.V. Somlyo, Bridging structures spanning the junctional gap at the triad of skeletal muscle. J. Cell Biol., 80: 743-750 (1979).

45. A. Saito, S. Seiler, A. Chu and S. Fleischer, Preparation and morphology of sarcoplasmic reticulum terminal cisternae from rabbit skeletal muscle. J. Cell Biol., 99:875-885 (1984).

46. A.H. Caswell and J.P. Brunschwig, Identification and extraction of proteins that compose the triad junction of skeletal muscle. J. Cell Biol., 99:929-939 (1984).

DRUG BIOTRANSFORMATION AND DRUG TOXICITY STUDIED WITH INTACT MAMMALIAN CELLS

Sten Orrenius

Department of Toxicology
Karolinska Institutet
Stockholm, Sweden

INTRODUCTION

Since many years our laboratory has been actively engaged in studies of biochemical mechanisms associated with drug biotransformation and drug toxicity. Our studies were originally concerned with the properties and regulation of the major enzymes involved in drug metabolism, in particular the cytochrome P-450 monooxygenase system. More recently, our research has been focused on biochemical mechanisms of importance for development of drug toxicity, and the role of the glutathione system as a major cellular defense mechanism. In these studies we have used a variety of experimental model systems, including perfused tissues, suspensions of intact cells, subcellular organelle fractions, and purified enzymes. This presentation summarizes some of our more recent studies of drug biotransformation and drug toxicity, using predominantly freshly isolated rat hepatocytes to investigate the underlying biochemical mechanisms.

DRUG BIOTRANSFORMATION - AN OVERVIEW

The biotransformation of foreign lipid-soluble compounds almost invariably results in formation of more polar products which can be excreted by the organism. Most of the reactions involved in this process are enzyme-mediated, and in mammals, as well as in many other species, the liver is the predominant tissue for drug biotransformation. However, various extrahepatic tissues do contribute to this process, and there are several examples of biotransformation reactions that occur more actively in extrahepatic tissue than in liver.

Drug biotransformation is frequently termed detoxication and detoxication enzymes are often used to denote the catalysts involved in this process. These expressions are however not ideal, since the products formed may sometimes be biologically more active and toxic than the parent compound. In fact, most, if not all, of the reactions generally referred to as detoxication reactions can result in generation of reactive and toxic metabolites. This is particularly important for the cytochrome P-450-catalyzed reactions which are now known to be critical events in the metabolic activation of a variety of hepatotoxic drugs and chemical carcinogens.

Most often a series of consecutive reactions operate to convert a lipophilic drug to polar and excretable products. For convenience, the reactions involved in drug biotransformation have been divided into phase I and phase II reactions. Phase I reactions include oxidation, reduction and hydrolysis, whereas phase II reactions denote synthetic reactions in which conjugates are formed between a drug or drug metabolite and endogenous acid or carbohydrate; the relationship between phase I and phase II reactions is illustrated schematically in Fig. 1. Several of the conjugates formed in phase II reactions may undergo further metabolic transformation prior to excretion; the conversion of glutathione conjugates to the corresponding mercapturic acid derivatives may serve to illustrate this.

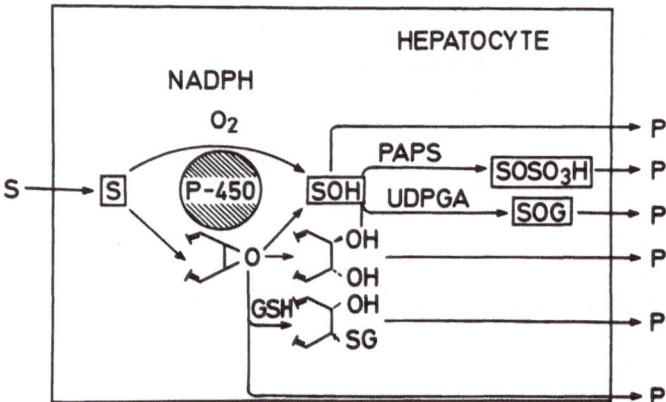

Fig. 1. Schematic representation of drug biotransformation by phase I (oxidation) and phase II (conjugation) reactions in hepatocytes. S, substrate; SOH, hydroxylated product; PAPS, 3'-phosphoadenosine-5'-phosphosulfate; UDPGA, uridine-5'-diphosphoglucuronic acid; SOG, glucuronide; GSH, glutathione, reduced form; SG, glutathione conjugate; P, product.

The detoxication enzymes consist of families of inducible isozymes with broad and overlapping substrate specificities[1,2]. The cytochrome P-450-linked monooxygenase system and the glucuronosyl transferases are membrane-bound enzymes which are located predominantly in the endoplasmic reticulum, whereas the sulfotransferases are found in the cytosol and the epoxide hydrolases and the glutathione transferases occur in both soluble and membrane-bound forms. In particular, the cytochrome P-450-linked monooxygenase system has been studied intensely during recent years and many of its properties are now known[2,3].

Most of the enzymes involved in drug biotransformation depend on the availability of cytosolic cofactors for activity. NADPH, uridine-5'-diphosphoglucuronic acid, 3'-phosphoadenosine-5'-phosphosulfate, and glutathione (GSH) are required as cofactors in the monooxygenation, glucuronidation, sulfation, and glutathione conjugation reactions, respectively. The various cofactors are all formed in energy-dependent reactions, and their regeneration may become rate-limiting for drug biotransformation under certain conditions[2]. In particular, the capability of the cell to regenerate GSH during drug biotransformation has attracted much attention[4], because of the important role of this thiol in the cellular defense against electrophilic drug and carcinogen metabolites.

GSH is the most important non-protein thiol found in animal cells[5]. It is characterized by its reactive thiol group and its γ-glutamyl bond which makes it resistant to normal peptidase activity. In many mammalian cells the concentration of GSH is very high, i.e. in the millimolar range. Most of the intracellular GSH is present in the cytosol, but the existence of a minor mitochondrial pool has also been demonstrated. The rate of GSH turnover varies markedly between different tissues. Under conditions of enhanced GSH utilization, the availability of intracellular cysteine appears to be rate-limiting for GSH resynthesis[4].

Among the multiple functions of intracellular GSH are those related to its role in drug biotransformation. Thus, GSH is known as a strong nucleophile, and electrophilic drug and carcinogen metabolites are often inactivated by formation of glutathione conjugates. In addition, GSH serves as a reductant in the metabolism of various hydroperoxides and free radicals[6]. These reactions may occur spontaneously, but are usually catalyzed by glutathione transferases and glutathione peroxidases, respectively.

The cytotoxicity of many xenobiotics is preceded by depletion of intracellular GSH and prevented under conditions of stimulated GSH biosynthesis. GSH depletion may occur as result of excessive utilization of GSH during formation of glutathione conjugates or oxidation of GSH to glutathione disulfide (GSSG) during metabolism of hydroperoxides and free radicals. Although the bulk of the GSSG formed intracellularly is readily re-reduced to GSH by NADPH-dependent glutathione reductase, a minor fraction of the GSSG is actively secreted into the extracellular fluid by a translocase which appears also to be responsible for the excretion of glutathione conjugates from the cell[7].

Even though many tissues are capable of forming glucuronic acid, sulfate and glutathione conjugates of various xenobiotics, the liver is the predominant site of formation of these conjugates. Once formed, they are readily excreted into bile and plasma, and there is no clear evidence for either accumulation or further metabolism of any of these conjugates within the liver. Whereas the glucuronides and glutathione conjugates, once formed in the liver, are released preferentially into the bile, a major fraction of the sulfate conjugates appears in the plasma.

USE OF ISOLATED CELLS IN METABOLIC STUDIES

During recent years the use of suspensions of freshly isolated cells has become frequent in metabolic studies. Suspensions are normally obtained by collagenase perfusion of the tissue and they usually retain excellent viability as well as the metabolic properties characteristic of their tissue of origin[8]. Suspensions of freshly isolated cells from liver, kidney, intestinal mucosa, heart, and several other tissues have now been characterized in detail and are in daily use in many laboratories.

Suspensions of hepatocytes and renal cells freshly isolated from the male rat have been used as experimental model systems in several of our studies of drug biotransformation and drug toxicity[4,8]. They can be obtained in good yields by collagenase perfusion of the liver and kidney, respectively, and may be incubated for several hours without significant loss of viability. It should be noted that the suspension of liver cells obtained is normally rather homogeneous and consists almost exclusively of hepatocytes because most nonparenchymal cells are removed during the isolation procedure. The renal cell preparation also consists almost exclusively of epithelial cells, but is clearly less homogeneous and probably contains tubular cells from all the different parts of the nephron. However, the biochemical characteristics of the renal cell preparation strongly suggest a predominance of cells derived from the proximal tubular epithelium.

The isolated cells maintain metabolic integrity for several hours of incubation, and can be prepared under sterile conditions for establishment of primary cultures. Cytochrome P-450 content in isolated liver cells is constant for several hours, but declines during longer incubation[9]. Glutathione level in hepatocytes is stable provided that either cysteine or methionine is present in the medium[4]. Addition of other amino acids or carbohydrates to the medium is typically not necessary with cells from fed animals for experiments lasting less than 2 hours, unless a particular metabolic stress is created. Cells from starved animals are more dependent upon extracellular metabolite sources, and have been used for study of supply of reducing equivalents and metabolic intermediates[8]. In liver cell incubations, ATP, ADP, AMP, NADH and NADPH levels and rates of glucose synthesis, albumin synthesis and endogenous respiration have been found to be constant up to 4 hours, and to be similar to values obtained for intact liver or isolated perfused liver.

Pretreatment of animals with various agents has been used to obtain hepatocytes with altered metabolic activities. Phenobarbital, 3-methylcholanthrene, β-naphthoflavone and ethanol have been used to increase the concentrations of various forms of cytochrome P-450[9]. Clofibrate has been administered to increase the content of peroxisomes, selenium-deficient diet and aminotriazole pretreatment have been used to decrease the levels of glutathione peroxidase and catalase, respectively and diethylmaleate pretreatment has been employed to lower GSH content in liver and kidney cells[4,8,9]. Incubations of isolated cells in different media, such as sulfate-free Krebs-Henseleit buffer and Krebs-Henseleit buffer with varying concentrations of calcium, have been used to study specific aspects of drug metabolism[8]. In general, with proper selection of conditions, these manipulations can be made such that there is no significant effect on the viability of the cells.

Metabolism of Paracetamol

Although there are important examples of endogenous compounds that are metabolized by glutathione conjugation (e.g. the leukotrienes), most electrophilic metabolites that react with GSH are of exogenous origin and formed in reactions catalyzed by the cytochrome P-450-linked monooxygenase system. Paracetamol, a widely used analgesic drug, is one such exogenous compound that is metabolized by the cytochrome P-450 system to an electrophilic intermediate which can form a conjugate with GSH (Fig. 2). The metabolism and hepatotoxicity of this drug has been intensely studied in several laboratories, including our own[4,6].

Paracetamol is mainly metabolized to the sulfate and glucuronide conjugates, but can also undergo cytochrome P-450-dependent oxidation to yield the reactive N-acetyl-p-benzoquinone imine[10], which is inactivated by conjugation with GSH; it is still uncertain whether, in liver, the conjugation with GSH occurs spontaneously

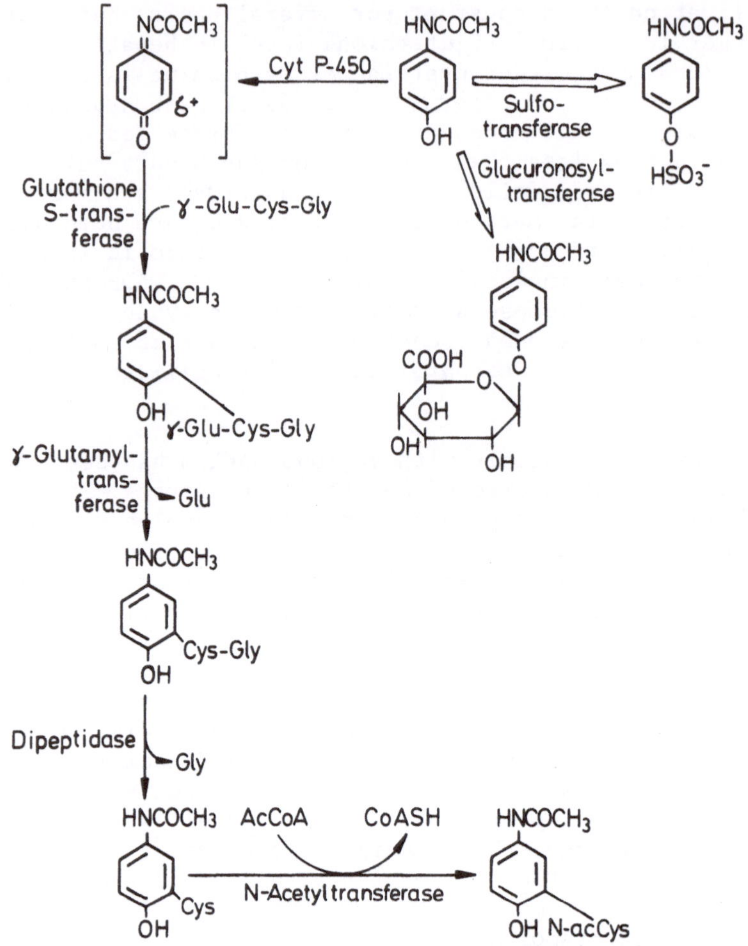

Fig. 2. Schematic illustration of major pathways of paracetamol
biotransformation in the rat. Formation of the glucuronide,
sulfate and glutathione conjugates occurs mainly in the
liver, whereas further metabolism of the glutathione
conjugate to the mercapturate occurs predominantly in
the kidney.

or is mediated by glutathione transferase(s). Incubation of isolated hepatocytes with paracetamol results in depletion of GSH
with concomitant formation of a glutathione conjugate which is
not further metabolized by the hepatocytes; depletion is followed
by rapid loss of cell viability. If the medium is supplemented
with either cysteine, methionine or N-acetylcysteine, stimulation

of GSH resynthesis occurs, and there is no loss of cell viability. Since both methionine and N-acetylcysteine are rapidly metabolized to cysteine in hepatocytes, and their protective effect against paracetamol toxicity is abolished in presence of inhibitors of glutathione synthesis, it appears safe to conclude that stimulation of glutathione biosynthesis is the biochemical mechanism underlying the protective effect of the sulfur amino acids against paracetamol hepatotoxicity.

Metabolism of Glutathione Conjugates

The kidney appears to play a prominent role in the metabolism of glutathione conjugates to mercapturates. In all species investigated, the brush border of the proximal tubular epithelium contains abundant γ-glutamyltransferase activity, which is known to catalyze the initial step in the degradation of extracellular glutathione and glutathione conjugates. Moreover, the tubular epithelium also exhibits dipeptidase and N-acetyltransferase activities, and thus contains all catalysts required for conversion of glutathione conjugates to the corresponding mercapturates. This can be illustrated by in vitro experiments[11], in which isolated renal tubular cells were found to catalyze the metabolism of the glutathione conjugate of paracetamol to the corresponding cysteine and N-acetylcysteine conjugates, with intermediate production of the cysteinylglycine derivative (Fig. 2); N-acetylation of the cysteine conjugate was rate-limiting for the overall conversion of the glutathione conjugate to the mercapturate. The observation that the cysteinylglycine derivative was recovered only in trace amounts, and did not accumulate during incubation, is consistent with high activity of cysteinylglycine dipeptidase relative to γ-glutamyltransferase. This could be further substantiated by including a competitive inhibitor of dipeptidase activity, phenylalanylglycine, in the incubation which caused a considerable increase in the level of the cysteinylglycine derivative[11]. As expected, the entire metabolic sequence was blocked in presence of γ-glutamyltransferase inhibitors, such as serine·borate or Anthglutin (1-γ-L-glutamyl-2-(carboxyphenyl)hydrazine), providing further evidence for the crucial role of this enzyme in initiating the degradation of glutathione conjugates.

USE OF ISOLATED CELLS IN STUDIES OF DRUG TOXICITY

It is now well established that the toxic effects of a wide variety of chemicals are mediated by reactive products formed during their biotransformation in the organism. It is equally clear that there exists a number of protective systems which can trap, or inactivate, toxic metabolites and thereby prevent their accumulation within the tissues and subsequent toxic effects.

Although phase I reactions, in particular those mediated by the cytochrome P-450-linked monooxygenase system, are most often responsible for the production of toxic metabolites, this is not always true and there are now many examples of metabolic activation via phase II reactions, although the latter normally serve a protective function. Hence, it is obvious that the formation of toxic metabolites cannot be attributed to any single enzyme or enzyme system, and that the balance between metabolic activation and inactivation is critical in deciding whether exposure to a potentially toxic compound will result in toxicity, or not.

During the last decade we have used freshly isolated cells from different mammalian tissues to study the metabolism of various foreign compounds and the balance between metabolic activation and inactivation[4,8]. Our studies have emphasized the protective function of the glutathione system, and provided several examples where depletion of intracellular glutathione results in a loss of the balance between metabolic activation and inactivation, and is rapidly followed by signs of toxicity. A recent study of the effects of oxidative stress induced by menadione (2-methyl-1,4-naphthoquinone) metabolism in isolated hepatocytes may serve to illustrate this.

Metabolism and Toxicity of Menadione in Hepatocytes

The metabolism of quinones by flavoenzymes can occur by either one- or two-electron reduction routes which often differ greatly in cytotoxicity. As illustrated schematically in Fig. 3, one-electron reduction results in formation of semiquinone free radicals which can readily reduce dioxygen, forming the superoxide anion radical, O_2^-, and regenerating the quinone. Dismutation of O_2^- and production of other highly reactive species of oxygen quickly lead to conditions of oxidative stress and toxicity as redox cycling of the quinone continues[12]. However, quinones can also undergo two-electron reduction, forming hydroquinones without production of free semiquinone intermediates. This reaction, which is catalyzed by DT-diaphorase[13], may serve an important function by competing with the single-electron pathway, since hydroquinones are often less reactive, and more easily excreted by the cell, than semiquinone radicals.

Although menadione can be excreted from liver cells as the free hydroquinone, conjugates with glucuronic acid, sulfate and glutathione appear to constitute the predominant metabolites. Whereas conjugation with glucuronic acid and sulfate occurs with menadiol, it is not yet clear to what extent the formation of glutathione conjugates also is preceded by reduction of the quinone. Quinones are able to form thiol conjugates nonenzymatically, and a

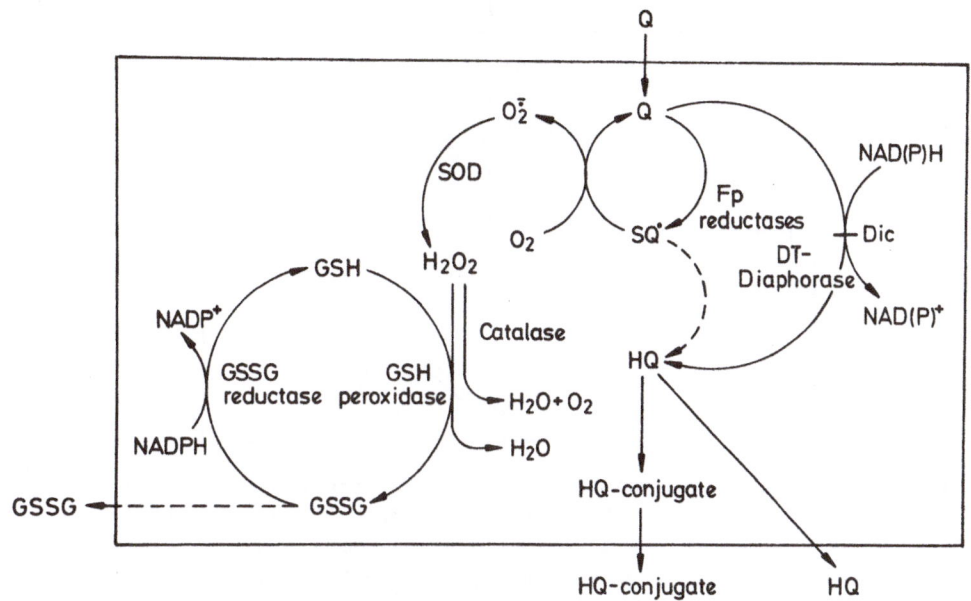

Fig. 3. Schematic representation of quinone metabolism in hepatocytes. Q, quinone; SO·, semiquinone radical; HQ, hydroquinone; Fp, flavoprotein; Dic, dicoumarol; SOD, superoxide dismutase; GSH, glutathione, reduced form; GSSG,
glutathione disulfide.

glutathione conjugate of menadione has been reported to contribute
to the overall redox cycling of the quinone in the perfused liver[14].

In liver cells, menadione metabolism involves both one- and
two-electron reduction pathways; the relative contribution of the
two routes depends on menadione concentration, and can be manipu-
lated by selective induction of either NADPH-cytochrome P-450
reductase or DT-diaphorase. Thus, hepatocytes isolated from
phenobarbital-treated rats exhibit increased redox cycling and
toxicity when exposed to menadione, whereas the reverse is true
for hepatocytes from 3-methylcholanthrene-treated rats[15].

As mentioned above, the cytotoxicity of menadione and other
quinones has been related to the oxidative stress caused by the
redox cycling of these agents in their target cells. During the
one-electron reduction of menadione to the semiquinone radical and
its subsequent reoxidation by molecular oxygen, as well as by the
direct interaction of menadione with intracellular thiols, O_2^- is
produced[16] (Fig. 4). Dismutation of O_2^- yields H_2O_2 which can be

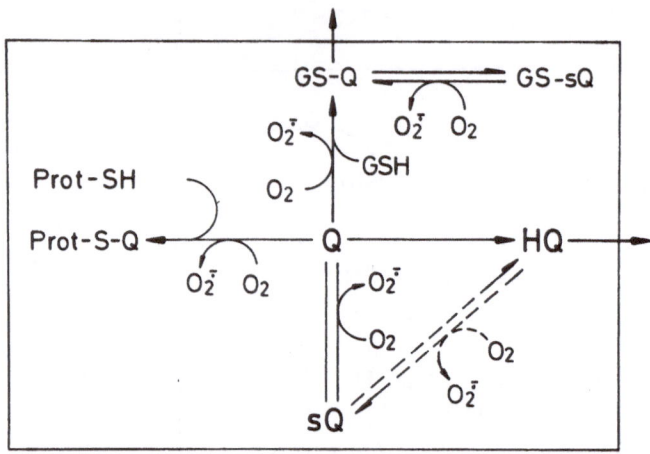

Fig. 4.   Mechanisms of menadione-induced superoxide radical formation in hepatocytes. Q, menadione; sQ, menasemiquinone;
HQ, menadiol; GSH, glutathione, reduced form; Prot-SH,
protein thiol; GS-Q, menadione-glutathione conjugate;
Prot-S-Q, protein-bound menadione.

further metabolized to H_2O by the glutathione peroxidase system
at the expense of GSH and NADPH[17].

Exposure of hepatocytes to toxic concentrations of menadione
is associated with extensive formation of O_2^- and H_2O_2, and the
oxidation of glutathione and pyridine nucleotides[15]. Treatments
which promote one-electron reduction and redox cycling of the
quinone potentiate menadione toxicity. Conversely, toxicity is
diminished when resynthesis of glutathione is facilitated. It thus
appears that glutathione depletion is a critical event in the initiation of menadione toxicity in isolated hepatocytes, and that
the metabolism of large amounts of H_2O_2 by glutathione peroxidase
is responsible for the loss of the major fraction of reduced
glutathione.

A recent analysis has shown that menadione metabolism by
isolated hepatocytes affects the status of both soluble (GSH) and
protein thiols[16]. In addition to oxidation of GSH to GSSG, formation of mixed disulfides and glutathione conjugate contribute to
GSH depletion during menadione metabolism. Following the decrease
in soluble thiols there is also a time- and dose-dependent loss of
protein sulfhydryl groups which is due to both oxidation and aryla-
tion of the protein thiols, oxidation being the quantitatively
more important mechanism[18]. This loss of protein thiols correlates

very closely with the onset of toxicity and invariably precedes cells death. It therefore seems that the depletion of protein thiols, chiefly through oxidative mechanisms, is a critical event which follows GSH depletion in oxidative cell injury caused by menadione.

Perturbation of Intracellular Calcium Ion Homeostasis during Oxidative Stress

The concentration of intracellular Ca^{2+} is a critical factor in the regulation of numerous metabolic processes. Ca^{2+} is known to be distributed throughout the cell in various functional compartments, including the mitochondria, endoplasmic reticulum, and binding proteins such as calmodulin[19]. The interdependent action of these compartments in controlling the free Ca^{2+} level is unquestioned, with the mitochondria playing a pivotal role by virtue of their ability to accumulate large amounts of Ca^{2+} during coupled respiration[20]. Studies with isolated mitochondria indicate that redox transitions, especially the oxidation of reduced pyridine nucleotides, are crucial in controlling mitochondrial Ca^{2+}[21]. Recent work by Lötscher et al.[22], and in our laboratory[23], has shown that the glutathione redox cycle may also contribute to this regulation by affecting the redox state of mitochondrial pyridine nucleotides. Relatively little is known, however, about the maintenance of Ca^{2+} homeostasis, and the interaction of the various Ca^{2+} compartments, in the intact cell.

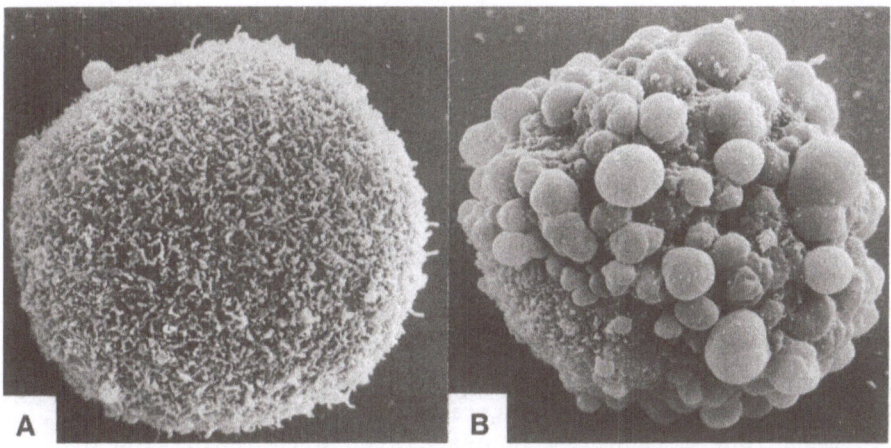

Fig. 5.   Scanning electron micrographs of hepatocytes incubated in absence (A) or presence (B) of menadione. Magnification: X 3000. The figure illustrates the formation of numerous surface blebs as result of menadione toxicity.

We have previously reported that menadione and several other
toxic agents cause the formation of numerous small blebs on the
surface of isolated hepatocytes[24], (Fig. 5). The same type of
surface blebbing can be induced by the calcium ionophore A23187,
suggesting that the alterations in surface structure caused by
menadione and other toxins are due to a redistribution of intracellular Ca^{2+}. This hypothesis is further supported by the observation that GSH depletion is followed by a redistribution of intracellular Ca^{2+} in hepatocytes incubated with menadione[15]. Detailed
analysis has shown that this effect is due to inhibition of Ca^{2+}
sequestration in both the mitochondria and endoplasmic reticulum
as result of menadione metabolism. Whereas the effect of menadione
on mitochondrial Ca^{2+} sequestration seems to be related to the
oxidation of pyridine nucleotides[23], the inhibitory effect on
endoplasmic reticular Ca^{2+} sequestration appears to be due to
oxidation/arylation of thiol group(s) critical for Ca^{2+}-ATPase
activity[25]. The effects of menadione metabolism on regulation of
Ca^{2+} compartmentation in hepatocytes are illustrated schematically
in Fig. 6.

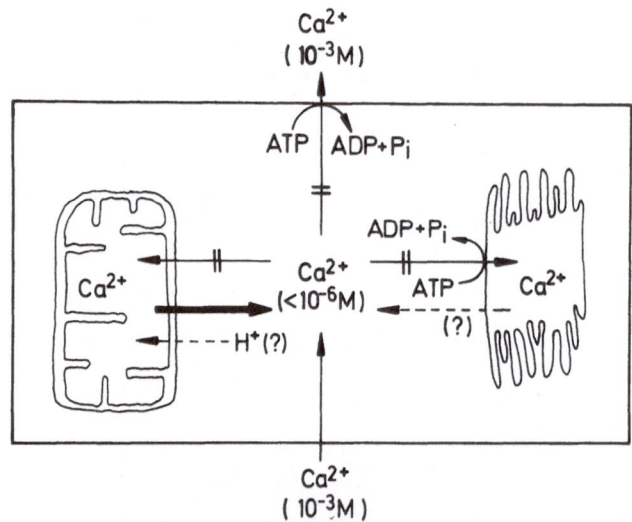

Fig. 6.   Schematic illustration of the effects of menadione metabo-
          lism on Ca^{2+} compartmentation in hepatocytes. As shown,
          menadione metabolism is associated with inhibition of
          Ca^{2+} sequestration in both the mitochondria and endoplas-
          mic reticulum, stimulation of Ca^{2+} release by the mito-
          chondria, and inhibition of active Ca^{2+} efflux from the
          cell.

Because of the inhibition of Ca^{2+} sequestration in the endoplasmic reticulum and mitochondria, incubation of hepatocytes with menadione results in a release of Ca^{2+} sequestered in these compartments into the cytosol. Normally, this would probably only cause a transient rise in the cytosolic Ca^{2+} concentration, because the plasma membrane Ca^{2+} pump would remove this Ca^{2+} from the cell, and the cytosolic Ca^{2+} concentration would return to its usual very low level. Recent studies in our laboratories have shown, however, that the hepatic plasma membrane Ca^{2+}-ATPase is inhibited by menadione and other agents which oxidize protein thiol groups[18,26].

The redox cycling of menadione in hepatocytes could therefore inhibit all three Ca^{2+} translocases present in the mitochondria, endoplasmic reticulum and plasma membrane. This would undoubtedly lead to a sustained rise in cytosolic Ca^{2+} level which could cause surface blebbing by altering microfilament organization. Obviously, measurements of cytosolic Ca^{2+} concentration following exposure of hepatocytes to menadione are required to substantiate this hypothesis, but this is practically very difficult. We have therefore used phosphorylase activity to monitor alterations in cytosolic free Ca^{2+} concentration in hepatocytes subjected to oxidative stress[27]. Phosphorylase activity has previously been demonstrated to be a valid indicator of fluctuations in cytosolic Ca^{2+} level; under appropriate experimental conditions, it is strictly dependent on a Ca^{2+}-requiring phosphorylase kinase[28]. Using this approach we have found that exposure of hepatocytes to toxic levels of menadione causes prolonged phosphorylase activation[18]. Thus, it appears that menadione-induced oxidative stress in hepatocytes can cause depletion of GSH and protein thiols, and a perturbation of Ca^{2+} homeostasis which may lead to an uncontrollable rise in cytosolic free Ca^{2+} concentration resulting in cell death. A schematic representation of these events is given in Fig. 7.

CONCLUDING REMARKS

Isolated mammalian cells have become an extremely useful experimental model for the study of many different aspects of drug metabolism and drug toxicity. Improved methodology has allowed the direct quantitation of various hemoproteins and detailed analysis of interactions between drugs and cytochrome P-450 in intact cells. The generation of cosubstrates for drug biotransformation is also readily studied and appears to be one of the areas that can be favorably approached in future studies with isolated cells. Recent studies employing model toxins, such as menadione, and isolated cells have also been described in this presentation, and it is hoped that these studies illustrate the

usefulness of the isolated cell model in studies of drug toxicity.
It should be borne in mind, however, that suspensions of isolated
cells are not necessarily an ideal experimental model in every
situation. One important consideration that is often overlooked is
that in such a preparation the spatial heterogeneity of the cells
is not the same as in the whole organ. This can be particularly
important in studies on cells from complex organs, such as the
kidney, where the metabolic zonation of biochemical reactions
probably plays a significant role.

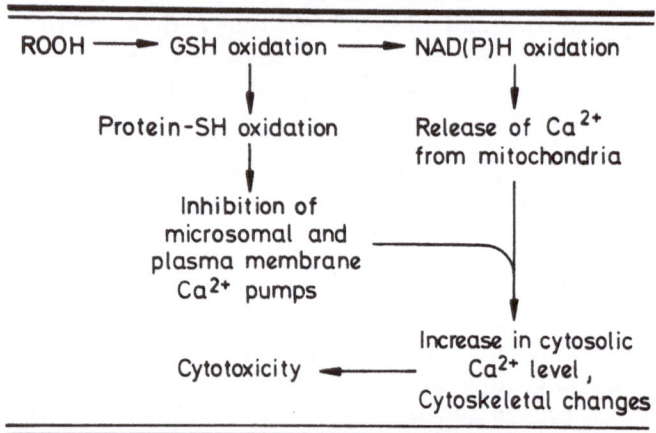

Fig. 7. Development of oxidative cell injury in hepatocytes.
Proposed sequence of events. ROOH, hydroperoxide; GSH,
glutathione, reduced form; Protein-SH, protein thiol.
It is suggested that formation of hydrogen peroxide as
result of redox cycling of menadione causes GSH and NADPH
oxidation and redistribution of intracellular Ca^{2+} leading
to an increase in cytosolic Ca^{2+} concentration which is
critically involved in the development of cytotoxicity.

Despite such limitations, freshly isolated cells are widely
used today in studies of cellular metabolism as well as drug bio-
transformation· and toxicity. Moreover, as further progress is made
in the development of techniques for studies on intact cells, it
is quite clear that such preparations will become increasingly
used in areas in which isolated cells constitute a good experimental model.

REFERENCES

1. W. B. Jakoby, "Enzymatic Basis of Detoxication", Vol. I & II, Academic Press, New York (1980).
2. S. Orrenius and H. Sies, Compartmentation of detoxication reactions, in: "Metabolic Compartmentation", H. Sies, ed., Academic Press, New York (1982).
3. M. D. Burke and S. Orrenius, Isolation and comparison of endoplasmic reticulum membranes and their mixed function oxidase activities from mammalian extrahepatic tissues, Pharmacol. Therap. 7:549 (1979).
4. S. Orrenius, K. Ormstad, H. Thor and S. A. Jewell, Turnover and functions of glutathione studied with isolated hepatic and renal cells, Federation Proc. 42:3177 (1983).
5. N. S. Kosower and E. M. Kosower, The glutathione status of cells, Int. Rev. Cytol. 54:109 (1978).
6. S. Orrenius and P. Moldéus, The multiple roles of glutathione in drug metabolism. Trends Pharmacol. Sci. 5:432 (1984).
7. H. Sies, Reduced and oxidized glutathione efflux from liver, in: "Glutathione: Storage, Transport and Turnover in Mammals", Y. Sakamoto, T. Higashi and N. Tateishi, eds, Japan Scientific Societies Press, Tokyo (1983).
8. M. T. Smith and S. Orrenius, Studies on drug metabolism and drug toxicity in isolated mammalian cells, in: "Drug Metabolism and Drug Toxicity", J. R. Mitchell and M. G. Horning, eds, Raven Press, New York (1983).
9. D. P. Jones, S. Orrenius and H. S. Mason, Hemoprotein quantitation in isolated hepatocytes, Biochim. Biophys. Acta, 576:17 (1979).
10. J. A. Hinson, S. D. Nelson and J. R. Mitchell, Studies on the microsomal formation of arylating metabolites of acetaminophen and phenacetin, Mol. Pharmacol. 13:625 (1977).
11. D. P. Jones, P. Moldéus, A. H. Stead, K. Ormstad, H. Jörnvall and S. Orrenius, Metabolism of glutathione and a glutathione conjugate by isolated kidney cells, J. Biol. Chem. 254:2787 (1979).
12. H. M. Hassan and I. Fridovich, Intracellular production of superoxide radical and of hydrogen peroxide by redox active compounds, Arch. Biochem. Biophys. 196:385 (1979).
13. L. Ernster, DT-Diaphorase, 10:309 (1967).
14. H. Wefers and H. Sies, Hepatic low-level chemiluminescence during redox cycling of menadione and the menadione-glutathione conjugate. Relation to glutathione and NAD(P)H: quinone reductase (DT-diaphorase) activity, Arch. Biochem. Biophys. 224:568 (1983).
15. H. Thor, M. T. Smith, P. Hartzell, G. Bellomo, S. A. Jewell and S. Orrenius, The metabolism of menadione (2-methyl-1,4-naphthoquinone) by isolated hepatocytes. A study of the implications of oxidative stress in intact cells, J. Biol. Chem. 257:12419 (1982).

16. D. Di Monte, D. Ross, G. Bellomo, L. Eklöw and S. Orrenius, Alterations in intracellular thiol homeostasis during the metabolism of menadione by isolated rat hepatocytes, Arch. Biochem. Biophys. 235:334 (1984).
17. A. Wendel, Glutathione peroxidase, in: "Enzymatic Basis of Detoxication", Vol. I, W. B. Jakoby, ed., Academic Press, New York (1980).
18. D. Di Monte, G. Bellomo, H. Thor, P. Nicotera and S. Orrenius, Menadione-induced cytotoxicity is associated with protein thiol oxidation and alteration in intracellular Ca^{2+} homeostatis, Arch. Biochem. Biophys. 235:343 (1984).
19. W. Y. Cheung, Calmodulin plays a pivotal role in cellular regulation, Science, 207:19 (1980).
20. E. Carafoli, Mitochondrial uptake of calcium ions and the regulation of cell function, Biochem. Soc. Symp., 39:89 (1976).
21. A. L. Lehninger, A. Vercesi and E. Bababunmi, Regulation of Ca^{2+}-release from mitochondria by the oxidation-reduction state of pyridine nucleotides, Proc. Natl. Acad. Sci. USA, 75:1690 (1978).
22. H. R. Lötscher, K. H. Winterhalter, E. Carafoli and C. Richter, Hydroperoxides can modulate the redox state of pyridine nucleotides and the calcium balance in rat liver mitochondria, Proc. Natl. Acad. Sci. USA, 76:4340 (1979).
23. G. Bellomo, S. A. Jewell and S. Orrenius, The metabolism of menadione impairs the ability of rat liver mitochondria to take up and retain calcium, J. Biol. Chem. 257:11558 (1982).
24. S. A. Jewell, G. Bellomo, H. Thor, S. Orrenius and M. T. Smith, Bleb formation in hepatocytes during drug metabolism is caused by alterations in intracellular thiol and Ca^{2+} homeostasis, Science, 217:1257 (1982).
25. D. P. Jones, H. Thor, M. T. Smith, S. A. Jewell and S. Orrenius, Inhibition og ATP-dependent microsomal Ca^{2+} sequestration during oxidative stress and its prevention by glutathione, J. Biol. Chem. 258:6390 (1983).
26. G. Bellomo, F. Mirabelli, P. Richelmi and S. Orrenius, Critical role of sulfhydryl group(s) in ATP-dependent Ca^{2+} sequestration by the plasma membrane fraction from rat liver, FEBS Lett. 163:136 (1983).
27. G. Bellomo, H. Thor and S. Orrenius, Increase in cytosolic Ca^{2+} concentration during t-butyl hydroperoxide metabolism by isolated hepatocytes involves NADPH oxidation and mobilization of intracellular Ca^{2+} stores, FEBS Lett. 168:38 (1984).
28. D. A. Malencik and E. H. Fisher, Structure, function and regulation of phosphorylase kinase, in: "Calcium and Cell Function", W. Y. Cheung, ed., Vol. III, Academic Press, New York (1982).

300

CHARACTERISATION OF THE NADPH - DEPENDENT SUPEROXIDE - GENERATING OXIDASE OF PHAGOCYTIC LEUKOCYTES

John F. Parkinson, Andrew R. Cross and Owen T.G. Jones

Department of Biochemistry
The Medical School, University of Bristol
University Walk, Bristol, BS8 1TD. U.K.

INTRODUCTION

Phagocytic leukocytes play a major role as part of the immune system in providing protection against infection. As part of their microbicidal activity, neutrophils, the most abundant blood phagocytes, convert molecular oxygen to highly reactive and toxic derivatives such as the superoxide anion, O_2^-, hydrogen peroxide, H_2O_2, hypochlorite, OCl^-, the hydroxyl radical, OH, and possibly singlet oxygen, $^1O_2$, (1).

A key process in the oxidative killing by neutrophils is the conversion of O_2 to O_2^-, a one - electron reduction. The enzyme complex involved, referred to as the NADPH - oxidase, is associated with the plasma membrane of neutrophils and accounts for nearly all the O_2 consumed by stimulated cells during the so-called "respiratory burst"; the oxidase is inactive in resting cells, (2). Other toxic derivatives can be produced from O_2^-, either spontaneously or by enzyme and metal-catalysed reactions, as shown in Table 1.

It is now generally accepted that the physiological electron donor for the oxidase complex is NADPH. This is in agreement with the observation that in stimulated cells the activities of the enzymes of the hexose monophosphate shunt are markedly increased, (3). Also, kinetic studies on the oxidase in cell - free extracts have shown a lower K_m for NADPH (approx. 50µM) than for NADH (approx. 500µM), (4,5).

Early investigations suggested the involvement of a flavoprotein in the oxidase since its activity in particulate and sol-

TABLE 1. Reactions which may be involved in O_2 uptake
by stimulated neutrophils.

1. $NADPH + 2O_2 \rightarrow NADP^+ + H^+ + 2O_2^{\overline{\cdot}}$ (catalysed by NADPH oxidase)

2. $2O_2^{\overline{\cdot}} + 2H^+ \rightarrow H_2O_2 + O_2$ (catalysed by superoxide dismutase)

3. $O_2^{\overline{\cdot}} + H_2O_2 \rightarrow OH^- + O_2 + OH^\bullet$ (Haber - Weiss reaction)

4. $Fe^{2+} + H_2O_2 \rightarrow Fe^{3+} + OH^- + OH^\bullet$ (Fenton reaction)

5. $O_2^{\overline{\cdot}} + H_2O_2 \rightarrow OH^- + {}^1O_2 + OH^\bullet$

6. $H_2O_2 + Cl^- \rightarrow OCl^- + H_2O$ (catalysed by myeloperoxidase)[*]

[*]Myeloperoxidase is a haemoprotein present in large quantities in neutrophils.

ubilised extracts of stimulated neutrophils could be stabilised by the addition of FAD, (6). The oxidase is also strongly inhibited by flavin analogues such as quinacrine and 5 - Deaza - FAD, (7,8). This laboratory has proposed a role for a b-type cytochrome in this oxidase complex, (9). The cytochrome, originally detected in rabbit neutrophils (10), is located in both the plasma membrane and granule fractions of neutrophils and is incorporated into the phagolysosomal membrane during phagocytosis, (11). The cytochrome becomes reduced when neutrophils are stimulated under anaerobic conditions, (12) and it can be readily observed in difference absorbance spectra of whole cell homogenates, with characteristic maxima at 428 and 559 nm, (see Fig. 1).

An important series of observations has been made, both with regard to the cytochrome - b and FAD, in studies on neutrophils from patients suffering from Chronic Granulomatous Disease (CGD). In this disease, which exhibits both X - linked and autosomal recessive modes of inheritance, the neutrophils are unable to mount a respiratory burst in response to stimulation. Toxic O_2 derivatives are not produced and these neutrophils cannot kill bacteria effectively, (13).

In patients with the X-linked form of CGD, cytochrome - b has been shown to be absent, (see Fig. 1) and also to be present at diminished levels in female carriers, (14). In the autosomal recessive form of CGD, cytochrome - b is present but fails to become reduced in whole cells stimulated under anaerobic conditions, (15). In this latter case the cells have been shown to have altered levels of FAD, (16,17) and may thus lack the reductase for cytochrome - b. Fusion of autosomal recessive cells with X - linked cells can result in the restoration of the normal respiratory burst response, (18).

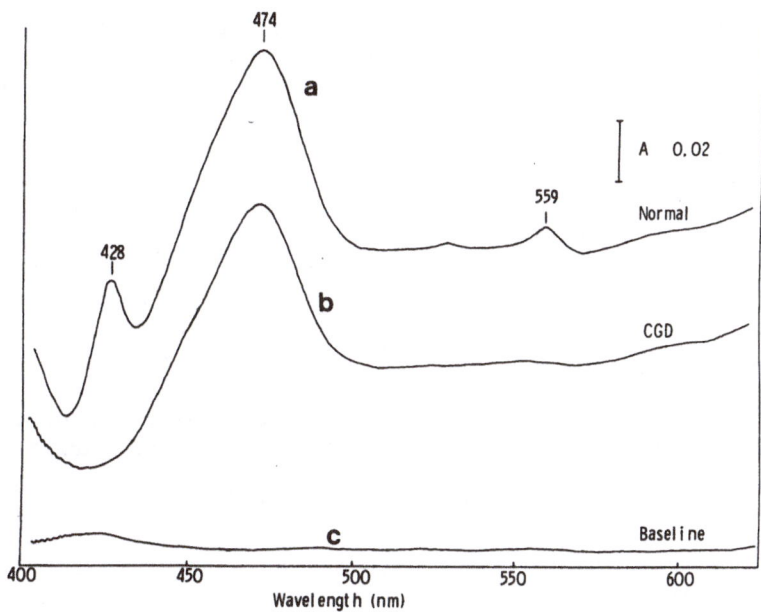

Fig. 1. Reduced - minus - oxidised defference spectra of neutrophil homogenates.
Spectrum (a) was obtained from normal neutrophils.
Spectrum (b) was obtained from neutrophils from a patient
with X - linked Chronic Granulomatous Disease. Spectrum
(c) was the instrument baseline. The peaks at 559 and
428nm are due to cytochrome - b absorbance. The peak at
474nm is due to myeloperoxidase absorbance.

Further evidence that the cytochrome - b is likely to be part
of the oxidase complex is provided by the fact that chemically
reduced cytochrome - b is rapidly oxidised by air with $t_{1/2}$ = 4.7
msec., (19). It also has an unusually low oxidation - reduction
mid-point potential ($Em_{7.0}$) of -245mV, (20), which is sufficiently
low to catalyse the reduction of O_2 to O_2^- at rapid rate, the $Em_{7.0}$
of the O_2/O_2^- couple being -160mV, (21).

Like other O_2 - binding haemoproteins, the reduced cytochrome
binds small ligands such as CO and n-butyl isocyanide, (19,22).
The cytochrome, now designated cytochrome - b_{-245}, has been shown
to be present in other phagocytes, including eosinophils, monocytes and macrophages, (23).

Taken as a whole, the evidence outlined above indicates that
the NADPH oxidase contains at least two proteins. These are
proposed to be cytochrome -b_{-245} and an FAD-linked flavoprotein,
either of which may be missing or malfunctioning in various forms
of CGD. Our current proposal is that the NADPH oxidase is an
electron transfer complex organised as shown in Scheme 1.

Cytochrome - b_245 has recently been purified to homogeneity from human neutrophils. It is a glycoprotein of molecular weight 68 - 78 Kd and is extremely hydrophobic, tending to form large aggregates, (24). The flavoprotein portion of the oxidase has not been well characterised. In detergent extracts of membranes from stimulated neutrophils, the ratio of FAD to cytochrome - b_245 is typically 1.3 : 1 (22). However, in partially purified preparations this ratio drops to 0.2 : 1 or even less, indicating that not all the FAD associated with neutrophil plasma membranes is associated with the oxidase, (25). Extraction of stimulated neutrophil plasma membranes with 1% deoxycholate causes complete loss of oxidase activity; the cytochrome - b_245 remains in a 100,000g membrane pellet whilst the supernatant contains a fluorescent flavoprotein which is reduced by added NADPH under anaerobic conditions, (26). This flavoprotein has not been purified.

The oxidase complex has been obtained in detergent solubilised extracts, but it is unstable, having a half-life of only a few minutes at room temperature. It also has a tendency to form large aggregates which make purification by conventional methods difficult. These aggregates may contain denatured protein and other protein contaminants which are difficult to remove. Several laboratories have obtained detergent solubilised preparations of the oxidase which were found to contain cytochrome - b_245 and FAD, (22, 25,27). In this article we describe a method for solubilisation of the oxidase from plasma membranes of stimulated pig neutrophils using a mixture of detergents. With this preparation we have been able to carry out various studies with inhibitors of the oxidase. More importantly, we have carried out kinetic experiments under both anaerobic and aerobic conditions. We believe that our results confirm that FAD and cytochrome - b_245 are organised within the oxidase complex as shown in Scheme 1.

$$NADPH \longrightarrow FLAVOPROTEIN \longrightarrow CYTOCHROME - b_{245} \begin{cases} O_2 \\ O_2^{\bullet-} \end{cases}$$

Scheme 1 : Organisation of the electron transfer components of the $O_2^{\bullet-}$ - generating NADPH - oxidase complex.

METHODS AND MATERIALS

Isolation of neutrophils: Whole pig blood (4 litres) was collected in plastic buckets containing EDTA in 0.9% saline at a final EDTA concentration of 0.15% (w/v), to prevent coagulation. Catering gelatine was then added to a final concentration of 1.25% (w/v) and the mixture allowed to stand for 1 hour, during which time erythrocytes aggregate and settle leaving an upper layer containing all the white blood cells. This layer was siphoned off and the cells harvested by centrifugation at 500g for 15 mins.

After washing once in 0.9% saline containing 5 units/ml heparin
and 1mM EDTA, the cells were subjected to several osmotic shocks
in order to remove contaminating erythrocytes.   This was achieved
by suspending the cells in 0.2% saline containing 5 units/ml
heparin and 1mM EDTA for 30 secs.   Isotonicity was then restored
by the addition of an equal volume of 1.6% saline containing 5
units/ml heparin and 1mM EDTA and the cells harvested by centrifugation at 400g for 7 mins.   When all traces of haemoglobin had
been removed, the cells were suspended in 0.9% saline and layered
on top of an equal volume of Ficoll-Hypaque and centrifuged at
800g for 20 mins.   The neutrophil pellet was collected and washed
twice with 0.9% saline.   Prior to stimulation the cells were
treated with 1mM di-isopropylfluorophosphate, left on ice for 5
mins, washed twice in 0.9% saline and finally suspended in a modified Krebs - Ringer buffer, pH 7.4, containing 50mM HEPES and 1mM
NaH_2PO_4 in place of $KHCO_3$ and supplemented with 2mM glucose and 5
units/ml heparin.

   Stimulation of neutrophils and oxidase preparations: The
cells, suspended in 100ml buffer, were incubated at 39°C (pig
blood temperature) and stimulated by the addition of 1μg/ml phorbol 12-myristate 13-acetate (PMA) for 6 mins.   Stimulation was
then stopped within seconds by the addition of frozen buffer and
the cells collected by centrifugation at 400g for 7 mins at 4°C.
The cells were then suspended in disruption buffer, pH 7.4, containing the following: 8.6% (w/v) sucrose, 2mM EGTA, 10mM HEPES,
4mM $MgSO_4$, 1mM NaN_3, 1mM Ca Cl_2, 10μg/ml DNAse 1, 0.01% (w/v)
silicon antifoam and 5μg/ml phenylmethylsulphonylfluoride;   the
protease inhibitors chymostatin, leupeptin, antipain, pepstatin
and 7-amino-1-dichloro-3-L-tosylamido-heptan-2-one were all added
at 1μg/ml.

   The cell suspension was disrupted by N_2-cavitation after
stirring on ice for 20 mins at 500 p.s.i. in a nitrogen bomb.
(Parr Instruments, U.K.).   Cell debris was removed by centrifugation at 500g for 7 mins and the post-nuclear supernatant thus
obtained was layered onto 40% (w/v) sucrose containing 10mM HEPES
and 1mM EGTA, pH 7.4.   After centrifugation for 1 hour at 100,000g
at 2°C, the plasma membrane fraction of the stimulated neutrophils
was collected from the interface between the soluble fraction and
the 40% sucrose, diluted with disruption buffer and pelleted by
centrifugation for 30 mins at 100,000g at 2°C.   The membrane
pellet was then suspended in extraction buffer, pH 8.0, containing
the following: 20% (v/v) glycerol, 20mM glycine, 1mM $MgSO_4$, 0.5mM
$CaCl_2$, 1mM NaN_3, 0.25% (v/v) Lubrol-PX, 0.25% (w/v) sodium deoxycholate and the same mixture of protease inhibitors as was used in
the disruption buffer.   This suspension was then sonicated 3 times
for 20 secs. at 100W using a Branson sonifier, with cooling in an
ice-water bath between treatments and centrifuged for 1 hour at

100,000g at 2°C.   The oxidase-enriched supernatant was pipetted
into aliquots and frozen in liquid N_2, in which it was also stored.
The green pellet was discarded.

Determination of $O_2^{\bullet -}$ - generating activity: NADPH and NADH -
dependent production of $O_2^{\bullet -}$ was measured as the superoxide dismutase sensitive rate of cytochrome c reduction at 20°C, using a dual
wavelength spectrophotometer set at 550 - 540nm, (19).

Absorbance spectroscopy: All absorbance spectra were recorded
on a rapid-scanning spectrophotometer constructed using a Hilger
D330 monochromator.   Signals were recorded at 0.05nm intervals at
12-bit resolution and processed in an ITT 2020 microcomputer with
a floppy disc drive unit.   Spectra were displayed on a visual
display unit and hard copies taken on an X - Y plotter (Hewlett
Packard).

Assay of Cytochrome - b_{-245} content: The cytochrome was
assayed by dithionite reduced-minus-oxidised difference absorbance
spectroscopy, using $\triangle E_{559-540nm} = 21.6$ $cm^{-1}mM^{-1}$ (19).

Assay of FAD content: FAD was assayed by dithionite reduced-
minus-oxidised difference absorbance spectroscopy, using
$\triangle E_{450-500} = 11.3$ $cm^{-1}mM^{-1}$ (28).   These values were confirmed by
extraction of total flavins which were assayed by fluoroscence
spectroscopy, (29).

Determination of $Em_{7.0}$:   The method, based on measurement of
absorbance spectra at known oxidation - reduction potentials was
as described previously, (20).

Spectra under anaerobic conditions: Sample was added to the
main well of a Thunberg cuvette with appropriate reagents pipetted
into the side-arm.   The assembly was repeatedly (20 times) evac-
ated and then flushed with N_2. N_2 was made O_2-free by flushing
through a solution containing proflavin hemisulphate (224mg/l),
methyl viologen (270mg/l) and EDTA (7.5g/l) in 250mM sodium phosphate buffer, pH 6.5, which was activated by light during use,
(30).   Anaerobiosis by this method was calibrated using myoglobin.
Incubations were carried out at 14°C.

Spectra under aerobic conditions: 0.4 ml sample was placed
in a 1ml cuvette in the rapid - scanning spectrophotometer.   After
spectra of the oxidised sample had been recorded, suitable additions of reagents were made and further spectra obtained.   The
sample was maintained in a state of aerobiosis by stirring between
spectral measurements.   Incubations were carried out at 20°C.

Protein assay: The method of Bramhall et al was used, (31),

306

Materials: PMA, Ficoll-Hypaque, di-isopropylfluorophosphate, all other protease inhibitors, proflavin hemisulphate, methyl viologen, p-chloromercuribenzoate, quinacrine, Lubrol-PX and sodium deoxycholate were all from Sigma Chemical Co. U.K.

Silicon antifoam was from BDH Ltd. U.K.
NADPH and NADH were from Boehringer Mannheim U.K.
All other reagents were of purest grade available.

Results and Discussion

The solubilised preparation of NADPH-oxidase from stimulated pig neutrophils obtained by our method has high superoxide-generating

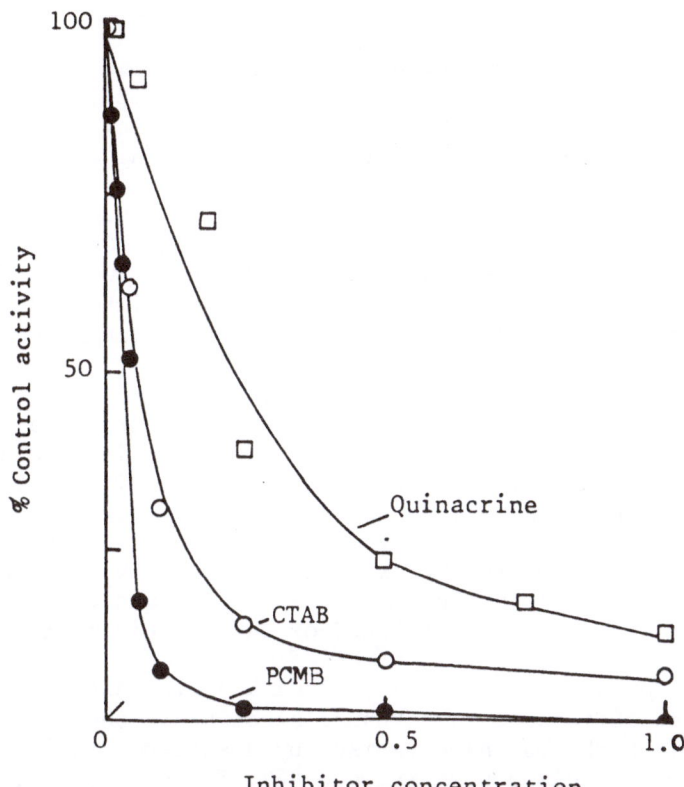

Fig. 2.   Effects of inhibitors on the $O_2^{\overline{\bullet}}$ -generating activity of solubilised oxidase.
The inhibitor concentration range 0 – 1.0 shown on the abscissa represents 0 – 4mM quinacrine, (□), or 0 – 0.01% w/v cetyltrimethylammonium bromide, (○), or 0 – 20μM p-chloromercuribenzoate, (●).

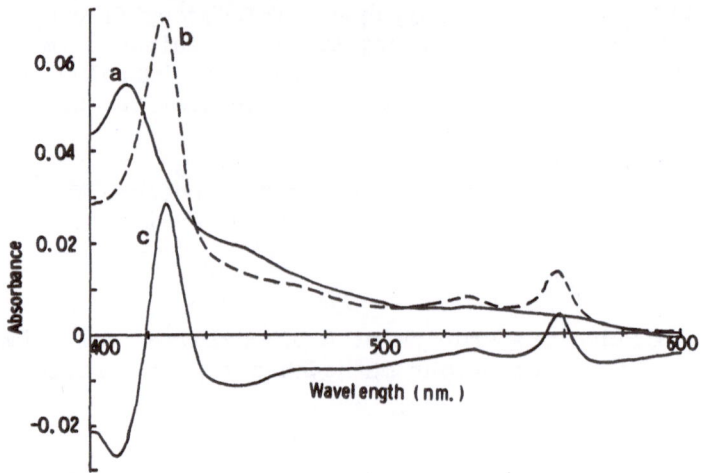

Fig. 3. Spectroscopic properties of the solubilised oxidase.
Spectrum (a), absolute air - oxidised sample. Spectrum
(b); absolute dithionite - reduced smaple. Spectrum (c);
dithionite - reduced - minus - air - oxidised sample.

activity with a yield of 40% from the postnuclear supernatant. The
specific activity of a typical extract, 365 nmoles $O_2^{\overline{\bullet}}$/min/mg
protein, is 12-fold that of whole cells (30 nmoles $O_2^{\overline{\bullet}}$/min/mg
protein) and is much greater than that obtained with oxidase extracted using exactly the same procedure from unstimulated cells,
2.0 nmoles $O_2^{\overline{\bullet}}$/min/mg protein.

The sensitivity of the oxidase to low concentrations of the
inhibitors p-chloromercuribenzoate (PCMB), quinacrine and cetyltrimethylammonium bromide (CTAB) was investigated. The results,
summarised in Fig. 2 agree with those previously reported (7).
PCMB is known to interact with cytochrome b-245 since it raises the
$Em_{7.0}$ of the cytochrome to approximately -180mV (32): Quinacrine
is an inhibitor of flavin-linked reactions (7) and CTAB, a cationic
detergent, is known to inhibit NADPH-oxidase of pig neutrophils (7).

Absolute oxidized, absolute reduced and also dithionite reduced-
minus-oxidised difference absorbance spectra are shown in Fig. 3.
The cytochrome b-245 is characterised by absorbance maxima at 417nm
in the absolute oxidised spectrum and 427, 531 and 559nm in the
absolute reduced and reduced-minus-oxidised difference spectra.
Absorbance due to FAD is seen as a broad band centred around 450nm
in the absolute oxidised spectrum and is seen as a trough centred
around 450nm in the difference spectrum .

The $Em_{7.0}$ of the cytochrome in this preparation, as determined by potentiometric titrations, was found to be -238mV, slightly higher than that for the cytochrome in intact membranes. Such titrations also showed that less than 8% of the cytochrome b of this preparation was reduced at oxidation-reduction potentials of -100mV and above. This indicates an insignificant contamination of this preparation by microsomal cytochrome b_5 and mitochondrial b-type cytochromes.

Detergent solubilised NADPH-oxidase was unstable at room temperature. Stability was increased by the addition of glycerol and protease inhibitors to the extraction buffer and storage in liquid nitrogen retained activity over a period of several months.

The $O_2^{\bar{}}$ generating oxidase activity of these preparations was found to have a K_m for NADPH of 45μM and for NADH of 460μM, in close agreement with previously reported figures for particulate preparations (27).

Reduction of FAD and cytochrome b-245 in anaerobic conditions

Addition of NADPH to the solubilised oxidase under anaerobic conditions gave rise to reduction of both cytochrome b-245 and FAD, shown by increases in absorbance at 427 and 559nm due to cytochrome b-245 and decreased absorbance at 450nm for FAD. This is similar to results obtained previously in whole cells and in isolated membranes. (12).

The results shown in Fig. 4 indicate that a substantial proportion of cytochrome b-245 (24%) and of FAD (34%) were reduced within 1 min. This initial rapid reduction was followed by a slow phase which led to almost complete FAD reduction and about 60% cytochrome b-245 reduction.

The effects of NADPH-oxidase inhibitors on these reduction rates is also shown in Fig. 4. PCMB inhibited cytochrome b reduction whilst having less effect on FAD reduction. CTAB and quinacrine also inhibited cytochrome b-245 reduction. The effect of quinacrine on FAD reduction could not be determined as quinacrine is highly coloured, but CTAB inhibited FAD reduction by NADPH.

Oxidase prepared from unstimulated cells exhibited reduction rates very different from those of stimulated cells (see Fig. 5). The initial fast reduction of both cytochrome b-245 and FAD was absent, but after 30 minutes extents of reduction were similar to those found with oxidase extracted from stimulated cells. The addition of methyl viologen, a mobile electron carrier, to the oxidase from stimulated cells resulted in a rapid and almost complete reduction of both cytochrome b-245 and FAD.

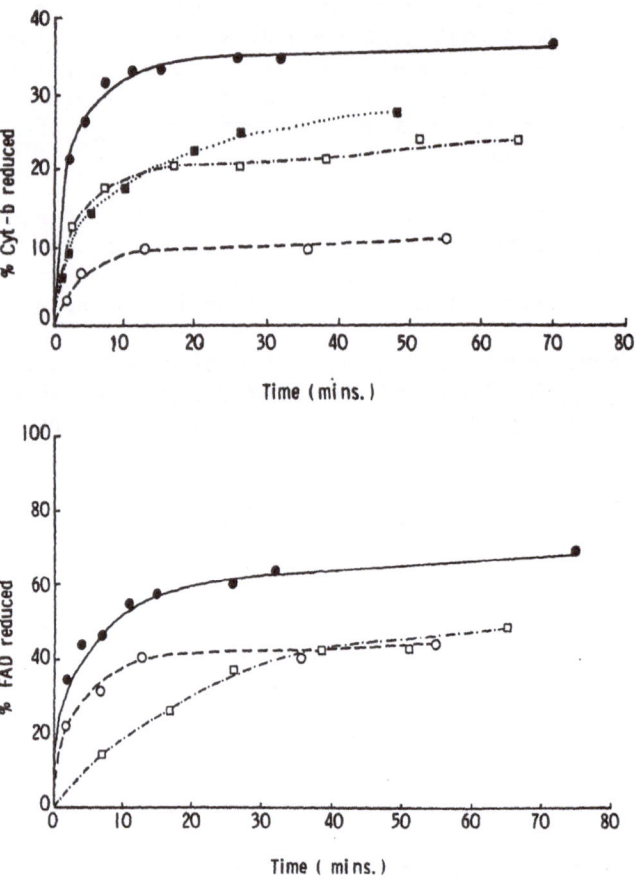

Fig. 4. Effects of inhibitors on the NADPH-dependent reduction of
cytochrome-b and FAD during anaerobic incubations. Inhibitors were used at the following concentrations; p-
chloromercuribenzoate, 12.5μM (○); cetyltrimethylammonium
bromide, 0.005% (w/v) (□); quinacrine, 5mM (■); control
(no inhibitor) (●). NADPH was present at 1mM.

      Whilst these anaerobic incubations indicate that the reduction
of cytochrome $\underline{b}$-245 and FAD are NADPH-dependent and sensitive to
oxidase inhibitors in a manner compatible with organisation of an
electron transport system according to scheme 1, they do not satisfy
kinetic considerations.   Simple calculations show that the rate of
cytochrome $\underline{b}$-245 reduction under these conditions is less than 10%
of that required to produce the rate of $O_2^{\bar{}}$-generation measured for
this preparation.   Since 24% of the cytochrome $\underline{b}$-245 was reduced
before we could take measurements we carried out a thorough search,
using anaerobic stopped-flow spectroscopy for any fast-reacting
cytochrome $\underline{b}$ component.   No such component was found.   This is an

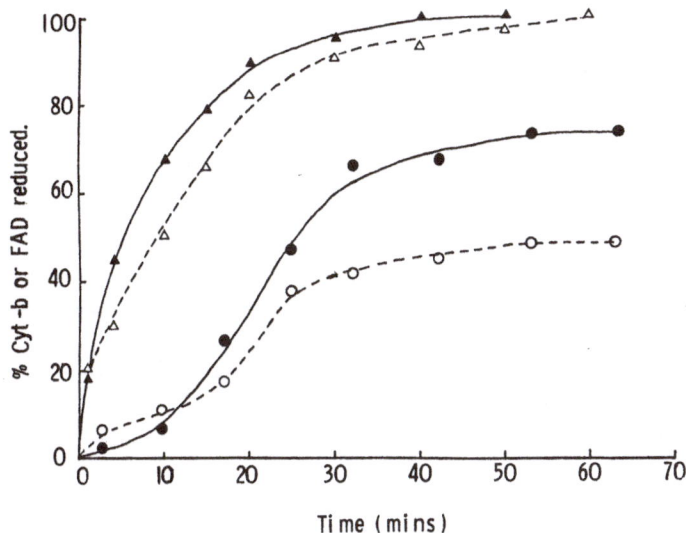

Fig. 5. Kinetics of cytochrome-b and FAD reduction under anaerobic conditions in the presence of Methyl Viologen and in oxidase from unstimulated cells.
Reduction of cytochrome-b ($\triangle$) and FAD ($\blacktriangle$) in the presence of 12.5μM Methyl Viologen and 1mM NADPH. Reduction of cytochrome-b ($\bigcirc$) and FAD ($\bullet$) in oxidase from unstimulated cells in the presence of 1mM NADPH.

unexpected result in view of the fact that there is much evidence to support the role of cytochrome $\underline{b}$-245 as the terminal electron donor in this oxidase complex.

The stimulatory effect of Methyl Viologen on both FAD and Cytochrome $\underline{b}$-245 reduction rates indicates that electron transfer between these components is limited under anaerobic conditions and that this is overcome by a mobile electron carrier. There may be build up of an electron transfer intermediate in anaerobic conditions which prevents electrons passing to cytochrome $\underline{b}$-245.

Reduction of Cytochrome $\underline{b}$-245 in aerobic conditions

For these experiments an oxidase preparation with $O_2^{\top}$-generating activity of 348 nmoles $O_2^{\top}$/min/mg protein, cytochrome $\underline{b}$-245 content of 445 nmoles/mg protein and FAD content of 492 pmoles/mg protein, was used.

Upon addition of 1mM NADPH under aerobic steady state conditions, approximately 9% of cytochrome $\underline{b}$-245 became reduced. (Fig. 6). An oxidase preparation from unstimulated cells showed no measurable cytochrome $\underline{b}$-245 reduction. This steady state level of reduction was abolished by leaving the oxidase preparation to stand at room temperature for 30 minutes prior to addition of NADPH. Addition of

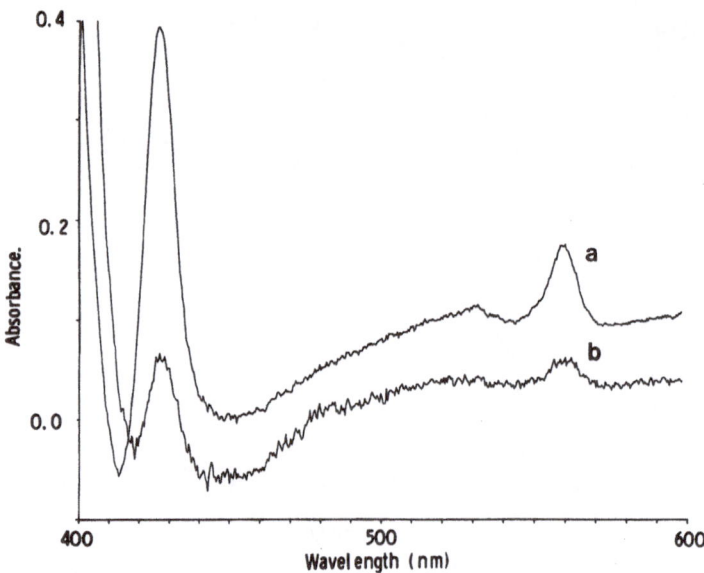

Fig. 6. Steady state cytochrome-b reduction by added NADPH under
aerobic conditions.
Spectrum (a), dithionite-reduced-minus-oxidised difference
spectrum of solubilised oxidase. Spectrum (b), difference
absorbance spectrum showing the steady state level of cytochrome-b reduction in the presence of 1mM NADPH under
aerobic conditions. The absorbance of spectrum (b) has
been expanded 4 times for clarity.

10µM PCMB which completely abolished $O_2^{\bar{}}$ generation by this preparation also abolished the steady state level of cytochrome b-245
reduction. NADPH-oxidase is insensitive to 1mM KCN (3) and
addition of this had no effect on the steady state level of reduction
of cytochrome b-245.

The steady state level of cytochrome b-245 reduction on the
addition of 50µM NADPH was 65% of that obtained with 100µM NADPH.
Addition of 1mM NADH gave similar levels of reduction to that
obtained by the addition of 100µM NADPH. Addition of 100µM NADH
gave no measurable cytochrome b-245 reduction.

These results indicate that the steady state level of reduction
of cytochrome b-245 has a much lower K_m for NADPH than for NADH, the
values being similar to those for $O_2^{\bar{}}$-generating activity. The
insensitivity to 1mM KCN and the complete abolition by 10µM PCMB
are also characteristics of NADPH oxidase and show that the cytochrome b that is reduced does not arise from microsomes or mitochondria.

The maximum rate of $O_2^{\bar{}}$-generation by this oxidase preparation is 348 nmoles O_2^-/min/mg protein. Since the cytochrome $\underline{b}_{-245}$ content is 445 pmoles/mg protein, this rate can be expressed as 13.03/sec/cytochrome $\underline{b}_{-245}$.

We have shown that the rate of oxidation of reduced cytochrome $\underline{b}_{-245}$ has $t_{\frac{1}{2}} = 4.7$ ms (19), giving a first order rate constant of 147/sec. Since the level of reduction of cytochrome $\underline{b}$ reduction under steady state conditions has now been measured, the rate of cytochrome $\underline{b}_{-245}$ reduction can be calculated as follows:

$$\text{rate of cytochrome } \underline{b}_{-245} \text{ reduction} = 147/\text{sec} \times \frac{[\text{reduced cytochrome}]}{[\text{oxidised cytochrome}]}$$

$$= 147/\text{sec} \times 0.09$$

$$= 13.23/\text{sec/cytochrome } \underline{b}_{-245}$$

This value of 13.23/sec/cytochrome $\underline{b}_{-245}$ calculated from our experimental data is very close to the measured rate for overall NADPH dependent O_2^--generation catalysed by this oxidase preparation. We believe this confirms the role of cytochrome $\underline{b}_{-245}$ in the oxidase complex as described in Scheme 1.

From a mechanistic point of view these results also show that electron flow through the NADPH oxidase is activated by O_2. The slow kinetics of cytochrome $\underline{b}_{-245}$ reduction by NADPH under anaerobic conditions are similar to those obtained when mammalian cytochrome-c oxidase is reduced anaerobically by reduced cytochrome-c, (33). In this latter case, "pulsed" cytochrome-c oxidase, which has been reacted with O_2, oxidises reduced cytochrome-c at a much faster rate than the cytochrome-c oxidase which has been left under anaerobic conditions, (34). It is likely that electron transfer within the NADPH oxidase complex requires a conformational change to occur upon binding of O_2, which allows electron transfer to take place. These results also explain why previous attempts to demonstrate the kinetic competence of cytochrome $\underline{b}_{-245}$ in this oxidase system, by measuring reduction rates of cytochrome $\underline{b}_{-245}$ under anaerobic conditions, have been unsuccessful.

A role for quinones within the oxidase complex has been proposed by other laboratories, (35). We have demonstrated previously that the quinone content of neutrophils is associated with mitochondrial but not with NADPH oxidase-containing subcellular fractions of neutrophils and also that the ratio of ubiquinone-10 to cytochrome-c oxidase in neutrophils is in agreement with published results obtained with mitochondria from other cell types, (36). High quinone content in neutrophils reported by some workers (37) may be due to contamination of neutrophil preparations with lymphocytes, which are a good source of mitochondria. Analysis for quinones in the solubilised oxidase preparation proved negative. Whilst we do not preclude a role for quinones in the oxidase complex we feel that published data do not provide a convincing argument for their involvement. Further investigations are required to settle this issue.

The mechanism of activation of the NADPH oxidase in stimulated neutrophils is not well understood. Since reduced cytochrome $\underline{b}_{-245}$ is rapidly oxidised by air whether it is prepared from resting or stimulated neutrophils, it is likely that the regulatory point within the oxidase lies within the NADPH/flavoprotein portion of this electron transfer system. As we have mentioned, little is known about this portion of the oxidase. The flavoprotein has not been purified, nor has it been identified on SDS-PAGE. Whether this flavoprotein is a substrate for protein C kinase, a kinase proposed to mediate many of the responses of stimulated neutrophils (38), has yet to be established. Future research will necessarily concentrate on the characterisation of this flavoprotein and we hope that our present method for preparing solubilised NADPH oxidase will provide us with a good starting point for this purpose.

Acknowledgements: We would like to thank Dr. R.N. Thorneley of the ARC Nitrogen Fixation Unit, Brighton Sussex, U.K., for use of his anaerobic stopped-flow spectrophotometer. Computer programmes for use with the rapid-scanning spectrophotometer were designed by Mr. J.N. Sturgis, of this department. The work described in this article was supported by grants from the Medical Research Council, the Wellcome Trust Foundation and the Arthritis and Rheumatism Research Council.

REFERENCES

1. A.J. Badwey and M.L. Karnovsky, Active oxygen species and the functions of phagocytic leukocytes. Ann. Rev. Biochem. 49: 695-726 (1980).
2. R.K. Root and J.A. Metcalf, H_2O_2 release from human granulocytes during phatocytosis: relationship of superoxide anion formation and cellular catabolism of H_2O_2: studies with normal and cytochalasin B treated cells. J. Clin. Invest. 60: 1266-1279 (1977).
3. A.J. Sbarra and M.L. Karnovsky, The biochemical basis of phagocytosis , 1: Metabolic changes during the ingestion of particles by polymorphonuclear leukocytes. J. Biol. Chem. 234: 1355-1362 (1959).
4. B.M. Babior, J.T. Curnette and B.J. McMurrich, The particulate superoxide-forming system from human neutrophils: properties of the system and further evidence supporting its role in the respiratory burst. J. Clin. Invest. 58: 989-996 (1976).
5. F. Rossi and M. Zatti, Changes in the metabolic pattern of

polymorphonuclear leukocytes during phagocytosis. Brit. J. Exp. Path. 45: 548-559 (1964).

6. B.M. Babior and R.S. Kipnes, Superoxide-forming enzyme from human neutrophils: evidence for a flavin requirement. Blood 50: 517-524 (1977).

7. H. Wakeyama, K. Takeshiga and S. Minakami, NADPH-dependent 2,6-dichlorophenolindophenol reudction by the phagocytic vesicles of pig polymorphonuclear leukocytes. Biochem. J. 210: 577-581 (1983).

8. D.B. Light, C. Walsh, A.M. O'Callaghan, E.T. Goetzl and A.I. Tauber, Characteristics of the cofactor requirements for the superoxide-generating NADPH oxidase of human polymorphonuclear leukocytes. Biochemistry 20: 1468-1476 (1981).

9. A.W. Segal, O.T.G. Jones, D. Webster and A.C. Allison, Absence of a newly described cytochrome-b from neutrophils of patients with Chronic Granulomatous Disease. Lancet (2), 446-449 (1978).

10. Y. Shinagawa, Y. Shinagawa, C. Tanaka and A. Teraoko, Electron microscopic and biochemical study of the neutrophilic granules from leukocytes. J. Elec. Micros. 15: 81-85 (1966).

11. A.W. Segal and O.T.G. Jones, Novel cytochrome-b system in phagocytic vacuoles of human granulocytes. Nature 276: 515-517 (1978).

12. A.W. Segal and O.T.G. Jones, Reduction and subsequent oxidation of a cytochrome-b of human neutrophils after stimulation with phorbol myristate acetate. Biochem. Biophys. Res. Commun. 88: 130-134 (1979).

13. M.L. Karnovsky, Chronic Granulomatous Disease – pieces of a cellular and molecular puzzle. Fed. Proc. 32: 1527-1533 (1973)

14. A.W. Segal and O.T.G. Jones. The subcellular distribution and some properties of the cytochrome-b component of the microbicidal oxidase system of human neutrophils. Biochem. J. 182: 181-188 (1979).

15. A.W. Segal and O.T.G. Jones. Absence of cytochrome-b reduction in stimulated neutrophils from both male and female patients with chronic granulomatous disease. FEBS Lett. 110: 111-114 (1980).

16. A.R. Cross, O.T.G. Jones, R. Garcia and A.W. Segal. The association of FAD with cytochrome b_{-245} of human neutrophils. Biochem. J. 208: 759-763 (1982).

17. T.G. Gabig. The NADPH-dependent O_2^--generating oxidase from human neutrophils. Identification of a flavoprotein that is deficient in a patient with chronic granulomatous disease. J. Biol.Chem. 258: 6352-6356 (1983).

18. M.N. Hamers, M. de Boer, L.J. Meerhof, R.S. Weening and D. Roos, Complementation in monocyte hydrids revealing genetic heterogeneity in chronic granulomatous disease. Nature 307: 553-555 (1984).

19. A.R. Cross, F.K. Higson, O.T.G. Jones, A.M. Harper and A.W. Segal, The enzymic reduction and kinetics of oxidation of cytochrome b_{-245} of neutrophils. Biochem. J. 204: 479-485 (1982).

20. A.R. Cross, O.T.G. Jones, A.M. Harper and A.W. Segal, Oxidation-reduction properties of the cytochrome-b found in the plasma membrane fraction of human neutrophils. Biochem. J. 194: 599-606 (1981).

21. P.M. Wood, The redox potential of the system oxygen-superoxide, FEBS Lett. 44: 22-24 (1974).

22. A.R. Cross, J.F. Parkinson and O.T.G. Jones, The superoxide generating oxidase of leukocytes·NADPH-dependent reduction of flavin and cytochrome-b in solubilised preparations. Biochem. J. 223: 337-344 (1984).

23. A.W. Segal, R. Garcia, A.H. Goldstone, A.R. Cross and O.T.G. Jones, Cytochrome b_{-245} of neutrophils is also present in human monocytes, macrophages and eosinophils. Biochem. J. 196: 363-367 (1981).

24. A.M. Harper, M.J. Dunne and A.W. Segal, Purification of cytochrome b_{-245} from human neutrophils. Biochem. J. 219: 519-527 (1984).

25. P. Bellavite : Biochemistry Journal (in press).

26. T.G. Gabig and B.A. Lefker, Catalytic properties of the resolved flavoprotein and cytochrome-b components of the NADPH dependent O_2^--generating oxidase from human neutrophils Biochem. Biophys. Res. Commun. 118: 430-436 (1984).

27. H. Wakeyama, K. Takeshiga, R. Takenagi and S. Minakami, Superoxide-forming NADPH oxidase preparation of pig polymorphonuclear leukocytes. Biochem. J. 205: 593-601 (1982).

28. R.M.C. Dawson, D.C. Elliot, W.H. Elliot and K.M. Jones, "Data for Biochemical Research - Clarendon press, Oxford (1969).

29. E.J. Faeder and L.M. Siegel, A rapid micromethod for determination of FMN and FAD mixtures. Anal. Biochem. 53: 332-336 (1976).

30. P.B. Sweetzer, Colorimetric determination of trace oxygen levels with the photochemically generated methyl viologen radical cation. Anal. Chem. 39: 979-982 (1967).

31. S. Bramhall, N. Noack, M. Wu and J.R. Lowenberg, A simple colorimetric method for determination of protein. Anal. Biochem. 31: 146-148 (1969).

32. P. Bellavite, A.R. Cross, M.C. Serra, A. Davoli, O.T.G. Jones and F. Rossi, The cytochrome-b and flavin content and properties of the O_2^--forming NADPH oxidase solubilised from activated neutrophils. Biochim. Biophys. Acta. 746: 40-47 (1983).

33. E. Antonini, M. Brunori, A. Colosimo, C. Greenwood and M.T. Wilson, Oxygen "pulsed" cytochrome c oxidase: functional properties and catalytic relevance. Proc. Natl. Acad. Sci. 74: 3128-3132 (1977).

34. L.E. Andreasson, B.G. Malmstrom, C. Stromberg and T. Vanngard, The reaction of ferrocytochrome c with cytochrome oxidase. A new look. FEBS Lett. 28: 297-301 (1972).

35. D.R. Crawford and D.L. Schneider, Evidence that a quinone may be required for the production of superoxide and hydrogen peroxide in neutrophils. Biochem. Biophys. Res. Commun. 99: 1277-1286 (1982).

36. A.R. Cross, O.T.G. Jones, R. Garcia and A.W. Segal, The subcellular localisation of ubiquinone in human neutrophils. Biochem. J. 216: 765-768 (1983).

37. F. Mollinedo and D.L. Schneider, Subcellular localisation of cytochrome-b and ubiquinone in a tertiary granule of resting human neutrophils and evidence for a proton pump ATPase. J. Biol. Chem. 259: 7143-7150 (1984).

38. A. Fujita, Diacylglycerol, 1-oleoyl-2-acetyl-glycerol, stimulates superoxide generation from human polymorphonuclear leukocytes. Biochem. Biophys. Res. Commun. 120: 318-324 (1984).

MECHANISMS OF MEMBRANE LIPID PEROXIDATION

Larry G. McGirr and Peter J. O'Brien

Department of Biochemistry
Memorial University of Newfoundland
St. John's, Newfoundland, Canada A1B 3X9

Lipid peroxidation is associated with CCl_4-induced hepato-toxicity and various pathological changes associated with the toxicity of some metals, herbicides, drugs, carcinogens, photo-sensitizers and ozone or NO_2. It has also been associated with atheroschlerosis, tumorigenesis and metastases, red cell disorders, inflammation, burns, eye cataracts, skin disorders, neurological disturbances, lung fibrosis, aging, etc. Dietary antioxidants and metal chelators are often protective. However, it remains to be proven whether lipid peroxidation is a direct cause of these toxic effects rather than a consequence (1-8).

In the following, new developments in the methodology for determining membrane lipid peroxides by noninvasive vs invasive techniques are outlined. The various activating species involved in initiating membrane lipid peroxidation are also reviewed. Finally, research results are presented showing that hydroxyl radicals and/or higher oxidation states of iron are involved in the initiation of lipid peroxidation by anticancer drugs.

NEW DEVELOPMENTS IN THE METHODS FOR DETERMINING MEMBRANE LIPID PEROXIDATION

Numerous methods have been used as an index of lipid peroxidation partly as a result of the large number of products which arise from lipid hydroperoxides on degradation. These include aldehydes, ketones, alcohols, hydrocarbons, epoxyesters, dihydroxyesters, various dimers and high molecular weight polymers (9). A detailed account of some of these methodologies

319

has recently been reviewed (8,10) and hence we will only provide a brief overview of some of these methods.

These methodologies can be divided generally into invasive and non-invasive techniques but the bulk of these procedures fall into the invasive category. The problem with invasive methodologies is the continued presence of handling artifacts as a result of the bringing together of unsaturated fatty acids and peroxidation catalysts like metal ions.

NON-INVASIVE TECHNIQUES

Breath Alkane Measurement

Of the numerous volatile compounds formed on peroxidation, only ethane and pentane have been found to be good indices of lipid peroxidation. The first report of ethane production in intact animals was made in 1974 where the authors (11) found increased whole body ethane expiration determined by gas chromatography in response to carbon tetrachloride, a known initiator of lipid peroxidation. Since then, ethane production has been found to respond to other treatments which result in lipid peroxidation, namely dietary vitamin E deficiency (12-14) and iron overload (12). Pentane production also responds to the same type of treatment (13, 15,16). Ethane was produced from linolenic acid (ω-3 fatty acid) while pentane was formed from linoleic acid and arachidonic acid (ω-6 fatty acids) on peroxidation (17).

It was recently suggested that measurement of breath ethane is a better index of lipid peroxidation than pentane since pentane was further metabolized by the intact animal (18,19). Furthermore, the role of bacterial lipid peroxidation in producing ethane and pentane in the intestinal tract (20) and its contribution to whole body or breath gas ethane/pentane production has yet to be clarified. This technique has been extended and used with perfused rat liver (21), hepatocytes (22) and liver microsomes (23) and has suggested that the liver is the source of most breath ethane upon treatment with various xenobiotics.

Urinary Malondialdehyde Values

It has recently been found that rat urinary malondialdehyde (MDA) determined as the thiobarbituric acid adduct by high pressure liquid chromatography (HPLC) responded to various treatments which cause in vivo lipid peroxidation; dietary vitamin E deficiency and iron nitrilotriacetate administration (24). Urinary MDA represents only a small percentage of administered MDA after intraperitonal injection or ingestion of MDA, 80% of dose is metabolized to CO_2 via a mitochondrial aldehyde dehydrogenase (25). Since free MDA does not exist in the urine to any large extent, the thiobarbituric acid adduct must result from a bound form of MDA. One bound form of MDA in rat urine has been identified as N-α-acetyl-ϵ-(2-propenal) lysine which presumably results from MDA reacting with the lysine groups of protein (McGirr, Hadley and Draper, unpublished results).

INVASIVE TECHNIQUES

Loss of Lipid Substrate

On peroxidation of microsomes in the presence of oxygen and NADPH, the progressive loss of specific fatty acids occurred as determined by gas chromatography (26-28). The phospholipids remained intact and became more polar. Changes occur in the polyunsaturated fatty acids, primarily the $C_{20:4}$ and $C_{22:6}$ acids but some decrease also occurs in the $C_{18:2}$ acids. It is important to note that MDA production only accounted for approximately 10-33% of the lipid loss.

Oxygen Uptake

During lipid peroxidation, oxygen is taken up to form the lipid hydroperoxides. The measurement of oxygen uptake can only be carried out on in vitro systems where the level of respiration is small. This method had been used with microsomal (29) and mitochondrial peroxidizing systems (30).

Detection of Lipid Hydroperoxides

Lipid hydroperoxides can be detected by numerous procedures which all involve some type of spectrophotometric determination. Iodometric assay originally developed for the determination of hydroperoxides in autoxidized fats has been modified to increase sensitivity and ease of use (31). In general, lipid hydroperoxides oxidize iodide under acidic conditions to release I_2 which reacts with iodide to give the triodide anion which can be determined spectrophotometrically (360 nm). The reaction must be kept under anaerobic conditions to prevent O_2 from oxidizing the iodide. This reaction is specific for lipid hydroperoxides and H_2O_2. Another reaction of this type involves the oxidation of dichlorofluorescin to the fluorescent dichlorofluorescein in the presence of hematin (32). This procedure is more sensitive than the iodometric assay, but appears to have a similar substrate specificity.

The use of glutathione peroxidase coupled to the enzyme glutathione reductase has led to another spectrophotometric method where the loss of NADPH is measured (33).

$$ROOH + 2GSH \longrightarrow ROH + GSSG + H_2O$$

$$GSSG + 2NADPH \longrightarrow 2GSH + 2NADP^+$$

This procedure has been modified by measuring GSSG formed or GSH left in the reaction by complexing with o-phthalaldehyde and measuring the fluorescent adduct (34).

Polyunsaturated fatty acids readily donate a hydrogen from the methylene group between double bonds; the resulting radical rearranges to form a conjugated double bond system on extraction of a hydrogen from another molecule or in the presence of O_2 giving the hydroperoxy radical. This conjugated system shows a large absorption at 233 nm with a smaller absorption in the 260-280 nm region. Although this method was originally developed for the food industry, it has found large use with toxicology, specifically in the response of the liver cell to various xenobiotics (35-37). Since the level of diene was small in relation to the nonconjugated lipids present, its UV absorption was only a small shoulder on the large end absorption given by the other lipids. To improve sensitivity, the difference spectrum between the peroxidised and non-peroxidised samples was used to quantitate the diene peak. Although ultraviolet absorption at 233 nm is non-specific, it does correlate very well with the amount of conjugated diene determined by reaction of [^{14}C]tetracyano-ethylene with diene to give the [^{14}C] Diels-Alder product (38). The use of double derivative spectroscopy (39) may improve the accuracy of measuring the 233 nm peak in biological systems. The detection of lipid hydroperoxides by the measurement of diene conjugation is not always quantitative. Dienes can form under anaerobic conditions, when lipid peroxidation does not occur, as a result of free radical attack (40). Also, non-conjugated lipid hydroperoxides can form upon oxidation by singlet oxygen (41).

The separation of the isomeric hydroperoxides of methyl linoleate by HPLC (42) may allow the detection of specific hydro-peroxides resulting from peroxidation of biological systems.

Lipid hydroperoxides have also been determined by the use of the thiobarbituric acid reaction (43). Peroxides decompose under the acid and heat conditions of the TBA test possibly as a result of Fe^{+3} contamination to give MDA (44).

Malondialdehyde Determination

One of the most popular methods of measuring lipid peroxida-tion has been the thiobarbituric acid reaction due to its sensitivity and ease of use. In the reaction, TBA combines with the three carbon dialdehyde malondialdehyde under acid conditions to form a pink adduct with a visible absorption at 533 nm (45).

The TBA test has been criticized for its lack of specificity as numerous other compounds not related to lipid peroxidation react with TBA to give interfering colored complexes, but most of these can be separated from TBA-MDA by chromatographic procedures (46,47). Also, numerous products of lipid peroxidation break down under the conditions of the TBA test to give MDA which then reacts with TBA to give the colored complex (48). In this case, one is not measuring free MDA but it still results in an index of lipid peroxidation. Since MDA is a reactive species and reacts with other biological compounds (protein, nucleic acids), other forms of bound MDA can exist which also result in a positive TBA test due to hydrolysis under the conditions of the TBA procedure (49).

In spite of these problems, it has been well demonstrated in the in vitro peroxidizing microsome systems that the TBA reaction is measuring only free MDA (50,51). Numerous other limitations of this method have been raised. Other sources of MDA besides lipid peroxidation exist: from deoxyribose (52) as a result of free radical damage to DNA and as a product of the prostaglandin/thromboxane biosynthetic pathway (53). Also, TBA does not react with numerous aldehyde products of peroxidation which have potent biological toxicity (4-hydroxynonenal) and hence the TBA test may not reflect the total biological consequences of peroxidation (54).

Free MDA can be determined by high pressure liquid chromatography (51) and has been used to detect free MDA in peroxidized liver microsomes.

Fluorescent Chromophores

During peroxidation of microsomes and mitochondia, chloroform-methanol soluble products showing fluorescent excitation/emission spectrum at 360-460 nm were formed (55). The same type of fluorescence developed during the autoxidation of arachidonic acid in the presence of phosphatidyl ethanolamine; yet, in the presence of phenylalanine the fluorescence material became water soluble (56). Similar fluorescent chromophores developed on inactivation of RNAase in the presence of peroxidizing arachidonate or MDA (57). The fluorescence was attributed to conjugated Schiff base formation between MDA and amino acid side chains based on similar fluorescence shown by MDA-amino acid Schiff bases (58). Although the fluorescent chromophores which developed during lipid peroxidation were thought to be a result of the reaction of MDA and amines to give the RN=CHCH=CHNHR Schiff base linkages, it has been suggested that the reaction of MDA and various amines under milder conditions results in fluorescent compounds of the 1,4-dihydropyridine-3,5-dicarbaldehyde type (59). Reaction of methylamine with oxidized fatty acids produced numerous fluorescent compounds but the major products were not 1,4-dihydropyridine-3,5-dicarbaldehydes nor Schiff bases indicating that MDA may not be responsible for a large portion of the fluorescence development

in peroxidizing lipid systems (60). It should be kept in mind that
the amount of MDA in autoxidizing fat experiments was a relatively
minor product in terms of total aldehyde in comparison to
peroxidizing microsomal systems (61). Regardless of the types of
bonds involved in the development of the fluorescence, the
measurement of fluorescent chromophores has been well correlated
with in vivo and in vitro lipid peroxidation (16).

Aldehydes

A large number of aldehydes were produced during microsomal
lipid peroxidation as determined by gas chromatography (62) or by
derivatization with 2,4-dinitrophenylhydazine (54) followed by
HPLC separation. Although individual measurement of one aldehyde
was possible, the relative amounts of a specific aldehyde varies
depending on the peroxidizing conditions. It may be possible,
using the 2,4-dinitophenylhydrazine procedure, to estimate total
aldehyde production. MDA is not measured by this reaction as it
gives the pyrazole derivative and not the hydrazone (63).

Hydroxy Fatty Acids

It is well known that lipid hydroperoxides decompose to
hydroxy fatty acids enzymatically and non-enzymatically. Numerous
hydroxy fatty acids have been identified in mouse liver after
treatment with CCl_4 by a combination of HPLC and GC/MS techniques
(64). Hydroxy fatty acids can also be formed by various lipoxygenase systems acting primarily on arachidonic acid (65) or
directly by the action of the cytochrome P450 system (60). In both
systems HPLC was used to separate the hydroxy acids.

Lipid and Lipid Oxy Radicals

It has recently been found that lipid derived carbon and
oxygen centered radicals can be detected by electron spin
resonance techniques by trapping the radicals with the spin trap
phenyl t-butylnitrone after treatment of rat liver microsomes
with carbon tetrachloride (67). In similar studies a lipid carbon
centered radical was trapped in the reaction of lipoxygenase and
linoleic acid (68). Although these methods would not be used as a
routine detection methods for lipid peroxidation, they may be used
in the future for specialized studies.

Chemiluminescence

Various organs, cells and subcellular fractions emit a low-
level chemiluminescence or 'dark' chemiluminescence which can
not be detected by the human eye (69,70). The chemiluminescence
emitted by mitochondrial and microsomal fractions exposed to lipid
peroxidizing conditions correlated very well with MDA production

(69-72). This observation has been extended to various organs which have increased chemiluminescence when exposed to oxidative stress; hyperbaric O_2, organic hydroperoxides or toxic chemicals. The species responsible for the light emission have not been fully explored. One possibility is singlet oxygen which releases a photon on return to the triplet state, $^3O_2$. Its formation possibly results from self reaction of lipid peroxy radicals (ROO,) to give the alcohol (ROH), ketone (RO) and $^1O_2$ (73). Singlet oxygen also reacts with double bonds to form dioxetanes which decompose to give activated carbonyls which also release light. The exact nature of the light emitting species can only be determined by examining the wavelength of the emitted light and the effect of specific scavengers.

Concluding Remarks

It is clear that there are several limitations for each method of determining lipid peroxidation which makes interpretation of the data difficult. For this reason, several methods are normally used in concert in order to allow cross-checking of the methods.

The measurement of various degradation products of hydro-peroxides are the easiest methods. However, there are numerous problems associated with these methods since the products are formed in small amounts (due to the large number of products) and the amount of each product varies according to the conditions of peroxidation. The various products are also further metabolized. For these reasons, it is better to measure primary products: the formation of hydroperoxides or the disappearance of unsaturated fatty acids. Finally, methods indicating toxicological consequences need to be developed. In this case, adduct formation with informational macromolecules would be useful eg. evidence of fluorescent DNA cross-linking.

2 THE ACTIVATING SPECIES INVOLVED IN INITIATING MEMBRANE LIPID PEROXIDATION

Initiation of lipid peroxidation is believed to involve abstraction of a methylene hydrogen from a polyunsaturated fatty acid by a variety of activated oxygen species or organic free radicals or as a result of the decomposition of preformed lipid hydroperoxides to free radicals and activated oxygen species. Metal complexes act as catalysts, particularly in their reduced form, by activating oxygen or decomposing formed lipid hydro-peroxides. The fatty acid alkyl radical formed then enters into the propagation cycles by the addition of dioxygen, resulting in the formation of lipid hydroperoxides. Table 1 summarizes the initiating oxidant species proposed for a variety of lipid peroxidizing systems including microsomal membranes, liposomes and polyunsaturated fatty acid micelles. Enzymes can play a role by catalyzing the reduction of the ferric ion. This may be direct in the case of NADPH; cytochrome P450 reductase or as a result of forming the reducing species superoxy radical in the case of xanthine oxidase. The initiating species proposed for microsomal lipid peroxidation include higher oxidation states of iron: perferryl for non-heme iron complexes and ferryl for heme iron. However, alkoxy and peroxy radicals may also be involved in each of these cases. In the case of Cu^{2+}/benzoyl peroxide a higher oxidation state of Cu^{2+} eg. $Cu^{+}.OOH$ could also be involved. Some photosensitizers, when light activated to the triplet state (^3P), can react with oxygen to form singlet oxygen ($^1O_2$) which can also initiate lipid peroxidation. Singlet oxygen has also been implicated in Cu^{2+} catalyzed lipid peroxidation. The singlet oxygen could be formed following dimerisation of the peroxy radicals (73). With liposomes and polyunsaturated fatty acid micelles the initiating species can be hydroxyl radicals formed by Fenton's reaction

$$Fe^{2+} + H_2O_2 \longrightarrow Fe^{3+} + OH^- + OH^\bullet$$

$$2O_2^- + 2H^+ \longrightarrow H_2O_2 + O_2$$

Fe-EDTA complexes are more effective in this type of initiation. Hydroxyl radicals, however, do not seem to be involved in the initiation by Fe^{2+} ADP or Cu^{2+}/H_2O_2. Finally, microsomal lipid peroxidation catalyzed by CCl_4 in the presence of NADPH, may be unusual in not requiring a metal catalyst. In this case the proposed initiator is the trichloromethyl radical or trichloromethyl peroxy radical formed by enzymatic reduction of the CCl_4.

THE NATURE OF THE INITIATING CATALYSTS IN LIPID PEROXIDATION INDUCED BY ANTICANCER DRUGS

Clinical use of the anticancer drug adriamycin is limited by its toxic side effects, particularly cardiomyopathy. Prevention of cardiac toxicity by vitamin E (89), the free radical scavenger N-acetylcysteine (90) or iron chelators (91) and the reported increase in whole animal toxicity and lipid peroxidation following the coadministration of iron and adriamycin (92) suggests that lipid peroxidation is involved. Little lipid peroxidation occurred with adriamycin or iron alone. Rat liver and heart microsomes also underwent lipid peroxidation with adriamycin in the presence of NADPH (93-95). However, recently peroxidation of phospholipid micelles has been demonstrated with adriamycin: ferric complexes without the need for enzymic activation. Furthermore, the peroxidation was unaffected by hydroxyl

Table 1 Mechanisms for initiating lipid peroxidation

Peroxidising system	Initiating catalyst	Reference
(A) Microsomal membrane		
Lipoxygenase	$Fe^{3+}.O_2^-$ (perferryl)	(74)
ADP-Fe^{3+} / NADPH	$Fe^{3+}.O_2^-$	(75)
Heme / H_2O_2	Heme-$Fe^{4+}O$ (ferryl)	(76)
Cytochrome P450 / ROOH	Heme-$Fe^{4+}O$, $ROO^\bullet$	(77)
Fe^{2+} / ROOH	$RO^\bullet$, $ROO^\bullet$	(78)
Cu^{2+} / Benzoyl peroxide	$ROO^\bullet$	(80)
Cu^{2+} / ROOH	$^1O_2$	(79)
Light / Photosensitizer (P) eg. dyes, chlorophyll, porphyrins, bilirubin	$^1O_2$, ^3P	(81)
CCl_4 / NADPH	$CCl_3^\bullet$, $CCl_3O_2^\bullet$	(67,82)
(B) Liposomes		
Fe^{2+} + H_2O_2	$OH^\bullet$	(83)
Fe^{2+} + O_2	FeO^{2+}	(84)
Fe^{2+} . ADP + O_2	$Fe^{2+}-O_2-Fe^{3+}$	(85)
Fe^{3+}.ADP + reductase/NADPH	$Fe^{3+}-O_2^-$	(76,86)
Cu^{2+} + H_2O_2	$Cu^+.OOH$	(87)
(C) Linolenate		
Fe^{3+} - EDTA + O_2 + Xanthine oxidase + Acetaldehyde	$OH^\bullet$	(88,89)
Fe^{2+} - EDTA + O_2	$OH^\bullet$	(88)
NO_2 + methyl linolenate	$NO_2^\bullet$	(120)
H_2O_2 + chloride + lactoperoxidase	$Cl^\bullet$, HOCl (?)	(121)
O_2^- + 0.1N H_2SO_4 + 85% EtOH	HO_2	(122)

radical scavengers, catalase and superoxide dismutase (96).
Other workers found no inhibition by benzoate, a hydroxyl radical
scavenger, superoxide dismutase or catalase and suggested that
perferryl complexes were responsible (97). It is possible that
hydroxyl radicals were formed in this system as the degradation of
deoxyribose by adriamycin: ferric complexes was substantially
inhibited by some hydroxyl radical scavengers (98). Membrane
lipid peroxidation by adriamycin: ferric complexes was found to
involve binding of the complex to the membrane and require glutathione (99). Hydrogen peroxide was formed and hydroxyl radicals
could be traced with a spin trap (100). They suggested that the
H_2O_2 was formed by autoxidation of the Fe^{2+} -adriamycin formed
following reduction of the ferric complex by the GSH and that the
hydroxyl radicals were formed from the H_2O_2 by the ferrous complex.
Furthermore, in contrast to the results of Gutteridge (96) and
Nakano (97), the lipid peroxidation was inhibited 50% by mannitol,
a hydroxyl radical scavenger (99). It has also been shown that
the adriamycin-Fe^{3+} complex forms ternary complexes with
phospholipid bilayers (101) and DNA (102).

In these systems, it appears that hydroxyl radicals may be
formed in the reaction mixture, but addition of scavengers of
hydroxyl radicals can have little effect on the rate of peroxidation. There is, therefore, a lot of confusion with regard to
whether hydroxyl radicals or ferryl ions are involved in the lipid
peroxidation or DNA damage induced by anticancer drugs.

The anticancer drug, mitimycin C, as well as the antibiotic,
nitrofurantoin or the herbicide, paraquat can induce life-threat-
ening pulmonary toxicity in humans. The pathophysiology of this
damage includes inflammation, hemorrhage, edema and interstitial
fibrosis. They are also readily reduced by microsomal NADPH:
cytochrome P-450 reductase to radical species which can react
with oxygen to form superoxide. These drugs have been shown to induce lipid peroxidation in rat or mouse lung microsomes as determined
by malonaldehyde and chemiluminescence (104). A level of 80-90
nmoles malonaldehyse / mg protein was approached only in the case
of paraquat and mouse lung microsomes. This is believed to represent 100% peroxidation. Inhibition by superoxy dismutase,
catalase and the hydroxyl radical scavenger dimethylurea was also
reported.

The anticancer antibiotic, bleomycin can also induce life-
threatening interstitial plumonary fibrosis in its clinical use.
A ferrous iron-bleomycin complex is 40% more effective than ferrous ion in catalyzing the peroxidation of a multi-lamella liposomal membrane (105). However, microsomal lipid peroxidation was
inhibited by this complex (106).

Table 2 Microsomal Lipid Peroxidation Catalyzed by the Reductive
Activation of Anticancer Drugs

Drug	Malonaldehyde (nmoles/mg protein/hr)	
	+ NADPH	+ NADH + xanthine oxidase
-----	19 ± 3	14 ± 3
Bleomycin + Cu^{2+} (20 μM)	24 ± 2	84 ± 2
Adriamycin	42 ± 5	73 ± 5
Daunomycin	16 ± 3	62 ± 4
Mitomycin C	43 ± 4	83 ± 7
Paraquat	51 ± 5	27 ± 5
Nitrofurantoin	35 ± 2	29 ± 5

Reaction Conditions: 3 ml of 0.1 M Tris-HCl buffer pH 7.4
containing 3 mg rat liver microsomes, 0.1 mM anticancer drug,
1 mM NADPH or 1 mM NADH and 10 units xanthine oxidase were
incubated at 37°C for 1 hour. The mean of three experiments is
given.

Hydroxyl radical formation has been reported during the reduction of mitomycin C and adriamycin by purified NADPH-cytochrome
P-450 reductase/NADPH (107) and during the autoxidation of mitomycin C, adriamycin, streptomycin and daunorubicin following reduction by NaBH$_4$ (108). Hydroxyl radicals were also formed during
the autoxidation of a ferrous ion-bleomycin complex (109,110) or a
cuprous ion-bleomycin complex (111). However, the latter complex
differs from the ferrous complex in being readily dissociated with
thiol compounds or traces of ferric ion and is unable to cause DNA
strand breakage. However, it has not yet been established whether
more complex systems such as liver microsomes and NADPH can
catalyze hydroxyl radical formation with anticancer drugs.

In the above systems, it appears that hydroxyl radicals may
be formed in the reaction mixture which can result in DNA damage.
However, hydroxyl radical scavengers in some systems does not

Table 3   Hydroxyl Radical Formation by the Reductive Activation of
Anticancer Drugs

Drug	NADH + Xanthine Oxidase (p moles ethylene)
-------	360 ± 42
Adriamycin	8,200 ± 562
Daunomycin	9,600 ± 437
Mitomycin C	2,000 ± 189
Bleomycin-Cuprous	8,300 ± 620
Paraquat	560 ± 36
Nitrofurantoin	1,680 ± 210

Reaction Conditions: 3 ml of 0.1 M phosphate buffer
pH 7.4; 10 mM 2-keto-4-thiomethyl-butyric acid; 0.1
mM anticancer drug; 1 mM NADH and 10 units xanthine
oxidase were incubated at 37°C for 10 minutes in 25 ml
Erlenmeyer flasks stopped with gas-sealed rubber caps.
The mean of three experiments is given.

affect the rate of lipid peroxidation.   There is therefore a lot
of confusion with regard to whether hydroxyl radicals or ferryl-
perferryl complexes are involved in the lipid peroxidation by
anticancer drugs.

In the following, we have investigated the nature of the initiating species involved in microsomal lipid peroxidation catalyzed
by various anticancer drugs, etc.   The formation of hydroxyl
radicals in this system was also measured by following the oxidation of 2-keto-4-thiomethyl-butyric acid to ethylene (112).   The
results show that there is a correlation between hydroxyl radical
formation and the rate of lipid peroxidation.   Both were affected
to the same degree by a variety of inhibitors.

As shown in Table 2, the bleomycin-Cu^{2+} complex readily catalyzed microsomal lipid peroxidation.   NADPH was required suggesting that reduction to bleomycin-Cu^+ was required.   Lipid peroxidation with bleomycin or the bleomycin-Fe^{3+} complex was only 30%
of that of the Cu^{2+} complex.   More extensive lipid peroxidation
occurred with xanthine oxidase and NADH as the reactant. Furthermore, as shown in Table 3, xanthine oxidase: NADH also catalyzed

Table 4 Effect of inhibitors on adriamycin catalysed microsomal lipid peroxidation.

INHIBITOR	MALONALDEHYDE (n moles/ mg protein/ hr)			
	+ NADPH	+ NADPH + adriamycin	+ NADH + xanthine oxidase	+ NADH + xanthine oxidase + adriamycin
None	19 ± 3	42 ± 5	9 ± 2	73 ± 5
Superoxy dismutase (10^{-5} M)	8 ± 1	32 ± 3	4 ± 1	23 ± 6
Azide (10^{-3} M)	29 ± 3	75 ± 6	13 ± 2	106 ± 9
Catalase (10^{-6} M)	14 ± 3	35 ± 5	5 ± 1	52 ± 5
Dimethylurea (10 mM)	12 ± 2	14 ± 4	7 ± 2	14 ± 2
Dimethylurea (1 mM)	14 ± 4	35 ± 3	8 ± 1	44 ± 5
Mannitol (10 mM)	17 ± 2	33 ± 4	8 ± 2	52 ± 2
Desferrioxamine (10^{-5} M)	2 ± 1	2 ± 1	2 ± 1	2 ± 1

Reaction Conditions: 3 ml of 0.1 M Phosphate buffer pH 7.4 containing 3 mg rat liver microsomes, 0.1 mM adriamycin, 1 mM NAD(P)H and 10 units xanthine oxidase were incubated at 37°C for 1 hour. The mean of three experiments is given.

extensive hydroxyl radical formation. Microsomal lipid peroxidation catalyzed by xanthine oxidase: NADH and the bleomycin-Cu^{2+} complex was also drastically inhibited by superoxide dismutase, catalase and hydroxyl radical scavengers.

As shown in Table 2, it can also be seen that microsomal lipid peroxidation readily occurs in the presence of NADPH and adriamycin, daunomycin, mitomycin C, paraquat and nitrofurantoin. More extensive lipid peroxidation occurred when NADH and xanthine oxidase was used to catalyze the reductive activation of these drugs. As shown in Table 4, the xanthine oxidase catalyzed lipid peroxidation required the presence of adriamycin. Xanthine oxidase and NADH did not catalyze lipid peroxidation. The

Table 5   The involvement of reductase and cytochrome P450 in
hydroxyl radical formation

Addition	Ethylene (n mol / mg protein / 10')	
		+ adriamycin (0.1 mM)
None	4.8 ± 0.2	14.1 ± 1.1
+ Catalase (10^{-5} M)	0.6 ± 0.1	1.2 ± 0.6
+ Azide (1 mM)	14.7 ± 1.1	23.2 ± 2.3
+ EDTA (0.2 mM)	16.2 ± 0.6	21.1 ± 2.2
+ EDTA (5.0 mM)	1.2 ± 0.2	1.2 ± 0.2
+ Desferrioxamine (0.3 mM)	1.2 ± 0.3	2.2 ± 0.3
+ Fe^{3+} / EDTA (0.2 mM)	31.0 ± 1.8	32.1 ± 2.1
+ Hexobarbital (5 mM)	9.8 ± 0.9	9.9 ± 1.0
+ Coumarin (1 mM)	10.7 ± 0.8	12.2 ± 1.1
+ Coumarin (1 mM) + dimethylurea (20 mM)	2.2 ± 0.2	4.3 ± 0.3
* Lysolecithin (0.4%)	0.8 ± 0.1	12.3 ± 1.4
* Linoleic acid hydro-peroxide (0.04%)	0.8 ± 0.1	13.4 ± 1.5

Reaction Conditions: 3 ml of 0.1 M phosphate buffer (pH 7.4),
10 mM 2-keto-4-thiomethyl-butyric acid; 3 mg microsomal protein
and 1 mM NADPH were incubated in 25 ml Erlenmeyer flasks for 10
minutes at 37°C. The mean of three experiments is given.
* Microsomes were pretreated with lysolecithin or linoleic acid
  hydroperoxide.

adriamycin dependent lipid peroxidation by xanthine oxidase and
NADH was more readily inhibited by superoxy dismutase or catalase
than the microsomal NADPH : cytochrome P-450 reductase system.

Both systems were completely inhibited by desferrioxamine and strongly inhibited by the hydroxyl radical scavenger, dimethylurea, and to a lesser extent, mannitol. Hydrogen peroxide seems to be involved as, in addition to the inhibition by catalase, a marked activation occurred when azide was added to inactivate endogenous microsomal catalase.

As shown in Table 3, hydroxyl radical formation readily occurred when adriamycin, daunomycin, mitomycin C, paraquat and nitrofurantoin was reductively activated by xanthine oxidase and NADH. Furthermore, as shown in Table 5, the adriamycin enhanced hydroxyl radical formation by liver microsomes was also stimulated by azide and inhibited by catalase. Complete inhibition occurred with desferrioxamine or high concentrations of EDTA. Low concentrations of EDTA or the ferric-EDTA complex markedly stimulated hydroxyl radical formation. Reagents which convert cytochrome P-450 to P-420, eg. lysolecithin or linoleic acid hydroperoxide or cytochrome P450 substrates, eg. hexobarbital or coumarin had little effect on hydroxyl radical formation in contrast to hydroxyl radical formation in the absence of adriamycin. This suggests that NADPH: cytochrome P-450 reductase but not cytochrome P-450 is involved in the reductive activation of adriamycin.

The above results suggest that hydroxyl radicals are involved in the stimulation of lipid peroxidation by certain drugs. Thus, if R represents the "drug" in its starting quinone form, xanthine oxidase or NADPH: cytochrome P-450 reductase could reduce R to its hydroquinone form RH_2.

$$R + NAD(P)H \longrightarrow RH_2 + NADP^+$$

H_2O_2 formation could then result from the dismutation of superoxy radicals formed by autoxidation of the hydroquinone.

$$RH_2 + O_2 \longrightarrow RH^{\cdot} + O_2^- + H^+$$
$$2 O_2^- + 2H^+ \longrightarrow H_2O_2 + O_2$$

Hydroxyl radicals can then be formed by a Fenton reaction between H_2O_2 and non-heme ferrous iron formed by reduction of non-heme ferric iron. This reduction could occur with either NADPH and reductase or by superoxy radicals. The Fe^{3+}-EDTA complex is particularly readily reduced by reductase or superoxy radicals.

$$Fe^{3+} + O_2^- \longrightarrow Fe^{2+} + O_2$$
$$Fe^{2+} + H_2O_2 \longrightarrow OH^{\cdot} + OH^- + Fe^{3+}$$

Xanthine oxidase and NADH also readily form superoxy radicals in the absence of drug substrates which can reduce the non-heme iron required for hydroxyl radical formation.

Adriamycin (doxorubicin DXH_2) complexes ferric ions with a stoichiometry of three adriamycin molecules to one ferric ion.

This complex can also catalyse microsomal lipid peroxidation in the absence of NADPH, presumably as a result of an intramolecular electron transfer.

$$DXH_2 + Fe^{3+} \longrightarrow DX^- - Fe^{3+} + 2H^-$$

$$DX^- - Fe^{3+} \rightleftharpoons DX^\cdot - Fe^{2+} \rightleftharpoons Fe^{2+} + DX^\cdot$$

$$DX^\cdot - Fe^{2+} + O_2 \longrightarrow Fe^{3+} - O_2^-$$

$$DX^\cdot - Fe^{2+} + H_2O_2 \longrightarrow DX - Fe^{3+} + OH^- + OH^\cdot$$

It is not known whether the following reaction occurs:

$$DX^- - Fe^{3+} + O_2^- \longrightarrow DX - Fe^{2+} + O_2$$

We have found that microsomal lipid peroxidation catalyzed by a ferric complex of adriamycin or bleomycin in the presence of NADPH was resistant to hydroxyl radical scavengers and superoxy dismutase suggesting that a ferryl complex is the initiating species.

$$DX - Fe^{3+} + H_2O_2 \longrightarrow OH^- + DX - Fe^{4+}.O \text{ (ferryl)}$$

Hydroxyl radical formation however still occurred and could explain the deoxyribose degradation by this complex (99).

$$DX - Fe^{2+} + H_2O_2 \longrightarrow DX - Fe^{3+} + OH^- + OH^\cdot$$

Clearly, ferryl or perferryl complexes are more effective than hydroxyl radicals in initiating lipid peroxidation. The reverse is true, however, for DNA damage.

Winterbourn et al (113,114) also found that hydroxyl radicals can also be formed from the reaction between adriamycin semi-quinone or paraquat radical and H_2O_2. Oxygen was not required but a ferric catalyst may be required.

$$RH^\cdot + H_2O_2 \longrightarrow RH^+ + OH^\cdot + OH^-$$

This reaction seems to be important in cardiac tissue and may explain the cardiotoxicity of anthracycline antibiotics (115). Elstner et al (116) have suggested that a crypto hydroxyl radical with a similar reactivity to the hydroxyl radical is formed by this reaction. A similar reaction between H_2O_2 and reduced nitrofurantoin was also found.

We also found that microsomal lipid peroxidation catalyzed by NADPH-ADP/Fe^{3+} was only slightly increased by adriamycin. Lipid peroxidation of both systems was also unaffected by catalase, superoxy dismutase and hydroxyl radical scavengers. The mechanism with ADP/Fe^{3+} may be very different as there is recent evidence that the ADP/Fe^{3+} complex is reduced by superoxy radicals but not by the reductase (117). Nakano et al (98) have obtained evidence for an adriamycin-ADP-Fe^{3+} complex and have concluded that an adriamycin-ADP-Fe^{2+} complex initiates lipid peroxidation in its perferryl oxidation state.

The reactions involved in the formation of hydroxyl radicals catalyzed by the bleomycin-Cu^{2+} complex could be explained as follows:-

$$2\ Cu^{2+} - BLM + NADPH \longrightarrow 2\ Cu^+ - BLM + NADP^+$$

$$Cu^+ - BLM + O_2 \longrightarrow Cu^{2+} - BLM + O_2^{\bar{}}$$

$$2\ O_2^{\bar{}} + 2H^+ \longrightarrow H_2O_2 + O_2$$

$$Cu^+ - BLM + H_2O_2 \longrightarrow Cu^{2+} - BLM + OH^{\bullet} + OH^-$$

Finally, as shown in Tables 4 and 5, microsomal lipid peroxidation and hydroxyl radical formation in the absence of adriamycin involves cytochrome P450 in contrast to that in the presence of adriamycin. Thus hydroxyl radical formation is stimulated by substrates of cytochrome P-450 which are poorly coupled, eg. hexobarbital and coumarin, and result in superoxy radical formation (118) and is inhibited by the conversion of cytochrome P-450 to cytochrome P-420 with lysolecithin or the destruction of cytochrome P-450 by linoleic acid hydroperoxide (119). This suggests that cytochrome P-450 is responsible for the formation of superoxy radicals by microsomes and NADPH. Another difference between the two systems is that the lipid peroxidation in the absence of adriamycin is less clear cut as to the nature of the initiating species, as the lipid peroxidation is less affected by hydroxyl radical scavengers. With ADP-Fe^{3+} as the catalyst, recent evidence suggests that a perferryl or ferryl complex is the initiating species (86,98).

CONCLUSIONS

The toxicity of some anticancer drugs could be mediated by lipid peroxidation or hydroxyl radical or other activated oxygen species. Thus microsomal lipid peroxidation readily occurred during the reductive activation, by liver microsomes and NADPH or xanthine oxidase and NADH, of the anticancer drugs adriamycin, daunomycin, bleomycin and mitomycin C as well as nitrofurantoin and paraquat. Hydroxyl radicals, or a species with a similar reactivity, are also formed. Their formation correlates with the rate of lipid peroxidation and both were affected similarly by various inhibitors. Non-heme ferrous or cuprous ions were required for the lipid peroxidation and hydroxyl radical formation. The formation of hydroxyl radicals could explain the DNA strand breakage, deoxyribose degradation and membrane lipid peroxidation.

ACKNOWLEDGEMENTS

This work was supported by the National Research Council of Canada and the National Cancer Institute of Canada. Technical assistance was provided by Wayne Redmond. The manuscript was prepared by Cindy O'Brien.

REFERENCES

1. P. J. O'Brien, Oxidation of Lipids in Biological Membranes and
    Intracellular Consequences, in:"Autoxidation of Unsaturated
    Lipids," H. Chan, ed., Academic Press, London (in press).
2. D. C. H. McBrien and T. F. Slater, eds.,in: "Free Radicals
    Lipid Peroxidation and Cancer," pp.1 Academic Press,
    London (1982).
3. G. L. Plaa and H. Whitshi, Chemicals, drugs and lipid peroxi-
    dation, Ann. Rev. Pharmacol. Tox. 16:125 (1976).
4. M. T. Smith, H. Thor and S. Orrenius, The role of lipid
    peroxidation in the toxicity of foreign compounds to liver
    cells, Biochem. Pharmacol. 32:763 (1983).
5. K. Yagi, ed., in:"Lipid Peroxides in Biology and Medicine,"
    p.1 Academic Press, New York (1982).
6. J. S. Bus and J.E. Gibson, Lipid peroxidation and its role in
    toxicology, in:"Reviews in Biochemical Toxicology," ed.,
    E. Hodgson, J. R. Bend and R. M. Philpot,
    North Holland, (1979).
7. M. A. Trush, E. G. Mimnaugh and T. E. Gram, Activation of
    pharmacologic agents to radical intermediates, Biochem.
    Pharmacol. 31:3335 (1982).
8. M. K. Logani and R. E. Davies, Lipid oxidation: Biological
    effects and antioxidants, Lipids 15:485 (1980).
9. E. N. Frankel, Lipid oxidation, Prog. Lipid Res. 19:1 (1980).
10. T. F. Slater, Overview of methods used for detecting lipid
    peroxidation, in: "Methods in Enzymology,"Vol.105,
    L. Packer, ed., Academic Press, New York (1984).
11. C. A. Riely, G. Cohen and M. Lieberman, Ethane evolution: a
    new index of lipid peroxidation, Science 183:208 (1974).
12. C. J. Dillard and A. L. Tappel, Volatile hydrocarbon and
    carbonyl products of lipid peroxidation: a comparison of
    pentane, ethane, hexanal and acetone as in vivo indices,
    Lipids 14:989 (1979).
13. C. J. Dillard, E. E. Dumelin and A. L. Tappel, Effect of
    dietary vitamin E on expiration of pentane and ethane by
    the rat, Lipids 12:109 (1977).
14. D. G. Hafeman and W. G. Hoekstra, Protection against carbon-
    tetrachloride-induced lipid peroxidation in the rat by
    dietary vitamin E, selenium and methionine as measured by
    ethane evolution, J. Nutr. 107:656 (1977).
15. C. J. Dillard, J. E. Downey and A. L. Tappel, Effect of anti-
    oxidants on lipid peroxidation in iron-loaded rats,
    Lipids 19:127 (1984).
16. A. L. Tappel, Measurement of and protection from in vivo
    lipid peroxidation, in: "Free Radicals in Biology,"Vol.IV,
    W. A. Pryor, ed., Academic Press, New York (1980).
17. E. E. Dumelin and A. L. Tapel, Hydrocarbon gases produced
    during in vitro peroxidation of polyunsaturated fatty
    acids and decomposition of preformed hydroperoxides,

336

Lipids 12:894 (1977).

18. H. Frank, T. Hintze, D. Bimboes and H. Remmer, Monitoring lipid peroxidation by breath analysis: endogenous hydrocarbons and their metabolic elimination, *Toxicol. Appl. Pharmacol.* 56:337 (1980).

19. G. D. Lawrence and G. Cohen, Ethane exhalation as an index of in vivo lipid peroxidation: concentrating ethane from a breath collection chamber, *Anal. Biochem.* 122:283 (1982).

20. D. Gelmont, R. A. Stein and J. F. Mead, The bacterial origin of rat breath pentane, *Biochem. Biophys. Res. Commun.* 102: 932 (1981).

21. A. Muller and H. Sies, Role of alcohol dehydrogenase activity and of acetaldehyde in ethanol induced ethane and pentane production by isolated perfused rat liver, *Biochem. J.* 206:153 (1982).

22. M. T. Smith, H. Thor, P. Hartzell and S. Orrenius, The measurement of lipid peroxidation in isolated hepatocytes, *Biochem. Pharmacol.* 31:19 (1982).

23. R. Reiter and A. Wendel, Drug-induced lipid peroxidation in mice-IV. In vitro hydrocarbon evolution, reduction of oxygen and covalent binding of acetaminophen, *Biochem. Pharmacol.* 32:665 (1983).

24. H. H. Draper, L. Polensek, M. Hadley and L. G. McGirr, Urinary malondialdehyde as an indicator of lipid peroxidation in the diet and in the tissues, *Lipids* (in press) (1984).

25. G. M. Siu and H. H. Draper, Metabolism of malonaldehyde in vivo and in vitro, *Lipids* 17:349 (1982).

26. R. A. Jordan and J. B. Schenkman, Relationship between malondialdehyde production and arachidonate consumption during NADPH supported microsomal lipid peroxidation, *Biochem. Pharmacol.* 31:1393 (1982).

27. H. E. May and P. B. McCay, Reduced triphosphopyridine nucleo-tide oxidase-catalyzed alterations of membrane phospho-lipids, I. Nature of lipid alterations, *J. Biol. Chem.* 243:2288 (1968).

28. H. E. May and P. B. McCay, Reduced triphosphopyridine nucleo-tide oxidase-catalyzed alterations of membrane phospho-lipids. II. Enzymic properties and stoichiometry, *J. Biol. Chem.* 243:2296 (1968).

29. P. Hochstein and L. Ernster, ADP-activated lipid peroxidation coupled to the TPNH oxidase system of microsomes, *Biochem. Biophys. Res. Commun.* 12:388 (1963).

30. R. K. Schneider, E. E. Smith and F. E. Hunter, Correlation of oxygen consumption with swelling and lipid peroxide forma-tion when mitochondria are treated with the swelling-inducing agents Fe^{+2}, glutathione, ascorbate or phosphate, *Biochemistry* 3:147 (1964).

31. M. Hicks and J. M. Gebicki, A spectrophotometric method for the determination of lipid hydroperoxides, *Anal. Biochem.* 99:249 (1979).

32. R. Cathcart, E. Schwiers and B. N. Ames, Detection of picomole levels of hydroperoxides using a fluorescent dichlorofluorescein assay, Anal. Biochem. 134:111 (1983).

33. R. L. Heath and A. L. Tappel, A new sensitive assay for the measurement of hydroperoxides, Anal. Biochem. 76:184 (1976).

34. T. Miura, J. Hiraizumi and M. Kimura, A micromethod for the fluorometric determination of lipid hydroperoxide by using glutathione peroxidase, J. Pharmacobio-Dyn. 3:5 (1980).

35. R. O. Recknagel and A. K. Ghoshal, Lipoperoxidation as a vector in carbon tetrachloride hepatotoxicity, Lab. Invest. 15:132 (1966).

36. R. O. Recknagel and A. K. Ghoshal, Quantitative estimation of peroxidative degeneration of rat liver microsomal and mitochondrial lipids after carbon tetrachloride poisoning, Exp. Mol. Pathol. 5:413 (1966).

37. K. S. Rao and R. O. Reckangel, Early onset of lipoperoxidation in rat liver after carbon tetrachloride administration, Exp. Mol. Pathol. 9:271 (1968).

38. R. L. Waller and R. O. Recknagel, Determination of lipid conjugated dienes with tetracyano-ethylene-^{14}C: Significance for study of the pathology of lipid peroxidation, Lipids 12:914 (1977).

39. F. Corongiu and A. Milia, An improved and simple method for determining diene conjugation in autoxidized polyunsaturated fatty acids, Chem. Biol. Interact. 44:289 (1983).

40. C. L. Wood, A. J. Gandolfi and R. A. VanDyke, Lipid binding of a halothane metabolite. Relationship to lipid peroxidation in vitro, Drug Metab. Dispos. 4:305 (1976).

41. M. J. Thomas and W. A. Pryor, Singlet oxygen oxidation of methyl linoleate: Isolation and characterization of the $NaBH_4$-reduced products, Lipids 15:544 (1980).

42. H. W. S. Chan and G. Levett, Autoxidation of methyl linoleate. Separation and analysis of isomeric mixtures of methyl linoleate hydroperoxides and methyl hydroxylinoleates, Lipids 12:99 (1977).

43. H. Ohkawa, N. Ohishi and K. Yagi, Assay for lipid peroxides in animal tissues by thiobarbituric acid reaction, Anal. Biochem. 95:351 (1979).

44. T. Asakawa and S. Matsushita, Thiobarbituric acid test for detecting lipid peroxides, Lipids 14:401 (1979).

45. R. O. Sinnhuber, I. C. Yu and Te. C. Yu, Characterization of the red pigment formed in the 2-thiobarbituric acid determination of oxidative rancidity, Food Res. 23:620 (1958).

46. J. M. C. Gutteridge and T. R. Tickner, The characterization of thiobarbituric acid reactivity in human plasma and urine, Anal. Biochem. 91:250 (1978).

47. R. P. Bird, S. S. O. Hung, M. Hadley and H. H. Draper, Determination of malonaldehyde in biological materials by high pressure liquid chromatography, Anal. Biochem.

128:240 (1983).

48. J. M. C. Gutteridge, J. Stocks and T. L. Dormandy, Thiobarbituric acid-reacting substances derived from auto-oxidizing linoleic and linolenic acids, Anal. Chim. Acta 70:107 (1974).

49. H. Buttkus and A. J. Rose, Amine-malonaldehyde condensation products and their relative color contribution in the thiobarbituric acid test, J. Am. Oil Chem. Soc. 49:440 (1972).

50. L. Poyer and P. B. McCay, Reduced triphosphopyridine nucleotide oxidase-catalyzed alterations of membrane phospholipids. IV. Dependence on Fe^{+3}, J. Biol. Chem. 246:263 (1971).

51. H. Esterbauer and T. F. Slater, The quantitative estimation by high performance liquid chromatography of free malonaldehyde produced by peroxidizing microsomes, I.R.C.S. Med. Sci. 9:749 (1981).

52. J. M. C. Gutteridge, Identification of malondialdehyde as the TBA-reactant formed by bleomycin-iron free radical damage to DNA, FEBS. Lett. 105:278 (1979).

53. M. Hamberg and B. Samuelsson, Oxygenation of unsaturated fatty acids by the vesicular gland of sheep, J. Biol. Chem. 242:5344 (1967).

54. H. Esterbauer, K. H. Cheeseman, M. U. Dianzani, G. Poli and T. F. Slater, Separation and characterization of the aldehydic products of lipid peroxidation stimulated by $ADP-Fe^{+2}$ in rat liver microsomes, Biochem. J. 208:129 (1982).

55. C. J. Dillard and A. L. Tappel, Fluorescent products of lipid peroxidation of mitochondria and microsomes, Lipids 6:715 (1971).

56. C. J. Dillard and A. L. Tappel, Fluorescent products from reaction of peroxidizing polyunsaturated fatty acids with phosphatidylethanolamine and phenylalanine, Lipids 8:183 (1973).

57. K. S. Chio and A. L. Tappel, Inactivation of ribonuclease and other enzymes by peroxidizing lipids and malonaldehyde, Biochemistry 8:2827 (1969).

58. K. S. Chio and A. L. Tappel, Synthesis and characterization of the fluorescent products derived from malonaldehyde and amino acids, Biochemistry 8:2821 (1969).

59. K. Kikugawa, Y. Machida, M. Kida and T. Kurechi, Studies on peroxidized lipids. III. Fluorescent pigments derived from reaction of malonaldehyde and amino acids, Chem. Pharm. Bull. 29:3003 (1981).

60. K. Kikugawa, S. Watanabe and T. Kurechi, Studies on peroxidized lipids: Fluorescent products derived from the reaction of unsaturated fatty acids and methylamine, Chem. Pharm. Bull. 32:638 (1984).

61. H. Esterbauer, Aldehydic products of lipid peroxidation, in: "Free Radicals, Lipid Peroxidation and Cancer," D. C. McBrien and T. F. Slater, ed., Academic Press, London (1982).

62. E. N. Frankel, W. E. Neff and E. Selke, Analysis of auto-xidized fats by gas chromatography-mass spectrometry: VII. Volatile thermal decomposition products of pure hydro-peroxides from autoxidized and photosensitized oxidized methyl oleate, linoleate and linolenate, Lipids 16:279 (1981).

63. T. W. Kwon and B. M. Watts, Malonaldehyde in aqueous solu-tion and its role as a measure of lipid oxidation in foods, J. Food Sci. 29:294 (1964).

64. H. Hughes, C. V. Smith, E. C. Horning and J. R. Mitchell, High-performance liquid chromatography and gas chroma-tography-mass spectrometry determination of specific lipid peroxidation products in vivo, Anal. Biochem. 130:431 (1983).

65. W. C. Hubbard and D. F. Taber, Analysis of hydroxy acids, in: "Leukotrienes and Prostacyclin," F. Bertii, G. Folio, and G. P. Velo, ed., Plenum Press, New York (1981).

66. J. Capdevilla, L. J. Marnett, N. Chacos, R. A. Prough and R. W. Estabrook, Cytochrome P450 dependent oxygenation of arachidonic acid to hydroxyicosatetraenoic acids, Proc. Natl. Acad. Sci. U.S.A. 79:767 (1982).

67. P. B. McCay, E. K. Lai, J. E. Poyer, C. M. Dubose and E. G. Janzen, Oxygen and carbon centered free radical formation during carbon tetrachloride metabolism, J. Biol. Chem. 259:2135 (1984).

68. J. J. M. C. DeGroot, G. J. Garssen, J. F. G. Vliegenthart and J. Boldingh, The detection of linoleic acid radicals in the anaerobic reaction of lipoxygenase, Biochem. Biophys. Acta. 326:276 (1973).

69. E. Cadenas, A. Boveris and B. Chance, Low-Level chemi-luminescence in biological systems, in: "Free Radicals in Biology," Vol. IV, W. A. Pryor, ed., Academic Press, New York (1984).

70. Yu. A. Vladimorov, V. I. Olenev, T. B. Suslova and Z. P. Cheremisina, Lipid peroxidation in mitochondrial membrane, Adv. Lipid Res. 17:173 (1980).

71. J. R. Wright, R. C. Rumbaugh, H. D. Colby and P. R. Miles, The relationship between chemiluminescence and lipid peroxidation in rat liver microsomes, Arch. Biochem. Biophys. 192:344 (1979).

72. N. R. DiLuzio and T. E. Stege, Enhanced hepatic chemi-luminescence following carbon tetrachloride or hydrazine administration, Life Sci. 21:1457 (1977).

73. F. Hawco, C. R. O'Brien and P. J. O'Brien, Singlet oxygen production during hemoprotein catalyzed lipid peroxide decomposition, Biochem. Biophys. Res. Comm. 76:354 (1973).

74. J. F. G. Vliegenthart and G. A. Veldink, Lipoxygenases, in: "Free Radicals in Biology," Vol. V, W. A. Pryor, ed., Academic Press, New York (1982).

75. S. D. Aust and B. A. Svingen, The role of iron in enzymatic lipid peroxidation, in: "Free Radicals in Biology," Vol. V, W. A. Pryor, ed., Academic Press, New York (1982).

76. F. Ursin, N. Maiorino, L. Ferri, N. Valente and C. Gregolin, Hydrogen peroxide and hematin in microsomal lipid peroxidation, J. Inorg. Biochem. 15:163 (1981).

77. P. J. O'Brien and A. D. Rahimtula, Involvement of cytochrome P-450 in the intracellular formation of lipid peroxides, J. Agric. Food Chem. 23:154 (1975).

78. J. M. C. Gutteridge, Lipid peroxidation initiated by superoxide dependent hydroxyl radicals using complexed iron and hydrogen peroxide, FEBS Lett. 172:245 (1984).

79. A. Ding and P.C. Chan, Singlet oxygen in copper catalyzed lipid peroxidation in erythrocyte membranes, Lipids 191: 278 (1984).

80. J. Silva and P.J. O'Brien, Benzoyl peroxide catalyzed membrane lipid peroxidation (submitted for publication).

81. A. Rahimtula, F. J. Hawco and P. J. O'Brien, The photodynamic inactivation of mixed function oxidase and peroxidation of membrane lipids, Photochem. Photobiol. 28:811 (1978).

82. L. G. Forni, J. E. Packer, T. F. Slater and R. L. Willson, Reaction of the trichloromethyl and halothane-derived peroxy radicals with unsaturated fatty acids: a pulse radiolysis study, Chem. Biol. Interact. 45:171 (1983).

83. M. Tien, B. A. Svingen and S. D. Aust, An investigation into the role of hydroxyl radical in xanthine oxidase dependent lipid peroxidation, Arch. Biochem. Biophys. 216:142 (1982).

84. B. Halliwell and J. M. C. Gutteridge, Oxygen toxicity, oxygen radicals, transition metals and disease, Biochem. J. 219: 1 (1984).

85. J. R. Bucher, M. Tien and S. D. Aust, The requirement for ferric ions in the initiation of lipid peroxidation by chelated ferrous ion, Biochem. Biophys. Res. Comm. 111:777 (1983).

86. L. A. Morehouse, C. E. Thomas and S. D. Aust, Superoxide generation by NADPH-cytochrome P450 reductase: the effect of iron chelators and the role of superoxide in microsomal lipid peroxidation, Arch. Biochem. Biophys. 232:366 (1984).

87. P. C. Chan, O. G. Peller and L. Kesner, Copper (II)-catalyzed lipid peroxidation in liposomes and erythrocyte membranes, Lipids 17:331 (1982).

88. M. Tien, L. A. Morehouse, J. R. Bucher and S. D. Aust, The multiple effects of ethylene diaminetetraacetate in several model lipid peroxidation systems, Arch. Biochem. Biophys. 218:450 (1982).

89. S. E. Fridovich and N. A. Porter, Oxidation of arachidonic acid in micelles by superoxide and H_2O_2, J. Biol. Chem. 256:260 (1981).

90. E. G. Mimnaugh, A. H. Siddik, R. Drew, B. I. Sikic and T. E. Gram, The effects of tocopherol on the toxicity, disposition and metabolism of adriamycin in mice, Toxic. Appl. Therap. 49:119 (1979).

91. J. H. Doroshow, G. Y. Locker, J. Ifrim and C. E. Myers, Prevention of doxorubicin cardiac toxicity in the mouse by N-acetylcysteine, J. Clin. Invest. 68:1053 (1981).

92. E. Herman, B. Ardalan, C. Rier, V. Waravdekar and S. Krof, Reduction of daunorubicin lethality and myocardial cellular alterations by pretreatment with ICRF-187 in Syrian hamsters, Cancer Treat. Rep. 63:89 (1979).

93. H. Kappus, H. Muliawan and M. E. Schenlen, in: "Mechanisms of Toxicity and Hazard Evaluation", B. Holmstedt, R. Lauwerys, M. Mercier and M. Roberfroid, ed., Elsevier/North Holland Biomedical Press, New York (1980).

94. J. Goodman and P. Hochstein, Generation of free radicals and lipid peroxidation by redox cycling of adriamycin and daunomycin, Biochem. Biophys. Res. Comm. 76:705 (1977).

95. E. G. Mimnaugh, M. A. Trush and T. E. Gram, Stimulation by adriamycin of rat heart and liver microsomal NADPH-dependent lipid peroxidation, Biochem. Pharmacol. 30:2797 (1981).

96. H. Muliawan, M. E. Scheulen and H. Kappus, Adriamycin stimulates only the iron ion induced, NADPH-dependent microsomal alkane formation, Biochem. Pharmacol. 31:3147 (1982).

97. J. M. C. Gutteridge, Adriamycin-iron catalyzed phospholipid peroxidation: a reaction not involving reduced adriamycin or hydroxyl radicals, Biochem. Pharmacol. 32:1949 (1983).

98. K. Sugioka and M. Nakano, Mechanism of phospholipid peroxidation induced by ferric ion-ADP-adriamycin coordination complex, Biochim. Biophys. Acta 713:333 (1982).

99. J. M. C. Gutteridge, Lipid peroxidation and possible hydroxyl radical formation stimulated by the self-reduction of a doxorubicin-iron (III) complex, Biochem. Pharmacol. 33: 1725 (1984).

100. C. E. Myers, L. Gianni, C. B. Simone, R. Klecker and R. Green, Oxidative destruction of erythrocyte ghost membranes catalyzed by the doxorubicin-iron complex, Biochemistry 21:1707 (1982).

101. J. R. F. Muindi, B. K. Sinha, L. Gianni and C. E. Myers, Hydroxyl radical production and DNA damage induced by anthracycline-iron complex, FEBS Lett. 172:226 (1984).

102. E. J. Demant, Binding of adriamycin-Fe^{3+} complex to membrane phospholipids, Europ. J. Biochem. 142:571 (1984).

103. H. Eliot, L. Gianni and C. Myers, Oxidative destruction of DNA by the adriamycin-iron complex, Biochemistry 23:928 (1984).

104. M. A. Trush, E. G. Mimnaugh, E. Ginsburg and T. E. Gram, Studies on the in vitro interaction of mitomycin c, nitrofurantoin and paraquat with pulmonary microsomes, Biol. Pharmacol. 31:805 (1982).

105. J. M. C. Gutteridge and F. Xiao-Chang, Enhancement of bleomycin-iron free radical damage to DNA by antioxidants and their inhibition of lipid peroxidation, FEBS Lett. 123:71 (1981).

106. W. M. Tom and M. R. Montgomery, Bleomycin toxicity. Alterations in oxidative metabolism in lung and liver microsomal fractions, Biochem. Pharmacol. 29:3239 (1980).

107. T. Komiyama, T. Kikuchi and Y. Sugiura, Generation of hydroxyl radical by anticancer quinone drugs in the presence of NADPH-cytochrome P-450 reductase, Biochem. Pharmacol. 31:3651 (1978).

108. J. W. Lown and H. H. Chen, Evidence for the generation of free hydroxyl radicals from certain quinone antitumor antibiotics upon reductive activation in solutions, Can. J. Chem. 59:390 (1981).

109. L. W. Oberley and G. R. Buettner, The production of hydroxyl radical by bleomycin and iron (II), FEBS Lett. 97:47 (1979).

110. Y. Suguira and T. Kikuchi, Formation of superoxide and hydroxy radicals in iron (II)-bleomycin-oxygen system: electron spin resonance detection by spin trapping, J. Antibiotics 31:1310 (1978).

111. Y. Suguira, The production of hydroxyl radical from copper (I) complex systems of bleomycin and tallysomycin: comparison with copper (II) and iron (II) systems, Biochem. Biophys. Res. Comm. 90:375 (1979).

112. G. Cohen and A. I. Cederbaum, Microsomal metabolism of hydroxyl radical scavenging agents: relationship to the microsomal oxidation of alcohols, Arch. Biochem. Biophys. 199:438 (1980).

113. C. C. Winterbourn, Evidence for the production of hydroxyl radicals from the adriamycin semiquinone and H_2O_2, FEBS Lett. 136:89 (1981).

114. D. A. Bates and C. C. Winterbourn, Deoxyribose breakdown by the adriamycin semiquinone and H_2O_2: evidence for hydroxyl radical participation, FEBS Lett. 145:137 (1982).

115. H. Noll and W. Jordan, Hydroxyl radical generation by adriamycin semiquinone and H_2O_2: an explanation for the cardiotoxicity of anthracycline antibiotics, Biochem. Biophys. Res. Comm. 114:197 (1983).

116. E. Paur, R. J. Youngman, E. Lengfelder and E. F. Elstner, Mechanisms of adriamycin-dependent oxygen activation catalyzed by NADPH-cytochrome c-oxidoreductase, Z.

<u>Naturforsch</u> 396:261 (1984).

117. G. W. Winston, D. E. Feierman and A. I. Cederbaum, The role of iron chelates in hydroxyl radical production by rat liver microsomes, NADPH-cytochrome P-450 reductase and xanthine oxidase, <u>Arch. Biochem. Biophys.</u> 232:378 (1984).

118. I. Johansson and M. Ingelman-Sundberg, Hydroxyl radical-mediated, cytochrome P-450 dependent metabolic activation of benzene in microsomes and reconstituted enzyme systems from rabbit liver, <u>J. Biol. Chem.</u> 258:7311 (1983).

119. E. G. Hrycay and P. J. O'Brien, The peroxidase nature of cytochrome P-450 utilizing a lipid peroxide substrate, <u>Arch. Biochem. Biophys.</u> 147:28 (1971).

120. W. A. Pryor and L. W. Lightsey, Mechanisms of nitrogen dioxide reactions: Initiation of lipid peroxidation and the production of nitrous acid, <u>Science</u> 214:435 (1981).

121. J. Kanner and J. C. Kinsella, Initiation of lipid peroxidation by a peroxidase/hydrogen peroxide/halide system, <u>Lipids</u> 18:204 (1983).

122. B. H. Bielski, R. L. Arudi and M. W. Sutherland, A study of the reactivity of HO_2/O_2^- with unsaturated fatty acids <u>J. Biol. Chem.</u> 258:4759 (1983).

FLUORESCENCE STUDIES OF LIPID ORDER IN PROTEOLIPOSOMES CONTAINING CYTOCHROME P-490 AND ITS REDUCTASE

Barbara C. Kunz*, Mike Rehorek**,
and Christoph Richter*

*Eidgenoessische Technische Hochschule, Laboratorium
 für Biochemie I, ETH-Zentrum, CH-8092 Zurich
**Biocenter of the University of Basel
 CH-4056 Basel, Switzerland

INTRODUCTION

Cytochrome P-450 and NADPH-cytochrome P-450 reductase are key enzymes of the hepatic microsomal monooxygenase system. They catalyze the oxidative and reductive metabolism of endogenous substrates and various xenobiotics (1-3). These enzymes are also involved in the biotransformation of physiologically important compounds such as fatty acids, prostaglandins, leukotrienes, and steroids. The reductase ($M_r \sim 78,000$) is anchored to the membrane of the endoplasmatic reticulum via a small ($M_r \sim 6,000 - 10,000$) hydrophobic segment (4,5). The large hydrophilic part, which contains 1 molecule of FMN and FAD, protrudes from the membrane into the cytoplasmic space. The topology of P-450 ($M_r \sim 50,000$) in membranes, on the other hand, is currently unknown (6). Limited proteolysis of microsomes does not release a soluble catalytic domain but results in conversion of P-450 to the catalytically inactive form P-420 (7). It therefore appears that P-450 is not simply anchored to the membrane by a single hydrophobic domain as its reductase. Recent sequencing studies support this view. P-450 and reductase can be isolated and reconstituted in a enzymatically active form.

Proposing the "fluid mosaic" model of the structure of cell membranes, Singer and Nicolson (8) emphasized that membrane proteins and phospholipids are amphipatic molecules, and that the rotational diffusion of these molecules across the plane of membrane would be very slow or negligible. There is now evidence that protein molecules retain their sidedness in natural membranes, and that phospholipids are distributed unevenly between the bilayer halves (9,10).

The arrangement of cytochrome P-450 and reductase in the microsomal membrane and their odd stoichiometry has raised questions as to the mechanism of electron transfer from the reductase to the cytochrome, and the functional interaction of the proteins. In microsomes there are up to 20 - 30 cytochrome P-450 molecules per one reductase molecule.

Reconstitution of membrane proteins has the advantage that side effects caused by other proteins present in intact microsomes can be avoided. Furthermore the stoichiometry of the two components can be altered. A suitable reconstitution method for the monooxygenase system is described in detail in ref. (11). Briefly, commercially available lipids or lipids extracted from microsomes are dried, suspended in phosphate buffer and solubilized with the detergent cholate. The proteins are added and the mixture is incubated for 15 h at 4° C. Then cholate is removed by dialysis at room temperature. The proteoliposomes will form during four hours of dialysis.

Rotational mobility of P-450 was recently demonstrated in microsomes and proteoliposomes (11-13) by flash-induced absorption anisotropy techniques of the heme-CO complex, and in proteoliposomes by delayed fluorescence polarization technique (14) or by magnetic CD spectroscopy (15). The absorption anisotropy technique was successfully applied to investigate intermolecular interactions of P-450 with itself or with its reductase in both microsomal membranes and reconstituted vesicles (11,16,17). These studies indicated the existence of a long-lived heterodimeric complex between one reductase and one cytochrome molecule in membranes. Evidence for significant molecular interactions between the two proteins in a membraneous reconstituted system was also recently provided (18) by the use of reductase specifically labelled with a fluorescent probe.

While these studies have given information about protein-protein interactions of the enzymes in membranes, little is known about lipid-protein interactions in the monooxygenase system. The study of lipid-protein interactions is of great interest in membrane biology. Such interactions affect the fluidity and order of the bilayer, modulate enzymatic activity of membrane proteins, and are responsible for changes in the melting temperature and cooperativity of the gel to liquid-cristalline phase transition. Biochemical studies give evidence for the importance of acidic phospholipids for reconstitution of the monooxygenase system (11,19,20). There are also indications that the phospholipid environment can affect the conformation and/or spin state of P-450 (21-23). Much less is known about the perturbation caused by the incorporation of proteins of the monooxygenase system on the dynamics of the lipid motion in the bilayer.

Diphenylhexatriene (DPH) is the most frequently used fluorescent probe in studies of lipid chain dynamics (24). It is a lipophilic molecule and will arrange parallel to the lipid hydrocarbon chains in the bilayer. DPH is fluorescent only in a hydrophobic environment (25). The similarity of size between DPH and a lipid chain excludes that pronounced perturbations of the lipid domain will be induced by DPH.

By determining the steady-state and time-dependent fluorescence depolarization of the probe DPH we investigate here the perturbation induced by P-450 and reductase as a function of lipid composition and lipid to protein ratio in reconstituted liposomes. The results show a considerable disordering of phospholipid acyl chains upon incorporation of the proteins into the membrane.

Principle of Optical Anisotropy

As pointed out in the excellent rewiew of Kinosita et al. (26), rotational motion is the change in orientation of an object with time. The simplest way of measuring rotation is to determine the orientation of the object twice, at times zero and t, and ask how much the orientation has changed during the time interval. When the sample contains a large number of objects, as is always the case for measurements at the molecular level, one would attempt to select a subpopulation that has a particular orientation at time zero and "mark" it with a tag. Later at time t, the orientation of the tagged objects will be determined for the second time (Fig.1). This is the basic principle common to all the optical anisotropy decay measurements.

For the optical measurement, the object must have a chromophore, either intrinsic or extrinsic. The anisotropic interaction of the chromophore with polarized light provides the basis for the two determinations of the orientation.

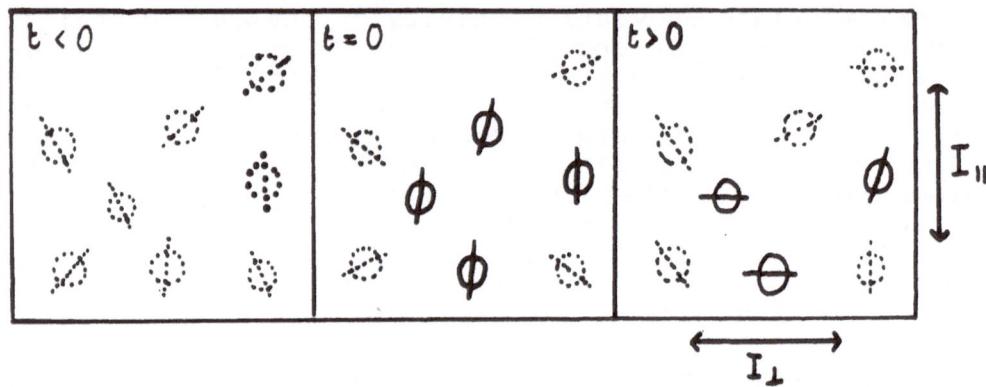

Figure 1: Principle of optical anisotropy decay.
At $t < 0$, the molecules with their chromophores are randomly distributed. At $t = 0$, a polarized light pulse excites those molecules that have their chromophores in the proper orientation relative to the exciting light as to absorb energy and undergo fluorescence, delayed fluorescence or phosphorescence. At $t > 0$, in principle three possibilities can be considered. First, the chromophore has not moved. In this case, the transition moment will still lay parallel to the exciting light, and the resulting optical anisotropy, r, does not change with time. Second, the chromophore has moved <u>isotropically</u>. In this situation, the anisotropy decreases with time and goes to zero. Third, the chromophore moves <u>anisotropically</u>. In this case, r decreases and reaches a constant value greater than zero. Instead of measuring the position of each chromophore, the number of chromophores with their transition moments parallel and perpendicular to the exciting light are determined. Total fluorescence intensity I_T, and anisotropy r are calculated from
$I_T = I_{\parallel} + 2I_{\perp}$ and $r = I_{\parallel} - I_{\perp} / I_T$.

Chromophores such as eosin will produce phosphorescence and delayed fluorescence with a long lifetime (10^{-3} sec). They serve as probes for the relatively slow rotational motions of proteins. The heme-CO complex of P-450 undergoes a photochemical reaction and therefore carries a long lived tag, suitable to measure rotational motion of cytochrome P-450. The fluorescent probe DPH has a much smaller lifetime (in the nanosecond range) and serves as probe for the relatively fast rotational motion of the lipids in a bilayer.

Steady-state Fluorescence Anisotropy

Steady-state fluorescence anisotropy under constant illumination of the probes has been widely used in the studies of membrane dynamics (27). The measurement is quick and relatively easy. The two fluorescence intensities, $I_\parallel$ and $I_\perp$ are alternately recorded by rotating the analyzer over 90°. Anisotropies are calculated from

$$r_s = I_\parallel - I_\perp / I_\parallel + 2 I_\perp \quad [1]$$

A major disadvantage of steady-state measurements is the fact that they normally do not separate dynamic and structural information due to the fact that the lifetime of the fluorescence can not be determined.

Fluorescence Decay Measurements

Under constant illumination the signals from all the different populations will overlap. To look at one single population a nanosecond-gated flashlamp has to be used. The probe is illuminated at time zero with one single light pulse. Then the fluorescence decay is recorded before the next population will become tagged. Exponential decays of the following form are assumed:

$$I_T^\delta(t) = \sum_{i=1}^{N} I_i \exp(-t/\tau i) \quad [2]$$

$$r^\delta (t) = \sum_{j=1}^{M} r_j \exp(-t/\varphi j) \quad [3]$$

where δ stands to indicate an infinitely short excitation pulse. The fluorescence intensities for vertical and horizontal polarization at time t after excitation, $I_V (t)$ and $I_H (t)$, are related to $I_T^\delta (t)$ and $r^\delta (t)$ by the following equations:

$$I_V (t) + 2I_H (t) = \int_0^t g(t') I_T^\delta (t-t') \, dt' \quad [4]$$

$$I_V (t) - I_H (t) = \int_0^t g(t') r^\delta (t-t') I_T^\delta (t-t') \, dt' \quad [5]$$

$$r (t) = I_V (t) - I_H (t) / I_V (t) + 2I_H (t) \quad [6]$$

where $g (t')$ is the response function of the apparatus (28).

Fluorescence Energy Transfer

Fluorescence energy transfer is a useful technique in determining microscopic parameters, such as intermolecu-

lar distances and interactions in biological systems
(29, and references cited therein). The application of
the technique for membranes and phospholipid vesicles
range from studies on membrane fusion and on antibody-
receptor clustering to the determination of lipid-protein interactions and of bilayer spacing.

Fluorescence energy transfer from donors to acceptors,
both randomly distributed within the plane of a bilayer,
can be described by theories for energy transfer in two-
dimensional systems (30-34). In such systems, where the
distances between donors and acceptors can cover an
extended range, a large portion of acceptors may be
located within R_o, the distance of 50% transfer
efficiency. The number of acceptors that are located
within an area, R_o^2, around a donor, is determined by
the concentration of acceptors and by the distance of
closest approach, R_c, between the donor and the acceptors (30). The extent of energy transfer can be described by the relative quantum yield, the ratio of the
areas under the normalized fluorescence decay curves of
the donor in the presence and absence of receptors (32).
The relative quantum yield is related to the transfer
efficiency by

$$E = 1 - q_{rel} \qquad [7]$$

Strong energy transfer from DPH to the chromophore of a
protein was observed in lipid vesicles containing bacteriorhodopsin (35) and cytochrome oxidase (36). The
spectral overlap between the fluorescence emission
spectrum of DPH and the absorption spectrum of
cytochrome P-450 is large enough to expect strong energy
transfer from DPH to the heme group of P-450.

RESULTS

Coreconstitution of P-450 and its reductase into liposomes results in a functionally competent monooxygenase
system (16). The interaction of the protein and the
lipid components of this reconstituted system was
investigated by the steady-state and time-resolved fluorescence of DPH.

Temperature Dependence of Steady-State Fluorescence
Anisotropy

Fluorescence anisotropy of DPH was determined as a
function of temperature and composition of liposomes

(Figure 2). Incorporation of P-450 alone or together
with reductase increased the steady-state fluorescence
anisotropy (r_s) of DPH in dioleoyl-PC (DOPC) and
phosphatidylcholine-phosphatidylethanolamine-phosphati-
dylserine (PC-PE-PS) (10:5:1) liposomes. In dimyristoyl-
PC (DMPC) and dipalmitoyl-PC (DPPC) vesicles, the
proteins decreased r_s significantly below the transition
temperature (Tc) of the gel to liquid-cristalline phase
transition. In all types of liposomes studied, reductase
coreconstituted with P-450 increased the changes in r_s
induced by the cytochrome alone. Changes in r_s between
four and 40°C were reversible in all experiments.

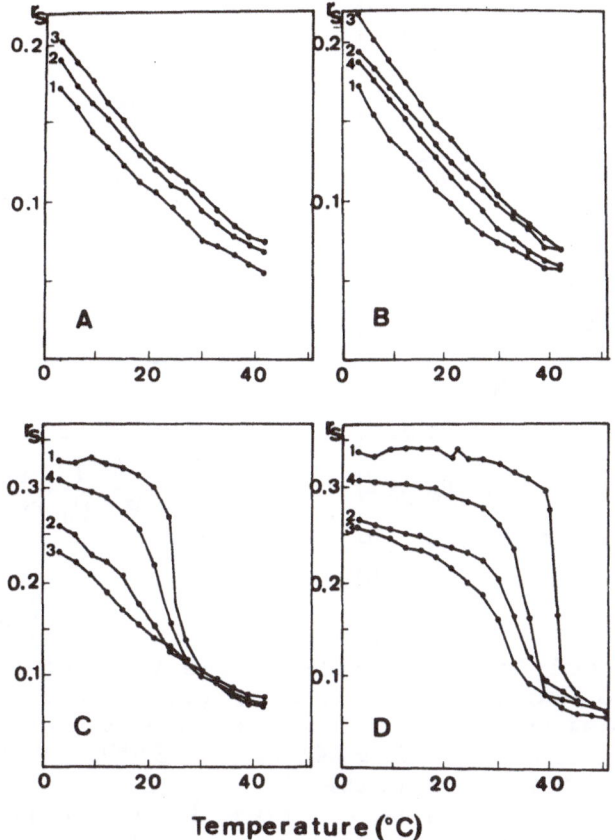

Temperature (°C)

Figure 2: Temperature dependence of steady-state
fluorescence anisotropy r_s of DPH incorporated into PC-
PE-PS = 10:5:1 (panel A), DOPC (B), DMPC (C), and DPPC
(D) vesicles. Lipid and cytochrome P-450 were 0.2 mg/ml
each. NADPH-cytochrome P-450 reductase was present in
amounts equimolar to cytochrome P-450 (2.8 μM). Concentration of DPH was 2 μM. Vesicles were in 20 mM HEPES pH

7.4 with 1.0 mM EDTA and 20% glycerol. Curve 1: Vesicles without proteins, curve 2: Vesicles containing P-450, curve 3: Vesicles containing P-450 and reductase, curve 4: vesicles containing reductase alone. In PC-PE-PS vesicles (A) reductase had no influence on r_s of DPH.

Time-Resolved Fluorescence Measurements

In contrast to the steady-state technique, time-resolved fluorescence measurements can give unambiguous informa-

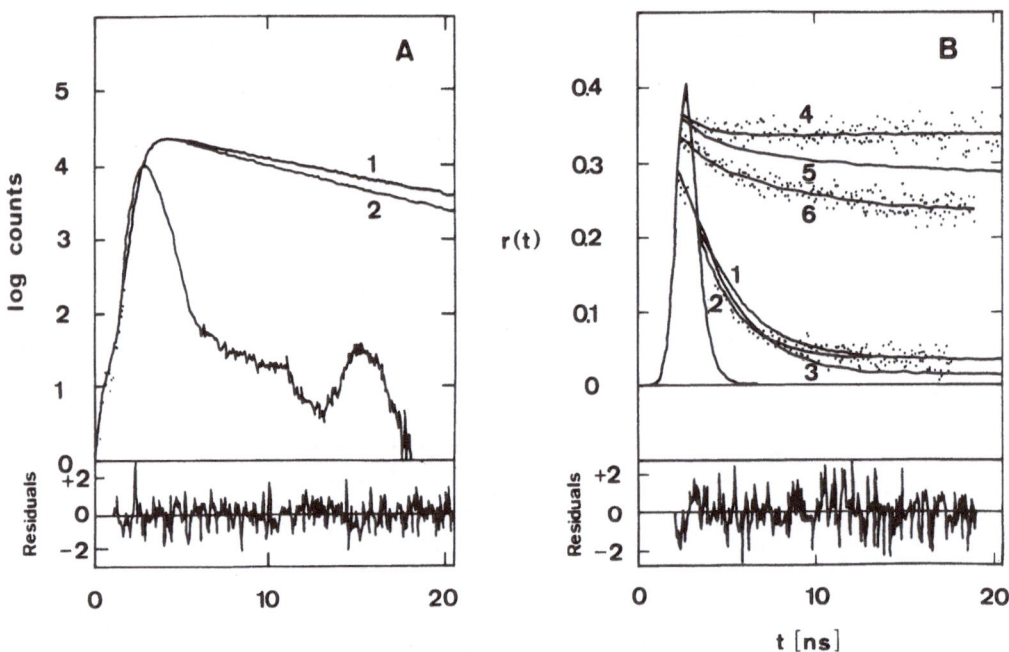

Figure 3: Panel A: Lamp pulse (curve without number) and fluorescence decay curves of DPH, calculated from eq.[1] for DMPC vesicles in the absence (curve 1) and in the presence (curve 2) of P-450, at 35°C. Bottom, the weighted residuals for curve 2.
Panel B: Experimental (·······) and calculated (———) aniso-tropy decay curves:
1 DMPC + P-450 (w/w =1/1), 35°C
2 pure DMPC, 35°C
3 DMPC + Reductase (w/w = 1/0.65), 35°C
4 pure DMPC, 10°C
5 DMPC + Reductase (w/w = 1/0.65), 10°C
6 DMPC + P-450 (w/w = 1/1), 10°C
For clarity, the data points for curves 1, 3, and 5 have been omitted. Bottom, the weighted residuals for curve1.

tion about lipid order and dynamics (37). Time courses of fluorescence intensity and depolarization of DPH in PC-PE-PS = 10:5:1 and in DMPC vesicles were measured at 35 and 10°C. The influence of P-450 and/or reductase on the dynamics and order of the lipid phase was investigated after reconstitution into these two vesicle systems. As an example, fluorescence decay curves of DPH in DMPC and DMPC containing P-450 are shown in Figure 3, panel A. Panel B shows the corresponding anisotropy decay curves. The fluorescence signals were analyzed according to equations [2]-[6], using least squares algorithm with x^2 to test the quality of the fits. Mean lifetime $\langle \tau \rangle$, relaxation time $\langle \varphi \rangle$, and order parameter S, obtained in the various systems are listed in Table I. The x^2 values were all close to one, indicating the good quality of the fits.

In both vesicle systems either protein lowered the order parameter S, the only exception being P-450 in DMPC proteoliposomes above Tc, where S remained unchanged. The lifetime of DPH fluorescence in the present liposomes is decreased only very little by cytochrome or reductase and is independent of the specific heme content of P-450 between eight and 14 nmol/mg. Appreciable energy transfer between the prosthetic groups of the proteins, i.e. heme or flavins, and DPH is therefore ruled out in contrast to the situation in liposomes containing bacteriorhodopsin and cytochrome oxidase for which pronounced energy transfer has been reported (35, 36). The absence of energy transfer between DPH and cytochrome P-450 allows a rough estimation of the location of the heme in the reconstituted cytochrome as outlined in the next section.

Fluorescence Energy Transfer

The overlap integral between the fluorescence emission spectrum of DPH and the absorption spectrum of P-450 was calculated by numerical integration to be $1.1 \cdot 10^{-13}$ cm^3 $lmol^{-1}$. For random orientations (orientation factor $K^2 = 2/3$) of the two chromophores this leads to a value for R_o of 5.0 nm. A similar value for R_o (5.1 nm) was calculated for heme $\underline{a}$ in DMPC vesicles containing cytochrome oxidase (36). In contrast to the cytochrome oxidase vesicles, where the energy transfer from DPH to heme $\underline{a}$ leads to a decrease of the average lifetime of DPH fluorescence by a factor of four, the energy transfer in the P-450 vesicles is very small. The average lifetime of DPH fluorescence in DMPC vesicles at 35°C decreases from 8.0 ns to 6.1 ns in the presence of P-450. This corresponds to a efficiency, E, of 0.24. As

Table I: Fluorescence Anisotropy Decay Parameters of DPH in PC/PE/PS Vesicles containing Cytochrome P-450 and/or NADPH-cytochrome P-450 Reductase

Lipid	Protein	10°C			35°C		
		$\langle\tau\rangle$ [ns]	$\langle\varphi\rangle$ [ns]	S	$\langle\tau\rangle$ [ns]	$\langle\varphi\rangle$ [ns]	S
PC/PE/PS	−	7.37±0.06	1.1±0.1	0.41±0.01	6.87±0.07	0.48±0.08	0.25±0.01
	P-450	6.32±0.05	0.5±0.1	0.33±0.01	6.0±0.1	0.68±0.08	0.18±0.01
	P-450/Red	6.52±0.05	0.8±0.1	0.32±0.02	6.0±0.1	0.82±0.08	0.15±0.01
	Red	6.69±0.05	1.1±0.1	0.25±0.02	6.3±0.1	0.68±0.08	0.0
DMPC	−[a]	11		0.9	8.0±0.1	0.44±0.06	0.29±0.01
	P-450	7.74±0.08	0.4±0.1	0.76±0.01	6.16±0.08	0.55±0.07	0.29±0.01
	P-450/Red	8 ±1	0.6±0.2	0.72±0.01	6.20±0.05	0.63±0.09	0.20±0.01
	Red	8.20±0.05	0.9±0.1	0.82±0.02	6.72±0.06	0.58±0.08	0.17±0.01

a The parameters for pure DMPC vesicles at 35°C and 10°C are taken from Rehorek and Heyn (unpublished), and ref. (36), respectively.

the protein radius of P-450, determined by rotational diffusion measurements (16,17), is ~2.0 nm, and as the molar protein to lipid ratio is high (1:70, 1-2 layers of lipid around one protein), the absence of strong energy transfer can only be explained by (i) a low, i.e. a non-random, orientation factor, K^2, or (ii) by a large vertical displacement of the heme group to the outside of the bilayer. The orientation factor for this system was calculated on the basis of eq. [21] of Dale et al. (38), with DPH wobbling around the membrane normal, and the heme group being tilted by 55° from the normal (17), and displaced from the bilayer midplane by a distance, d. The low order parameter, S, of DPH in these vesicles (0.287) and the planar symmetry of the heme group yields values for K^2 between 0.6 and 0.8, depending on d. For d > 5.0 nm, R_O = 5.2 nm and is almost independent of the relative orientation of the two chromophores.

The lateral distribution function of the acceptors (P-450) and the donors (DPH) at such high protein to lipid ratio (1:70) is unknown, since there might be some aggregates of cytochrome P-450 present (17). Therefore it is not possible to calculate analytically the distance, d, from the measured transfer efficiency. From numerical calculations we estimate, however, that the heme group in cytochrome P-450 is positioned at a distance d $\geqslant$ 6.0 nm from the bilayer midplane. This estimation provides the first experimental evidence for location of P-450's heme outside the membrane bilayer.

CONCLUSION AND OUTLOOK

The fluorescence measurements mentioned above illustrate two aspects of this powerful technique. First, useful qualitative information about lipid mobility can be obtained by the relatively simple and inexpensive steady-state technique. However, comparison of results in Figure 2 and Table I underline the fact that unambiguous information about lipid order and dynamics can only be obtained by time-resolved fluorescence decay measurements. Second, electronic excitation energy can be efficiently transferred between a fluorescent energy donor and a suitable energy acceptor over distances as large as 70 Å, making energy transfer a spectroscopic ruler in the 10-60 Å range.

The data show a decreased order of the phospholipid acyl chains in proteoliposomes containing cytochrome P-450 and its reductase above as well as below Tc. No other protein studied so far disordered acyl side chains both above and below Tc; rather, most proteins increase the order of their surrounding phospholipids. Future experiments will investigate if also the polar part of the phospholipids is mobilized by cytochrome P-450 and/or its reductase.

Fluorescence energy transfer experiments place the heme of cytochrome P-450 outside the membrane bilayer. It is hoped that current protein sequencing data and future crystallographic investigations will provide additional information concerning the structure of this important membrane-bound heme protein.

REFERENCES

(1)   Estabrook, R.W., Werringloer, J., and Peterson, J.A. (1979) in Xenobiotic Metabolism: In Vitro Methods (Paulson, G.D., Frear, D.S., and Marks, E.P., eds) Symposium Series No. 97, pp. 149-179, American Chemical Society, Washington D.C.
(2)   White, R.E., and Coon, M.J. (1980) Annu. Rev. Biochem. 49, 315-356
(3)   Omura, T. (1978) in Cytochrome P-450 (Sato, R., and Omura, T., eds) Chap. 1, Kodanska, Tokyo
(4)   Gum, J.R., and Strobel, H.W. (1979) J. Biol. Chem. 254, 4177-4185
(5)   Black, S.D., French, J.S., Williams, C.H. Jr., and Coon, M.J. (1979) Biochem. Biophys. Res. Commun. 91, 1528-1535
(6)   De Pierre, J.W., and Ernster, L. (1977) Annu. Rev. Biochem. 46, 201-262
(7)   Sato, R., Nishibayashi, H., and Ito, A. (1969) in Microsomes and Drug Oxidations (Gillette, J.R., Conney, A.H., Cosmides, G., Estabrook, R.W., Fouts, J.R., and Mannering, G.J., eds) pp. 111-132, Academic Press, New York
(8)   Singer, S.J., and Nicolson, G.L. (1972) Science 175, 720-731
(9)   Etemadi, A.H., (1980) Biochim. Biophys. Acta. 604, 347-422
(10) Etemadi, A.H., (1980) Biochim. Biophys. Acta. 604, 423-475
(11) Kawato, S., Gut, J., Cherry, R.J., Winterhalter, K.H., and Richter, C. (1982) J. Biol. Chem. 257, 7023-7029
(12) Richter, C., Winterhalter, K.H., and Cherry, R.J. (1979) FEBS Lett. 102, 141-154
(13) Mc Intosh, P.R., Kawato, S., Freedman, R.B., and

Cherry, R.J. (1980) FEBS Lett. 122, 54-58

(14) Greinert, R., Staerk, H., Stier, A., and Weller, A. (1979) J. Biochem. Biophys. Methods 1, 77-83

(15) Boesterling, B., and Trudell, J.R. (1982) J. Biol. Chem. 257, 4783-4787

(16) Gut, J., Richter, C., Cherry, R.J., Winterhalter, K.H., and Kawato, S. (1982) J. Biol. Chem. 257, 7030- 7036

(17) Gut, J., Richter, C., Cherry, R.J., Winterhalter, K.H., and Kawato, S.(1983) J. Biol. Chem. 258, 8599-8594

(18) Nisimoto, Y., Kinosita, K., Jr., Ikegami, A., Kawai, N., Ichihara, I., and Shibata, Y. (1983) Biochemistry 22, 3594-3603

(19) Haaparanta, T., Rydstroem, J., and Ingelman-Sundberg, M. (1980) in Microsomes, Drug Oxidations, and Chemical Carcinogenesis (Coon, M.J., Conney, A.H., Estabrook, R.W., Gelboin, H.V., Gillette, J.R., and O'Brien, P.J. eds) pp. 533-536, Academic press, New York

(20) Ingelmann-Sundberg, M.J., Haaparanta, T., and Rydstroem, J. (1981) Biochemistry 20, 4100-4106

(21) Taniguchi, H., Imai, Y., and Sato, R. (1980) in Microsomes, Drug Oxidations, and Chemical Carcinogenesis (Coon, M.J., Conney, A.H., Estabrook, R.W., Gelboin, H.V., Gillette, J.R., and O'Brien, P.J. eds) pp. 537-540, Academic press, New York

(22) Gibson, G.G., Sligar, S.G., Cinti, D.L., and Schenkman, J.B. (1980) in Microsomes, Drug Oxidationsand Chemical Carcinogenesis (Coon, M.J., Conney, A.H., Estabrook, R.W., Gelboin, H.V., Gillette, J.R., and O'Brien, P.J., eds.) pp.119-122, Academic Press, New York

(23) Yang, C.S., and Tsong, T.Y. (1980) in Microsomes, Drug Oxidations and Chemical Carcinogenesis (Coon, M.J., Conney, A.H., Estabrook, R.W., Gelboin, H.V., Gillette, J.R., and O'Brien, P.J., eds.) pp.199-202, Academic Press, New York

(24) Shinitzky, M., and Barenholz, Y. (1978) Biochim. Biophys. Acta 515, 367-394

(25) Shinitzky, M., and Barenholz, Y. (1974) J. Biol. Chem. 249, 2652-2657

(26) Kinosita, K., Jr., Kawato, S., and Ikegami, A. (1984) Adv. Biophys. 17, 147-202

(27) Shinitzky, M., and Barenholz, Y. (1978) Biochim. Biophys. Acta 515, 367

(28) Kawato, S., Kinosita, K., Jr., and Ikegami, A. (1977) Biochemistry 16, 2319-2324

(29) Kellerer, H., and Blumen, A. (1984) Biophys. J. 46, 1-8

(30) Wolber, P.K., and Hudson, B.S. (1979) Biophys, J. 28, 197-210
(31) Koppel, D.E., Fleming, P.J., and Strittmatter, P. (1979) Biochemistry 18, 5450-5457
(32) Dewey, T.G., and Hammes, G.G. (1980) Biophys. J. 32, 1023-1036
(33) Fung, B.K., and Stryer, L. (1978) Biochemistry 17, 5241-5248
(34) Estep, T.N., and Thompson, T.E. (1979) Biophys. J. 26, 195-208
(35) Rehorek, M., Dencher, N.A., and Heyn, M.P. (1983) Biophys. J. 43, 39-45
(36) Kinosita, K., Jr., Kawato, S., Ikegami, S., Yoshida, S., and Orii, Y. (1981) Biochim. Biophys. Acta 647, 7-17
(37) Heyn, M.P. (1979) FEBS Lett. 108, 359-364
(38) Dale, R.E., Eisinger, J., and Blumberg, W.E. (1979) Biophys. J. 26, 161-194

AKNOWLEDGEMENT

This work was supported by the Swiss National Science Foundation, Grant Nr. 5.521.370.646/0, and the very generous financial help of Solco AG, Basle.

THE MECHANISMS INVOLVED IN THE REMOVAL OF OXYGEN AND OXYGEN RADICALS

IN THE N_2-FIXING CYANOBACTERIUM NOSTOC MUSCORUM

Leah Karni and Elisha Tel-Or

Dept. Of Agricultural Botany
The Hebrew University of Jerusalem
Rehovot 76100, Israel

INTRODUCTION

Cyanobacteria (blue green algae) consist of the largest group of photosynthetic prokaryotes in nature. They differ from other photosynthetic bacteria by their oxygenic photosynthetic activity employing water as primary electron donor for CO_2 fixation, (Binder 1982). The cyamobacteria are therefore combining the simplicity of the prokaryotic cell structure, with the highly developed photosynthetic functions of the higher plants, (Papageorgiou and Tzani, 1980). The cyanobacterium Nostoc muscorum is also capable of fixing molecular nitrogen to ammonia under aerobic conditions. The cells are organized in a multicellular filament including the vegetative cells responsible for oxygenic photosynthesis and CO_2 fixation, and fewer heterocysts which accomodate the N_2-fixing enzyme, nitrogenase, and its supporting systems providing reductants, ATP and protecting mechanisms. Heterocysts are formed by differentiation of vegetative cells during grown in media deficient of combined nitrogen compounds, and frequency of heterocysts in air-grown cultures approaches 5-8 percent (Wolk 1982, Peterson and Burris 1976). The frequency of heterocysts in nitrate-grown cultures is lower, and is very low in ammonia-grown cultures (Wolk 1979, Neuer et al. 1983). The heterocyst is considered to be a specialized cell for N_2-fixation, under aerobic conditions, acquiring unique structure and functions throughout the process of differentiation, providing the suitable conditions for the expression and activity of nitrogenase (Stewart et al. 1969, Haselkorn 1978). Nitrogenase catalyses the reduction of N_2 to ammonia, described in the following equation:

$$N_2 + 8e^- + 8H^+ + 16 \text{ ATP} \longrightarrow 2NH_3 + H_2 + 16(\text{ADP+Pi}).$$

The enzyme activity involves two nonheme iron proteins: dinitrogenase (Fe-Mo protein) and dinitrogenase reductase (Fe protein) each of which is extremely sensitive to oxygen. The nature of O_2 inhibition is irreversible, attacking the iron-sulphur active sites of both proteins (Palmer et. al. 1972). O_2 inhibits also the de novo synthesis of nitrogenase polypeptides, via the regulatory H and L genes within the nif genome in the heterocysts (Stewart 1982).

The heterocyst is the most suitable cell to acccomodate nitrogenase synthesis and function. O_2 concentration is lower in the heterocyst in comparison with the neighboring vegetative cells and the cell is protected by few layer of extracellular envelope, composed of glycolipids which eliminate the diffusion of oxygen into the heterocyst (Lambian and Wolk 1973). Furthermore, the water splitting manganese-containing complex is inactive in the heteroocyst, eliminating O_2 production by the photosynthetic apparatus in the cell (Tel-Or and Stewart 1977). However, O_2 diffusion into the heterocysts from the vegetative cells may occur through the microplasmodesmata bridging the heterocyst with its adjacent vegetative cells (Neuer et al. 1983). Peterson and Burris (1976) suggested that respiration provides active means to remove oxygen in heterocysts of Anabaena, where respiratory activity was enriched in the heterocysts in comparison with the vegetative cells.

I. REMOVAL OF O_2, SUPEROXIDE AND PEROXIDE

We found rather low respiration activity in isolated heterocysts of Nostoc muscorum; the rate of O_2 consumption did not exceed 4.5 nmol O_2.mg prot^{-1}. min^{-1}, while the rate of O_2 diffusion into heterocysts may approach 5-15 nmol.mg prot^{-1}.min^{-1}.

The oxyhydrogen activity of uptake hydrogenase may be therefore extremely important in the heterocysts employing the H_2 produced by nitrogenase to reduce the respiratory components, leading to the scavenging of O_2 by cytochrome oxidase (Stewart et al., 1978). Oxygen damage to the heterocysts may also derive from its radicals, superoxide and peroxides which are considered to be potent and highly reactive toxic compounds (Bothe 1980). Superoxide is produced in photosynthetic prokaryotes by the interaction of O_2 with reducing intermediates of the photosynthetic electron transport chain (i.e., ferredoxin) in the Mehler type reaction (Allen 1977). Superoxide may also be produced by the photosynthetic pigments, phycocyanin and chlorophyll, through a direct interaction between the excited pigment with O_2 in the light (Steinitz 1977). The enzyme superoxide dismutase (EC 1.15.1.1) is active in the removal of the toxic superoxide radical, following the dismutation reaction:

$$2O_2^- + 2H^+ \longrightarrow H_2O_2 + O_2$$

Table 1. The Reduction of H_2O_2 by GSH and Ascorbate

Reductant	H_2O_2[a] reduced (nmoles.min^{-1})
GSH 0.5 mM	47 ± 5.2
Ascorbate 0.5 mM	4.9 ± 1.0

[a]10 mM solution of H_2O_2 in 100mM Na_2HPO_4 pH 8.0

We have detected relatively high activities of superoxide dismutase
in heterocysts and vegetative cell preparations, 34.2 and 64.2
units. mg prot^{-1} . min^{-1}, respectively. The lower activity of
superoxide dismutase in the heterocysts may correspond with the
anticipated lower superoxide concentration in these cells where O_2
concentration is lower than in the vegetative cells. Superoxide
dismutase activity in Nostoc muscorum was inhibited by H_2O_2 and was
cyanide insensitive, suggesting the presence of the iron-containing
enzyme, similar to superoxide dismutase of Spirulina and Plectonema
(Asada et al. 1975).

Peroxide is generated from superoxide by the following reactions: (i) dismutation by superoxide dismutase; (ii) non-enzymic
reduction superoxide by ascorbate (Allen, 1975; Gallon, 1980).
Peroxide may be removed by catalase (EC 1.11.1.6) following the
reaction:

$$H_2O_2 + H_2O_2 \longrightarrow 2H_2O + O_2$$

Catalase activity was detected in few cyanobacterial preparations;
catalase activity induced by O_2 and light was demonstrated in Anabaena cylindrica (Mackey and Smith 1983) and a low catalase activity
was detected in heterocysts and vegetative cells of Anabaena cylindrica; 0.143 and 0.164 units . mg protein^{-1}, respectively (Henry
et al. 1978).

Glutathione peroxidase (EC 1.11.1.9) and ascorbate peroxidase
activities have been demonstrated in cell preparations of the unicellular N_2-fixing cyanobacterium Gloeotheca (Tozum and Gallon 1979)
suggesting that these enzymes are involved in the removal of peroxides in this organism. We were unable to observe either glutathione peroxidase or ascorbate peroxidase activities in cell free preparations of Nostoc muscorum. We have further determined the
glutathione content in filaments of Nostoc muscorum employing the
The total glutathione content (GSH + GSSG) in cell sonicates was
fluorimetric determination described by Hissin and Hilf (1976).

found to be 0.6-0.7 mM. This concentration is relatively high and
we therefore followed the nonenzymic reduction of H_2O_2 by GSH and
compared it with the nonenzymic reduction of H_2O_2 by ascorbate
(Table 1).

Glutathione was found to be a better reductant of H_2O_2 than
ascorbate, at the concentratyion of 0.5 mM which is similar to the
endogenous glutathione concentration.

II. GLUTATHIONE REDUCTASE ACTIVITY

The cells of Nosotc muscorum containing such high glutathione
concentration were expected to be provided with glutathione reductase to regenerate the reduced form of glutathione, as suggested
for Gloetheca (Tozum and Gallon 1979) and for chloroplasts (Halliwell et al. 1981). We have detected high activities of NADPH-
dependent glutathione reductase (EC 1.6.4.2) in cell free preparations of heterocysts and vegetative cells of Nostoc muscorum as
shown in Figs 1-a,b. The activities of the glutathione reductase
did vary between preparations at the range of 30-200 nmoles GSH.mg
prot^{-1}.min^{-1}. The kinetic analysis of glutathione reductase described by the double reciprocal plots exhibits lower apparent

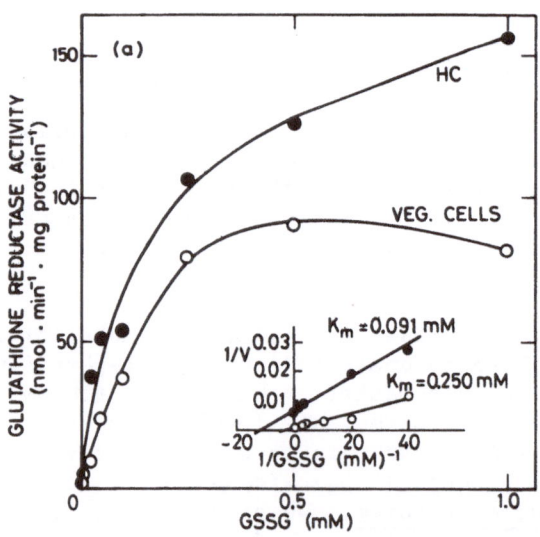

Fig. 1a. Glutathione reductase activity as a function of GSSG
concentration.

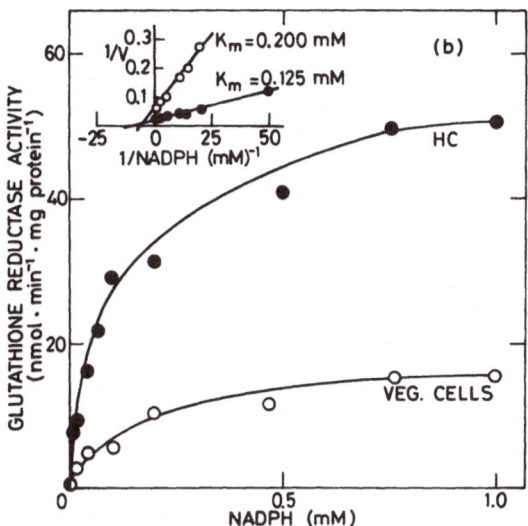

Fig. 1b.   Glutathione reductase activity as a function of NADPH
concentration.

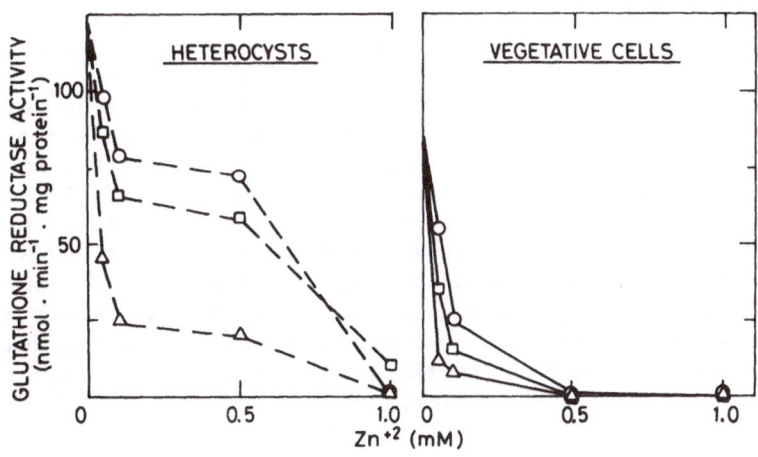

Fig. 2.   The effect of Zn^{+2} on glutathione reductase activity: o,
Zn^{+2} added without preincubation; □ , Zn^{+2} added after preincubation
with GSSG 1 mM for 5 min; Δ , preincubation with Zn^{+2} for 5 min.

Table 2. GSSG Reduction[a] in Heterocysts and Vegetative Cells

Electron donor[b]	Enzymes Involved	Heterocysts	Vegetative Cells
NADPH	Glutathione Reductase	35.6	82.0
Isocitrate + NADPH	Glutathione Reductase + Isocitrate Dehydrogenase	18.6	17.0
Glucose-6-P + NADPH	Glutathione Reductase + Glucose-6-P Dehydrogenase	3.7	0

[a] Activity is expressed in nmoles $GSH.mg\ prot^{-1}.min^{-1}$.
[b] 1mM conc. of each.

Km values for GSSG and NADPH in the heterocyst preparation, 0.091mM and 0.125 respectively, than for the vegetative cell preparations, 0.250 mM and 0.200 mM respectively. Glutathione reductase activity in Nostoc muscorum was sensitive to Zn^{+2} ions in both cell preparations as shown in Fig 2. A brief preincubation for 5 min of the crude cell preparation with 1mM GSSG protected the enzyme against the Zn^{+2} inhibition, suggesting that thiol groups are present in the active site of glutathione reductase in Nostoc muscorum which interact with Zn^{+2} ion (Halliwell and Foyer 1978). The high gluta-thione reductase activity demonstrated is most significant for the potential employment of GSH to remove H_2O_2 in Nostoc muscorum. NADPH provided by the photosynthetic apparatus of the vegetative cells, or by isocitrate dehydrogenase and glucose-6-P dehydrogenase is the heterocysts could be employed for the regeneration of the GSH pool in either of the cell types (Karni et al. 1982, Karni and Tel-Or 1983). We have therefore studied the rates of glutathione reduction in heterocysts and vegetative cells with either NADPH as an electron donor or with isocitrate/NADPH and glucose-6P/NADPH as the electron donors, as shown in Table 2. The rates of GSSG reduc-tion in both cell preparations are much higher with isocitrate as the electron donor than with the glucose-6-P, suggesting that iso-citrate dehydrogenase is potentially more active as a regenerator of NADPH than glucose-6-P dehydrogenase, as proposed by Karni and Tel-Or (1982).

CONCLUSIONS

This report indicates the potential role of glutathione in the removal of peroxide. The high content of glutathione in the filaments of <u>Nostoc muscorum</u> and the high activity of glutathione reductase in the vegetative cells and the heterocysts should provide an active and useful means to remove H_2O_2. Although glutathione peroxidase activity was not detected, it may be present in the cells, or alternatively GSH could be employed to regenerate ascorbate which will be further employed to remove H_2O_2 by ascorbate peroxidase. A comprehensive model, describing the reductive pathways

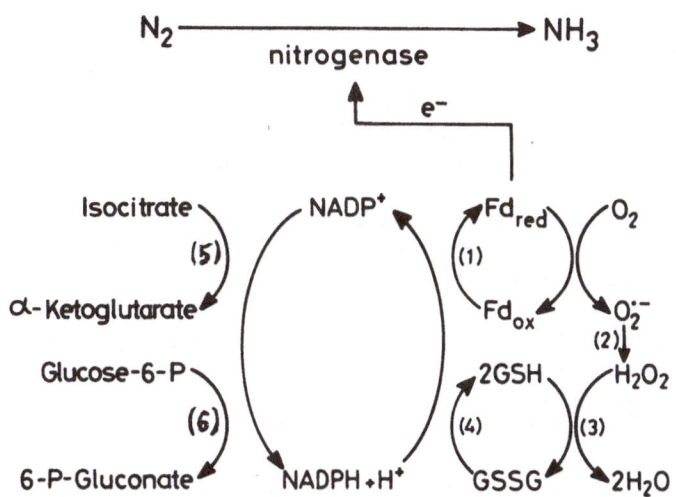

Fig 3: A suggested model for scavenging superoxide and peroxide in <u>N. muscorum</u>.

The enzyme involved: (1) Ferredoxin-NADP oxidoreductase; (2) Superoxide dismutase; (3) Non-enzymatic reaction of reduction of peroxide by reduced glutathione; (4) Glutathione reductase; (5) Isocitrate dehydrogenase; (6) Glucose-6-phosphate dehydrogenase.

leading to active N_2-fixation in Nostoc muscorum is presented in Fig 3. NADPH may be generated by isocitrate dehydrogenase and/or glucose-6-P dehydrogenase, leading to the reduction of ferredoxin which further reduces dinitrogenase reductase. The reduced ferredoxin may also produce superoxide radicals, dismutated by superoxide dismutase to peroxide. Glutathione, regenerated by glutathione reductse is involved in the removal of peroxide, providing the protection required for functional N_2-fixation in the filaments under aerobic conditions.

ACKNOWLEDGEMENT

We are grateful to Mrs. Yael Libal and Mr. Stephen J. Moss for their assistance. This research was supported by the Basic Research Fund of the Israel Academy for Sciences and Humanities.

REFERENCES

1. Allen, J.F. (1977) Superoxide and photosynthetic reduction of oxygen. In: Superoxide and Superoxide Dismutases (Michelson, A.M., McCord, J.M. and Fridovich, I., eds.), Academic Press, London, pp. 417-436.

2. Asada, K., Yoshikawa, K., Takahashi, M., Maeda, Y. and Enmanji, K. (1975) Superoxide dismutases from a blue-green alga Plectonema boryanum. J. Biol. Chem. 250:2801-2807.

3. Binder, A. (1982) Respiration and photosynthesis in energy-transducing membranes of cyanobacteria. J. Bioenergetics and Biomembranes 14:271-286.

4. Bothe, H., Neuer, G., Kalbe, I. and Eisbrenner, G. (1980) Electron donors and hydrogenase in nitrogen-fixing microorganisms. In: Nitrogen Fixation (Stewart, W.D.P. and Gallon, J.R., eds.), Academic Press, London, pp. 83-112.

5. Halliwell, B., Foyer, C.H. and Charles, S.A. (1981) The fate of hydrogen peroxide in illuminated chloroplasts. In: Fifth International Congress on Photosynthesis (Akoyunoglou, G., ed.), Balaban Int. Sci. Services, Philadelphia, pp. 279-283.

6. Haselkorn, R. (1978) Heterocysts. Ann. Rev. Plant Physiol. 29: 319-344.

7. Henry, L.E.A., Palmer, J.M. and Hall, D.O. (1978) The induction of histochemically-detectable superoxide dismutase (Cu/Zn type) band on acrylamide. FEBS Lett. 93:327-330.

8. Hissin, P.J. and Hilf, R. (1976) A fluorimetric method for determination of oxidized an reduced glutathione in tissues. Anal. Biochem. 74:214-226.

9. Karni, L. and Tel-Or, E. (1983) Isocitrate dehydrogenase as a potential electron donor to nitrogenase of Nostoc muscorum. In: Photosynthetic Prokaryotes: Cell Differentiation and Function (Papageorgiou, G.C. and Packer, L., eds.), Elsevier Scientific Publishing Co., Inc., New York, pp. 257-264.

10. Karni, L., Miller, N. and Tel-Or, E. (1982) Isocitrate dehydrog-

enase as a potential electron donor to nitrogenase of <u>Nostoc muscorum.</u> Isr. J. Botany 31: 190-198.

11. Lambien, F. and Wolk, C.P. (1973) Structural studies on the glycolipids from the envelope of the heterocyst of <u>Anabaena cylindrica.</u> Biochemistry 12:791-798.

12. Mackey, E.J. and Smith, G.D. (1983) Adaptation of the cyanobacterium <u>Anabaena cylindrica</u> to high oxygen tensions. FEBS Lett. 156:108-112.

13. Neuer, G., Papen, H. and Bothe, H. (1983) Heterocyst biochemistry and differentiation. In: <u>Photosynthetic Prokaryotes: Cell Differentiation and Function</u> (Papageorgiou, G.C. and Packer, L. eds.) Elsevier Sci. Publ. Co. Inc., New York. pp. 219-242.

14. Palmer, G., Multani, J.S., Cretney, W.C., Zumft, W.G. and Mortenson, L.E. (1972) Electron paramagnetic resonance studies on nitrogenase. Arch. Biochem. Biophys. 153:325-332.

15. Papageorgiou, G.C. and Tzani, H. (1980) The action of lysozyme on gluteraldehyde-treated filaments of the cyanobacterium <u>Phormidium luridum.</u> J. App. Biochem. 2:230-240.

16. Peterson, R.B. and Burris, R.H. (1976) Properties of heterocysts isolated with colloidal silica. Arch. Microbiol. 108:35-40.

17. Steinitz, Y. (1977) Study of photosynamic damage in cyanobacteria and mechanisms of their resistance to photooxidative death. Ph.D. thesis, The Hebrew University, Jerusalem.

18. Stewart, W.D.P. (1982) Nitrogen fixation--its current relevance and future potential. Isr. J. Botany 31:5-44.

19. Stewart, W.D.P., Hyastead, A. and Pearson, H.W. (1969) Nitrogenase activity in heterocysts of blue-green algae. Nature 224:226-228.

20. Stewart, W.D.P.., Rowell, P., Codd, G.A. and Apte, S.K. (1978) N_2fixation and photosynthesis in photosynthetic prokaryotes. In: <u>Proc. Fourth International Congress on Photosynthesis</u> (Hall, D.O., Coombs, J. and Goodwin, T.W., eds.), The Biochemical Society, London, pp. 133-146.

21. Tel-Or, E. and Stewart, W.D.P. (1977) Photosynthetic components and activities of nitrogen-fixing isolated heterocysts of <u>Anabaena cylindrica.</u> Proc. Roy. Soc. Lond. B. 198:61-86.

22. Tozum, S.R.D. and Gallon, J.R. (1979) The effects of methyl viologen on <u>Gloeocapsa</u> sp. LB 795, and their relationship to

the inhibition of actylene reduction (nitrogen fixation) by oxygen. J. Gen. Microbiol. 111:313-326.

23. Wolk, C.P. (1979) Intercellular interactions and pattern formation in filamentous cyanobacteria. In: 37th Symp. Soc. Developmental Biology (Subtelny, S. and Konigsberg, I.R., eds.), Academic Press, New York, pp. 247-266.

24. Wolk, C.P. (1982) Heterocysts. In: The Biology of Cyanobacteria (Carr, N.G. and Whitton, B.A., eds.), Blackwell Sci. Pbl., Oxford, pp. 359-386.

THE ROLE OF CATIONS IN THE

PHOTOINDUCED ELECTRON TRANSPORT OF CYANOBACTERIA

G.C. Papageorgiou, K. Kalosaka, T. Lagoyanni and
G. Sotiropoulou

Nuclear Research Center Demokritos, Department of
Biology, 153 10 Aghia Paraskevi, Athens, Greece

INTRODUCTION

The complex series of cooperative photophysical, physicochemical and biochemical processes, whose integrated outcome is oxygenic photosynthesis, takes place in the thylakoid membrane of
plants, which is electrically charged on both sides. Due to the
preponderence of ionized negative groups (protein carboxyls,
pKa = 3.8-4.8; phospholipid and sulfolipid headgroups, pKa = 2.0),
over positive groups (amino-, guanidino-, imidazolylo-; pKa = 9.2-
12.5), the net charge on the surfaces of the membrane is negative.
The overall surface charge density, taking into account both sides
of the membrane, is approx. 1-3 $\mu C/m^2$. The outer thylakoid surface
has a pKa of 4.3, and the inner of 4.1, indicating that all these
groups are ionized in the physiological pH range (reviewed in 1-4).
This surface electricity underlies an extensive body of phenomeno-
logy reported in the literature, concerning the regulatory effects
of electrolytes, and particularly of cations, on the partial processes of photosynthesis. Applying the Gouy-Chapman theory of the
diffuse double layer, Barber [1,2] made a serious attempt to syste-
matize many of the reported observations.

Barber differentiates between phenomena controlled by the

Abbreviations: Chl, chlorophyll; DBMIB, 2,5-dibromo-3-methyl-6-
isopropyl-p-benzoquinone; DCIP, 2,6-dichlorophenol indophenol;
DCMU, 3-(3',4'-dichlorophenyl)-1,1-dimethyl urea; DPC, diphenyl
carbazide; FD, ferredoxin; FeCN, $K_3[Fe(CN)_6]$; Hepes, N-2-hydro-
xyethyl piperazine-N-ethane sulfonic acid; LHC, light-harvesting
chlorophyll a/b protein; MV, methylviologen; PD, p-phenylene
diamine.

surface potential, on one hand, and those controlled by the space charge density and by hydrophobic interactions, on the other. The potential Ψ_x, defined as the electrostatic potential difference at a distance x from the membrane surface and the bulk aqueous phase, determines the concentration of ions along the normal to the membrane in accordance to Boltzmann's distribution law

$$C_x = C_b e^{-Z_i F \Psi_x / RT}$$

In this expression, C_x and C_b stand for the local and the bulk phase concentrations of the ith ionic species, while the other symbols have their usual meaning. The ions are considered to be nonbinding, i.e. they do not change the surface charge density by attaching themselves to ionized groups. In response to this distribution law, cations will accumulate near the electronegative membrane surface, while anions will tend to move away from it.

The space charge density and the intermembrane and intramembrane hydrophobic interactions control such phenomena as binding of extrinsic proteins to thylakoids [5,6], lateral movements of protein complexes [7], and thylakoid stacking [8]. The detachment of the coupling factor, for example, as a result of insufficient counterion screening of the negative surface charge of thylakoids is the cause for the long known uncoupling effect of low salt suspension media [9]. Likewise, translational (lateral and vertical) and rotational movements of protein-Chl complexes may underly such cation regulated phenomena, as excitation energy spillover from photosystem II to Photosystem I units [4,10,11], ΔpH-induced quenching of Chl a fluorescence [12-14], partition of hydrophilic and hydrophobic excitation quenchers to the Chl a domains of the membrane [15-17], and physiological adaptation to the light conditions of the ambience [State 1 - State 2 transitions; 18-20].

Nearly all the information we have on the cation regulation of the partial processes of photosynthesis derives from experiments with fragmented higher plant chloroplasts. In contrast, our knowledge of these phenomena in the cyanobacteria, another major class of oxygenic photosynthesizers, is quite meager. The main reasons for this information paucity are: (i) the impermeability of the thylakoid cell envelope to electrolytes; (ii) the impossibility to isolate cyanobacterial thylakoids of a degree of intactness, comparable to that of higher plant thylakoids; and (iii) the rapid structural and functional deterioration of the cyanobacterial thylakoid membrane, once it is extracted from the cytoplasm.

The cyanobacteria, evolutionary predecessors of algae and higher plants, share with them common mechanisms of electron and proton transport, oxygen evolution, and photophosphorylation, yet they differ from them in several key aspects. The cyanobacteria are prokaryotes. Therefore, only one permeability barrier, the cell envelope (cell wall plus cell membrane) intervenes between the suspension medium and the surface of thylakoid vesicles. In contrast, two such barriers, the cell envelope and the chloroplast envelope are present in eukaryotic photosynthetic cells.

The cyanobacteria are devoid of LHC, the light harvesting pigment-protein complex of higher plants, which supplies excitation to the reaction centers of photosystem II. Instead, they employ the phycobilisome, an organelle external to the membrane (recently reviewed in 21-23). The phycobilisome is a spheroidal or discoidal particle, 25-60 nm in diameter, depending on the cyanobacterium species, consisting of hexameric phycobiliprotein subunits, arranged in a layered structure. The phycobilisome core consists of allophycocyanin, in contact with a layer of phycocyanin subunits, which are covered with a layer of phycoerythrin subunits. Some species, (e.g. _Anacystis nidulans_) have only allophycocyanin and phycocyanin. The phycobiliprotein subunits are held together in the phycobilisome non-covalently, primarily by forces originating from hydrophobic interactions. Noncovalent is, also, the attachment of phycobilisomes to the outer surface of thylakoid vesicles, where they form rows spaced at 40-80 nm, depending on the species, with an intra-

TABLE 1. Ionizable and Non-ionizable Constituents of Thylakoid Membranes from Spinach[1] and _Anacystis nidulans_[2]

Constituent	Percent by weight	
	Spinach	Anacystis
Ionizable constituents	(75.9)	(76.3)
Protein	68.1	70.0
Diacyl sulfoquinovosyl glycerol	3.1	2.1
" glycerophospho glycerol	3.5	4.2
" glycerophospho choline	1.2	–
Non-ionizable constituents	(24.1)	(23.7)
Diacyl monogalactosyl glycerol	9.3	10.5
" digalactosyl	7.1	2.2
Chlorophyll plus carotenoids	7.7	11.0

Compiled from:

[1]Murphy, D.J. and Woodrow, I.E. (1983) Biochim. Biophys. Acta 725, 104-112.
[2]Omata, T. and Murata, N. (1983) Plant and Cell Physiol. 24, 1101-1112

row, center-to-center periodicity of approx. 10 nm.

Each phycobilisome exists in strict spatial association with a single membrane-embedded photosystem II complex (reaction center plus an antenna of approx. 20 Chl a molecules), to which it donates electronic excitation [24,25]. The regularity of phycobilisome arrangement on the thylakoid surface should impose a similar regularity on the arrangement of photosystem II complexes within the membrane phase, and most likely on photosystem I complexes. Extensive lateral displacements of membrane proteins should not be allowed in the cyanobacterial thylakoid, if efficient excitation transfer from phycobilisome to photosystem II complex is to be preserved. This stands in sharp contrast with the enormous lateral displacements of membrane proteins in higher plant thylakoids that are associated with such cation controlled phenomena as excitation spillover, thylakoid stacking, and light adaptive State 1 - State 2 transitions [2,8]. Our conclusion is further supported by the lower fluidity of the cyanobacterial thylakoid membrane, relative to the fluidity of higher plant thylakoids [26], which is attributed to its higher proportion of saturated fatty acids [27,28].

The mass composition of cyanobacterial [27] and of higher plant thylakoids [28] in ionizable and non-ionizable constituents is roughly the same (Table 1). The former, however, carry as much as 3-10 times more negative electricity per unit area, than the latter [29]. In this respect, the thylakoids of cyanobacteria resemble the chromatophores of the nonoxygenic photosynthetic bacteria [3].

Recent studies implicate Ca^{2+} as an indispensible factor for the normal function of pre-photosystem II electron transport, both in higher plants and in cyanobacteria. Substituted phenothiazines, which act as antagonists of calmodulin, the Ca^{2+}-binding protein of animal cells, inhibit electron transport at a site located between the DPC and DCIP coupling sites [30-32; see diagram below].

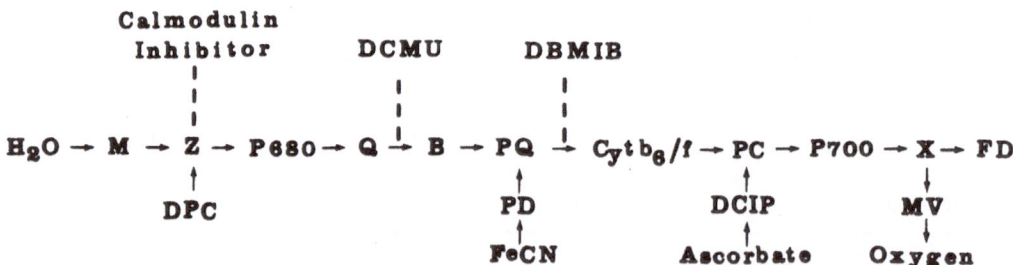

Although calmodulin itself has not been detected in plants, these results implicate a Ca^{2+}-binding protein as part of the thylakoid membrane.

Several lines of evidence are in agreement with that. A transient treatment of isolated spinach chloroplasts with glacial acetic acid extracts several polypeptides from the membrane, one of which (16 kDa) has been proposed to be the Ca^{2+}-binding protein, acting near the photosystem II reaction center [33]. Washing O_2-evolving photosystem II particles, isolated from spinach, with concentrated NaCl solutions (1-2 M) causes the release of two polypeptides, 17 and 23 kDa [34-36], inhibition of O_2 evolution [35,36], and retardation of electron donation to Z^+, the unidentified electron donor on the oxidizing side of the photosystem II reaction center [34]. O_2 evolution and electron donation to Z^+ is restored by the addition of Ca^{2+} ions, at a relatively high concentration (approx. 10 mM), or by reconstituting the particles with the 17 and 23 kDa polypeptides, and adding Ca^{2+} at a lower concentration. This suggests that Ca^{2+} binds to the membrane with high affinity, only when these polypeptides are in position.

Earlier, Piccioni and Mauzerall [37] and Brand [38] had demonstrated that cyanobacterial thylakoid fragments require approx. 5-20 5-20 equivalents of Ca^{2+} for efficient H_2O FeCN electron transport. Higher plant chloroplasts have no such requirement. Although no attempt was made to differentiate between the nonspecific charge screening effect of Ca^{2+}, which would bring more of the FeCN anion to the membrane surface, and a true Ca^{2+} catalytic effect, a specific Ca^{2+} requirement for the photosynthetic electron transport of cyanobacteria was claimed. In view, however, of the more recent experimental observations described above, it appears quite likely that upon mechanical cell disruption the cyanobacterial thylakoids release the high affinity Ca^{2+} binding polypeptides, setting thus the conditions for the observed Ca^{2+} cation requirement. Higher plant thylakoids, on the other hand, are much stabler in vitro, as testified so many other observations, and they release the Ca^{2+} binding polypeptides only upon concentrated saline treatment. Hence, they exhibit no specific requirement for Ca^{2+}.

Employing the unicellular cyanobacterium Anacystis nidulans, we have initiated in our laboratory a series of experiments with the following objectives: (i) To prepare ion-permeable cells (permeaplasts) with minimally damaged thylakoids, as assessed in terms of several structural and functional indices. (ii) To measure the surface charge density and the surface potential both at the outer and at the inner thylakoid membrane-H_2O interfaces. (iii) To differentiate between nonspecific, charge-screening cation effects, and catalytic effects due to specific binding of cations on negative sites of the membrane. Here, we shall outline some of our experimental observations.

MATERIALS AND METHODS

Anacystis nidulans was cultured photoautotrophically as in [39]. Late logarithmic phase cells were resuspended in 0.05 M Hepes NaOH, pH 7.5, at a cell density corresponding to 0.15 mg Chl $\underline{a}$/ml, and 10 mg/ml egg white lysozyme (3×recrystallized; Sigma grade I) and 1 mM EDTA were added. The mixture was incubated at 37°C. Permeabilization of the cell envelope to ions was assessed in terms of the Hill reaction rates, with ionic oxidants (FeCN, MV) serving as electron acceptors.

Photosynthetic O_2 evolution was measured with a Clark type O_2 electrode as in [39]. The reaction mixture, made in 0.25 M sorbitol, 0.25 M KCl, and 0.05 Hepes·NaOH, pH 7.5, contained cells equivalent to 20 μg Chl $\underline{a}$/ml, and depending on the particular assay the following additions: FeCN, 1 mM; DCIP, 34 μM; Na-ascorbate, 2 mM; MV, 0.1 mM; DPC, 0.5 mM; ADP, 2 mM; KH_2PO_4, 5 mM; and DCMU, 10 μM.

FeCN photoreduction was assayed spectrophotometrically, in terms of the absorbance decrease at 420 nm, using a millimolar coefficient of $\varepsilon=1.0$ $mM^{-1} \cdot cm^{-1}$. The reaction mixture (6 ml) was made in a medium consisting of 0.5 M sorbitol, 0.005 M Hepes·NaOH, pH 7.5, and contained 1.25 mM FeCN and 25 μg Chl $\underline{a}$/ml. The mixture was illuminated for 4 min with saturating white light, filtered through a 5 cm layer of 1% w/v $CuSO_4$, which absorbed most of the infra-red radiation. The illumination time was within the linear range of A420 bleaching. Subsequently, 0.8 ml of the illuminated reaction mixture was mixed with 5% w/v trichloroacetic acid, and the mixture was centrifuged in a clinical centrifuge. Absorption measurements were performed in the clear supernatant. Independent tests had shown that there is no decomposition of FeCN when illuminated samples were mixed with trichloroacetic acid.

Absorption and difference absorption spectra were measured with a microprocessor-controlled spectrophotometer (Hitachi Model 557). Dark reduction of P700, following light-induced oxidation, was measured in the dual wavelength mode using 700 nm and 720 nm ($\Delta\lambda = 2$ nm) as the measuring and the reference wavelengths. Actinic light, obtained from an incandescent source, passed in sequence through a heat reflecting mirror (Oriel No 5740), a heat absorbing filter (Ealing E26-3681), a color glass filter (Corning C.S. 4-72) and 60 cm of a fiber optics guide. It emerged as a blue beam of 93.5 $W \cdot m^{-2}$, which was directed to the sample cuvette (diffusing side) at right angles to the measuring and the reference beams. Corning filters C.S. 2-64 and C.S. 7-54 guarded the photomultiplier entrance against actinic light leaks. The low Chl $\underline{a}$ content of the samples (7.9 μg/ml), the blue actinic light, and the distal positioning of the sample rendered the Chl $\underline{a}$ fluorescence artifacts insignificant.

Fluorescence was measured with a Perkin-Elmer Model MPF-3L spectrofluorometer. Chl a was determined in methanolic extracts according to McKinney [40], and phycobiliproteins in aqueous extracts according to Kursar and Alberte [41]. Except for the lysozyme permeabilization reaction, all preparations and assays were performed at room temperature.

RESULTS

(a) <u>Preparation and characterization of Anacystis permeaplasts.</u>
Various techniques have been described, which succeed in permeabilizing the cyanobacteria cell envelope to electrolytes. They include lyophilization [42], treatment with aldehydes [43], digestion of the cell wall peptidoglycan with lysozyme [44-47], and transient exposure to EDTA and concentrated KCl [48]. The enzymatic procedure is the mildest, but has two complications: first,

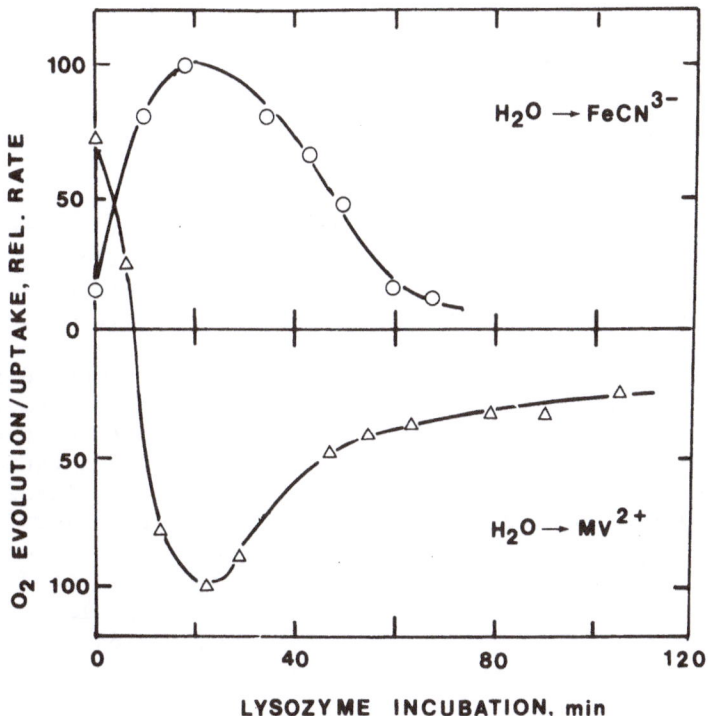

Fig. 1 The rate of light-induced O_2 evolution in the presence of ferricyanide (FeCN) and of methylviologen (MV)-mediated O_2 uptake by Anacystis nidulans cells, as a function of the duration of lysozyme treatment of the cyanobacterium.

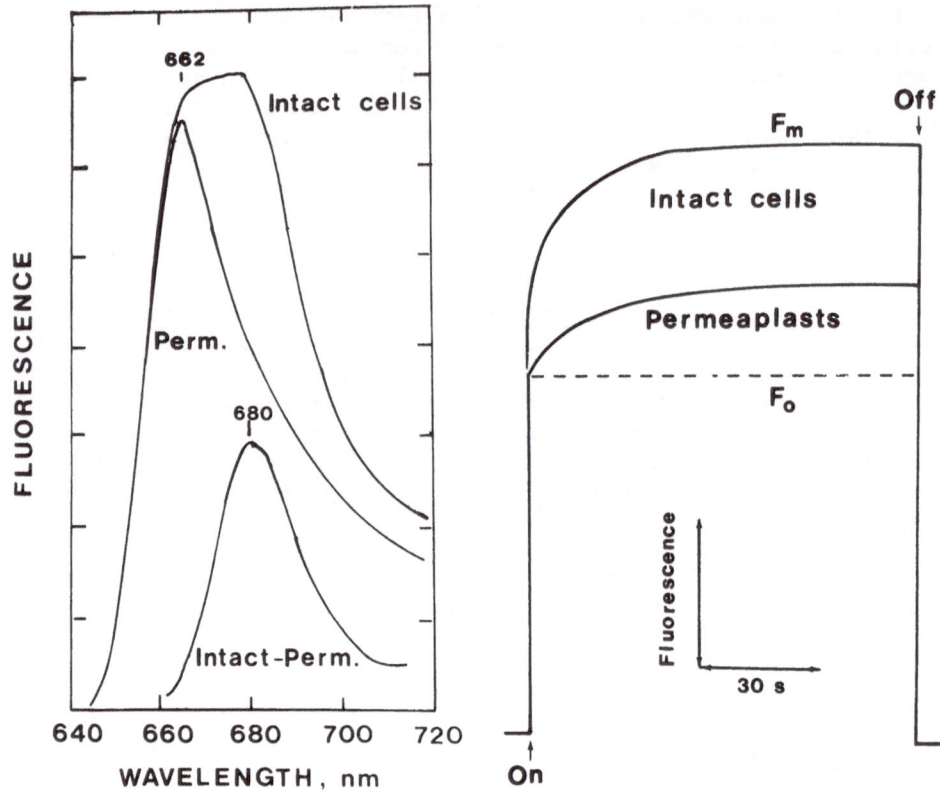

Fig. 2   Fluorescence emission
          spectra of intact Anacy-
          stis cells, and of permea-
          plasts obtained after 20
          min treatment with lyso-
          zyme. Excitation: λ=580
          nm; $\Delta\lambda$=2 nm; detection,
          $\Delta\lambda$=2 nm. The difference
          fluorescence spectrum
          (intact cells-permeaplasts)
          is also displayed.

Fig. 3   The slow induction of
          Chl a fluorescence of
          DCMU-poisoned intact
          Anacystis cells and
          Anacystis permeaplasts.
          Samples were dark adapted
          for 4 min, prior to the
          recording of the kine-
          tics. Excitation:
          λ=436 nm, $\Delta\lambda$=32 nm, and
          Corning glass filter
          C.S. 5-60; detection:
          λ=685 nm, $\Delta\lambda$=2 nm, and
          Corning glass filter
          C.S. 2.64.

some cyanobacteria, including Anacystis, are very resistant to lysozyme, requiring extraordinarily long incubations; second, such
long incubations often lead to the inactivation of the photosynthetic O_2 evolution [39,44,46], and to the appearance of a DCMU- and
cyanide-insensitive photoinduced O_2 uptake [49].

To prepare, therefore, ion-permeable and optimally active Anacystis cells, the lysozyme treatment should be made as short as
possible. To achieve this objective, we screened several incubation
media (unpublished experiments). Best results were obtained with
the hypoosmotic medium described in Materials and Methods. With
this medium, we achieved the fastest permeabilization ever reported
for Anacystis. The permeabilization kinetics with respect to an
anionic ($FeCN^{3-}$) and a cationic (MV^{2+}) are illustrated in Fig. 1.
Electron transport rates rise at first, indicating penetration of
the ionic acceptors to the cytoplasm, but then they fall off as a
result of the exportation of cytoplasmic solutes. Following the
terminology of Ward and Myers [44], we shall designate the ascending and the descending branches of such typical permeabilization
kinetics as phase 1 and phase 2, respectively. By examining the
effect of permeabilization on several partial electron transport
processes, we established that the site of phase 2 inhibition precedes the site of electron donation by DPC (see electron transport
diagram in Introduction). Since lack of Ca^{2+} ions blocks electron
transport between DPC and the photosystem II reaction center, phase
2 inhibition is unrelated to the loss of this cation, and indeed
addition $CaCl_2$ failed to reverse it.

Late phase 1, or early phase 2 permeaplasts were tested for
stability and integrity of the photosynthetic apparatus with respect
to several indices. Resuspended in 0.5 M sorbitol, 0.05 M Hepes.
NaOH, pH 7.5, at room temperature, they maintained steady photosystem II plus photosystem I electron transport activity for more
than 4 h. Electron transport is accelerated in the presence of
phosphorylation cofactors and uncouplers, indicating capacity for
photosynthetic control, and hence capacity to build up a pH difference (acid inside) across the permeaplast thylakoid membrane. On
the other hand, however, after cell permeabilization, phycobilins
can no longer sensitize the fluorescence of Chl $\underline{a}$, indicating detachment of phycobilisomes from the thylakoid surface, as a result
of the altered ionic conditions in the cytoplasm. (Fig. 2).

A sensitive index of membrane integrity is the ability to
undergo the so-called state 1-state 2 transition. The term denotes
a light adaptive process, which enables oxygenic photosynthetic
plants to distribute the electronic excitation between photosystem
I and II evenly. In the cyanobacteria, the transition from the
dark-adapted state (state 2) to the light-adapted state (state 1)
is reported by a rise in the Chl $\underline{a}$ fluorescence yield, as shown in
Fig. 3. In this experiment, we blocked electron transport with
DCMU in order to avoid interference by the Q-controlled fast transients of Chl $\underline{a}$ fluorescence [13]. Furthermore, in order to avoid
a strong background of inert fluorescence emitted by the detached
phycobilins, excitation was given at 436 nm, where only Chl $\underline{a}$
absorbs. Fig. 3 shows that after permeabilization, Anacystis loses
a substantial part of the variable fluorescence amplitude, signifying

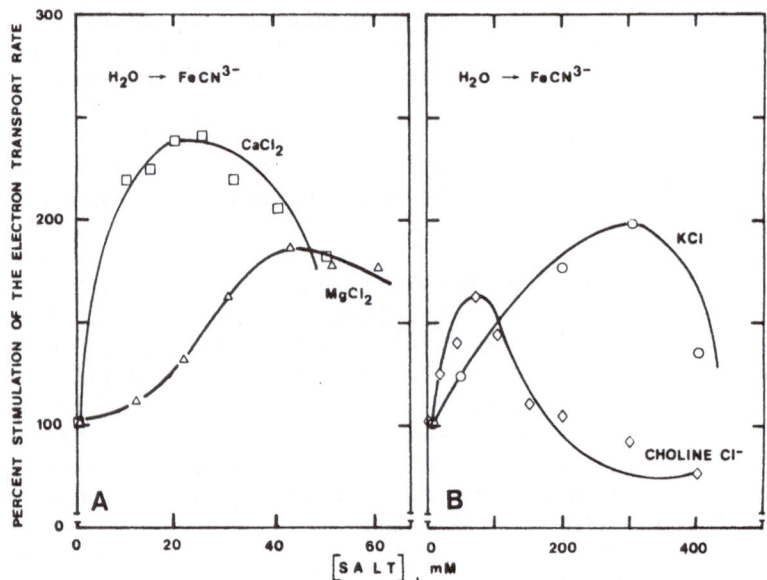

Fig. 4 Percent stimulation of the rate of ferricyanide (FeCN)-
photoreduction, measured as absorbance decrease at 420 nm,
as a function of the monovalent (A) and divalent (B) cation
chlorides in Anacystis permeaplast suspensions. Rates in
μequiv.mg Chl$\cdot$h^{-1}, at zero concentrations of KCl, choline
chloride, $CaCl_2$, and $MgCl_2$ are: 67, 58, 71 and 61.

an impairment of the ability to undergo state 1-state 2 transitions.
This experiment indicates that the change in the ionic conditions
in the cytoplasm, influences phenomena which occur within the membrane phase of thylakoids.

(b)   Interactions of cations with the outer thylakoid surface.
Cations influence the physicochemical properties of thylakoids
basically in two ways: (i) nonspecifically, by accumulating as
counterions in the vicinity of the H_2O-membrane interfaces; and
(ii) specifically, by binding to negative membrane sites. The
differentiation between nonspecific and specific cation effects is
often difficult. Furthermore, in the case of fragmented thylakoid
vesicles, the ambiguities with regard to the proper orientation of
the membrane, and with regard to the discontinuity of the outer
and the inner aqueous phases are always present. For studies of
ion interactions with thylakoids, therefore, cyanobacteria permeaplasts, with sealed and properly oriented thylakoid membranes,
offer definite advantages to the experimenter.

Figure 4 displays the effects of monovalents (A) and divalent

(B) cations on the rate of FeCN photoreduction. The bell-shaped
concentration curves indicate rate stimulation at low cation concentrations and inhibition at high cation concentrations. Similar
effects have been reported for isolated thylakoid fragments from
cyanobacteria [37,38,45]. A noteworthy feature of these curves is
the location of the maximum. In the case of the nonbinding cholinate cation, maximal rate stimulation is achieved at approx. 75 mM,
whereas 250-300 mM of the binding cation K^+ is required for the
same effect. On the other hand, maximal stimulation by divalent
cations occurs at lower concentrations, 20-25 mM for $CaCl_2$ and
40-50 mM for $MgCl_2$.

The electrostatic theory [1,2] predicts (i) increased FeCN
anion concentration at the membrane surface, when the surface potential shifts to less negative values upon increasing the electrolyte
concentration of the bulk aqueous phase, and an attendant stimulation
of the Hill reaction rate; and (ii) a more drastic rate stimulation

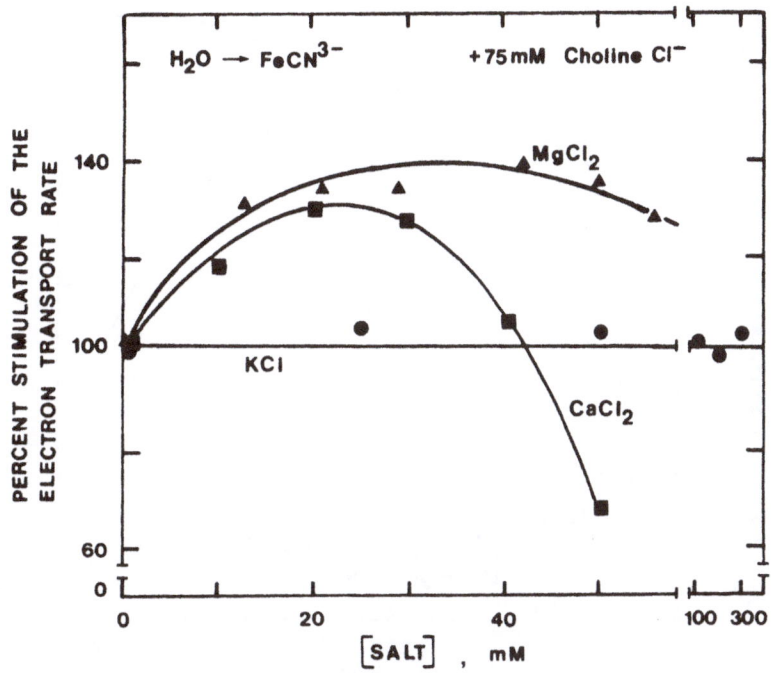

Fig. 5   Percent stimulation of the ferricyanide (FeCN)-photoreduction
         rate by Anacystis permeaplasts, as a function of the concen-
         tration of metal chlorides, added on top of a background
         concentration of 75 mM choline chloride. Actual rates of
         zero concentrations of KCl, $CaCl_2$ and $MgCl_2$, in μequiv.mg
         $Chl^{-1} \cdot h^{-1}$ are: 117, 131 and 98.

379

by divalent cations. Electrostatic interactions alone, however, do
not account for the concentration curves of Fig. 4. Were, for
example, the membrane surface potential the only determinant of the
FeCN photoreduction rate, then we should have not observed differences in the ascending portions of the concentration curves for cations
of the same valence. Electrostatic interactions cannot account,
also, for the Hill reaction inhibition at higher cation concentrations. It seems plausible, therefore, to postulate both cation-
specific and valence-specific effects as determinants of the FeCN
reduction rate.

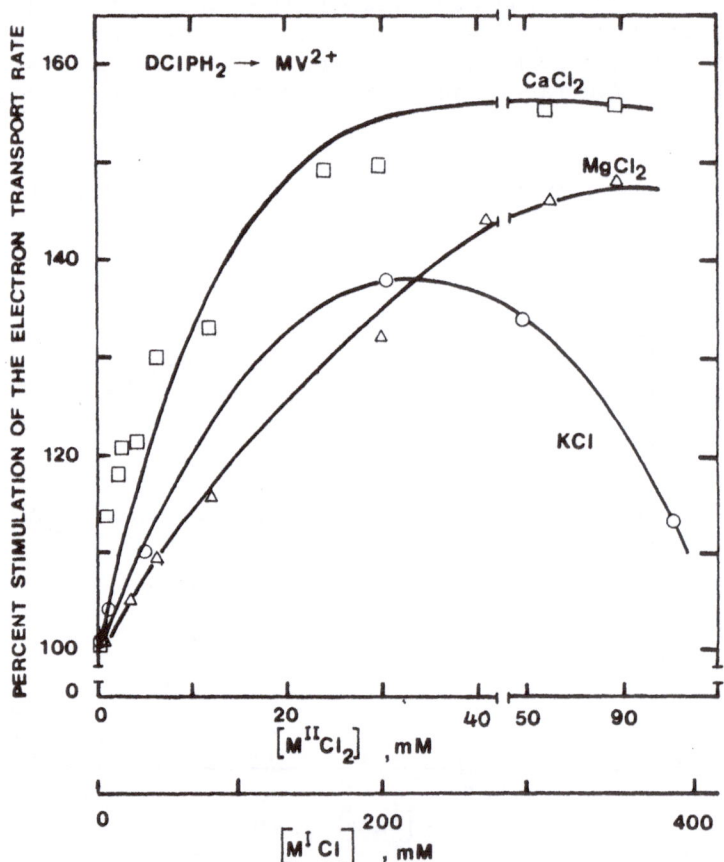

Fig. 6 Percent stimulation of methylviologen – (MV) – mediated O_2
uptake by Anacystis permeaplasts, as a function of the metal
chlorides concentration in the assay medium. Assays were
performed in the presence of 10 µM DCMU, with ascorbate-
reduced DCIP as the electron donor. Actual rates at zero
concentrations of KCl, $MgCl_2$ and CaCl are: 534, 482 and 87,
µequiv.mg $Chl^{-1} \cdot h^{-1}$.

What could then be the electrolyte-dependent factors, which in addition to the surface potential effects, govern the rate of photo-induced electron transfer from a nonmembrane donor, namely H_2O, to a nonmembrane acceptor, FeCN ? It appears that higher order effects should be taken into account, such as for example structural changes occurring upon charge neutralization, and chaotropic effects exerted by the added anions (notably Cl^-).

In an attempt to differentiate between specific and nonspecific cation phenomena, we investigated the effects of added metal chlorides on the FeCN Hill reaction of permeaplasts, in the presence of a constant background of 75 mM choline chloride. At this background concentration of the cholinate cation, it can be calculated that the negative electrostatic potential of the membrane surface is reduced by approx. 55%, allowing thus an 8fold rise in the local concentration of the FeCN anion. Under this charge screening by the nonbinding cation, we expect a clearer manifestation of the specific effects of the added metal cations.

In the presence of background cholinate, KCl no longer stimulates FeCN reduction, but Ca^{2+} and Mg^{2+} are still active (Fig. 5)

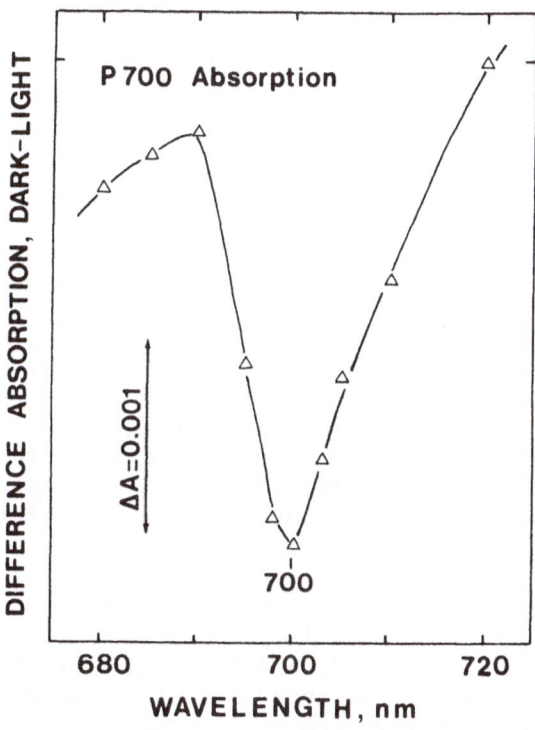

Fig. 7 The dark-light difference absorption spectrum of <u>Anacystis</u> nidulans P700, measured in a suspension containing 7.9 µg Chl <u>a</u>/ml, 10 µM DCMU, 50 µM MV, 0.4 M sorbitol, and 1 mM Hepes.NaOH, pH 7.5 .

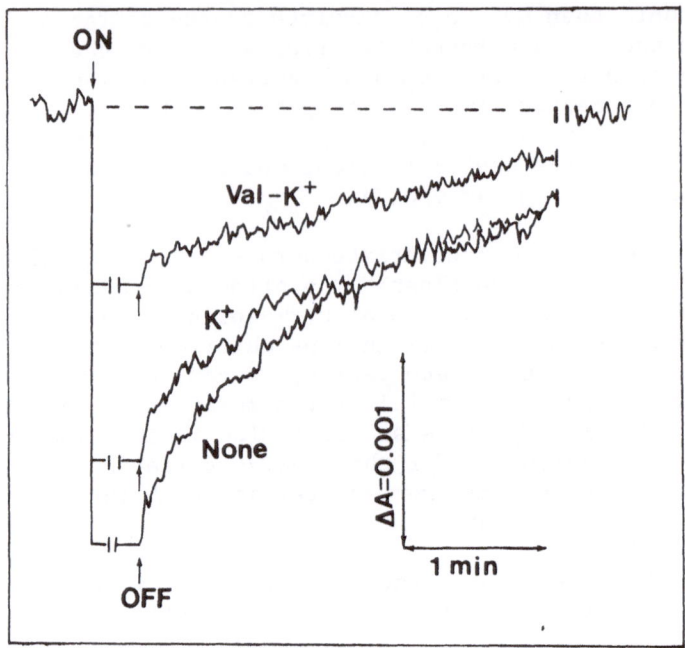

Fig. 8 Dark recovery kinetics of photooxidized P700 in 6-day old
        Anacystis permeaplasts. Control (none); plus 50 mM KCl
        (K^+); plus 50 mM KCl and 3 μM valinomycin (Val-K^+).

The ineffectiveness of K^+, in this case, may be explained as
follows: according to Fig. 4A, cholinate and K^+ do not differ
significantly in their rate stimulation effect up to a concentration of approx. 100 mM. Beyond 10 mM, however, they exert mutually
compensation stimulatory (K^+) and inhibitory (cholinate) effects,
whose sum total is the KCl curve shown in Fig. 5. Divalent cations,
on the other hand, are preferentially attracted to the membrane
surface, in accordance to Boltzmann's distribution law (see Introduction) and succeed in stimulating the reduction of FeCN, even in
the presence of the background cholinate.

Figure 6 displays cation concentration curves of photoinduced
MV-mediated O_2 uptake by Anacystis permeaplasts. Ascorbate-reduced
DCIP was the source of electrons. Rate stimulation both by monovalent and divalent cations is observed, although the acceptor is
a cation, and a more positive membrane potential would repel rather
than attract it. The experiment illustrates that direct interactions between the electrically charged membrane and diffusible ions
are not the only determinants of the rate of photosynthetic electron transport.

382

(c) <u>Interactions of cations with the inner thylakoid surface.</u>
The photooxidation of P700, the reaction center chromophore of
photosystem I, is reported by an absorption loss, whose dark-minus-
light difference spectrum is centered at 698-702 nm (Fig. 7).
After turning the actinic light off, the absorption loss is reversed,
reflecting the reduction of P700 by endogenous electron donors. In
cyanobacteria [50] and in higher plant chloroplasts [51-53], the
dark reduction of P700 occurs at rates which exceed the time resolution capability of double beam spectrophotometers. These instruments are slow because of the time delay the sharing of the photomultiplier by the two beams imposes.

Upon aging, however, of Anacystis the dark reduction of P700
becomes much slower (K. Kalosaka and G.C. Papageorgiou, unpublished
experiments), allowing detection with a double beam spectrophotometer. Such a recording is illustrated by the lowest trace in
Fig. 9. Using a difference absorption coefficient of e (697-720) =
58 mM^{-1}cm^{-1} [54], we may calculate from the amplitude of the dark
reversal a ratio of 180-220 Chl <u>a</u>/P700. Accordingly, the kinetic
trace accounts for all P700 present in the sample.

Kinetic evidence strongly suggests that plastocyanin is the
primary endogenous electron donor to P700 [51-53], although under
certain circumstances alternate electron donation routes may operate [55-57], and an intermediate may intervene between plastocyanin and P700 [58]. Plastocyanin is a Cu-containing protein,
having an isoelectric point at pH 4.0, which easily dissociates
from the lamella [59,60]. As a result of its negative electric
charge at the physiological pH range, charge compensation is required for the interaction of plastocyanin with the thylakoid surface. The possibility, therefore, exists that the slower dark
reduction of P700 in aged Anacystis reflects an insufficient compensation of negative charge of the membrane.

We tested this hypothesis by measuring the dark reversal kinetics of the P700 photobleaching in Anacystis permeaplasts in the
presence of 50 mM KCl, and in the presence of 50 mM KCl plus 3 μM
valinomycin. Fig. 8 shows that only in the presence of the cationophore we could observe a significant acceleration of the dark
reduction of P700. When intact cells were used, instead of permeaplasts, no acceleration was obtained, either with or without the
cationophore.

These results indicate the following: (i) Plastocyanin functions
in a space which is inaccessible to medium electrolytes, except
when the cell envelope is permeabilized by lysozyme and ionophores
are present to transport ions across the thylakoid membrane. It
should be noted here, that an ongoing controversy exists with
regard to the accessibility of plastocyanin to ionic solutes of
the medium (see Ho and Krogmann,ref. 57). (ii) Permeaplast

thylakoids are impermeable to K^+, indicating intactness of the vesicles. Valinomycin enables K^+ to cross into the inner space and neutralize the negative potential of the inner thylakoid surface, facilitating the interaction of plastocyanin with the membrane. Finally, (iii) as in other photosynthetic bacteria [61], the system valinomycin-K^+ cannot overcome the permeability barrier imposed by the Gram negative envelope of the intact cell.

DISCUSSION

As Gram negative microorganisms, the cyanobacteria have their cytoplasm enclosed within a complex envelope, consisting of the cell wall (outer membrane, periplasmic zone, and peptidoglycan layer) and the cell membrane. The envelope is impermeable to electrolytes, but becomes permeabilized after hydrolysis of the polysaccharide backbone of peptidoglycan with lysozyme, although the enzyme does not act on the cell membrane itself. The peptidoglycan layer serves as an exoskeleton to the cell, exerting morphological control over its cytoplasmic elements. Cells which have the peptidoglycan layer completely removed, but retain the outer membrane are called spheroplasts. Very often, however, the term is used incorrectly to identify protoplasts, namely cells deprived both of the peptidoglycan layer and the outer membrane, the latter as a result of EDTA treatment. Protoplasts and spheroplasts are osmotically responsive cells, assuming spherical shape in suspension. A partial enzymatic disintegration of peptidoglycan, on the other hand, results in osmoresistant, ion-permeable cells known as permeaplasts [44].

Enzymatically permeabilized cyanobacteria offer the experimenter two advantages, that are unattainable with isolated thylakoid fragments: virtually intact ion-accesible thylakoids and superior structural and functional stability. Some cyanobacteria, nevertheless, exhibit unusualy resistance to lysozyme. This may be attributed to excessive molecular sieving by the outer membrane, and to the steric hindrance the enzyme molecules encounter, once they reach the peptidoglycan network [62]. One unusually resistant cyanobacterium is <u>Anacystis nidulans</u>, which requires pretreatment with EDTA, excessive lysozyme, and long incubations in order to be permeabilized [39,44,49].

We have shown, however, in this work, that in hypotonic medium, consisting of 50 mM Hepes.NaOH, pH 7.5 and 1 mM EDTA, lysozyme converts Anacystis to highly active permeaplasts in relatively short time. These permeaplasts possess a fully functional electron transport chain (H_2O MV), and are capable of photosynthetic control, which indicate presence of intact, ion-impermeable thylakoids. Thus, unless transported by an ionophore, an ion added to the suspension medium interacts with the outer thylakoid surface only.

384

The unimpeded passage of ions through the cell envelope succeeds in communicating the low ionic strength conditions of the medium to the cytoplasm, with severe consequences on phycobilisomes, the extra-thylakoid light harvesting organelles of cyanobacteria. In the permeabilized cell, phycobilins cannot sensitize the fluorescence of Chl $\underline{a}$. This implies, at least, detachment of phycobilisomes from the thylakoid surface, and possibly further dissociation of the organelle into phycobilin subunits.

The light adaptive State 1 - State 2 transitions, which characterize the oxygenic two-photosystem photosynthetic organisms, involve displacements of chlorophyll-protein subunits within the membrane phase. Accordingly, they may serve as an ultrasensitive index of membrane integrity. In higher plant thylakoids, a balanced delivery of excitation to the subsets of photosystem I and photosystem II centers is believed to be achieved either by controlling the excitation spillover rate [19], or by controlling the absorption cross section of each photosystem [20]. The latter mechanism presupposes, however, extensive lateral movements of light-harvesting chlorophyll-protein subunits, which, as argued in the Introduction, are unlikely in the cyanobacteria. For these organisms, a model of noncommunicating "puddles" is more appropriate in describing the system antenna pigments-reaction center [63].

It is quite probable, therefore, that the light adaptive transitions in cyanobacteria involve slight displacements and/or orientation changes of membrane phase pigment proteins only. These changes influence critically the probability of resonance excitation transfer from photosystem II to photosystem I chromophores. The observed suppression of the slow induction of Chl $\underline{a}$ fluorescence (which reports on State 2 - State 1 transition; ref. 15), upon permeabilization of the Anacystis cell envelope to ions, witnesses therefore an impairment within the membrane phase. This impairment, however, does not affect the fundamental energy transducing properties of thylakoids, which manifest as photoinduced electron and proton transport.

Due to the impermeability of permeaplast thylakoids to ions, electrolytes present in the suspension medium interact with the outer thylakoid surface only. We have examined such interactions in terms of photoinduced electron transfer to ionic Hill oxidants ($FeCN^{3+}$; MV^{2+}).

Cations, which as counterions are attracted to the negatively charged thylakoid surface influence its physicochemical properties either by means of charge screening, or by charge neutralization [64]. In the first instance, the chemical nature of the counterion is immaterial, as it never comes into contact with the membrane, but resides at a distance from it, in the diffuse double layer. The most important factor is the valence of the counterion, and to a lesser degree its size, as determined by the hydration sphere, or other attached ligands. In the second istance, the chemical affinity of the counterion to specific negatively charged groups of the membrane is critically important, since the equilibrium binding constant determines the extent of charge neutralization.

The data displayed in Fig. 4 are in agreement with the results of Mohanty and coworkers [65-67] and the point out to an interplay of both charge screening and charge neutralization. Evidence for the former is the greater effectiveness of divalents in stimulating electron transport, relative to the monovalent cations. Evidence for the latter are the striking differences for cations within the same valence group, which we interpret as reflecting differences in the equilibrium binding constants.

The divalent/monovalent hierarchy in affecting the rate of FeCN photoreduction is expressed not only with regard to the stimulation of the rate at low cation concentrations, but also with regard to its inhibition at higher concentrations. Macromolecular polycations, such as polylysine, have been reported to inhibit photosynthetic electron transport near the plastocyanin-P700 couple [68]. This inhibition bears, however, no resemblance to the inhibition of FeCN reduction by cations, as observed here, since the latter inhibition occurs also in the presence of DBMIB, which blocks electron transport ahead of plastocyanin (Sotiropoulou and Papageorgiou, unpublished experiments; also Wavare and Mohanty, ref. 67).

We consider two expressions for the combined effect of cation screening and neutralization of membrane negative charges, with regard to the Hill reaction rates: an influence on electron exchanges between membrane phase intermediates, and aqueous phase Hill oxidants. Since the determinant factor in the latter case, according to the Gouy-Chapman theory [1,2], is the membrane potential, cations should stimulate the photoreduction of anionic acceptors (e.g. $FeCN^{3-}$) and should suppress the reduction of cationic acceptors (e.g. MV^{2+}). Contrary to expectation, however, cations stimulate the rate of photoinduced O_2 uptake in the presence of MV^{2+} (Fig. 7). We believe that this constitutes a clear indication of cations influencing directly electron exchanges between membrane phase intermediates.

The possibility of examining the functional consequences of cation interactions with the inner surface of permeaplast thylakoids is limited by the availability of specific ionophores for transporting the cations across the thylakoid membrane. We have demonstrated that K^+ accelerates the reduction of P700 by an endogenous donor (presumably plastocyanin) only in the presence of the K^+-specific cationophore valinomycin. This suggests that the interaction of the negatively charged donor with the inner thylakoid surface is electrostatically controlled, and is accelerated upon collapsing the inner surface potential upon importation of K^+. This experiment, performed with virtually intact permeaplast thylakoids, clearly

proves that plastocyanin operates on the inner portion of the thylakoid membrane.

In the present work, we examined the effects of metal and non-
metal cations on the photosynthetic electron transport of enzymatically permeabilized Anacystis cells. Our results demonstrate an
interplay both of electrostatic screening and neutralization of
membrane charges, as a source of the observed phenomena. On the
other hand, the relatively high (millimolar) cation concentrations,
required for maximal electron transport stimulation, rule out specific cation catalytic effects, at least for cations acting on the
outer thylakoid surface. Thus, differences among cations of the
same valence group reflect differences in binding constants, rather
than differences in specificity.

ACKNOWLEDGEMENT

Partial support by EEC Grant ESD-043-GR(B) is acknowledged.

REFERENCES

1. J. Barber, Membrane surface charges and potentials in relation
       to photosynthesis, Biochim. Biophys. Acta 594:253 (1980).
2. J. Barber, Influence of surface charges on thylakoid structure
       and function, Ann. Rev. Plant Physiol. 33:261 (1982).
3. W. Junge and J.B. Jackson, The development of electrochemical
       potential gradients across photosynthetic membranes, in
       "Photosynthesis: Energy conversion by plants and bacteria,"
       Govindjee, ed., Academic Press, New York (1982).
4. W. Junge, Electrogenic reactions and proton pumping in green
       plant photosynthesis, in "Current topics in membranes and
       transport," C.L. Slayman, ed., vol. 16, Academic Press,
       New York (1982).
5. A.T. Jagendorf, Mechanism of photophosphorylation, in "Bioenerge-
       tics of Photosynthesis," Govindjee, ed., Academic Press, New
       York (1975).
6. S. Lien and A. San Pietro, Interaction of plastocyanin and P700
       in photosystem I reaction center particles from Chlamydomonas
       reinhardi and spinach, Arch. Biochem. Biophys. 194:128 (1979).
7. G.K. Ozakian and P. Satir, Particle movements in chloroplast
       membranes: quantitative measurements of membrane fluidity
       by freeze-fracture technique, Proceed. Nat. Acad. Sci. USA
       21:2052 (1974).
8. S. Kaplan and C. Arntzen, "Photosynthetic membrane structure and
       function, in Photosynthesis: Energy conversion by plants and
       bacteria", Govindjee, ed., Academic Press, New York (1982).
9. A.T. Jagendorf and M. Smith, Uncoupling of phosphorylation in

spinach chloroplasts by absence of cations, <u>Plant Physiol</u>.
37:135 (1962).

10. N. Murata, Control of excitation transfer in photosynthesis.
    II. Magnesium ion dependent distribution of excitation
    energy between two pigment systems in spinach thylakoids,
    <u>Biochim. Biophys. Acta</u> 189:171 (1969).

11. N. Murata, Effects of monovalent cations on light energy distri-
    bution between two pigment systems in photosynthesis in
    isolated spinach chloroplasts, <u>Biochim. Biophys. Acta</u> 197:
    250 (1971).

12. C.A. Wraight and A.R. Crofts, Energy-dependent quenching of
    chlorophyll <u>a</u> fluorescence in isolated chloroplasts, <u>Eur</u>.
    <u>J. Biochem</u>. 17:319 (1970).

13. G.C. Papageorgiou, Chlorophyll fluorescence: an intrinsic
    probe of photosynthesis, <u>in</u> "Bioenergetics of Photosynthesis",
    Govindjee, ed., Academic Press, New York (1975).

14. G.H. Krause, J.-M. Briantais and C. Vernotte, Characterization
    of chlorophyll fluorescence quenching in chloroplasts by
    fluorescence spectroscopy at 77 K. I. ΔpH-dependent quench-
    ing, <u>Biochim.Biophys.Acta</u> 723:169 (1983).

15. G.C. Papageorgiou, M. Tsimili-Michael and J. Isaakidou, Quen-
    ching of excited chlorophyll <u>a</u> <u>in</u> <u>vivo</u> by nitrobenzene,
    <u>Biophys. J</u>. 15:83 (1975).

16. G.C. Papageorgiou, On the mechanism of the PMS effected quen-
    ching of chloroplast fluorescence, <u>Arch. Biochem. Biophys</u>.
    166:330 (1975).

17. G.C. Papageorgiou, Applications of spectroscopy to biological
    research, <u>in</u> "Problems of contemporary biophysics", Polish
    Scientific Publishing Company, Warsawa-todz (1979).

18. C. Bonaventura and J. Myers, Fluorescence and oxygen evolution
    from <u>Chlorella</u> pyrenoidosa, <u>Biochim. Biophys. Acta</u> 189:366
    (1969).

19. W.P. Williams, D. Furtado and A.R. Nutbeam, Comparison between
    state I/state II adaptation in a unicellular green alga and
    high energy state quenching in isolated intact spinach
    chloroplasts, <u>Photochem. Photobiophys</u>. 1:91 (1980).

20. J. Barber, An explanation for the relationship between salt-
    induced thylakoid stacking and chlorophyll fluorescence
    changes associated with changes in spillover of energy from
    photosystem II to photosystem I, <u>FEBS Lett</u>. 118:1 (1980).

21. E. Gantt, Structure and function of phycobilisomes: Light-
    harvesting pigment complexes in red and blue-green algae,
    <u>Intern. Rev. Cytol</u>. 66:45 (1980).

22. G. Cohen-Bazire and D.A. Bryant, Phycobilisomes: Composition
    and structure <u>in</u> "The Biology of Cyanobacteria", N.G. Carr
    and B.A. Whitton, eds., Blackwell, Oxford (1982).

23. A.N. Glazer, Phycobilisomes: Structure and dynamics, <u>Ann. Rev</u>.
    <u>Microbiol</u>. 36:173 (1982).

24. M. Mimuro and Y. Fujita, Excitation energy transfer between
    pigment system II units in blue-green algae, <u>Biochim. Biophys</u>.
    <u>Acta</u> 504:406 (1978).

25. A. Manodori and A. Melis, Photochemical apparatus organization in Anacystis nidulans (cyanophyceae), Plant Physiol. 74:67 (1984).
26. N. Murata, The lipid phase of photosynthetic membranes, in Advances in photosynthesis research, C. Sybesma, ed., vol. 3, M. Nijhoff/Dr. W. Junk, The Hague (1984).
27. T. Omata and N. Murata, Isolation and characterization of the cytoplasmic membranes from the blue-green alga (cyanobacterium) Anacystis nidulans, Plant and Cell Physiol. 24:1101 (1983).
28. D.J. Murphy and I.E. Woodrow, Lateral heterogeneity in the distribution of thylakoid membrane lipid and protein components and implications for the molecular organization of photosynthetic membranes, Biochim. Biophys. Acta 725:104 (1983).
29. K. Kalosaka and G.C. Papageorgiou, Surface electric properties of thylakoid fragments isolated from vegetative and heterocystous cyanobacteria, in Advances in photosynthesis research, C. Sybesma ed., vol. 2, M. Nijhoff/Dr. W. Junk, the Hague (1984).
30. R. Barr, K.S. Troxel and F.L. Crane, Calmodulin antagonists inhibit electron transport in photosystem II of spinach chloroplasts, Biochem. Biophys. Res. Commun. 104:1182 (1982).
31. R.R. England and E.H. Evans: A requirement for Ca^{2+} in the extraction of O_2-evolving photosystem 2 preparations from the cyanobacterium Anacystis nidulans. Biochem. J. 210: 473 (1983).
32. E.K. Pistorius, Effects of Mn^{2+}, Ca^{2+} and chlorpromazine on photosystem II of Anacystis nidulans: An attempt to establish a functional relationship of aminoacid oxidase to photosystem II. Eur. J. Biochem. 135:218 (1983).
33. R. Barr, K.S. Troxel and F.L. Crane, A calcium selective site in photosystem II of spinach chloroplasts, Plant Physiol. 73:309 (1973).
34. D.F. Ghanotakis, G.T. Babcock and C.F. Yocum, Calcium reconstitutes high rates of O_2 evolution in polypeptide-depleted photosystem I preparations, FEBS Lett. 167:127 (1984).
35. D.F. Ghanotakis, J.N. Topper, G.T. Babcock and C.F. Yocum, Water-soluble 17 and 23 kDa polypeptides restore O_2 evolution activity in creating a high affinity binding site for Ca^{2+} on the oxidizing side of photosystem II, FEBS Lett. (in press 1984).
36. M. Miyao and N. Murata, Calcium ions can be substituted for the 24 kDa polypeptide in photosynthetic oxygen evolution, FEBS Lett. 168:118 (1984).
37. R.G. Piccioni and D.C. Mauzerall, Calcium and photosynthetic O_2 evolution in cyanobacteria, Biochim. Biophys. Acta 504: 384 (1978).
38. Brand, J.J., The effect of Ca^{2+} on oxygen evolution in membrane preparations from Anacystis nidulans. FEBS Letters 103:114 (1979).

39. G.C. Papageorgiou, Photosynthetic activity of diimidoester-modified cells permeaplasts and cell-free membrane fragments of the blue-green alga Anacystis nidulans, Biochim. Biophys. Acta 461:379 (1977).

40. G. MacKinney, Absorption of light by chlorophyll solutions, J. Biol. Chem. 140:315 (1941).

41. T.A. Kursav and R.S.Alberte, Photosynthetic unit organization in a red alga, Plant Physiol. 42:409 (1983).

42. B. Gerhardt and A. Trebst, Photosynthetische Reaktionen in lyophilisierten Zellen der Blaualge Anacystis, Z. Naturforsch. 20B:879 (196).

43. U.W. Hallier and R.B. Park, Photosynthetic light reactions in chemically fixed Anacystis nidulans, Chlorella pyrenoidosa and and Porphyridium cruentum, Plant Physiol. 44:535 (1969).

44. B. Ward and J. Myers, Photosynthetic properties of permeaplasts of Anacystis, Plant Physiol. 50:547 (1972).

45. A. Binder, E. Tel-Or and M. Auron, Photosynthetic activities of membrane preparations of the blue-green alga Phormidium luridum, Eur. J. Biochem. 67:187 (1976).

46. G.C. Papageorgiou and H. Tzani, Action of lysozyme on glutaraldehyde-treated filaments of the cyanobacterium Phormidium luridum, J. Appl. Biochem. 2:230 (1980).

47. H. Spiller, Photophosphorylation capacity of stable spheroplast preparations of Anabaena, Plant Physiol. 66:446 (1980).

48. S.J. Robinson, C.S. De Roo and C.F. Yocum, Photosynthetic ele electron transfer in preparations of the cyanobacterium Spirulina platensis, Plant Physiol. 70:154 (1982).

49. G. Sotiropoulou, T. Lagoyanni, and G.C. Papageorgiou, Effects of Ca^{2+} ions on the light-induced electron-transport activities of Anacystis nidulans permeaplasts and spheroplasts, in Advances in photosynthesis research, C. Sybesma, ed., vol. 2, M. Nijhoff/Dr. W. Junk, The Hague (1984).

50. T. Hiyama and B. Ke, Laser-induced reactions of P700 and cytochrome f in a blue-green alga, Plectonema boryanum, Biochim. Biophys. Acta 226:320 (1971).

51. W. Haehnel, The reduction of chlorophyll a as an indicator for proton uptake between the light reactions of chloroplasts, Biochim. Biophys. Acta 440:506 (1976).

52. W. Haehnel, Structural and quantitative analysis of membrane-bound components of the photosystem I complex of spinach ch chloroplasts by immunological methods, in "Bioenergetics of membranes", L. Packer, G.C. Papageorgiou, and A. Trebst, eds., Elsevier/North-Holland, Amsterdam (1977).

53. W. Hoehnel, A. Prüpper, H. Krause, Evidence for completed plastocyanin as the immediate electron donor of P700, Biochim. Biophys. Acta 593:384 (1980).

54. S. Lien and A. San Pietro, Interaction of plastocyanin and P700 in photosystem I reaction center particles from Chlamydomonas reinhardi and Spinach, Arch. Biochem. Biophys. 194:128 (1979).

55. Wood, P.M., Interchangeable copper and iron proteins in algal photosynthesis, Europ. J. Biochem. 87:9 (1978).

56. Böhme H. and P. Böger, Reciprocal formation of cytochrome c-553 and plastocyanin in Scenedesmus, FEBS Letters 85:337 (1978).

57. K.K. Ho and D.W. Krogmann, Photosynthesis in "The biology of cyanobacteria", N.G. Carr and B.A. Whitton, eds., Blackwell, Oxford (1982).

58. B. Bouges-Bocquet and R. Delosme, Evidence for a new electron donor to P700 in Chlorella pyrenoidosa, FEBS Letters 94:100 (1978).

59. S. Katoh, Plastocyanin, Meth. Enzymol. 23:408 (1971).

60. W.L. Ellefson, E.A. Ulrich, and D.W. Krogmann, Plastocyanin, Meth. Enzymol. 69:223 (1980).

61. K. Matsuura and M. Nishimura, Sidedness of membrane structures in Rhodopseudomonas spheroides. Electrochemical titration of the spectrum changes of carotenoid in spheroplast, spheroplast membrane vesicles, and chromatophores, Biochim. Biophys. Acta 459:483 (1977).

62. J.W. Costerton, J.M. Ingram and K.-J. Cheng, Structure and function of the cell envelope of Gram negative bacteria, Bacteriol. Rev. 38:87 (1974).

63. K.K. Karukstis and K. Sauer, Fluorescence decay kinetics of chlorophyll in photosynthetic membranes, J. Cellular Biochem. 23:131 (1983).

64. S. McLaughlin, Electrochemical potentials at membrane-solution interfaces, in Current Topics in Membranes and Transport, F. Bronner and A. Kleinzeller, eds., vol. 9, Academic Press, New York (1977).

65. R.A. Wavare and P. Mohanty, Aluminum stimulation of photoelectron transport in spheroplasts of cyanobacterium Synechococcus cedrorum, Photobiochem. Photobiophys. 3:327 (1982).

66. R.A. Wavare and B. Subbalakshmi and P. Mohanty, Effect of Al^{3+} on electron transport catalysed by photosystem I and II of photosynthesis in cyanobacterium Synechococcus spheroplasts and beet-spinach chloroplasts, Indian J. Biochem. Biophys. 20:301 (1983).

67. R.A. Wavare and P. Mohanty, Cations stimulate electron transport associated with photosystem II and inhibit electron flow linked with photosystem I in spheroplasts of the cyanobacterium Synechococcus cedrorum, Photobiochem. Photobiophys. 6:189 (1983).

68. M.L. Richter and P.H. Homann, Surface-charge related actions of polylysine on thylakoid membranes, Arch. Biochem. Biophys. 222:67 (1983).

HALORHODOPSIN, THE LIGHT-DEPENDENT CHLORIDE TRANSPORT SYSTEM

OF HALOBACTERIA

Janos K. Lanyi

Department of Physiology and Biophysics
University of California
Irvine, CA 92717

INTRODUCTION

The halobacteria are microorganisms, adapted to growth and survival in nearly saturated brines (Larsen, 1967; Kushner, 1978). Because their intracellular salt concentration is also very high but consists mainly of KCl rather than NaCl, it might be expected that the exclusion of sodium ions and the uptake of potassium ions is an important function of their cytoplasmic membrane. Indeed, a powerful sodium/proton antiport system has been described (Lanyi and MacDonald, 1976; Eisenbach et al, 1977) in these membranes, driven by protonmotive force created by outward proton transport during respiration (under aerobic conditions) or during illumination though the operation of bacteriorhodopsin, a retinal protein with proton translocating activity. The respiratory chain is not very unusual for a bacterial membrane (Lanyi, 1968; Cheah, 1969), but bacteriorhodopsin is unique for the halobacteria, and its study has contributed greatly to our understanding of how proton pumps function (Stoeckenius et al, 1979). Another aspect of the membrane bioenergetics of these organisms is the transport of chloride. Since the concentration of this anion in the cytoplasm is nearly the same as in the medium, and a large membrane potential (negative inside) exists during growth, chloride in these organisms is far from its electrochemical equilibrium and must be imported by active transport. It is now evident (Schobert and Lanyi, 1982; Bamberg et al, 1984a) that a second retinal protein, halorhodopsin, is a light-driven, inward directed chloride pump, in the same sense that bacteriorhodopsin is an outward directed proton pump.

SOME PRINCIPLES OF SOLUTE TRANSPORT

From a thermodynamic point of view, we distinguish passive and active transport on the basis of whether the solute flux across the membrane causes the absolute value of the net electrochemical potential for that solute to decrease or increase, respectively. Thus, passive transport is downhill and will decrease the size of the electrochemical gradient of the solute across the membrane. For charged solutes this might mean that both electrical and concentration gradients approach zero, or that the electrical and concentration terms are in the opposite sense and balance one another. Passive transport may be non-facilitated (and therefore dependent on the solubility of the substance in question in the lipid bilayer), or facilitated, either by small organic molecules such as valinomycin, gramicidin, etc., or by protein-type transport systems. An illustration of passive transport is the efflux of potassium ions from a KCl-filled liposome upon addition of valinomycin, a potassium-specific ionophore. In large liposomes the loss of a small amount of potassium will produce a membrane potential, negative inside, so as to stop the net potassium flux. In the absence of permeability to other ions, the membrane potential will endure for some time, and will be nearly equivalent to the original concentration difference for the potassium.

Active transport is uphill in that it proceeds against an electrochemical potential, and therefore it requires energy input. This energy might come from the downhill flux of another solute, for coupled transport systems, such as symport or antiport depending on the direction of the two fluxes relative to one another. Active transport is always facilitated. An example for active transport is the uptake of protons into KCl-filled liposomes upon addition of nigericin, a potassium/proton exchange ionophore. In this system the potassium moves downhill, as with valinomycin, but this flux is coupled by the antibiotic to the uphill movement of the protons, which establishes a pH difference (acid inside) during the process. Electrical potential difference is not produced in this case. It should be noted, of course, that in the absence of other permeable ions this process will also lead to an equilibrium, but one in which some of the potassium gradient will have been used up to produce a net gradient for the protons. In this way we may speak of energy coupling between the gradients of potassium and protons. A special class of active transport systems derives energy for the transport from the breaking and forming of chemical bonds or from light. These systems, which transduce energy, are called pumps. Examples include the ionmotive ATPases, redox systems such as the mitochondrial complex III and cytochrome oxidase, and the bacterial transport rhodopsins.

THE LIGHT-DRIVEN TRANSPORT OF CHLORIDE

Among the large variety of existing strains of <u>Halobacterium</u> <u>halobium</u>, there are some which produce no bacteriorhodopsin, yet exhibit light-dependent transport. This phenomenon, now established to be due to the second retinal protein, halorhodopsin, are easiest to observe in cell envelope vesicles, produced by sonic treatment of whole cells (MacDonald and Lanyi, 1975; Lanyi and MacDonald, 1979). The envelope vesicles are mostly devoid of cytoplasmic components, they are at least 85% right-side-out by a number of criteria, and their internal contents can be readily adjusted by either osmotic shock or simple long-term equilibration with the desired salt mixture (Lanyi and MacDonald, 1979). They are stable for months.

When a suspension of the bacteriorhodopsin-deficient envelope vesicles were illuminated, extensive proton uptake was observed as detected by a glass electrode in the buffer and by flow dialysis, particularly when uncouplers (protonophores) were included in the system (Lindley and MacDonald, 1979; MacDonald et al, 1979). The proton uptake was also shown by fluorescence quenching of 9-amino acridine, which detected internal acidification (Greene and Lanyi, 1979). A membrane potential developed, negative inside, as detected by the uptake of a radioactively labeled membrane-permeant cation (MacDonald et al, 1979), or by fluorescence quenching for a cyanine dye, commonly used to follow membrane potentials (Greene and Lanyi, 1979). The magnitudes of these gradients were estimated by calibrating the fluorescence changes with KCl-filled envelopes suspended in NaCl-KCl mixtures and adding valinomycin (for membrane potential) or nigericin (for pH difference). Using the calibration curves, it was confirmed that these oppositely directed gradients for protons cancelled each other quantitatively during the steady state established by the illumination (Greene and Lanyi, 1979), as expected if light caused the active transport of an ion which is not a proton (Lindley and MacDonald, 1979; MacDonald et al, 1979). Stated in another way, in these vesicles the electrical potential was not created by proton pumping. Rather, protons moved passively, in response to the electrical potential. Abolishing the light-dependent potential, for example with sodium flux facilitated by gramicidin, indeed abolished the proton movements.

The passive nature of the proton uptake was further demonstrated by treating the envelopes with low concentrations of dicyclohexylcarbodiimide, an agent which in many other membranes blocks proton permeability through free Fo channels. As expected, the light-dependent internal acidification was abolished although the membrane potential was still produced. Adding an uncoupler restored the light-dependent proton influx, and at rates more rapid than in the untreated envelopes. Hence, it could be concluded that proton uptake is a secondary, and not necessary, process in the generation of membrane potential (Greene and Lanyi, 1979).

It was proposed (Lindley and MacDonald, 1979) that it is the active extrusion of sodium ions which generates the membrane potential in this system. However, we had been unable to demonstrate, by using sodium-dependent amino acid transport, net export of sodium from the envelopes (Luisi et al, 1980). On reexamining this point, it became clear that the active transport had to be for chloride, rather than for sodium. These experiments were carried out (Schobert and Lanyi, 1982) by comparing cell envelope vesicles containing only halorhodopsin or only bacteriorhodopsin, as light-driven transport systems. The former were produced from H. halobium strain L-33 (Wagner et al, 1981), which has a genetic defect in bacterio-opsin, the latter were constructed by retinal-reconstitution of envelopes prepared from a retinal-negative, but bacterio-opsin positive strain, which produces very little halorhodopsin under these condition.

With bacteriorhodopsin-vesicles, suspended in NaCl, we expected light-energized proton extrusion, proton recirculation via the sodium/proton antiporter causing active sodium extrusion, and passive chloride extrusion in response to the membrane potential, negative inside, created by the cation efflux. The net loss of sodium and chloride was expected to cause loss of water during the illumination, i.e. energized volume decrease. This change was expected to be sensitive to uncouplers, as well as to cationic iopnophores, such as gramicidin. With halorhodopsin-vesicles, there were two alternatives. An outward-directed sodium pump would produce a membrane potential, negative inside, and the observed inward proton movement, but it would also result in passive chloride loss and energized volume decrease. This change was expected to be uncoupler insensitive, but gramicidine-sensitive. An inward-directed chloride pump, on the other hand, would create the observed potential and proton uptake, but also passive sodium ion influx, and therefore energized volume increase. This change was expected to be uncoupler insensitive, as for the sodium pump model, but also gramicidin insensitive. Tests of the volume changes in these vesicles, determined by light-scattering (Schobert and Lanyi, 1982) and later with a spin-probe of volume (Mehlhorn, Schobert, Packer and Lanyi, unpublished), fulfilled the expectations for the bacteriorhodopsin-vesicles as controls, and unequivocally supported the chloride pump model. In fact, gramicidin (as well as valinomycin plus potassium) not only did not inhibit, but accelerated the volume increase of the halorhodopsin-vesicles during the illumination.

The chloride-pump model for halorhodopsin was suggested also by the requirement of chloride (or bromide) for all of the light-dependent bioenergetic phenomena (membrane potential, pH change, volume change) measured with envelopes suspended in sodium sulfate or potassium phosphate (Schobert and Lanyi, 1982). The chloride requirement was on the vesicle exterior, as expected for inward chloride transport, and showed an apparent affinity constant of about 40 mM.

Direct demonstration of chloride uptake was by illuminating the envelope vesicles in sulfate plus a limited amount of chloride, separating aliquots of envelopes by rapid Agarose chromatography, and analyzing them for total contained chloride. Chloride uptake in the light (but not in the dark) proceeded for over 30 mins, and amounted to more than 3 orders of magnitude more chloride translocated than halorhodopsin molecules present. The transport was in-different to whether sodium or potassium was the cation present. Sodium contamination of the potassium phosphate amounted to about 5% of the chloride transported.

Active transport of chloride requires that it be transported against a concentration gradient. Chloride concentration gradients across the vesicle membrane were determined in the presence of a large amount of sodium sulfate by adding low concentrations of triphenyl tin (TPT), which will function as a chloride/hydroxyl exchanger (Selwyn et al, 1970), and measuring pH changes in the vesicle suspensions. Calibration curves, obtained in the dark with envelopes loaded with known amounts of chloride and suspended in chloride-sulfate mixtures, allowed the estimation of the size of the chloride gradients, and indicated furthermore that preformed chloride gradients collapse with a half-life of only about 30 mins. The light-driven chloride gradients were measured with this method after the pH changes but not the chloride gradient, caused by the illumination, had relaxed (about 10 mins). Bacteriorhodopsin-vesicles showed light-dependent chloride extrusion and an inward directed chloride gradient developed, as expected. Halorhodopsin-vesicles, in contrast, took up chloride, and an outward directed chloride gradient of at least 5-fold developed.

These experiments established that halorhodopsin is a light-driven chloride pump by four criteria: a) chloride is transported against an electrical or concentration gradient, with energy input (i.e. it is active transport), b) the translocation is insensitive to, or enhanced by cationic ionophores (i.e. it is not a secondary coupled system), and c) the light-induced membrane potential depends on the presence of chloride but is indifferent to cations (i.e. chloride is the substrate of the pump). In addition, d) the spectroscopic and photochemical properties of halorhodopsin were greatly influenced by those anions which are transported, as explained below.

Action spectra for the creation of membrane potential and the passive proton uptake yielded a putative absorption maximum for halorhodopsin of 585-590 nm (Greene and Lanyi, 1979; Greene et al, 1980). This maximum is 15-20 nm red-shifted from that of bacterio-rhodopsin.

While bacteriorhodopsin-containing cells contain so much of this pigment that colonies and suspensions appear purple, the halo-rhodopsin-containing cells do not appear obviously colored. Detection and identification of the small amounts of the latter pigment was therefore difficult and had to be indirect. Cell envelope vesicles of a carotenoid-deficient strain were bleached by extensive illumination in the presence of hydroxylamine, a treatment which would remove retinal from bacteriorhodopsin (Oesterhelt at al, 1974; Oesterhelt and Schuhmann, 1974). Adding retinal to such bleached preparations produced difference spectra where the disappearance of absorption at 380 nm (free retinal) occurred concurrently with the appearance of an absorption maximum at 585-590 nm (Lanyi and Weber, 1980). Transport activity was restored partially upon the retinal reconstitution (Lanyi and Weber, 1980; Weber and Bogomolni, 1981), but in proportion to the amount of absorption increase produced. From the absorption changes the extinction coefficient of halorhodopsin was estimated to be about 50,000 1/M.1/cm.

It is now clear (Bogomolni and Spudich, 1982; Lanyi and Schobert, 1983) that the membranes used in these experiments contained another rhodopsin as well, "slowly cycling rhodopsin," whose existence was not known at the time. The retinal reconstitutions thus yielded a mixture of the two pigments. In possession of this information, the real absorption maximum of halorhodopsin is now estimated to be at 580 nm (Spudich and Bogomolni, 1983).

The retinal reconstitution was used also to identify the opsin molecule, by employing radioactively labeled retinal, and reducing the reconstituted pigment with cyanoborohydride in the presence of trace amounts of diethyl ether to produce a stable product (Lanyi and Oesterhelt, 1982). Fluorography of SDS gels indicated that the molecular weight of halo-opsin is somewhat less than that of bacterio-opsin, but clearly distinguishable from it on gels where both opsins are labeled. Since the molecular weight of bacteriorhodopsin is 26,000 Da (it runs anomalously on SDS gels), that of halorhodopsin should be about 25,000 Da. This value was confirmed once halorhodopsin could be produced in purified form (Hegemann et al, 1982; Steiner and Oesterhelt, 1983; Taylor et al,

1983; Sugiyama and Mukohata, 1984; Ogurusu et al, 1984).

PURIFICATION OF HALORHODOPSIN, FUNCTIONAL RECONSTITUTION

The purification strategies developed in several laboratories are rather similar, and consist of a preliminary extraction of the membranes, followed by detergent solubilization, and hydroxylapatite and hydrophobic column chromatography. The initial extraction of 30-50% of the proteins could be achieved by lowering the salt concentration to 0.1 M and/or adding a mild detergent, Tween 20. The remaining still membrane-like material, which contains all of the halorhodopsin is called "Tween membranes." It can be dissolved at 1,2 or 4 M NaCl in Lubrol PX (Steiner and Oesterhelt, 1983), octylglucoside (Taylor et al, 1983), C12E9 (Sugiyama and Mukohata, 1984), or Triton X-100 (Ogurusu et al, 1984). Elution with the detergent solutions from hydroxylapatite and phenyl or octylsepharose columns produces pure halorhodopsin, with no contamination by slowly cycling rhodopsin or proteins giving extra bands on SDS gels. An additional criterion of purity is that the absorption ratio at 580 nm and 280 nm is no higher than 1.6 to 1.7. Octyl-glucoside can be directly removed by dialysis, the other detergents have to be first exchanged for a dialyzable detergent.

Amino acid analyses of purified halorhodopsin have revealed significant differences from bacteriorhodopsin. Halorhodopsin contains a cysteine residue essential for activity (Ariki and Lanyi, 1984), and two histidine residues (Ariki and Lanyi, 1984; Sugiyama and Mukohata, 1984), which bacteriorhodopsin lacks. The protein is particularly rich in arginine residues, but poor in lysine. Partial sequence information now available (Hegemann and Oesterhelt, unpublished; Ariki and Lanyi, unpublished) indicates that the two opsins have little homology. Immunologically they are not cross-reactive (Hegemann et al, 1982).

Liposomes reconstituted with pure halorhodopsin by conventional means were functionally only marginally active, but after adsorption onto planar lipid films produced chloride-dependent sustained photocurrents (Bamberg et al, 1984b). The requirement for a chloride ionophore for this activity in this system, and in similar experiments with Tween membranes (Bamberg et al, 1984a), was consistent with the role of halorhodopsin as a chloride pump. When octylglucoside-halorhodopsin was first extensively dialyzed to aggregate it, and then reconstituted into liposomes, the proteoliposomes were active, and illumination in the presence of uncoupler produced chloride dependent pH lowering, attributed to passive proton efflux in response to chloride extrusion (Bogomolni et al, 1984). The halorhodopsin in these reconstituted liposomes are thus reversed in orientation from the cells and cell envelope vesicles. The reconstitution experiments indicate that the halorhodopsin chromoprotein (retinal plus opsin) is sufficient for the chloride transport activity.

ANION BINDING SITES IN HALORHODOPSIN

As expected for a chloride pump, halorhodopsin shows various chloride and anion effects in its chromophoric properties, which can be interpreted as consequences of occupying binding sites of differing affinities and specificities. Most of these effects are readily observable for halorhodopsin in both cell envelope vesicles (i.e. in the original membrane-bound form), and in detergent. The most significant difference between these systems is that the affinity constant for chloride for two of the sites, as well as for the transport, is about 5-fold higher in the envelopes. It should be noted that as a halophilic protein, halorhodopsin seems to require high ionic strength for intergrity and stability, like many other enzymes from the halobacteria (Lanyi, 1974), although not bacteriorhodopsin. We have been able to get consistent results only at high concentrations (above 0.5 M) of sodium sulfate or potassium phosphate as background salts.

Site I is identified from the effect of anions on the pKa of the Schiff-base (Lanyi and Schobert, 1985). The latter can be titrated in sodium sulfate between the protonated (580 nm) and the deprotonated (410) state to yield a pKa of 7.4. Adding a variety of anions at relatively low concentrations raises this pKa to 8.7, accounting for earlier observations that chloride and other anions caused a negative peak at 410 nm and a positive peak near 560-580 nm (Lanyi and Schobert, 1983; Steiner et al, 1984). In purified halorhodopsin the apparent affinity of this site for chloride is about 10 mM (Steiner et al, 1984; Lanyi and Schobert, 1985), its specificity is chloride,bromide,nitrate,iodide,thiocyanate,azide > nitrate > acetate ≫ others. Access to Site I appears to be on the exterior of envelope vesicles, i.e. on the cis side of the chloride translocation.

Of the anions listed above, only chloride, bromide and to some extent iodide, shift the absorption all the way to 578 nm (Steiner et al, 1984). If these anions are added to halorhodopsin already in a solution of one of the other Site I anions, e.g. nitrate, a shift in the absorption maximum from about 565 nm to 578 nm is observed (Ogurusu et al, 1984; Lanyi, 1984a). This more specific effect is termed Site II. Occupancy of Site II has dramatic effects on the photochemistry of halorhodopsin. It has been known for some time that in NaCl the principal photointermediate of halorhodopsin absorbs near 500 nm (Weber and Bogomolni, 1981; Tsuda et al, 1982; Schobert et al, 1983; Steiner et al, 1984), an intermediate we refer to as HR(520). It arises with very rapid (usec) kinetics, and decays on a 10-15 msec timescale. In the absence of Site II anions this intermediate is completely absent, and another one absorbing in the red region (Schobert et al, 1983), referred to as HR(660), is produced, which decays on a 1-2 msec timescale. As the chloride (or bromide) concentration is increased more HR(520) and less HR(660) is produced, both with intermediate decay kinetics. An exact description of the photocycle consistent with these observations is not yet available. It is clear, however, that Site II is involved in the light-dependent events in this pigment. This site is also obviously involved in the transport, since its affinity for chloride of about 10 mM, and its specificity agree with that of the light-dependent generation membrane potential. Binding of chloride to Site II can be competitively inhibited by a diuretic drug (Schobert et al, 1983), which inhibits transport and lowers the yield of HR(520). Occupancy of Site II has no effect on the pKa of the Schiff-base. Access to Site II is also on the cis side of the chloride translocation (Schobert et al, 1983).

Site III was first identified from chloride-37 NMR line-broadening (Falke et al, 1984), which suggested a binding site with an affinity of 110 mM, and a specificity (from competition with chloride binding) of iodide > thiocyanate > bromide, nitrate > chloride ≫ others. The other sites were not visible by the NMR method. This site corresponds to one later described for the effect of azide on the light-dependent deprotonation of the Schiff-base (Hegemann et al, 1984; Lanyi and Schobert, 1985). During the normal photocycle of halorhodopsin, a small fraction of the pigment ends up in the deprotonated state, i.e. in the form HR(410). The return reaction (reprotonation) is in the order of hours, hence sustained illumination will cause the accumulation of the blue-shifted species, particularly at alkaline pH. The reaction is termed "light-dependent deprotonation reaction," and it is not observable in single turn-over (i.e. flash) experiments. In the presence of azide, however, both forward and reverse reactions proceed with greatly increased rate, and HR(410) is now observable with high yield even during a single turn-over of HR(520). Since the steady state value of the ratio HR(410)/HR(578) is unchanged by the azide, it was suggested that the deprotonation is a side-reaction at the HR(520) and/or HR(660) level (Hegemann et al, 1984). Indeed, the decay kinetics of these intermediates and the rise kinetics of HR(410) roughly agree (Lanyi and Schobert, 1985). The acceleration of the protonation reaction of the Schiff-base by the azide is probably due to the conduction of the proton to and from the Schiff-base nitrogen, since other anions with moderate pKa values, such as cyanate and sulfide, were also found to be effective (Hegemann et al, 1984). Competition of other anions with azide in the light-dependent production of HR(410) is in a pattern consistent with that in the NMR experiments (Lanyi and Schobert, 1985).

It should be understood that the anion binding sites described above refer to distinct effects of anions on the halorhodopsin chromophore. Binding of ions which does not produce easily observable consequences, is not excluded, but such binding might not be particularly relevant to the photochemistry and thus to the transport.

LIGHT-DEPENDENT MOLECULAR EVENTS IN HALORHODOPSIN

In bacteriorhodopsin the light-dependent absorption changes in the near UV have revealed trans-to-13-cis isomerization on a usec time-scale (Kuschmitz and Hess, 1982), and deprotonation of a tyrosine on a msec time-scale (Bogomolni et al, 1978; Hess and Kuschmitz, 1979). The former event coincides with the K intermediate, the latter with the M intermediate. The configurational change of the retinal produces a distinct "cis-peak" near 340 nm, with spectral complexities well reproduced in a dark-adapted vs. light-adapted difference spectrum (Kuschmitz and Hess, 1982). The tyrosine deprotonation is modeled simply with an alkaline vs. neutral tyrosine difference spectrum. These events in bacteriorhodopsin are thought to be directly involved in the proton transport. For halorhodopsin the absorption changes in the near UV consist only of the trans-to-13-cis isomerization, with no evidence for any deprotonation of tyrosine (Lanyi, 1984b). The kinetics of the rise of the cis-peak is very fast, and either coincides or preceeds that of HR(520). The decay of this feature agrees well with the decay of HR(520).

A MODEL OF CHLORIDE TRANSPORT BY HALORHODOPSIN

As expected from their functions as a proton pump and a chloride pump, bacteriorhodopsin and halorhodopsin appear to be very different proteins. Their similarity as rhodopsins does not, thus, conceal a real photochemical dissimilarity. Both pigments undergo a photocycle on a msec time-scale, but unlike in bacterio-rhodopsin, the Schiff-base in halorhodopsin is deprotonated only with very low efficiency unless azide is added. This is somewhat surprizing, in view of the fact that the pKa of this Schiff-base is already near that observable in free solution (latter about 7), while that for bacteriorhodopsin is above 13 until light is absorbed (Druckmann et al, 1982). Furthermore, the pKa of the halorhodopsin Schiff-base is readily influenced by low concentrations of a variety of anions (Site I). It would appear, therefore, that in halorhodopsin the Schiff-base is near the aqueous phase and might not be directly involved in the translocation. If this idea is correct, any functional model for the chloride transport should be fundamentally different from proposed models for proton transport by bacteriorhodopsin. It should be mentioned, however, that it is possible to construct analogous functional models for the two retinal pigments, and one featuring chloride binding and translocation directly by the Schiff-base nitrogen itself in halorhodopsin was proposed recently (Schulten and Oesterhelt, presented at the 3rd EMBO Workshop on Molecular Biology of Retinal Proteins, 1984).

A model of halorhodopsin (Lanyi, 1984b), in which the retinal isomerization indirectly drives chloride transport is shown in Fig. 1. The protein is represented by a box containing retinal (straight line for trans configuration, bent line for the cis isomer) and two binding sites for chloride. The upper of these sites communicates with the membrane surface facing the exterior, the lower with that facing the cytoplasm. The upper site is a high affinity site for chloride, and might be Site II. There is, in addition, a movable charge in the protein, driven by the charge of the Schiff-base. Light causes retinal isomerization and the movement of the Schiff-base, hence that of the movable charge also (species II). The movable charge modulates the affinity of the lower site to chloride, thus in species II this site has become a high affinity site. The upper site exists in an equilibrium with external chloride (species I and VI), both species exhibiting the affinity change described above. It is in species II only, however, the chloride in the upper site can migrate to the lower site, provided that the affinity of the latter is increased over that of the former. The species III so created can decay in three ways. The first of these is the reverse pathway back to species I, via III – II – I, a nonproductive process. The second of these is to lose chloride to the interior, and return to species I via the III – VII – VI – I or the the III – VII – II – I pathway, replacing the chloride in the upper site in the process. The third possibility is to regain a chloride in the upper site directly, and return to I via III – IV – V – I. The second and third possibilities are productive transport pathways. The probability that these occur, and not the first, is determined by the relative affinities of the upper and lower sites to chloride in species II and III.

The model might explain how external chloride appears (Schobert et al, 1983) to increase the life-time of HR(520) during the photocycle, since species III is driven to species IV, which has a longer reisomerization time, by external chloride. However, by the same token, the model neeeds to invoke protein conformational changes to drive the IV – V transition.

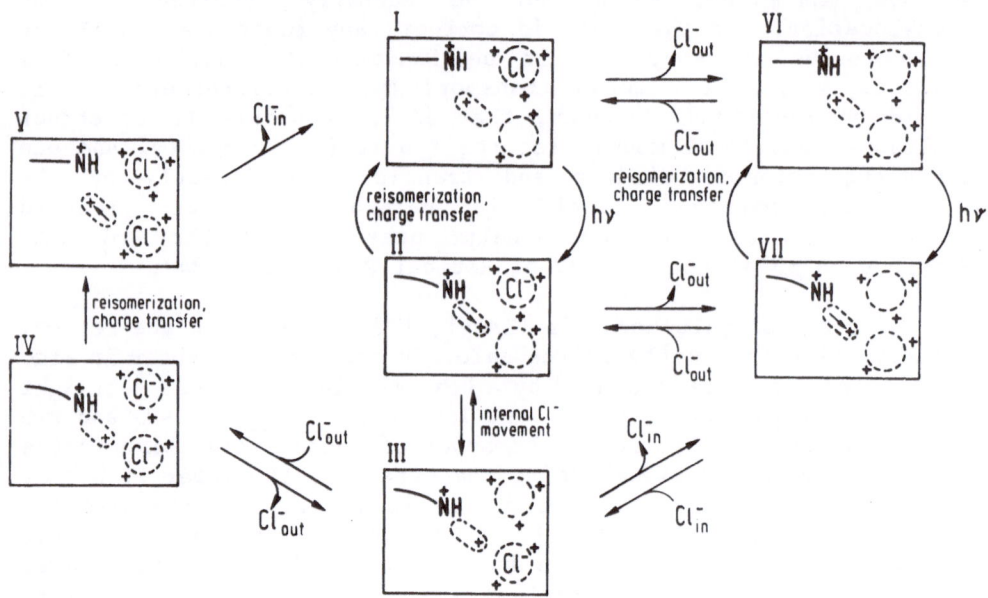

Figure 1. A model for chloride translocation by halorhodopsin. Details are explained in the text. Reproduced with permission from Lanyi (1984b).

REFERENCES

Ariki, M. and Lanyi, J.K. (1984) J.Biol.Chem. <u>259</u>, 3504-3510.

Bamberg, E., Hegemann, P. and Oesterhelt, D. (1984a) Biochim.Biophys. Acta (in press)

Bamberg, E., Hegemann, P. and Oesterhelt, D. (1984b) Biochemistry (in press)

Bogomolni, R.A. and Spudich, J.L. (1982) Proc.Nat.Acad.Sci.U.S.A. <u>79</u>, 6250-6254.

Bogomolni, R.A., Stubbs, L. and Lanyi, J.K. (1978) Biochemistry
    17, 1037-1041.

Bogomolni, R.A., Taylor, M.E. and Stoeckenius, W. (1984)
    Proc.Nat. Acad.Sci.U.S.A. (in press)

Cheah, K.S. (1969) Biochim.Biophys.Acta 180, 320-333.

Druckmann, S., Ottolenghi, M., Pande, J. and Callender, R.H.
    (1982) Biochemistry 21, 4953-4959.

Eisenbach, M., Cooper, S., Garty, H., Johnstone, R.M.,
    Rottenberg, H. and Caplan, S.R. (1977) Biochim.Biophys.Acta
    465, 599-613.

Falke, J.J., Chan, S.I., Steiner, M., Oesterhelt, D., Towner,
    P. and Lanyi, J.K. (1984) J.Biol.Chem. 259, 2185-2189.

Greene, R.V. and Lanyi, J.K. (1979) J.Biol.Chem. 254,
    10986-10994.

Greene, R.V., MacDonald, R.E. and Perreault, A.J. (1980) J.Biol.
    Chem. 255, 3254-3257.

Hegemann, P., Steiner, M. and Oesterhelt, D. (1982) EMBO J.
    1, 1177-1183.

Hess, B. and Kuschmitz, D. (1979) FEBS Lett. 100, 334-340.

Kuschmitz, D. and Hess, B. (1982) FEBS Lett. 138, 137-140.

Kushner, D.J. (1978) in Microbial Life in Extreme Environments
    (ed. D.J. Kushner), pp. 317-368, Academic Press, London

Lanyi, J.K. (1968) Arch.Biochem.Biophys. 128, 716-724.

Lanyi, J.K. (1974) Bacteriol. Rev. 38, 272-290.

Lanyi, J.K. and Schobert, B. (1985) J.Biol.Chem. (in press)

Lanyi, J.K. (1984a) Biochem.Biophys.Res.Comm. 122, 91-96.

Lanyi, J.K. (1984b) FEBS Lett. (in press)

Lanyi, J.K. and MacDonald, R.E. (1979) Methods Enzymol. 56,
    398-407.

Lanyi, J.K. and Oesterhelt, D. (1982) J.Biol.Chem. <u>257</u>, 2674-2677.

Lanyi, J.K. and Schobert, B. (1983) Biochemistry <u>22</u>, 2763-2769.

Lanyi, J.K. and Weber, H.J. (1980) J.Biol.Chem. <u>255</u>, 243-250.

Larsen, H. (1967) Adv.Microbial Physiol. <u>1</u>, 97-132.

Lindley, E.V. and MacDonald, R.E. (1979) Biochem.Biophys.Res.Comm. <u>88</u>, 491-499.

Luisi, B.F., Lanyi, J.K. and Weber, H.J. (1980) FEBS Lett. <u>117</u>, 354-358.

MacDonald, R.E., Greene, R.V., Clark, R.D. and Lindley, R.V. (1979) J.Biol.Chem. <u>254</u>, 11831-11838.

MacDonald, R.E. and Lanyi, J.K. (1975) Biochemistry <u>14</u>, 2882-2889.

Oesterhelt, D., Schuhmann, L. and Gruber, H. (1974) FEBS Lett. <u>44</u>, 257-261.

Oesterhelt, D. and Schuhmann, L. (1974) FEBS Lett. <u>44</u>, 262-265.

Ogurusu, T., Maeda, A. and Yoshizawa, T. (1984) J.Biochem. (Tokyo) <u>95</u>, 1073-1082.

Schobert, B. and Lanyi, J.K. (1982) J.Biol.Chem. <u>257</u>, 10306-10313.

Schobert, B., Lanyi, J.K. and Cragoe, E.J., Jr. (1983) J.Biol. Chem. <u>258</u>, 15158-15164.

Selwyn, M.J., Dawson, A.P., Stockdale, M. and Gains, N. (1970) Eur.J.Biochem. <u>14</u>, 120-126.

Spudich, J.L. and Bogomolni, R.A. (1983) Biophys.J. <u>43</u>, 243-246.

Steiner, M. and Oesterhelt, D. (1983) EMBO J. <u>2</u>, 1379-1385.

Sugiyama, Y. and Mukohata, Y. (1984) J.Biochem. (Tokyo) <u>96</u>, 413-420.

Steiner, M., Oesterhelt, D., Ariki, M. and Lanyi, J.K. (1984)
J.Biol. Chem. 259, 2179-2184.

Stoeckenius, W., Lozier, R.H. and Bogomolni, R.A. (1979) Biochim.
Biophys.Acta 505, 215-278.

Taylor, M.E., Bogomolni, R.A. and Weber, H.J. (1983)
Proc.Nat.Acad. Sci.U.S.A. 80, 6172-6176.

Tsuda, M., Hazemoto, N., Kondo, M., Kobatake, Y. and Terayama,
Y. (1982) Biochem.Biophys.Res.Comm. 108, 970-976.

Wagner, G., Oesterhelt, D., Krippahl, G. and Lanyi, J.K. (1981)
FEBS Lett. 131, 341 345.

Weber, H.J. and Bogomolni, R.A. (1981) Photochem.Photobiol.
33, 601-608.

THE DETERMINATION OF THE ELECTROCHEMICAL POTENTIAL DIFFERENCE

OF PROTONS IN BACTERIAL CHROMATOPHORES

Rita Casadio, Giovanni Venturoli and B. Andrea Melandri

Institute of Botany- University of Bologna
Via Irnerio 42
40126 Bologna, Italy

INTRODUCTION

In cells and in subcellular vesicles, a thermodynamic description of the energy-driven metabolic reactions is based on the evaluation of the free energy change which results from the vectorial proton movements across the coupling membrane. According to the chemiosmotic hypothesis for energy coupling introduced by P.Mitchell(1), the bulk to bulk phase electrochemical potential difference of protons ($\Delta\tilde{\mu}_{H^+}$) is the driving force for ATP synthesis and active transport in all energy conserving membranes. In the last decade, many experiments were performed in order to test the validity of the chemiosmotic hypothesis in energy conserving systems such as bacteria, mitochondria, chloroplast and bacterial chromatophores. A common feature of these studies was to determine the extent of $\Delta\tilde{\mu}_{H^+}$ sustained by substrate oxidation or light-driven electron flow, and to compare it to thermodynamic and kinetic parameters of the phosphorylation reaction and of metabolite transport across the coupling membrane. Many efforts were, therefore, done in order to find and develop methods and techniques suited to monitor $\Delta\tilde{\mu}_{H^+}$, especially in those vesicular systems, where a direct evaluation was prevented by the smallness of the internal lumen.

$\Delta\tilde{\mu}_{H^+}$ generated by the vectorial proton transport consists of two components: one due to charge separation across the coupling membrane or membrane potential difference ($\Delta\psi$), and one due to the proton concentration difference between the inner and outer compartment(ΔpH). Measurements of $\Delta\tilde{\mu}_{H^+}$ imply, therefore, an evaluation of both $\Delta\psi$ and ΔpH according to the equation:

$$\Delta \tilde{\mu}_{H^+} / F = \Delta \psi - 2.3RT/F(\Delta pH) \qquad (mV) \qquad Eqn.1$$

where R, T and F are the gas constant, the absolute temperature and the Faraday constant, respectively. If $\Delta \tilde{\mu}_{H^+}$ is considered as the transmembrane activity difference of protons between the inner and the outer compartment, the sign of $\Delta \tilde{\mu}_{H^+}$ depends on the orientation of the vectorial proton movements across the membrane, being positive when it is inwardly and negative when it is outwardly oriented, with respect to the internal volume.

A detailed description of the variety of the methods and techniques used to determine $\Delta \psi$ and ΔpH can be found elsewhere (2, 3, 4). As a general remark, however, it must be noticed that in spite of the large amount of data, the evaluation of both $\Delta \psi$ and ΔpH is still a matter of debate, since different methods gave different results, even when applied to the same vesicle system. This was obviuosly misleading when the results obtained were compared , for example, with the thermodynamic and kinetic parameters of the phosphorylation reaction, and interpreted in terms of the chemiosmotic hypothesis.

A general method for evaluating both components of $\Delta \tilde{\mu}_{H^+}$ relies on the changes of the equilibrium distribution of a suitable probe following either the establishment of $\Delta \psi$ and of ΔpH. The underlying assumption of this approach is that a true electrochemical equilibrium distribution of the probe is established, across the coupling membrane, in the steady state conditions for ion fluxes . Ions and weak amines or acids are used to detect $\Delta \psi$ and ΔpH respectively. The choice of the probe depends on the polarity of the vesicle system in respect to proton flow. Probes which accumulate into the inner compartment are generally preferred to maximize the observed change in the equilibrium distribution, given the low internal to external volume ratio(10^{-2};10^{-3}). Moreover the response of the probe must be fast in order to rapidly reequilibrate, following the establishment of ΔpH and $\Delta \psi$. In this respect, lipophilic probes are faster than hydrophilic ones; however, on the contrary of the formers, they are reported to affect the membrane permeability, being largely sequestered within the phospholipid bilayer. They may cause, therefore, uncoupling effects even at micromolar concentrations(5). Several techniques, including flow dialysis and ion-selective liquid membrane microelectrodes have been developed to follow changes of the equilibrium distribution of the probe during energy transducing events in the coupling membrane.

Dye-probes offer the advantage that changes in response to $\Delta \tilde{\mu}_{H^+}$ can be spectroscopically detected, often increasing the time resolution of the observed phenomena. Among the

410

spectroscopic probes, some are naturally occurring. It was indeed observed that in photosynthetic membranes, some endogenous pigments, such as carotenoids and chlorophylls, behave as molecular voltmeters, undergoing absorption changes that can be related to electrogenic events(6). Bacterial chromatophores have provided, so far, a well defined system suited to investigate the energy coupling processes. In these inside-out vesicles, obtained by mechanical rupture of photosynthetic cells, $\Delta\psi$ was determined from the ion distribution method (utilizing different ion probes, such as $T\phi B^-$,$^{14}[C]SCN^-$, OCC^-, ANS^-, $OX-VI$) and from the red shift of the endogenous carotenoid band. ΔpH was also measured with the ion distribution method using weak amines, such as $^{14}[C]$methylamine and the fluorescent probe 9-aminoacridine(for review, see Ref.2). Published $\Delta\tilde{\mu}_{H^+}$ values range, however, from 400 mV in chromatophores of Rps.capsulata(7) to about 130 mV in chromatophores of Rps.rubrum(8), depending on the methods applied and the experimental conditions. The highest value of $\Delta\tilde{\mu}_{H^+}$ was obtained when $\Delta\psi$ was evaluated from the red shift of the carotenoid band and ΔpH from the fluorescence quenching of 9-aminoacridine. The use of these spectroscopic methods was challenged purely on the basis of the different results obtained utilizing hydrophilic probes, the equilibrium distribution of which was detected mainly by means of flow dialysis. Each of the methods utilized is, however, based upon different assumptions, the validity of which must be tested in order to evaluate the reported values. We will focus therefore, on the present knowledge on ΔpH and $\Delta\psi$ measurements in bacterial chromatophores by spectroscopic techniques, and review the our latest results, which in our opinion add to the validity of the experimental data previuosly published.

THE EVALUATION OF ΔpH

Measurements with fluorescent amines

The proton concentration difference has been estimated in bacterial chromatophores, from the change in the equilibrium distribution of different amines(A), which are supposed to permeate the membrane in neutral form and to distribute between the inner and the outer compartment according to the pH difference. For $pHout > pHin \gg pKa$, the approximation holds:

$$\Delta pH = pHin - pHout \simeq \log[A]in/[A]out \qquad \text{(Eqn.2)}$$

The model, originally proposed by Schuldiner et al.(9), assumes that 9-aminoacridine, a fluorescent monoamine with pKa=10.0, is redistributed across the membrane in response to a proton activity difference and distributes according to ΔpH, as a result of diffusion and protonation coupled equilibria(Fig.1). When this fluorescent amine is used, it is

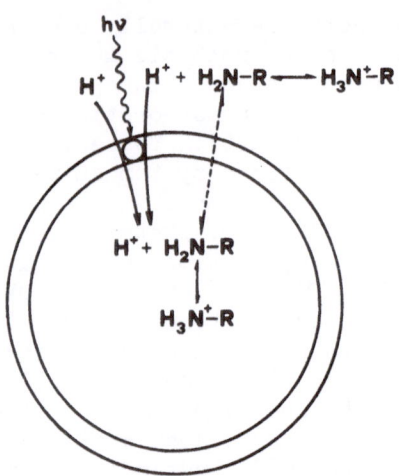

Fig. 1.   The coupling of protonation and diffusional equilibria
          for distribution of a weak amine in response to a pH
          difference between outer and inner compartment.   The
          R group can be different, according to the technique
          employed: an isotope-labeled aliphatic group, in flow
          dialysis; a fluorescent chromophore or a nitroxide
          label for ESR spectroscopy.

believed that the fluorescence detected after equilibration with
the protonic gradient, is that due to the probe remaining in the
outer phase of the vesicle.   In this case, the extent of ΔpH can
be mathematically related to the quenching (Q) of fluorescence
that follows energization, according to:

$$\log [A] \, in/[A] \, out = \log Q/(100-Q) + \log Vout/Vin \quad (Eqn.3)$$

where Vout and Vin are the external and internal aqueous
compartments, respectively.   This relation holds, provided that
the intensity of fluorescence is linearly related to the free
amine of the outer phase, and that the amine behaves ideally.

We, recently, reinvestigate the reliability of the use of
9-aminoacridine fluorescence quenching for measurements of Δ pH,
by using different approaches(10).   According to Eqn.3,
logQ/(100-Q) should be a linear function of the transmembrane proton
concentration difference, with an ideal slope of one.   Moreover,
the value of log Vout/Vin can be extrapolated from the curve at
Δ pH=0, and the apparent inner volume of the acidic compartment
can be estimated therefrom.   When artificial pH's of known extent
are imposed to bacterial chromatophores, results as those shown
in Fig.2, are obtained.   The relation between logQ/(100-Q) is
clearly linear with the ideal slope of one up to pH's as high as

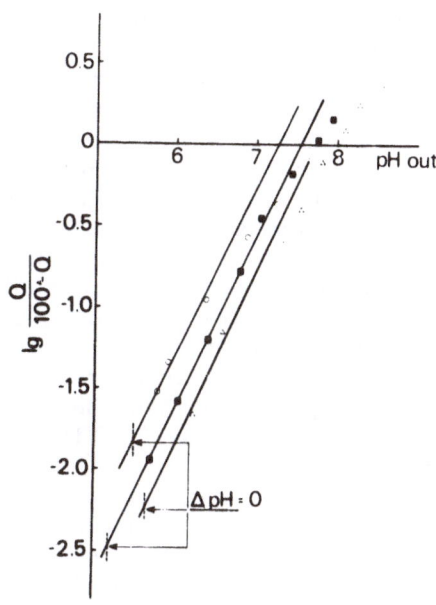

Fig. 2.   Calibration of the transitory quenching of
9-aminoacridine fluorescence, induced by an
artificially imposed Δ pH.
Δ pH's were artificially imposed by adding 5-30 μl of
1M NaOH to a fluorescence cuvette, equipped with
magnetic stirring and containing chromatophores
previously equilibrated at pH 5.5 against a 200mM
succinate-tricine buffer. Bchl and 9-aminoacridine
concentrations were 30 and 5 μM, respectively. KCl
concentration was 50 mM in all the experiments.
(O) in the presence of 5mM $MgCl_2$; (■) in the
presence of 5mM $MgCl_2$ and 0.95mM sucrose; (Δ) in
the presence of 100mM $MgCl_2$.
(Reproduced from Ref.10)

2-2.5 pH units. An apparent inner internal volume of about
0.7 ml/mg Bchl was obtained, when the ionic strength of the medium
was in the range of 0.1M, as detailed in the legend of Fig.2.
The presence of 5mM $MgCl_2$ and 50mM KCl is, in principle,
sufficient to buffer the negative charges distributed on the
external surface of the chromatophores (the surface charge
distribution was evaluated to be about -2×10^{-3} elementary
charges/$Å^2$ with pK=6.5, (11)), and to prevent possible dye
stacking phenomena on the membrane. The extrapolated Vin value
markedly decreases at increasing $MgCl_2$ concentrations, as it shown
also in Fig.2; the change observed, however, could not be
entirely attributed to osmotic shrinkage, since it was much more

413

reduced by increasing the $MgCl_2$ concentration than by adding a correspondent amount of sucrose. It is also worth to mention that the observed linearity and the inner volume value obtained for the chromatophore population, are independent of the absolute value of the inner pH. The Vin value, obtained from this method, is much higher, however, than that measured utilizing, for example, isotopic techniques(in the range of 0.1ml/mgBchl), and suggests that deviations from the ideal behaviour, expected in the proposed model, occur. These results can be explained by considering that the probe activity coefficient is lower in the inner than in the outer compartment, causing an accumulation of the probe larger than that expected in the ideal case. Obviously, this effect results in an overestimation of ΔpH as high as 1.3-1.5 units, especially when the inner concentration of the probe is calculated by assuming the Vin value obtained from the isotope distribution method. It is therefore suggested that correction for the non ideal volume can be made on the basis of the empirical calibration curve, which allows an estimation of the dye non-ideal accumulation, under the different experimental conditions used. The fluorescence quenching is, however, unanbiguously related to the energization of the chromatophore membranes.

The fluorescence intensity of 9-aminoacridine in buffer at pH 8 is highly reduced upon addition of chromatophores to the assay medium(Fig.3); it is possible, by correcting for inner filter effects due to absorption of the chromatophore pigments, to account for this reduction. A large decrease of the fluorescence yield is noticed when chromatophores are energized with infrared light; most noticeably, this phenomenon is prevented or abolished when uncoupling agents are added before or during steady state illumination, respectively. The parallel or crossed polarized fluorescence spectrum is also independent of the energization state of the added chromatophores, indicating that in both instances, 9-aminoacridine behaves as a freely rotating fluorescence emitter. These spectral observations are consistent with the notion that most of the fluorescence intensity, observed in all conditions, is only due to the probe in the outside aqueous buffer.

The probe response to a transmembrane ΔpH, as time resolved in stopped-flow experiments, occurs with a halftime of about 100ms at 25°C and in the presence of 5mM $MgCl_2$ and 50 mM KCl. When a fixed ΔpH is applied, the fluorescence is observed to decrease with a first order kinetics fitting the whole transition. The time constant of the fluorescence quenching is markedly dependent on the temperature; the activation energy between 5 and 30°C is constant, unaffected by increasing the ionic strenght and equal to about 17 Kcal/mole. This value is consistent with that for the diffusion of a hydrophobic molecule

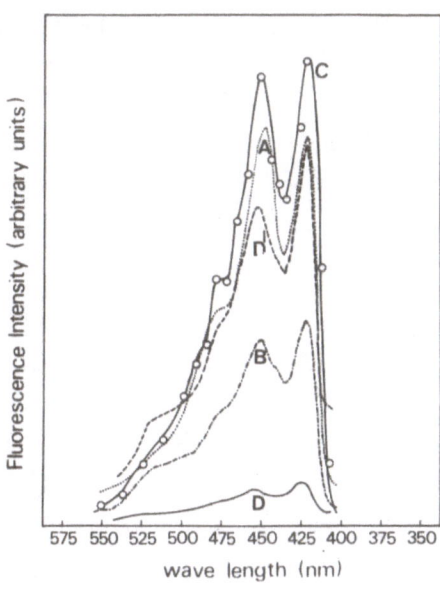

Fig. 3. Fluorescence emission spectra of 9-aminoacridine under different experimental conditions.
Spectrum A): 5μM 9-aminoacridine in 20mM glycylglycine buffer, pH 8.0, 5 mM $MgCl_2$, 50mM KCl and 0.2mM Na succinate; final volume 2ml. Spectrum B): same as A), but after addition of chromatophores corresponding to 27μg of Bchl. Spectrum C): same as B), but after correction for internal filter absorption by the chromatophore pigments. Spectra D and D'): same as B), but after 3 min of irradiation with infrared light (the sensitivity of the instrument was increased 10 folds for spectrum D). Spectrum B was recorded after addition of 2 uM nigericin, to prevent actinic effects of the exciting beam.
(Reproduced from Ref.10).

in a lipid lattice, and further supports the transmembrane translocation of 9-aminoacridine in response to a ΔpH. Following Schuldiner's model, the concentration of the unprotonated form of the fluorescent amine in the outer phase should limit the diffusion of the molecule through the bilayer and therefore the onset of the quenching. An increase of the final pH of the acid-base transition from 7.6 to 9.4, which raises the concentration of the uncharged amine of about 40 folds, brings about only a 11-fold increase in the rate constants, leaving, however, the activation energy practically unaffected. These results are also indicative of a deviation of the probe from the ideal behaviour.

In conclusion, although binding of the probe may occur in the inner volume of the chromatophore membrane, probably due to electrostatic effects and dye aggregation at high concentrations, the fluorescence quenching of 9-aminoacridine, when properly calibrated, can be utilized with a good degree of confidence to determine ΔpH in bacterial chromatophores.

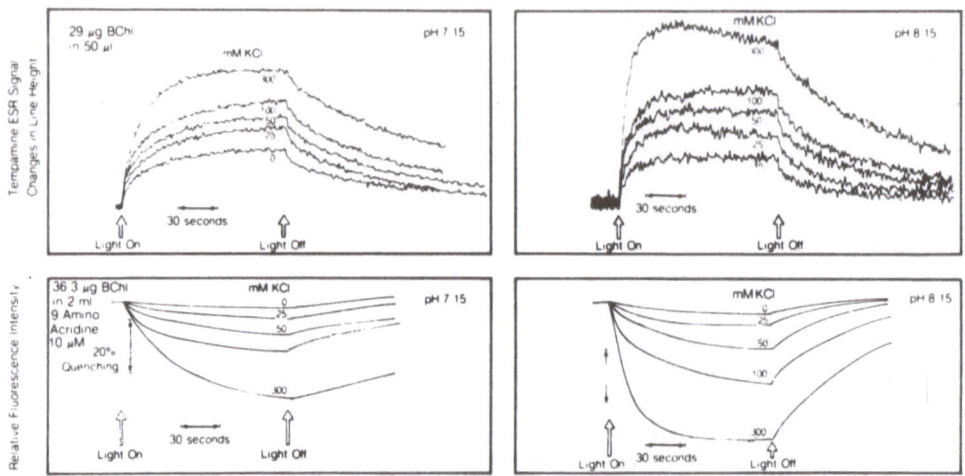

Fig. 4.   Comparison of the response to light-induced pH gradients of 9-aminoacridine fluorescence and tempamine ESR signal heights in chromatophores.
Assay conditions: 20mM Pipes-Tris, 5mM MgCl$_2$, 0.2mM Na succinate, 50μM Tempamine, 100mM $\left[(CH_3)_4N\right]_2$Mn-EDTA, 0.5μM valinomycin, KCl as indicated.
(Reproduced from Ref.13).

Measurements with ESR-labeled probes

The rationale behind the use of nitroxide labeled amine, detectable with ESR spectroscopy, can be also understood from Fig.1(12). Tempamine, (4-amino-2,2,6,6-tetramethyl-piperidine-N-oxyl, pKa=8.8, the probe most commonly utilized in this type of measurements), will distribute again according to Eqn.3, in response to a   pH. In this case, however, the spectral signal is detected only from the probe residing in the inner compartment, since the signal from the outside probe is totally eliminated by the addition of large concentrations of line-broadening agents impermeable to the inner volume. This experimental approach offers the clear advantage of giving direct information on the spectral characteristics of the inside probe; any strong interaction affecting its activity and its distribution will be quite likely detectable from distortions of the ESR spectral profile.

The response of the tempamine signal (50 μM) in the presence of bacterial chromatophores and of 100mM $[(CH_3)_4 N]_2$MnEDTA, added as external line-broadener, is shown in Fig.4 (13).The measurements were performed at two different pH's , in the presence of valinomycin and increasing concentrations of K^+(from 0 to 300mM); this latter addition establishes an increasing high Δ pH during steady state illumination. The values of Δ pH can be evaluated from Eqn.3, provided that the volume in which tempamine

Table 1. Comparison of ΔpH in illuminated chromatophores of Rhodopseudomonas sphaeroides, Ga, as measured with the spin-labeled amine tempamine and the fluorescent amine 9-aminoacridine.

Additions	External pH= 7.15		External pH= 8.15	
	9-amino acridine	Tempamine	9-amino acridine	Tempamine
none	1.01	1.35	1.07	1.16
25mM KCl	1.36	1.53	1.39	1.38
50mM KCl	1.60	1.53	1.70	1.62
100mM KCl	1.79	1.74	2.04	1.74
300mM KCl	2.12	2.01	2.36	2.16

Assay conditions as shown in the legend of Fig.4.
The Δ pH for tempamine measurements was computed using the osmotic volume determined with tempone; for 9-aminoacridine an internal volume of 253μg/mgBchl, as evaluated from the empirical calibration curves, was utilized.

is trapped and is inacessible to line-broadening, is determined under identical conditions. This can be easily done in parallel measurements, by substituting tempamine with another membrane permeable and non-protonizable probe like tempone (4-oxo-2,2,6,6-tetramethyl-piperidine-N-oxyl). Again, the signal, in the presence of the line-broadening agent, will be derived only from the inaccessible space, and, if diffusion equilibrium of tempone holds, will be proportional to the internal volume. The volumes, obtained from tempone determination, are exceedingly small (about $3\mu l$/mg Bchl, in the presence of 100mM $[(CH_3)_4N]_2$MnEDTA). At lower osmolarity, the inner volume values are higher but never larger than $15\mu l$/mg Bchl. They are, therefore, much smaller than those evaluated from the empirical calibration of the 9-aminoacridine fluorescence quenching. Nevertheless, the Δ pH values, measured from tempamine and tempone distribution are comparable to those determined with 9-aminoacridine(Fig.4 and Table 1), and , if run in parallel and under identical conditions, practically coincident.

These observations, obtained with techniques completely independent of each other, can be considered as an accurate test for the reliability of both methods, provided that proper calibrations of the 9-aminoacridine fluorescence are made. They also indicate that, in the case of tempamine, the signal, detected by ESR spectroscopy, is due to a free probe present in the inner compartment as a thermodynamically ideal solute; this is also confirmed by the lack of any distortion in the ESR spectrum, as compared to that of a solution in aqueous buffers.

THE CAROTENOID ELECTROCHROMIC SHIFT AS A PROBE OF $\Delta\Psi$ IN

CHROMATOPHORES

Reversible absorption changes, interpretable as red shifts of the carotenoid spectrum, can be observed following illumination of chromatophore suspensions. The spectral response of carotenoids, induced either by continuos light or by short(ns or us) flashes, was very soon recognized as related to the high energy state of the membrane, on the basis of the sensitivity of the signal to inhibitors of electron transfer, to phosphorylation and to agents affecting the ionic conductivity of the membrane. The most widely accepted view is that the carotenoid spectral shift reflects a Stark effect, i.e. a perturbation of the molecular energy levels, caused by light-induced transmembrane electric fields. This idea, originally put foreward by Junge and Witt(6), was most convincingly sustained by the observation(14) that absorption changes, indistinguishable from the light-induced ones, could be promoted in the dark by transmembrane diffusion potentials, caused by a salt jump in the presence of a cation-specific ionophore. This approach has been extensively

utilized for the calibration of the light-induced absorption
changes. According to the basic theory of electrochromism, the
shift ($\Delta\lambda$) of the absorption maximum of the pigment molecule,
caused by an electric field E, is given by

$$\Delta\lambda = (\lambda^2/hc)(\Delta\mu E \cos\vartheta + 0.5\Delta\alpha E \cos\varphi) \qquad \text{(Eqn.4)}$$

where h is Plank's constant, c is the speed of light, $\Delta\mu$ the
difference between the permanent dipole moments in the ground and
excited states and $\Delta\alpha$ the difference between the polarizability
of ground and excited states, ϑ and φ the angles between the
electric field and $\Delta\mu$ or $\Delta\alpha$, respectively. On the basis of
this equation, a mainly linear relation is expected for molecules
with a permanent dipole and a quadratic response for polarizable
chromophores with no permanent dipole. A quadratic dependence
has been demonstrate at high field intensities for intrisic
carotenoids in blebs, subjected to externally applied voltage
pulses. In contrast, the response observed in chromatophores has
been found essentially linear. Since carotenoids have no
permanent dipole, it has been proposed that they sense, in situ,
a permanent field, which induces a high background polarization
of the chromophores, causing a pseudo-linear response to
superimposed electric fields. If this interpretation is correct,
it can be shown that the direction of the spectral shift(blue or
red) should be determined by the orientation of the applied
electric field with respect to the direction of the permanently
induced dipole moment. This expected behaviour has been
experimentally verified by inducing fields of opposite polarity
in the same membrane system or in membrane with opposite
sidedness. Moreover, a number of observations indicate that only
a small fraction of the total carotenoid pool in the membrane,
undergoes electrochromical absorption changes, and that the
field-sensing carotenoids in chromatophores are associated with
one of the two light-harvesting, pigment-protein complexes(LHII).
Further evidence supporting the electrochromic interpretation of
the carotenoid spectral shift has come from the analysis of
difference spectra of the light induced absorption changes. In
chromatophores, the ionophore sensitive difference spectrum in the
carotenoid region could be, indeed, accurately reconstructed from
the electrochromic difference spectra of isolated carotenoids
measured in microcapacitors(for review, see Ref.2, 4, 15,16).

Calibration of the light-induced band shift is generally
accomplished by inducing diffusion potentials of known extent
with K^+ pulses in the presence of valinomycin. Assuming that an
electrochemical equilibrium for K^+ is very rapidly reached after
the pulse, and that the distribution of K^+ is not affected by
other ion fluxes, the following relation between the induced
$\Delta\psi$ and the amount of K^+ added in the pulses, has been obtained

$$\Delta\Psi = (RT/F)\ln(\ \left[K\right]_{o}^{+} + \left[K\right]^{+}\text{added}\ )/(\left[K\right]_{o}^{+} + C\ \Delta\Psi/FVin)\quad (\text{Eqn.5})$$

where Vin is the inner volume, $\left[K\right]^{+}$added is the concentration of K^{+} added, $\left[K\right]_{o}^{+}$ the initial concentration of endogenous K^{+}present in the chromatophores before the addition, C is the charging capacitance of the membranes contained in the assay and R, T, F have the usual meaning. It has been indeed observed that, as predicted by this relation, the pattern of the calibration curve($\Delta\Psi$ vs. log K^{+}added) deviates from linearity only at low concentrations of external K^{+}added, especially when the initial internal concentration of K^{+}before the pulse is low; at higher concentrations of K^{+}added (> 5mM),$\Delta\Psi$ becomes linear with a slope that approaches the Nernst dependence. The slope of the calibration curve can be affected by parameters such as pH, ionophore concentration and ionic strenght of the medium(2). Special care should be, therefore, taken to maintain constant these parameters during calibration, in order to avoid surface potential asymmetries caused by the umbalanced ion distribution, following salt pulses.

    The use of the electrochromic carotenoid band shift as a "molecular voltmeter" monitoring transmembrane   , and its calibration against diffusion potentials, relies primarily on the assumption that the light-induced absorption changes are caused by delocalized $\Delta\Psi$ across the membrane and not by dipole fields localized near to the pigment complexes, associated with the responding chromophores. A detailed kinetic analysis of the carotenoid shift induced by single turnover flash activation of chromatophores, has shown that the onset of the electrochromic signal comprises three phases, characterized by different time constants and clearly attributable to distinct and localized charge separations, promoted by light-induced electron transfer processes(4). The idea that the primary electrogenic events monitored by the carotenoid band shift, are rapidly delocalized over the entire dielectric membrane bilayer, through ionic conductance in the aqueous phases, is mainly supported by accelerating effects of ionophores on the decay of the carotenoid signal, at a concentration equivalent to one molecule of ionophore per membrane vesicle(4).

    To further test the ability of carotenoids to sense delocalized $\Delta\Psi$, we recently investigated the response of their spectral shift to electrogenic events in phospholipid-enriched as compared to native chromatophores(17). It is possible, indeed, to enrich selectively these vesicle systems in their phospholipid content, by freezing and thawing in the presence of presonicated liposomes. Morphologycal analysis shows that the intramembrane protein complexes are diluted, within the phospholipid bilayer, after this procedure. The primary electrogenic events within the reaction center complexes result unaffected in this system, whereas the cyclic electron flow through the b-c$_1$ complex is

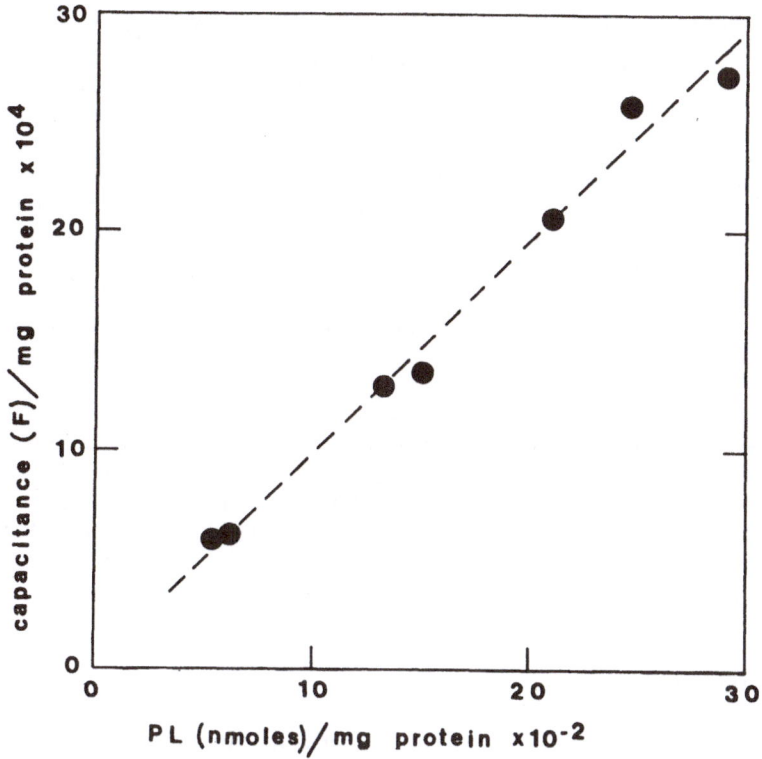

Fig. 5. Charging capacitance of the phospholipid enriched chromatophores as a function of the increasing surface area of the phospholipid bilayer. The charging capacitance ($C = Q/\Delta\psi$) was calculated from the membrane potential produced by the photooxidation of the reaction centers contained in the assay.

progressively impaired at increasing average phospholipid area per reaction center complex. The ionic permeability of the fused chromatophores is not substantially increased, at least up to a five-fold enrichment of the phospholipid content, as judged from the decay of the carotenoid signal induced by single turnover flashes. In the phospholipid-enriched chromatophores, spectral changes of the carotenoid band can also be detected by inducing diffusion potentials in the dark with K^{+}-valinomycin pulses, and, the slope of the calibration curves, relating absorption changes to the extent of $\Delta\psi$, are very similar when normalized per reaction center content of the sample. On the contrary, the $\Delta\psi$ values obtained upon light-induced charge separation within the reaction center complex(i.e. in the presence of antimycin, when only phase I plus phase II are detectable) are progressively reduced at increasing phospholipid content. The measured $\Delta\psi$, when normalized per photooxidizable reaction center, allows to estimate the charging capacitance(C) of the membrane bilayer. As it is shown in Fig.5, the C value, when expressed per mg of protein, is linearly related to the increasing surface area of the phospholipid bilayer. The absolute value of C, expressed per surface area, is about $1\mu F/cm^2$, namely a value comparable with that evaluated in black lipid planar membranes. The most reasonable interpretation of these data is that the carotenoids are responding to a charge separation which is rapidly delocalized over the membrane dielectric of the chromatophore vesicles.

As a general conclusion, therefore, the model proposed for interpreting the response of carotenoids to transmembrane fields, seems quite sound and in general agreement with most of the experimental results. The use of this technique for the evaluation of $\Delta\psi$ appears, therefore, to be totally justified, not only for its obvious kinetic advantages, but also for its sensitivity and accuracy in steady state measurements.

ACKNOWLEDGEMENTS

The fruitful collaboration of Dr. R. Melhorn and of Prof. L. Packer, University of California, Berkeley, in performing the ESR measurements is gratefully acknowledged.

This work was partially supported by Consiglio Nazionale delle Ricerche, grant no. 83.1997.04.

REFERENCES

1. P. Mitchell, Chemiosmotic Coupling in Oxidative and Photosynthetic Phosphorylation, Biol. Rev. Cambridge Phylos. Soc. 41:445 (1966).
2. A. Baccarini Melandri, R. Casadio and B. A. Melandri, Electron Transfer, Proton Translocation and ATP Synthesis in Bacterial Chromatophores, Curr. Top. Bioenerg. 12:197 (1982).
3. H. Rottenberg, The Measurements of Membrane Potential and pH in Cells, Organelles and Vesicles, Meth. Enzymol. 55:547 (1979).
4. W. Junge and B. J. Jackson, The Development of Electrochemical Potential Gradient across Photosynthetic Membranes, in: "Photosynthesis: Energy Conversion by Plants and Bacteria", Vol. 1, p. 589, Govindjee, ed., Academic Press, New York (1982).

5. R. Casadio, G. Venturoli and B. A. Melandri, Light-Induced
    Transmembrane Potential Difference in Chromatophores of
    Photosynthetic Bacteria: Measurements with a
    Ion-Selective Minielectrode, Photobiochem.
    Photobiophys. 2:245 (1981).
6. W. Junge and H. T. Witt, On the Ion Transport System of
    Photosynthesis. Investigation on a Molecular Level,
    Z. Naturforsch. 23B:244 (1966).
7. A. Baccarini Melandri, R. Casadio and B. A. Melandri,
    Thermodynamics and Kinetics of Photophosphorylation in
    Bacterial Chromatophores and their Relation with The
    Transmembrane Electrochemical Potential Difference of
    Protons, Eur. J. Biochem. 78:389 (1977).
8. D. B. Kell, P. John and S. J. Fergusson, Measurements by a
    Flow Dialysis Technique of the Steady State Proton
    Motive Force in Chromatophores from Rhodospirillum
    Rubrum. Comparison with Photophosphorylation Potential,
    Biochim. Biophys. Acta 502:111 (1978).
9. S. Schuldiner, H. Rottenberg and M. Avron, Determination
    of pH in Chloroplasts. Fluorescent Amines as a Probe
    for the Determination of Δ pH in Chloroplasts, Eur. J.
    Biochem. 25:64 (1972).
10. R. Casadio and B. A. Melandri, Calibration of the Response
    of 9-Aminoacridine Fluorescence to Transmembrane pH
    Differences in Bacterial Chromatophores,
    Archiv. Biochem. Biophys. in the press (1985)
11. K. Matsuura, K. Masamoto, I. Itoh and M. Nishimura,
    Effect of Surface Potential on the Transmembrane
    Electric Field Measured with Carotenoid Spectral Shift
    in Chromatophores from Rhodopseudomonas Sphaeroides,
    Biochim. Biophysic. Acta 547:91 (1979).
12. R. Melhorn, P. Candau and L. Packer, Measurements of
    Volumes and Electrochemical Gradients with Spin Probes
    in Membrane Vesicles, Meth. Enzyml. 88:751 (1981).
13. B. A. Melandri, R. Melhorn and L. Packer, Light-Induced
    Proton Gradient and Internal Volumes in Chromatophores
    of Rhodopseudomonas Sphaeroides, Archiv. Biochem.
    Biophys. 235:97 (1984).
14. J. B. Jackson and A. R. Crofts, The High Energy State in
    Chromatophores from Rhodopseudomonas Sphaeroides,
    FEBS Lett. 4:185 (1969).
15. C. A. Wraight, R. J. Cogdell and B. Chance, Ion Transport
    and Electrochemical Gradients in Photosynthetic
    Bacteria, in: "The Photosynthetic Bacteria", p.471,
    R. K. Clayton and W. R. Sistrom, eds., Plenum Press
    New York (1978).
16. W. A. Cramer and A. R. Crofts, Electron and Proton
    Transport, in: "Photosynthesis: Energy Conversion by
    Plants and Bacteria", Vol.1, p.387, Govindjee, ed.,
    Academic Press, New York (1982).

17. R. Casadio, G. Venturoli, A. Di Gioia, P. Castellani,
    L. Leonardi and B. A. Melandri, Phospholipid
    Enriched Chromatophores. A System Suited to Investigate
    Ubiquinone Mediated Interactions of Protein Complexes
    in Photosynthetic Oxidoreduction Processes, J. Biol.
    Chem. 259:9149 (1984).

KINETICS OF PROTON-TRANSPORT-COUPLED ATP SYNTHESIS

DRIVEN BY AN ARTIFICIALLY GENERATED ΔpH AND $\Delta\psi$

Ulrike Junesch, Gerlinda Thulke and
Peter Gräber

Max-Volmer-Institut für Biophysikalische und
Physikalische Chemie, Technische Universität
Berlin, Strasse des 17. Juni 135
1000 Berlin 12

INTRODUCTION

Coupling of vectorial transport phenomena and scalar chemical reactions occur at practically all biological membranes. Of special importance is the coupling between ion transport and ATP synthesis/ hydrolysis which is effected by membrane-bound enzymes, the ATPases. In photosynthetic vesicles of chloroplasts, this reaction is coupled with proton translocation across the thylakoid membrane and can be described by:

$$ADP + P_i + nH^+_{in} \rightleftharpoons ATP + nH^+_{out} \qquad (1)$$

where ADP, ATP and P_i (inorganic phosphate) stand for the total concentration of the respective compound, including all its ionization states that depend on Mg^{2+} and H^+ concentration; n is the number of protons which are translocated per ATP synthesized/hydrolyzed; H^+_{in}, H^+_{out} are the proton concentration inside and outside of the vesicle.

According to the chemiosmotic theory (1) it results quantitatively

$$\Delta G = -n(2.3\ RT\ \Delta pH + F\Delta\psi) + \Delta G_p \gtreqless 0 \qquad (2)$$

($\Delta G < 0$, ATP synthesis; $\Delta G > 0$, ATP hydrolysis, $\Delta G = 0$, equilibrium; $\Delta pH = pH_{out} - pH_{in}$; $\Delta\psi = \psi_{in} - \psi_{out}$, and ΔG_p, phosphate potential)

The aim of this work is a quantitative analysis of energetics and kinetics of ATP synthesis in chloroplasts and, thus, to obtain information about the mechanism of this reaction. For this purpose, the rate of ATP synthesis is measured, using artificially impressed ΔpH, $\Delta \psi$ and ΔG_p.

A transmembrane ΔpH can be generated by incubating chloroplasts in an acidic medium I (e.g., pH 4) for about 30 s (see scheme in Fig. 1, top). This leads to an equilibrium of the ion concentrations of the outer phase and the inner one. Injection of this acidic suspension into a strongly buffered basic medium II (e.g., pH 8) leads to an immediate neutralization at the outside; whereas, the ion concentration inside the vesicle remains identical to that in medium I for a short time. Then, the ions begin to move until both phases are in equilibrium again. The movement of the protons is connected with ATP synthesis (2).

A transmembrane $\Delta \psi$ can be generated either by an external electric field (3,4) or by generation of a diffusion potential (5). In both cases ATP synthesis has been observed. Fig. 1, bottom, shows a scheme for the generation of a diffusion potential across the thylakoid membrane. In this case the chloroplasts are incubated in a medium with a low KCl concentration. This suspension is injected into a solution with a high KCl concentration. If valinomycin is also present (valinomycin forms a lipid soluble complex with K^+ and thereby increases the K^+ permeability of the membrane), the K^+ permeates much faster than the other ions so that a diffusion potential is generated that is mainly due to the transmembrane K^+ imbalance.

Our experiments were performed in the ms time range. Therefore, ΔpH, $\Delta \psi$ and the concentrations of the substrates ADP and P_i remain constant at their initial values. Since the ATP concentration is measured, the phosphate-potential, ΔG_p, is also known.

MATERIALS AND METHODS

A transmembrane ΔpH and $\Delta \psi$ in the ms-time range was generated with a rapid mixing quenched flow apparatus (Durrum D 133). Fig. 2 shows a scheme of the experimental procedure: in syringe I the chloroplasts

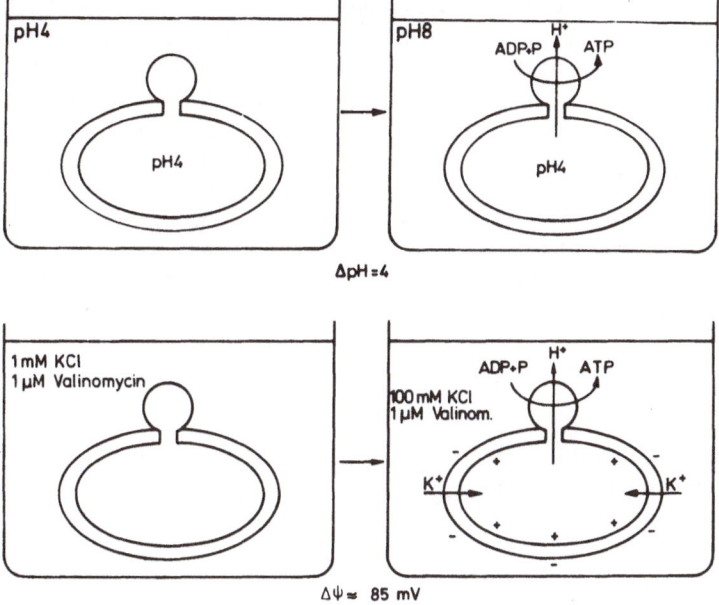

Fig. 1. Scheme of the procedure for artificial generation of a transmembrane proton concentration
difference, ΔpH (top), and of a transmembrane
electric potential difference, $\Delta\psi$ (bottom), in
a chloroplast suspension.

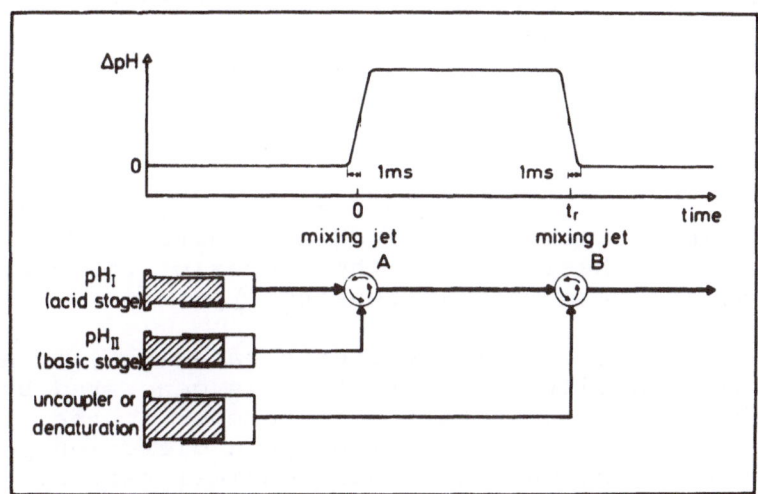

Fig. 2. Scheme of a rapid mixing quenched flow apparatus for rapid starting and stopping of proton
transport coupled ATP synthesis.

are incubated in the acidic medium I for 30 s (pH_I = 3.8 - 6.5). In mixing chamber A, this suspension is mixed with the basic solution II which is adjusted to give a final pH = 8.2. Thereby a transmembrane pH gradient is generated within the mixing time (about 2 ms), and the reaction is allowed to proceed for a preselected time. After this time, mixing with the quenching solution III (4% trichloroacetic acid, TCA) in mixing chamber B terminates the reaction. The composition of medium I was varied according to the desired ΔpH and $\Delta\psi$. For further details see ref. 6.

RESULTS

1. The Rate of ATP Synthesis as a Function of ΔpH and $\Delta\psi$ at constant pH_{out}

Fig. 3 shows the time course of ATP synthesis in an acid base transition experiment where ΔpH = 3.2 and additional K^+/valinomycin diffusion potentials of increasing magnitude (5; 44; 60 mV) were generated. The diffusion potentials were calculated from the Gold- mann-Hodgkin-Katz-equation as described in ref. 6. Up to a reaction time of about 200 ms a linear increase of the ATP yield was observed. Therefore, one can conclude that within this time ΔpH and $\Delta\psi$ remain practically constant within the accuracy of the measurements. The slopes of the curves, representing the rate of ATP synthesis, became smaller after this linear range and finally approached zero, since ΔpH and $\Delta\psi$ decayed completely. Fig. 3 shows clearly that ATP rate and ATP yield increase with increasing magnitude of the diffusion potential.

Fig. 4 shows the ATP yield as a function of the reaction time for different ΔpH at constant $\Delta\psi$. The rate of ATP synthesis, i.e., the slopes of the curves in Fig. 4, increase with increasing ΔpH.

Fig. 5 shows the ATP synthesis rate when, in addi- tion to ΔpH, a diffusion potential is generated by a transmembrane K^+ concentration difference in the presence of valinomycin. Obviously, a diffusion poten- tial displaces the curves at half maximal rate by approximately 0.7 (curve B) and 1.4 (curve C) ΔpH units to the lower values. The maximal rate 380 mM ATP/(M Chl · s) is for all three curves the same. Therefore, it represents the maximum turnover of the ATPase under these conditions.

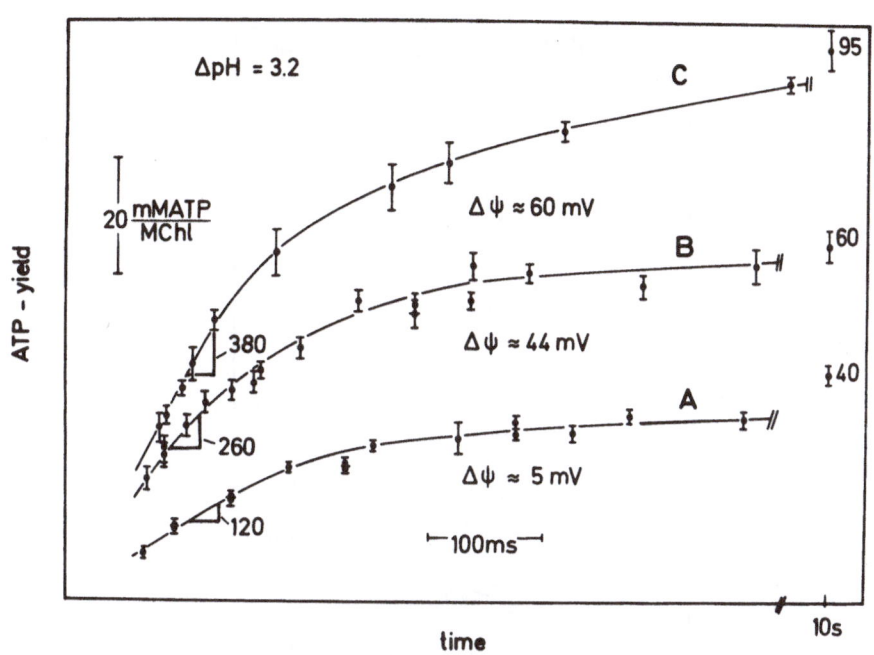

Fig. 3. ATP yield as a function of reaction time from quenched flow experiments at $\Delta pH = 3.2$, $pH_{out} = 8.20 \pm 0.05$, $Mg^{2+} = 2$ mM, $P_i = 5$ mM, $ADP = 100$ μM, with $\Delta\Psi \approx 5$ mV, $\Delta\Psi \approx 44$ mV and $\Delta\Psi \approx 60$ mV. For further details of reaction conditions see Materials and Methods and Ref. 6. The error bars represent the standard deviation from four determinations of the ATP concentration. The slopes are given in mM ATP/(M Chl·s).

429

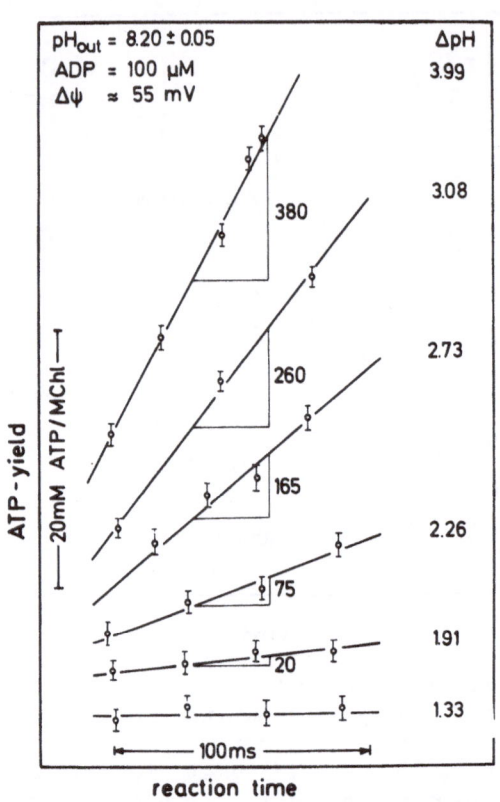

Fig. 4. ATP yield as a function of reaction time at
different Δ pH at constant $\Delta\Psi$ = 55 mV (Mg^{2+} =
2 mM, P$_i$ = 5 mM). For further details see
Ref. 6.[1] The slopes are given in mM ATP/(M Chl·s).

The number of ATPases per chlorophyll is 1:1085 (6). With this number the maximum turnover of the ATPase is k_{max} = 410 ATP/(ATPase · s). This corresponds to a time of about $\tau \approx$ 2.4 ms for the reaction cycle of the ATPase.

Also, in light-driven phosphorylation the maximal rate is of similar magnitude (7,8). This demonstrates that energization by artificially generated ΔpH and $\Delta\psi$ is equivalent to light-induced energization with regard to ATP synthesis.

The transmembrane difference of the electrochemical potential of protons, $\Delta\tilde{\mu}_H$+, can be calculated from ΔpH and $\Delta\psi$. If the data from Fig. 5 are plotted versus $\Delta\tilde{\mu}_H$+, one obtains only one curve, i.e., the rate of ATP synthesis is determined only by $\Delta\tilde{\mu}_H$+ under these conditions (6).

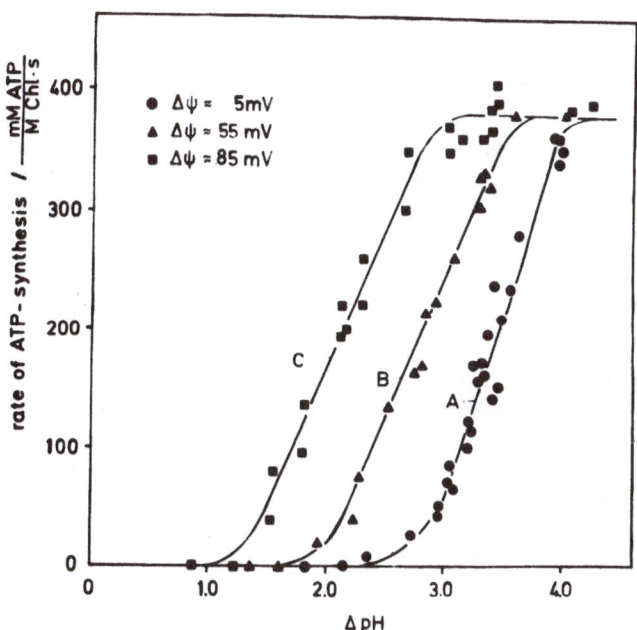

Fig. 5. Rate of ATP synthesis as a function of ΔpH with different diffusion potentials. Data from Figs. 3 and 4 and additional sets of experiments. pH_{out} = 8.2+0.05, Mg^{2+} = 2 mM, P_i = 5 mM, ADP = 100 µM.

2. The Rate of ATP Synthesis as a Function of P_i Concentration at Constant Δ pH

It has been proposed that in the catalytic cycle of the ATPase the energy input (ΔpH and $\Delta\Psi$) is used for the release of ATP and the binding of ADP and P_i (9). Therefore, the apparent Michaelis constants, K_M, for P_i and/or ADP might increase with decreasing ΔpH (10). In chloroplasts the membrane was energized by the light-driven electron transport and different ΔpH have been generated by (a) varying the light intensity, (b) addition of electron transport inhibitors, and (c) addition of uncouplers (11,12). It was found that the K_M-value for ADP decreases when the ΔpH is decreased by addition of an electron transport inhibitor or by decreasing the light intensity (11,12). If the ΔpH is decreased by addition of an uncoupler, the K_M-value increases (11) as reported earlier for submitochondria (10).

These investigations contain a systematic error: if at a constant light intensity the ADP concentration is changed from zero to about 100 µM, the ΔpH decreases by about 0.3-0.5 units, e.g., from ΔpH = 3.2 to 2.7 (13,14). The reason for this effect is that parallel to the basal proton efflux a phosphoylating proton efflux occurs which increases with increasing ADP concentration. This leads to a lowering of the steady state level of ΔpH. Therefore, the rate at low ADP concentration is measured at a higher ΔpH than that at high ADP concentration and a correction of this effect would decrease the rate at low ADP concentration considerably. For example, according to the results shown in Fig. 5, the rate of ATP synthesis decreases from about 130 mM ATP/(M Chl · s) at ΔpH = 3.2 to about 20 mM ATP/(M Chl · s) at ΔpH = 2.7 ($\Delta\Psi \approx$ 5 mV).

Using the technique described above, the rate of ATP synthesis was measured as a function of the P_i concentration at constant Δph, $\Delta\Psi$ and ADP concentration. Fig. 6 shows the rate of ATP synthesis at different P_i concentrations at Δph = 2.8 and $\Delta\Psi$ = 0. Fig. 7 shows a similar measurement at Δph = 2.8 with an additional diffusion potential $\Delta\Psi \approx$ 76 mV.

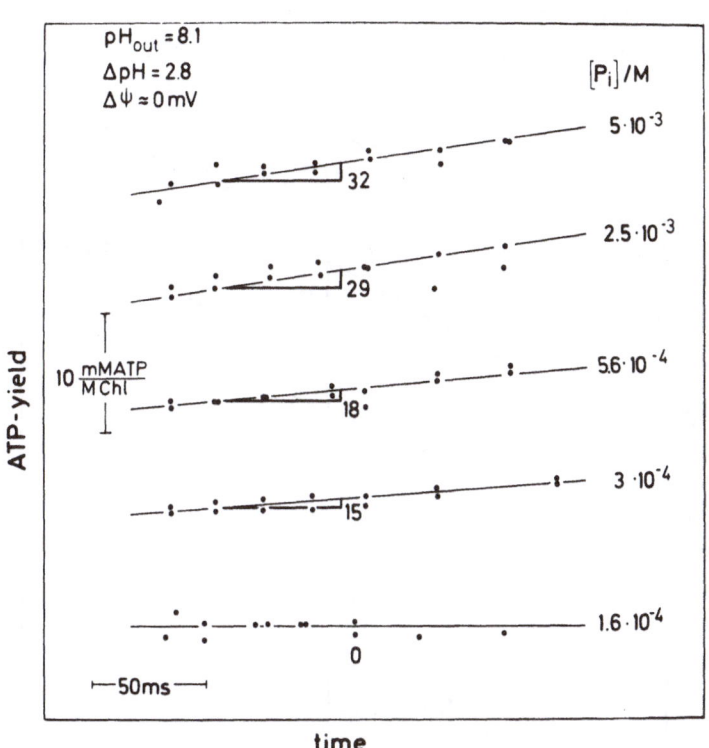

Fig. 6. ATP yield as a function of reaction time at different P_i concentrations (ΔpH = 2.8; $\Delta \Psi$ = 0; ADP = 100 μM, Mg^{2+} = 2 mM). The slopes are given in mM ATP/(M Chl·s).

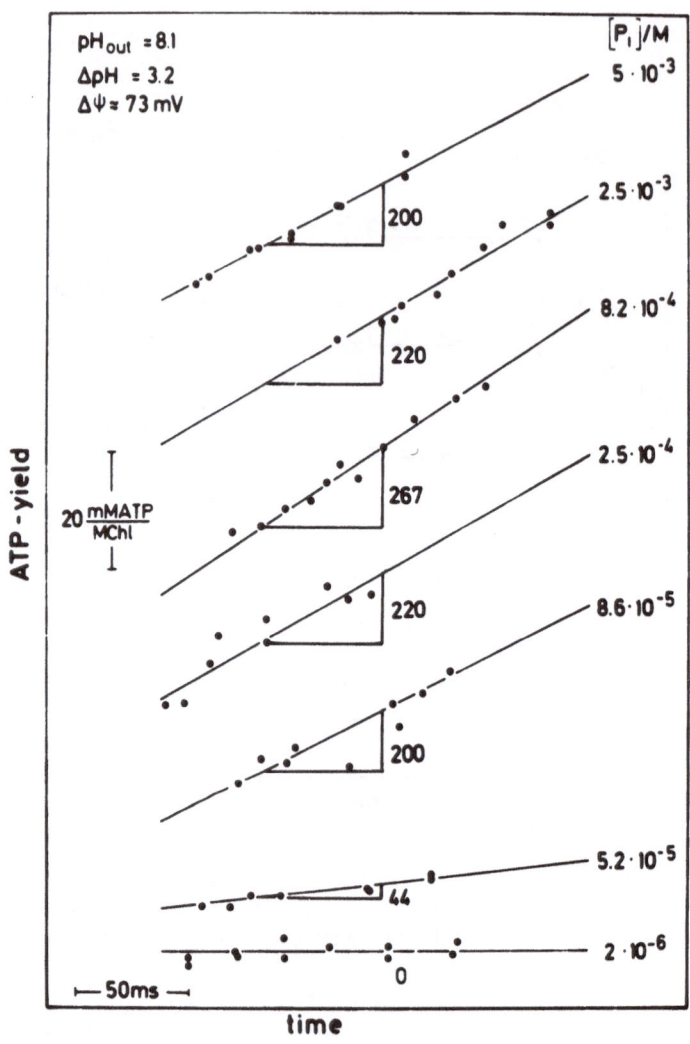

Fig. 7. ATP yield as a function of the reaction time at different P_i concentrations ($\Delta pH = 2.8$; $\Delta\Psi \approx 76$ mV, ADP = 100 µM, Mg^{2+} = 2 mM). The slopes are given in mM ATP/(M Chl·s).

434

Fig. 8 shows the rate of ATP synthesis as a function of the P_i concentration for $\Delta pH = 2.8$ with $\Delta\Psi = 0$ and $\Delta\Psi \approx 76$ mV. The curve at $\Delta\Psi = 0$ increases at low P_i in a sigmoidal way and reaches a saturation at high P_i. For $\Delta\Psi = 76$ mV there is first a rapid increase of the rate with increasing P_i to a maximum and then the rate decreases again.

Fig. 9 shows the rate of ATP synthesis as a function of the P_i concentration at a higher energization, i.e., $\Delta pH = 3.2$ with $\Delta\Psi = 0$ and $\Delta\Psi = 73$ mV. These curves have a very similar shape as those shown in Fig. 8, only the maximal rates are higher due to the higher energization.

It is evident from the results shown in Figs. 8 and 9 that the data cannot be described by a simple Michaelis-Menten kinetics. In table I different parameters describing the curves are listed.

Table I

Energization			max. Rate	$[P_i]$ for half-max. Rate
ΔpH	$\Delta\Psi$/mV	$\Delta\tilde{\mu}_H$+/mV	$V_{max}/\left(\dfrac{mM\ ATP}{M\ Chl\cdot s}\right)$	$P_i/\mu M$
2.8	0	165	29	300
3.2	0	189	75	250
2.8	76	241	170	120
3.2	73	262	275	100

It can be seen that with increasing energization the maximal rate increases and that the P_i concentration necessary for the half-maximal rate decreases. This result corresponds qualitatively with the observation that with increasing uncoupler concentration the K_M increases (10,11).

The decrease of the rate at high P_i concentration in the presence of a diffusion potential (Figs. 8 and 9) may be explained as follows: Whereas the transmembrane ΔpH is measured by glass electrodes, the diffusion potential is calculated from the Goldmann-Hodgkin-Katz-equation (6). This implies that the effect of bivalent

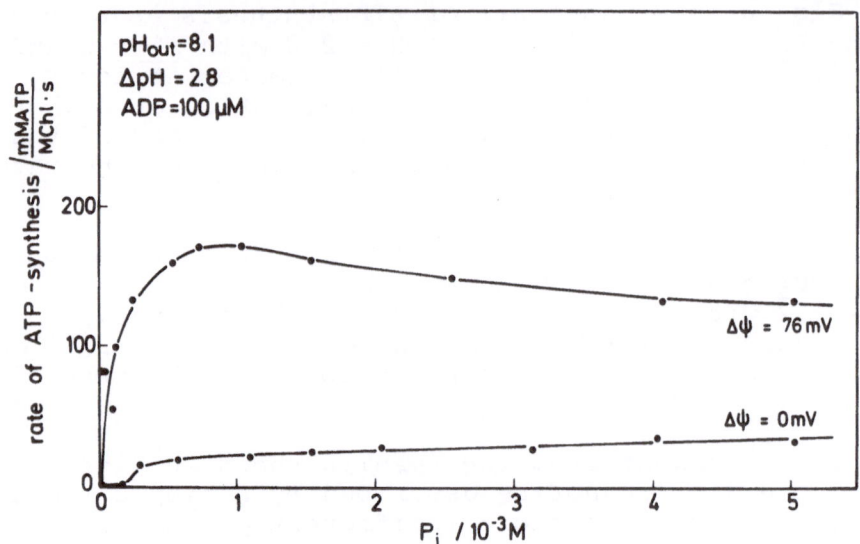

Fig. 8. Rate of ATP synthesis as a function of P_i concentration. ΔpH = 2.8 with $\Delta\Psi$= 0 and $\Delta\Psi$ = 76 mV. Data from Figs. 6 & 7 and additional sets of experiments

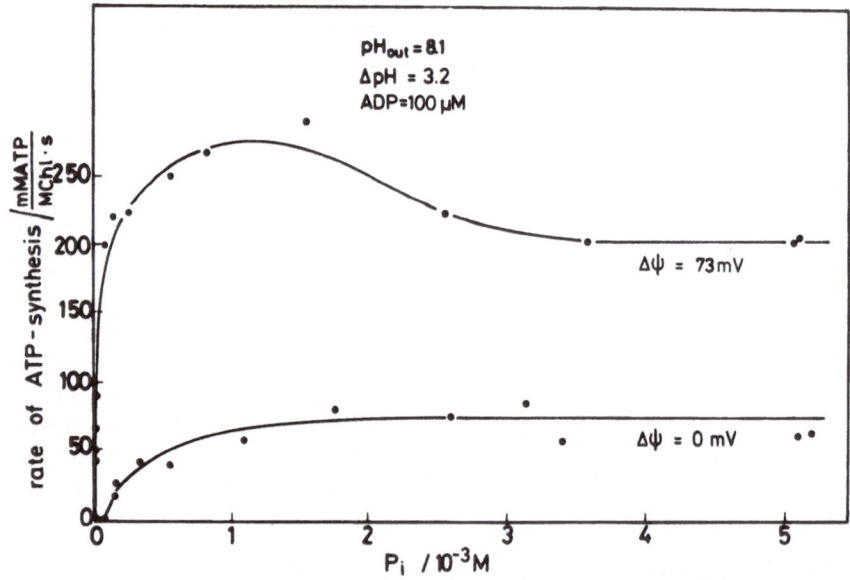

Fig. 9. Rate of ATP synthesis as a function of P_i concentration. ΔpH = 3.2 with $\Delta\Psi$= 0 and $\Delta\Psi$ = 73 mV.

436

ions is neglected. This approximation is correct at low P_i concentrations; however, at high P_i concentrations the calculated $\Delta\Psi$ is certainly higher than the true $\Delta\Psi$. In other words, increasing the P_i concentration may lead to a decrease of $\Delta\Psi$ and since the energization decreases also, a decrease of the rate should be observed.

These measurements show that with the described method useful data on the substrate dependence of the rate of ATP synthesis can be obtained. However, at the present state it is too early to draw final conclusions regarding the underlying enzymatic mechanism.

ACKNOWLEDGEMENT

This work was supported by grants from the Deutsche Forschungsgemeinschaft.

REFERENCES

1. P. Mitchell, Nature 191:144 (1961)
2. A.T. Jagendorf and E. Uribe, Proc. Natl. Acad. Sci. USA 55:170 (1966)
3. H.T. Witt, E. Schlodder, and P. Gräber, FEBS Lett. 69:272 (1976)
4. E. Schlodder, P. Gräber, and H.T. Witt, in: "Electron Transport and Phosphorylation", pp. 105-175, J. Barber, ed., Elsevier, Amsterdam (1982)
5. E. Uribe and B.C.Y. Li, Bioenergetics 4:435 (1973)
6. P. Gräber, U. Junesch, and G.H. Schatz, Ber. Bunsenges. Physik. Chemie 88:599 (1984)
7. M. Avron, Biochim. Biophys. Acta 40:257 (1960)
8. B. Rumberg and T. Heinze, Ber. Bunsenges. Physik. Chemie 84:1055 (1980)
9. P.D. Boyer, in: "Bionergetics of Membranes", C.P. Lee, ed., Addison-Wesley (1979)
10. C. Kayalar, J. Rosing, and P.D. Boyer, Biochem. Biophys. Res. Comm. 72:1153 (1976)
11. Ch. Winkler, Biochem. Biophys. Res. Comm. 99:1095 (1981)
12. S. Bickel-Sandkötter and H. Strotmann, FEBS Lett. 125:188 (1981)
13. B. Rumberg and U. Siggel, Naturwiss. 56:130 (1969)
14. U. Pick, H. Rottenberg, and M. Avron, FEBS Lett. 32:91 (1973)

LIGHT INDUCED GENERATION OF A PROTON MOTIVE FORCE AND Ca^{++}-TRANSPORT IN MEMBRANE VESICLES OF <u>STREPTOCOCCUS</u> <u>CREMORIS</u> FUSED WITH BACTERIORHODOPSIN PROTEOLIPOSOMES

Arnold J.M. Driessen, Klaas J. Hellingwerf and
Wil N. Konings

Department of Microbiology, University of Groningen
Kerklaan 30, 9751 NN Haren, The Netherlands

INTRODUCTION

The chemiosmotic hypothesis (Mitchell, 1966,1972) postulates that proton translocation by primary proton pumps generates an electrochemical proton gradient (Δp) across the cytoplasmic membrane. The electrochemical proton gradient or proton motive force (pmf) is composed of electrical and chemical parameters according to the equation:

$$\Delta p = \Delta \psi - 2.3 \; RT/F \; \Delta pH \quad (mV) \tag{1}$$

where $\Delta \psi$ represents the electrical potential and ΔpH the chemical potential of protons across the membrane (2.3 RT/F is equal to 58.8 mV at 25°C).

In bacteria, a pmf can be generated by proton translocation from the cytoplasm to the external medium coupled to (i) electron transfer in electron transfer chains (Konings and Michels, 1980), (ii) ATP hydrolysis via the membrane-bound Ca^{++}, Mg^{++} stimulated ATPase complex (Harold, 1977), (iii) efflux of metabolic endproducts (Otto et al., 1980; Ten Brink and Konings, 1980) and (iv) in Halobacteria to the light-induced proton pump bacteriorhodopsin (Bakker et al., 1976; Hellingwerf et al., 1979). These proton pumping processes convert redox, chemical or light energy into electrochemical energy and are termed primary transport processes.

It is well recognized that in intact bacteria or isolated cytoplasmic membrane vesicles the pmf is the main driving force and/or a regulator of a large number of different metabolic processes, such as ATP synthesis, motility and secondary transport of solutes across the cytoplasmic membrane (Kaback, 1982; Skula-chev, 1984; Hellingwerf and Konings, 1984). In addition to trans-port systems which catalyze the coupled movement of protons with solutes (H^+-solute symport or -antiport), transport systems have been described that catalyze Na^+-solute symport (Lanyi, 1979). In these cases, $\Delta\mu_{Na}+$ rather than the pmf is the driving force.

Strictly fermentative bacteria (for instance lactic acid streptococci) lack functional electron transfer chains (Ritchey and Seeley, 1976). As a consequence of their impairment in the syn-thesis of porphyrins, they are not able to synthesize cytochromes (Sijpensteijn, 1970). This excludes pmf generation across the cyto-plasmic membrane by electron transfer (Bragg, 1979). These organ-isms produce ATP via substrate level phosphorylation and generate a pmf by proton extrusion through the Ca^{++}, Mg^{++} stimulated membrane-bound ATPase (Harold et al., 1970; Maloney, 1982). Generation and maintenance of a pmf by ATP hydrolysis would consume a considerable fraction of the ATP formed by substrate level phosphorylation and this would greatly effect the amount of ATP available for biosyn-thetic purposes. To prevent this drain on ATP, these organisms have developed systems which mediate the electrogenic efflux of endpro-ducts across the cytoplasmic membrane in symport with protons (Otto et al., 1980, 1982; Ten Brink and Konings, 1980; Ten Brink, 1984). In this way the energy conserved in the chemical gradient of the endproducts can be converted into a pmf (Michels et al., 1979).

For the study of the properties of different pmf driven trans-port systems isolated cytoplasmic membrane vesicles are well suited (Kaback, 1974). The polarity of the membrane in these vesicles is the same as in intact bacteria. In the vesicular membrane energy transducing systems are present such as the Ca^{++}, Mg^{++} stimulated ATPase and the solute carrier proteins. All these systems have retained their functional properties. The vesicular membrane sepa-rates an inner compartment from the external medium and under appropriate conditions secondary transport of a large number of solutes can take place (Kaback, 1974; Konings and Michels, 1980). The primary transport systems can generate a pmf in the membrane vesicles. The addition of ionophores such as valinomycin and nigericin in the presence of K^+ makes it possible to vary the mag-nitude of the components of the pmf, the $\Delta\psi$ and the ΔpH.

In membrane vesicles of lactic acid streptococci it would in principle be possible to generate a pmf by ATP hydrolysis. However, in practice this is difficult to achieve since the cytoplasmic mem-brane is impermeable for ATP and thus ATP cannot be hydrolysed by the Ca^{++}, Mg^{++} stimulated ATPase located at the inner surface of the vesicular membrane. Since cytochrome-linked electron transport chains are absent electron flow also cannot be used to generate a pmf. In membrane vesicles of S. cremoris $\Delta\psi$ and ΔpH can therefore only be generated artificially. This can be achieved by imposing either an outwardly directed potassium diffusion gradient in the presence of valinomycin, which leads to the generation of $\Delta\psi$, interior negative, or by an outwardly directed acetate diffusion gradient, which leads to a ΔpH, interior alkaline. With these procedures the uptake of a number of amino acids could be energized in membrane vesicles of S. cremoris (A.J.M. Driessen, unpublished results; Otto et al., 1982). With these procedures, however, a $\Delta\psi$ or ΔpH can only be generated transiently which severely limits the information that can be obtained about the role of the pmf in solute transport. More detailed studies on the energetics of solute transport in membrane vesicles of streptococci are preferentially done in a modelsystem in which a pmf can be generated for a longer period of time.

Recently a method for the induction of fusion between liposomes and bacterial membrane vesicles was described (A.J.M. Driessen, D. Hoekstra, G. Scherphof and J. Wilschut, manuscript submitted for publication). In this study the pH dependent interaction between large unilamellar phospholipid vesicles (LUV) and membrane vesicles derived from Bacillus subtilis was investigated, with a fluorescent assay for mixing membrane phospholipids based on resonance energy transfer (RET) (Struck et al., 1981). Efficient interaction was observed when liposomes, containing negatively charged phospholipids were mixed with the bacterial membranes at low pH. The observed interaction was the result of membrane fusion since (i) the RET probes and a fluorescent phosphatidylcholine analog were diluted to similar extents into the bacterial membrane, (ii) the transfer of the RET probes and another non-exchangeable lipid marker, cholesteryloleate from the liposomes to the bacterial vesicles occurred simultaneously and proportionally, (iii) interaction products were formed with a intermediate buoyant density, and (iv) liposomes-encapsulated colloidal gold was transferred into the aqueous space of the bacterial vesicles as revealed by thin section electron microscopy. The demonstration that both the membrane phospholipids and the internal contents of the bacterial membrane vesicles and the liposomes were mixed virtually ruled out any other interaction than fusion. This fusion process was strictly dependent on a protein located at the outer surface of the bacterial membrane since it was extremely sensitive to proteolytic treatment of the bacterial membrane.

Low pH induced fusion of liposmes with membrane vesicles derived from Streptococcus cremoris can also take place. In order to incorporate a primary proton pump in membrane vesicles of S. cremoris these membrane vesicles were fused with liposomes containing the light-induced proton pump Bacteriorhodopsin (Brh). In this paper we demonstrate that S. cremoris membrane vesicles efficiently fuse with Brh proteoliposomes at low pH which leads to a functional incorporation of Brh into S. cremoris membrane vesicles. Part of this work has been published elsewhere (A.J.M.

Driessen, K.J. Hellingwerf and W.N. Konings, manuscript submitted for publication).

MATERIALS AND METHODS

Growth of Cells and Preparation of Membrane Vesicles

Streptococcus cremoris Wg2 (prt⁻) was obtained from the Dutch Institute of Dairy Research (Nederlands Instituut voor Zuivel Onderzoek, Ede, The Netherlands). S. cremoris was grown anaerobic- ally on MRS broth (Otto, 1981) at controlled pH of 6.4 in a 5-l fermenter. Membrane vesicles of S. cremoris were prepared as described previously (Otto et al., 1982), suspended in 50 mM potas- sium phosphate buffer pH 7.0 containing 10 mM $MgSO_4$ at a protein concentration of 10-15 mg/ml and stored in liquid nitrogen.

Halobacterium halobium NRL strain R1 (Stoeckenius and Kunau, 1968) was grown and purple membrane sheets were isolated as described by Danon and Stoeckenius (1974). The concentration of purple membranes was determined using a molar extinction coeffi- cient at 560 nm of 63,000 $M^{-1}cm^{-1}$ (Oesterhelt and Hess, 1973). Purple membranes were stored in liquid nitrogen at a protein con- centration of 12 mg/ml in distilled water.

Preparation of Bacteriorhodopsin Proteoliposomes

Brh proteoliposomes were prepared by co-sonication (Racker, 1973; Hellingwerf, 1979) of purple membrane sheets (4 mg/ml) and phospholipids (20 mg/ml) in 0.15 M KCl pH 6.0 or 50 mM potassium phosphate buffer pH 6.0, supplemented with 0.1 mM ethylenediamine- tetraacetate (EDTA). Sonication was performed with a 9 mm probe, at a frequency of 21 kHZ and an amplitude of 4 μm (peak to peak) under nitrogen atmosphere for in total 700 s at 4°C. Alternating inter- vals of 15 s sonication and 45 s rest were used. Protein free lipo- somes were removed by centrifugation (150,000 x g; 60 min; 4°C). Brh proteoliposomes were always used immediately after preparation.

Fusion between S. cremoris Membrane Vesicles and Brh Proteolipo- somes

S. cremoris membrane vesicles (2-4 mg protein/ml) and Brh pro- teoliposomes (1-2 mg protein/ml) were mixed at pH 5.0 in 25 mM potassium phosphate buffer supplemented with 25 mM Na-acetate. Mem- branes were mixed by rotation during 30 min at 4°C. Alternatively fusion was induced by lowering the pH to 5.0 by adding small amounts of 1 N HCl. After 10 min incubation at 25°C the pH was re- adjusted to 7.0 by adding small amounts of 1 N KOH. Fused membranes were collected by centrifugation (48,000 x g; 25 min; 4°C) and washed once in 50 mM potassium phosphate buffer pH 7.0.

Sucrose Density Gradient Centrifugation

Fused membranes were washed twice in 50 mM potassium phosphate pH 7.0 to remove non-fused Brh proteoliposomes and an aliquot of the resuspended pellet was mixed with 42% (w/v) sucrose in 50 mM potassium phosphate pH 8.0 containing 1 mM EDTA to a final sucrose concentration of 7%. Subsequently, 1 ml of this suspension (2 mg protein) was layered on top of a discontinuous sucrose gradient, composed of the following sucrose concentrations (w/v): 15% (6 ml), 30% (3 ml), 38% (3 ml), 46% (3 ml), 50% (3 ml), 54% (3 ml) and 65% (3 ml) in 50 mM potassium phosphate buffer pH 8.0 supplemented with 500 mM NaCl. After the addition of an overlay of buffer, the gradients were centrifuged in a Sorvall SS-90 vertical rotor at 34,500 x g during 3 h at 4°C. During fractionation the gradients were scanned at 280 nm with a Perkin Elmer doublebeam spectrophotometer, model 124, using a flow cell. Absorption spectra in the visible region were recorded with an Aminco Chance DW 2a spectrophotometer (American Instrument Company, Silver Spring MD).

Fusion Assays

Fusion was monitored with the resonance energy transfer (RET) fusion assay as described by Struck et al. (1981). For this assay purple membrane sheets were bleached by illumination with a 400 Watt lamp for 3 h in a thermostated vessel at 30°C in the presence of 2 M hydroxylamine (Ebrey et al., 1977) at a protein concentration of 1 mg/ml. After bleaching, hydroxylamine was removed by extensive washing with distilled water. Retinyl oxide was removed from lyophilized bleached purple membrane sheets by hexane extraction and brief sonication (Tokumaga and Ebrey, 1978). This treatment resulted in more than 90% bleaching of the pigment as judged from the absorbance at 578 nm.

The fluorescence donor N-(7-nitro-2,1,3-benzoxadiazol-4-yl)-phosphatidylethanolamine (N-NBD-PE) and the fluorescence acceptor N-(lissamine Rhodamine-ß-sulfonyl)dioleoylphosphatidylethanolamine (N-Rh-PE) were incorporated into Brh proteoliposomes each at a concentration of 0.5 mol% total phospholipid phosphorus. Brh proteoliposomes were prepared as described above in 10 mM 4-(2-hydroxyethyl)-1-piperethanesulphonic acid (HEPES) supplemented with 100 mM NaCl. A Brh to phospholipid ratio (molar) of 1 to 160 was used throughout. Protein free liposomes were removed by centrifugation (150,000 x g; 60 min; 4°C).

Brh proteoliposomes (50 nmol phospholipid) were preincubated during 1 min in 2 ml 10 mM potassium phosphate buffer supplemented with 10 mM Na-acetate and 100 mM NaCl at the indicated pH at 25°C, prior to the addition of S. cremoris membrane vesicles (50 nmol phospholipid; 55 µg protein). NBD fluorescence was monitored continuously using a Perkin Elmer MPF 43 fluorimeter, with a thermostatically controlled, magnetically, stirred cell holder, and a cutoff filter (> 520 nm) between sample and emission monochromator. N-NBD-PE was excited at 475 nm and monitored at 530 nm using 5 nm bandpass slits. Triton X-100 (1.0% (v/v) final concentration) was added to the vesicle-suspension to determine the NBD fluorescence at infinite probe dilution. This value represents 70% of the maximal NBD fluorescence, since Triton X-100 reduces the NBD quantum yield (Struck et al., 1981).

For measurements of the energy transfer efficiency between the fluorescence donor N-NBD-PE and acceptor Brh, non-bleached purple membranes instead of bleached purple membranes were used and only the fluorescence probe N-NBD-PE was incorporated into the Brh proteoliposomes at a concentration of 0.5 mol%. Preparation of the Brh proteoliposomes and fluorescence measurements were the same as above. For the determination of the relation between energy transfer efficiency and Brh concentration in the liposomal membrane, 1 ml of a Brh proteoliposomes suspension, containing 2.5 mg protein (10 mg phospholipid) in 150 mm KCl pH 6.0 was applied on a discontinuous Ficoll gradient in 10 mM potassium phosphate buffer 6.0 supplemented with 0.1 mM EDTA and 150 mM KCl, with 4.5 ml each of the following Ficoll concentrations (w/v): 2%, 3%, 4%, 5%, 7.5% and 15%. After addition of an overlay consisting of 10 mM potassium phosphate buffer pH 6.0 supplemented with 0.1 mM EDTA and 150 mM KCl, the gradients were centrifuged as described in the section sucrose density gradient centrifugation. After fractionation, fractions were assayed for the Brh and phospholipid content in the presence of 1% (v/v) Triton X-100, as determined by the absorbance at 568 nm and the NBD fluorescence. The energy transfer efficiency was calculated from the NBD fluorescence in the absence (ΔF) and presence (Fo) of 1% (v/v) Triton X-100 as described (Struck et al., 1981). After correction of Fo for the effect of Triton X-100 on the NBD fluorescence quantum yield, the energy transfer efficiency (E) is defined as:

$$E = 1 - Fo/\Delta F \tag{2}$$

Binding Assay

S. cremoris membrane vesicles were incubated for 10 min at 37°C with either Brh proteoliposomes (phospholipid/Brh ratio (mol/mol) 160/1), containing 6 mol% 14-C labeled egg PC (3.4 Ci/mol) or with liposomes containing either 0.5 mol% N-NBD-PE and 1.0 mol% cholesteryl-1(1-14C-)oleate (20 Ci/mol) or 0.5 mol% N-NBD-PE and 6 mol% 14C-egg PC (3.4 Ci/mol), in 10 mM potassium phosphate buffer pH 4.0, supplemented with 10 mM Na-acetate and 100 mM NaCl. Phospholipid phosphorus concentrations were 0.5 mM each of the bacterial membrane vesicles and the liposomes. Subsequently, the mixtures were centrifuged for 5 min in an Eppendorf microfuge. Radioactivity, NBD fluorescence and Brh absorbance at 568 nm were determined in the initial reaction mixture and in the supernatant after centrifugation.

Measurements of the Proton Motive Force

The $\Delta\psi$ (interior positive) was calculated from the distribution of tetraphenylboron (TPB$^-$) between the bulk phase of the external medium and the intervesicular fluid using the Nernst equation. The concentration of TPB$^-$ in the external medium was determined with a Tetraphenylphosphonium (TPP$^+$) selective electrode (Casadio et al., 1981) constructed according to Shinbo et al. (1978). A final concentration of 1 μM TPB$^-$ was used. Measurements were performed in the presence of 0.1 μM TPP$^+$ to increase the permeability of the membrane for TPB$^-$. Potassium ion free buffers and membrane vesicles preparations were used. The intravesicular concentration was calculated from the amount of TPB$^-$ which had disappeared from the external medium. Since TPB$^-$ binding was not saturated between 1 and 30 μM TPB$^-$, TPB$^-$ accumulations were corrected for concentration dependent binding according to the model of Lolkema et al. (1982). Alternatively, the $\Delta\psi$, interior positive, was calculated from the distribution of thiocyanate (SCN$^-$) measured by rapid filtration. Since no binding of SCN$^-$ was observed, no correction was applied. The $\Delta\psi$, interior negative, was calculated from the distribution of TPP$^+$, using a TPP$^+$ selective electrode as described above. TPP$^+$ was used at a concentration of 2 μM.

Measurements of light-dependent pH changes were performed in a 2.0 ml thermostated incubation vessel (25°C) equipped with a magnetic stirrer. The pH of the medium was continuously measured with an Orion pH electrode (pH 2000) connected to an amplifier and a recorder. Light from a slide projector (400 Watt) was provided to the membrane vesicle suspension by means of a fiber-optic light guide. Measurements were performed at a Brh concentration of 0.25 mg protein/ml in 0.15 M KCl pH 6.0. The pH changes upon illumination were calibrated by the addition of small amounts of 0.01 N KOH.

Calcium Transport Assay

Uptake studies of calcium were performed with the rapid filtration technique (Kaback, 1974). Fused membranes were resuspended into 100 μl 50 mM potassium phosphate buffer, containing 10 mM MgSO$_4$ pH 7.0 (unless indicated otherwise). The suspension was stirred and illuminated for 2 min prior to the addition of 500 μM 45 CaCl$_2$. Other additions were made as indicated in the legends to figures. At the times indicated uptake was stopped by addition of 2.0 ml cold 0.1 M LiCl. The mixture was subsequently filtered over cellulose acetate filter (pore size: 0.45 μM), presoaked in 0.5 mM CaCl$_2$ to lower binding, and washed once with the same amount of 0.1 M LiCl. A light intensity of about 1 kJ/m^2.s and a final protein concentration between 1 and 2 mg/ml was used throughout. All experiments were done at 25°C. Radioactivity was determined with a liquid scintillation counter.

Protein and Lipid Determination

Protein was measured by the method of Lowry et al. (1951), using bovin serum albumin as standard. Lipid concentrations were determined by analysis of lipid membrane phosphorus (Amesz and Dubin, 1960).

Materials

14C-Methylamine (2.22 TBq/mmol), 14C-KSCN (2.22 TBq/mmol) and 45 Calcium chloride (1.85 TBq/mmol) were obtained from the Radiochemical Centre, Amersham, Bucks. UK. N-NBD-PE, N-Rh-PE (both from Avanti Biochemicals Birmingham AL, 14C Egg PC (2.5 TBq/mmol) and 14C Cholesteryloleate (2.2 TBq/mmol) were generous gifts from Dr. J. Wilschut (Dept of Physiological Chemistry, University of Groningen, The Netherlands). Egg PC and Cardiolipin were obtained from Sigma Chemical Co. Ionophores and uncouplers were dissolved in pure ethanol. Addition of these compounds were made so that the final ethanol concentration in transport assay did not exceed 1% (v/v). All other chemicals were reagent grade and obtained from commercial sources.

RESULTS AND DISCUSSION

Fusion between Streptococcus cremoris Membrane Vesicles and Brh Proteoliposomes

Bacteriorhodopsin (Brh) can be reconstituted into phospholipid vesicles by various procedures. Depending on the reconstitution procedure employed, these vesicles either alcalify (Racker, 1983; Hellingwerf, 1979) or acidify (Happe et al., 1977; Hellingwerf, 1979) their suspending medium upon illumination. In our study we used Brh proteoliposomes that were prepared by cosonication of purple membranes and synthetic phospholipids. Using a phospholipid to Brh molar ratio of 160 and a total sonication time of 700 s, a preparation is obtained that shows a net alkalinisation of its surrounding medium upon illumination with an extent of 1.8-2.2 mol H^+/mol Brh .

In order to test fusion between S. cremoris membrane vesicles and Brh proteoliposomes we applied the resonance energy transfer (RET) fusion assay as described by Struck et al. (1981). This assay monitors changes in the spatial organization of two fluorescent lipid probes in the membrane. The excitation spectrum of N-(lissamine Rhodamine-ß-sulfonyl)dioleoyl phosphatidylethanolamine

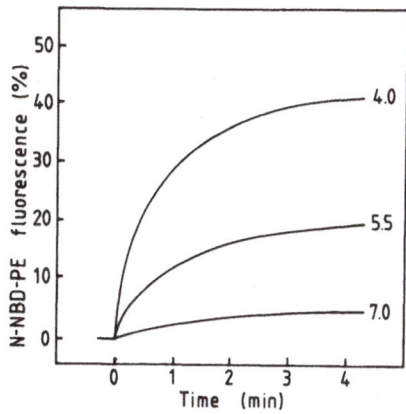

Fig. 1. Effect of the pH on the energy transfer between N-NBD-PE and N-Rh-PE upon fusion of S. cremoris membrane vesicles with Brh proteolipsomes, containing CL.

(N-Rh-PE, the acceptor) overlaps the emission spectrum of N-(7-nitro-2,1,3-benzoxadiazol-4-yl)phosphatidylethanolamine (N-NBD-PE, the donor). Incorporation of sufficient amounts of both probes into a single bilayer, results in efficient energy transfer that can be observed as a low level of NBD fluorescence. Dilution of both probes into a non-labeled membrane can be monitored as an increase in NBD fluorescence, due to the decrease in the energy transfer efficiency between the two fluorescent probes. Since the absorption spectrum of Brh strongly overlaps the emission spectrum of N-NBD-PE, purple membranes had to be bleached in order to be able to utilize the RET assay. This was achieved by illumination of purple membranes in the presence of 2 M hydroxylamine (Ebrey et al., 1977). This procedure resulted in more than 90% bleaching of the pigment. Subsequently the retinyl oxide was extracted from lyophilized bleached purple membrane sheets with hexane (Tokumaga and Ebrey, 1978). Bleached Brh was incorporated into cardiolipin (Cl) liposomes containing 0.5 mol % of both N-NBD-PE and N-Rh-PE. At this probe concentration the energy transfer efficiency is approximately linear to the N-NBD-PE probe concentration in the membrane, allowing an utilization of this assay for a quantitation of the fusion efficiency (Struck et al., 1981). Protein free liposomes were removed by centrifugation.

Fig. 1 shows that mixing of equal quantities (phospholipid) of S. cremoris membrane vesicles and Brh proteoliposomes at low pH resulted in a decrease of the fluorescence energy transfer efficiency, monitored as an increase in NBD fluorescence. At neutral pH only a low level of interaction was observed. As pointed out by Driessen et al. (A.J.M. Driessen, D. Hoekstra, G. Scherphof and J.

Wilschut, manuscript submitted for publication), dilution of N-NBD-PE and N-Rh-PE from labeled Brh proteoliposomes to unlabeled S. cremoris membrane vesicles could either occur through transfer of individual phospholipid molecules or by fusion of both membranes. Struck et al. (1981) have shown that N-NBD-PE and N-Rh-PE are non-exchangeable, suggesting that the observed interaction only involves membrane fusion.

In order to further substantiate this, binding of Brh proteoliposomes to S. cremoris membrane vesicles at low pH was studied. S. cremoris membrane vesicles were incubated at pH 4.0 with Brh proteoliposomes labeled with 14C- Egg PC. In a parallel experiment S. cremoris membrane vesicles were incubated with liposomes either labeled with 14C-Egg PC and N-NBD-PE or with 14C-Cholesteryloleate and N-NBD-PE. After centrifugation the labels left in the supernatant were determined. A simultaneous disappearance of the labels from the supernatant is observed (Table 1) suggesting that all the labels are part of one unit. In the absence of S. cremoris membrane vesicles no pelleting of Brh proteoliposomes or liposomes was observed. It should be stressed that in principle this experiment does not discriminate between (i) aggregation, (ii) transfer of individual molecules between both membranes that are irreversibly associated or (iii) membrane fusion. However, the same extent of association of the different labels and the completely different nature of the label used, virtually rules out exchange of individual lipid molecules, leaving aggregation or fusion as the only possible mechanisms of interactions. The lipid mixing observed in the RET assay (Fig. 1) cannot be reconciled with simple aggregation or liposomes. Therefore, the interaction must involve a process of membrane fusion. Furthermore, the results obtained are in full accordance with results obtained with Bacillus subtilis membrane vesicles and liposomes. For the latter system fusion as a mechanism for the observed interaction is well established (Driessen A.J.M., Hoekstra, D., Scherphof, G. and Wilschut, J., manuscript submitted for publication).

Table 1. Association of DOPC/CL (1/1), DOPC and Brh/DOPC/CL (1/49.5/49.5) liposomes with S. cremoris membrane vesicles. The amount of label pelleted is calculated from the concentration of labels in the initial reaction mixture and in the supernatant after centrifugation.

Liposome composition	Label pelleted (%): 14C-Egg PC	Cholesteryl 14C-oleate	Brh	N-NBD-PE
DOPC	9.1	11.6	–	8.8
DOPC/CL (1/1)	93.3	96.2	–	94.9
DOPC/CL/Brh (49.5/49.5/1)	75.8	n.d.	77.9	–

Since S. cremoris membrane vesicles and Brh proteoliposomes have a different buoyant density, fusion between both membranes should result in a decrease of the buoyant density of the S. cremoris membrane vesicles. When the bacterial membranes were mixed at pH 4.0 with Brh proteoliposomes, containing Cl, analysis of the interaction product on a sucrose density gradient revealed a peak with a density intermediate of those of S. cremoris membrane vesicles and Brh proteoliposomes (Fig. 2, panel A). This peak contained Brh as shown by its visible absorption spectrum (Fig. 2, panel B). Sucrose density gradient centrifugation was performed in the presence of 500 mM NaCl in order to prevent aggregation. Unfused Brh proteoliposomes were removed from the fused membranes prior to sucrose density gradient centrifugation.

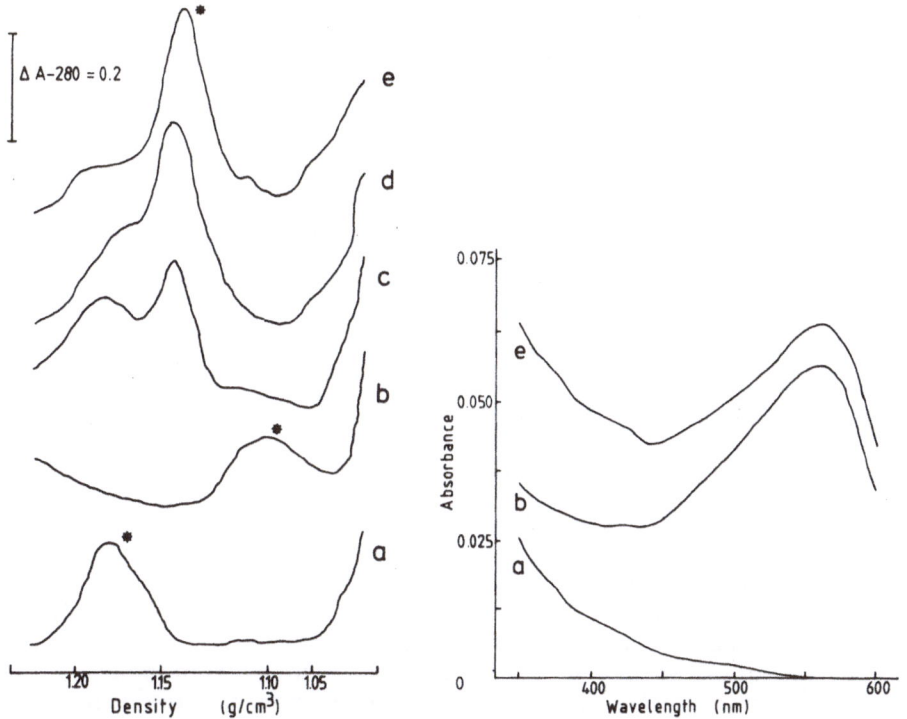

Fig. 2. Aalysis of S. cremoris membrane vesicles fused with Brh proteoliposomes by sucrose density gradient centrifugation. Panel A: Absorbance at 280 nm of the sucrose gradients. Panel B: Absorption spectra of the peak fractions. (a) S. cremoris membrane vesicles only; (b) Brh proteoliposomes reconstituted with CL/Egg PC (1/1); (c) S. cremoris membrane vesicles fused with Brh proteoliposomes containing Egg PC and (d) instead of Egg PC CL/Egg PC (1/1) and (e) instead of Egg PC CL only. Absorption spectra of the peak fractions are indicated with an * in panel A.

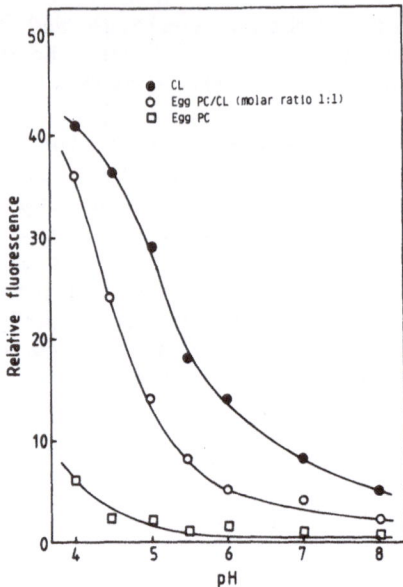

Fig. 3. Maximal NBD fluorescence development as a result of fusion between S. cremoris membrane vesicles and Brh proteoliposomes as a function of the pH and phospholipid composition of the Brh proteoliposomes.

To investigate some general properties of the fusion reaction, fluorescent labeled Brh proteoliposomes with different Cl to Egg PC ratios were prepared and examined for their fusogenic properties. Mixing of equal quantities (phospholipid) of S. cremoris membrane vesicles with Brh proteoliposomes containing a substantial amount of Cl, resulted in high NBD fluorescence yields (Fig. 3). However, Brh proteoliposomes containing only the neutral charged Egg PC showed poor fusogenic capacities, indicating the strict dependence on negative charge of the liposomes for the fusion reaction (see also Fig. 2, panel A). Fusion increased drastically with decreasing pH. It should be noted that at neutral pH hardly any association of the S. cremoris membrane vesicles and Brh proteoliposomes containing Cl, took place as could be deduced from binding experiments (data not shown). At pH 4.0, mixing of equal quantities (phospholipid) of S. cremoris membrane vesicles and Brh proteoliposomes resulted in a 42% increase in NBD fluorescence yield. Theoretically fusion of equal quantities of labeled and non-labeled membranes should result in a 50% increase in NBD fluorescence yield at the probe concentration used. This indicates that fusion between S. cremoris membrane vesicles and Brh proteoliposomes can be very efficient. As shown in Fig. 4 an increase in the steady state NBD fluorescence yield was observed with increasing ratios of S. cremoris membrane vesicles to Brh proteoliposomes. This implicates that continued fusion of fusion products with unfused S. cremoris membrane vesicles must have taken place.

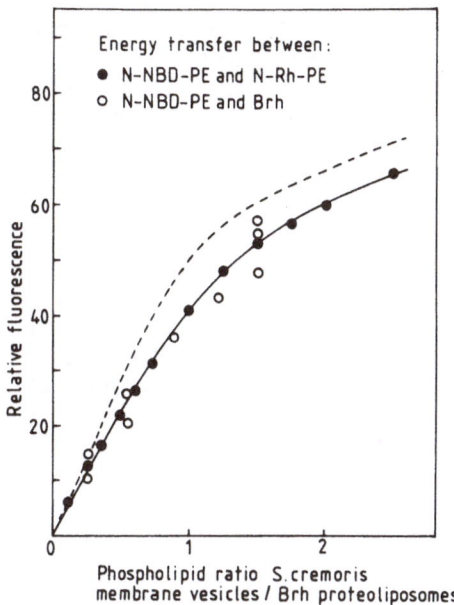

Fig. 4. Maximal NBD fluorescence development as a result of fusion
between S. cremoris membrane vesicles and Brh proteoliposomes as a function of the phospholipid weight ratio of
S. cremoris membrane vesicles and Brh proteoliposomes. The
theoretical derived curve is indicated by the broken line
(---). (●) energy transfer between N-NBD-PE and N-Rh-PE;
(o) energy transfer between N-NBD-PE and Brh.

The results described above show that an intermixing of membrane phospholipids takes place when both membranes are mixed at
low pH. In order to investigate whether in addition to dilution of
phospholipids also dilution of Brh into the membrane of S. cremoris
occurred, an experiment was performed in which the change in fluorescence energy transfer between the donor N-NBD-PE and the acceptor
(non-bleached Brh) was measured. Fig. 5 shows that the absorption
spectrum of Brh overlaps the emission spectrum of N-NBD-PE. Fractionation of Brh proteoliposomes prepared by cosonication of purple
membranes, cardiolipin and N-NBD-PE (0.5 mol % total phospholipid)
on a Ficoll gradient reveals a heterogeneous distribution of Brh
into phospholipid vesicles with respect to the Brh to phospholipid
ratio (not shown). Measurement of the energy transfer efficiency
between N-NBD-PE and Brh in these fractions showed the correlation
between the ratio Brh to phospholipid and the energy transfer efficiency as plotted in Fig. 6. A maximal efficiency of energy transfer of about 0.6 could be measured. At a phospholipid to Brh molar
ratio of 1000 or higher, the concentration of Brh in the membrane
and the energy transfer efficiency were approximately linearly
related, allowing a quantitative examination of the fusion reaction

with this assay. Mixing of <u>S. cremoris</u> membrane vesicles and Brh proteoliposomes at low pH resulted in an increase of NBD fluorescence yield (Fig. 7). Further analysis of the fusion reaction with this assay showed that identical steady state NBD fluorescence yields were attained as measured with the RET assay (Fig. 4). These results implicate that Brh molecules are able to diffuse freely in the <u>S. cremoris</u> bilayer.

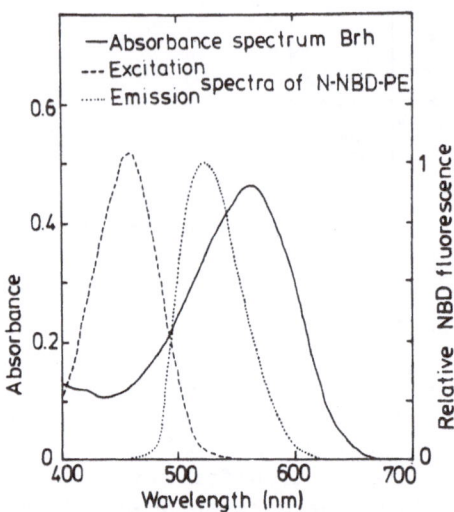

Fig. 5.   The fluorescence emission and excitation spectra of N-NBD-PE and the absorbance spectrum of Bacteriorhodopsin.

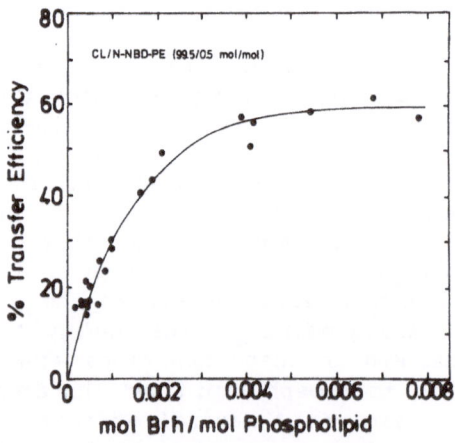

Fig. 6.   The energy transfer efficiency between N-NBD-PE and Brh as a function of the Brh to phospholipid molar ratio.

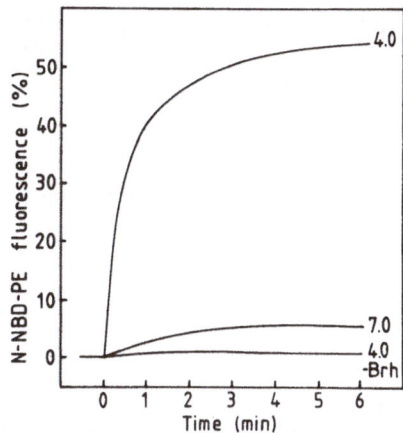

Fig. 7. Effect of the pH on the energy transfer between N-NBD-PE and upon fusion of S. cremoris membrane vesicles with Brh liposomes containing Cl. A S. cremoris membrane vesicles to Brh proteoliposomes ratio (phospholipid) of 1.5 was used.

Generation of a Proton Motive Force in the Fused Membranes

The direction of proton pumping of Brh proteoliposomes reconstituted from purple membrane sheets and synthetic phospholipids by cosonication is opposite to that in whole cells of Halobacterium halobium. This implicates that the majority of the vesicles generates an inside out oriented pmf (interior positive and acidic). Measurements of the accumulation of the lipophilic anion Tetraphenylboron (TPB$^-$), performed with a TPP$^+$ selective electrode (Shinbo et al., 1978), that also responds to TPB$^-$ (Casadio et al., 1981) indicate that upon illumination a maximal $\Delta\psi$ of about +80 mV is generated in these Brh proteoliposomes (not shown). Since binding of TPB$^-$ was not saturated at the probe concentration used, accumulation values of TPB$^-$ were corrected for concentration dependent binding of TPB$^-$ according to the model of Lolkema et al. (1982). However, under appropriate conditions (high TPP$^+$ concentration (A.J.M. Driessen, unpublished results)) also accumulation of the lipophilic cation tetraphenylphosphonium (TPP$^+$) can be observed, indicating that a small population of Brh proteoliposomes is present that generate a rightside out pmf (interior negative and alkaline). The net direction of proton translocation did not change upon fusion. Also in the fused membranes a $\Delta\psi$, interior positive, is generated, as shown by the accumulation of TPB$^-$ upon illumination (Fig. 8). The addition of triphenyltin chloride (TPT), a chloride/hydroxyl antiporter (Selwyn et al., 1970; Mokuhata and Kaji, 1981) which dissipates the ΔpH in the presence of chloride,

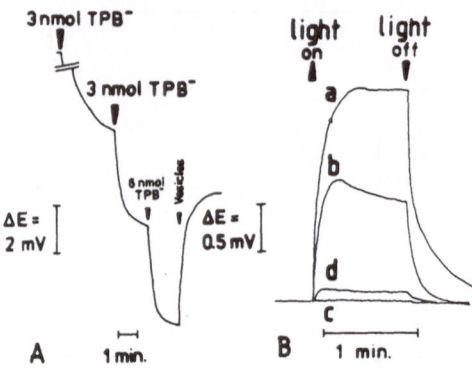

Fig. 8. Generation of Δψ (inside positive) by S. cremoris membrane vesicles fused with Brh proteoliposomes, upon illumination. Panel A: Calibration of the TPP$^+$ selective electrode by the addition of TPB$^-$. Panel B: Accumulation of TPB$^-$ upon illumination. S. cremoris membrane vesicles were fused with Brh proteoliposomes at pH 5.0. TPB$^-$ accumulation (a) in the presence of 4 nM TPT and (b) without TPT, (c and d) as (a and b) with S. cremoris membrane vesicles fused with Brh proteoliposomes at pH 7.0. Fused membranes were washed twice with 10 mM sodiumphosphate buffer pH 7.0 supplemented with 50 mM Na$_2$SO$_4$, prior to the determination of Δψ.

led to an increased accumulation of TPB$^-$. These results indicate that at pH 7.0 in addition to a Δψ, a ΔpH was generated in the fused membranes. This conclusion could further be substantiated by measurements of the ΔpH with the weak base Methylamine by flow dialysis (not shown). The addition of the Δψ dissipating ionophore valinomycin or the uncoupler 5 chloro-3-tert-butyl-2'-chloro-4'-nitro-salicylanilide (S-13) inhibited TPB$^-$ uptake. Similar results were obtained with radioactive labeled thiocyanate (SCN$^-$) as probe for a Δψ inside negative (not shown). Both with TPB$^-$ and SCN$^-$ a maximal Δψ of about +50 mV was measured. The pmf recorded by these procedures in the fused membranes do not unequivocally demonstrate that Brh is functionally incorporated in the S. cremoris membrane vesicles. In order to establish whether Brh was actually functioning in the S. cremoris membrane vesicles, an experiment was performed in which the passive proton permeability of the fused membranes was measured as a function of the amount of dicyclohexylcarbodiimide (DCCD), with which the S. cremoris membrane vesicles were pretreated. At low concentration DCCD selectively inhibits the proton translocating ATPase. The F1 component of the membrane bound ATPase is partially lost during the isolation of membrane vesicles from S. cremoris cells. This can be concluded from experiments in

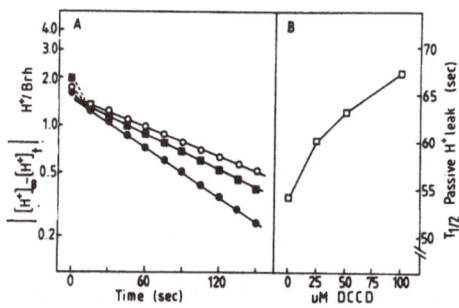

Fig. 9. Effect of DCCD pretreatment of S. cremoris membrane vesicles on the passive back diffusion of protons in S. cremoris membrane vesicles fused with Brh proteoliposomes.
Panel A: Semilogarithmic traces of proton back diffusion.
(o) Brh proteoliposomes only, (●) S. cremoris membrane
vesicles fused with Brh proteoliposomes and (■) S. cremoris membrane vesicles fused with Brh proteoliposomes
pretreated with 100 μM DCCD. Panel B: Halftime of the
passive back diffusion of protons as a function of the
concentration of DCCD, with which S. cremoris membrane
vesicles were pretreated.

potassium diffusion gradient in DCCD treated and nontreated S.
cremoris membrane vesicles were compared. A slow-down of the decay
of the Δψ is observed in the presence of DCCD (not shown). The
relative passive proton permeability of Brh proteoliposomes and
fused membranes can be determined by measurements of external pH
changes when illumination is switched off. The decay of the alkalinisation observed in the dark is a sum of two exponentials (Eisenbach et al., 1978), the first phase being faster than the second.
The slow phase represents net proton transport across the membrane,
and the rapid phase represents a Δψ dependent reaction (Arents et
al., 1981). Since the slow phase reflects back diffusion of protons
under influence of the pmf it will be limited by the flux of accompanying ions as well as by the proton permeability. Therefore it
can be used to determine the relative proton permeabilities of the
Brh proteoliposomes and fused membranes. A decrease of the halftime
of proton back diffusion could be observed when Brh proteoliposomes
were fused with S. cremoris membrane vesicles (Fig. 9, panel A),
indicating an increase in the relative proton permeability of the
membrane. This decrease, however, was less pronounced when S. cremoris membrane vesicles were pretreated with DCCD as shown in Fig.
9 (Panel B). It should be emphasized that DCCD treatment of the S.
cremoris membrane vesicles had no significant effect on the fusion
efficiency as measured by the RET assay (not shown). These results
indicate that Brh is capable of generating a pmf in the fused membranes upon illumination.

Calcium Uptake by the Fused Membranes

Since the pmf generated by Brh in the fused membranes upon illumination is inside out with respect to the orientation of the pmf in whole cells of S. cremoris, the uptake of calcium, a solute normally extruded by S. cremoris cells was examined in the fused membranes. When S. cremoris membrane vesicles fused with Brh proteoliposomes at pH 5.0 were illuminated (1 kJ/m^2.s), calcium uptake could be observed (Fig. 10). Ca^{++} uptake was stimulated about two-fold by the addition of the ionophore valinomycin (5 µM). In the presence of the uncoupler carbonylcyanide-p-trifluoromethoxyphenyl-hydrazone (CCCP) (10 µM) or S-13 (5 µM), or in the presence of both valinomycin and nigericin , only binding of calcium to the vesicles was observed. The inhibition of the calcium uptake by the ΔpH dissipating ionophore nigericin and the stimulation of Ca^{++} uptake by the Δψ dissipating ionophore valinomycin, which action is accompanied by an increase in the ΔpH (not shown), suggests that calcium uptake occurs via H$^+$/Ca^{++} antiport. Such a mechanism for the extrusion of calcium has been reported for a number of different bacteria (Rosen, 1983). The light dependency of the uptake of calcium is shown in Fig. 11. The uptake increased hyperbolically with increasing light intensity upto saturation. Particularly at high

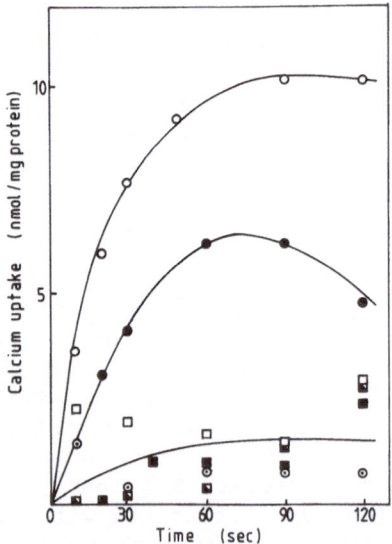

Fig. 10. Calcium uptake by S. cremoris membrane vesicles fused with Brh proteoliposomes at pH 5.0. (●) no additions; (□) plus 0.5 µM nigericin; (o) plus 5 µM valinomycin; (◉) plus 10 µM CCCP or (■) in the dark. (◪) Uptake in fused vesicles with bleached Brh.

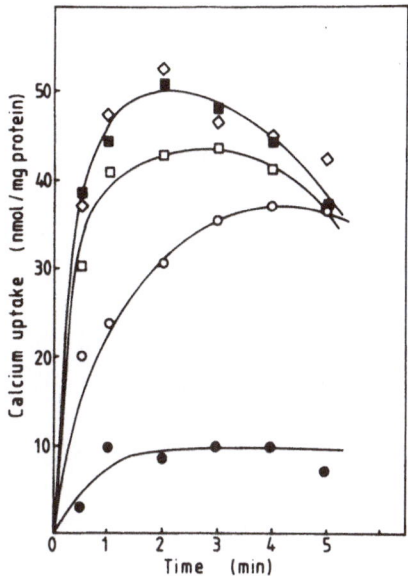

Fig. 11. Calcium uptake by S. cremoris membrane vesicles fused with Brh proteoliposomes at different light intensities. (●) dark; (o) 30%; (□) 45%); (■) 60% or (◇) 80% of the maximal intensity (1.5 kJ/m^2.s).

light intensities the Ca^{++} uptake reached its maximum level after 1 to 2 min after which the Ca^{++} leaked out. This leakage is partly due to an increased denaturation of Brh at the pH = 8.0, the pH at which the uptake experiments were performed and partly the result of an increased permeability of the fused membranes by a concentration dependent association of calcium with cardiolipin in the presence of phosphate (Fraley et al., 1980). In the dark only a low level of calcium uptake was observed (Fig. 10 and 11). Bleaching of Brh in the presence of hydroxylamine also abolished calcium uptake. These results show that the fused membranes at least have retained the capacity to accumulate a solute (calcium), driven by ΔpH generated by Brh in the light.

CONCLUDING REMARKS

With the system described in this paper (see Fig. 12) exclusively antiport systems can be studied. Experiments are in progress in which Brh is incorporated in a rightside out orientation in the S. cremoris membrane vesicles by the use of rightside out oriented Brh proteoliposomes (Happe et al., 1977; Hellingwerf, 1979). Such a system should result in a modelsystem in which H$^+$/amino acid symport in lactic acid streptococci and other anaerobic fermentative bacteria can be studied.

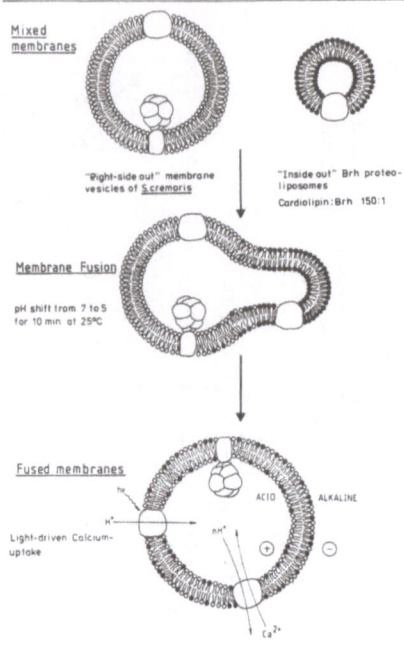

Scheme of the incorporation of Bacteriorhodopsin in
membrane vesicles of Streptococcus cremoris

Fig. 12. Scheme of fusion between S. cremoris membrane vesicles and
Brh proteoliposomes.

It is further of interest to note that bacterial membranes retain their pmf generating capacities when fused with liposomes by low pH treatment (A.J.M. Driessen, unpublished results) (Snozzi and Crofts, 1984; Pennoyer et al., 1984; Casadio et al., 1982) as opposed to the membranes obtained by low pH induced fusion of mitochondrial inner membranes with liposomes (Schneider et al., 1980a; 1980b). The molecular basis for this difference is unknown. Bacterial membranes fused with liposomes offer attractive experimental tools for the study of functional properties of bacterial membranes.

Acknowledgements

The authors wish to thank Drs J. Wilschut and D. Hoekstra for their help with the Resonance energy transfer fusion assay. This study has been made possible by the Stichting voor Biofysica with financial support from the Netherlands Organization for the Advancement of Pure Scientific Research (ZWO). The help of Marry Pras in preparing the manuscript is greatly appreciated.

Abbreviations

Brh, Bacteriorhodopsine; pmf, transmembrane electrochemical proton gradient, $\Delta\psi$ transmembrane electrical potential ; ΔpH, transmembrane pH gradient; N-NBD-PE, N-(7-nitro-2,1,3-benzoxadiazol-4-yl)phosphatidylethanolamine; N-Rh-PE, N-(lissamine Rhodamine-ß-sulfonyl)dioleoyl phosphatidylethanolamine; DOPC, dioleoyl phosphatidylcholine; Egg PC, egg phosphatidylcholine; CL, cardiolipin; HEPES, N-2-hydroxyethylpiperazine-N'-2-ethanesulfonic acid; EDTA, ethylenediamine tetraacetic acid; CCCP, carbonylcyanide p-trifluoromethoxyphenylhydrazone; TPT, triphenyltin chloride; S13, 5-chloro-3-tert-butyl-2'-chloro-4'-nitro-salicylanilide; TPP^+, tetraphenylphosphonium chloride; TPB^-, tetraphenylboron chloride.

REFERENCES

Amesz, B. N., and Dubin, D. T., 1960, The role of polyamines in the neutralization of bacteriophage deoxyribonucleic acid, J. Biol. Chem., 235:769.

Arents, J. C., van Dekken, H., Hellingwerf, K. J., and Westerhoff, H. V., 1981, Linear relations between proton current and pH gradient in Bacteriorhodopsin liposomes, Biochemistry, 20:5114.

Bakker, E. P., Rottenberg, H., and Caplan, S. R., 1976, An estimation of the light-induced electrochemical potential difference of protons across the membrane of Halobacterium halobium, Biochim. Biophys. Acta, 440:557.

Bragg, P. D., 1979, Electron transport and energy transducing systems in E. coli, in: "Membrane Proteins in Energy Transduction", R.A. Capaldi, ed., Marcel Dekker AG, Basel.

Casadio, R., Venturoli, G., and Melandri, B. A., 1981, Light-induced transmembrane potential difference in chromatophores of photosynthetic bacteria: Measurements with an ion-selective mini-electrode, Photobiochem. Photobiophys., 2:245.

Casadio, R., Venturoli, G., and Melandri, B. A., 1982, Studies on Phospholipid Enriched Chromatophores from Rps. sphaeroides, Characterization and Perspectives, in: "Short Reports of the 2nd European Bioenergetics Conference", Université Claude Bernard, Lyon.

Danon, A., and Stoeckenius, W., 1974, Photophosphorylation in Halobacterium halobium, Proc. Natl. Acad. Sci. USA, 71:1234.

Ebrey, T., Becker, B., Mao, B., Kilbride, P., and Honig, B., 1977, Exciton interactions and chromophore orientation in the purple membrane, J. Mol. Biol., 112:377.

Eisenbach, M., Garty, H., Bakker, E. P., Klemperer, G., Rottenberg, H., and Caplan, S. R., 1978, Kinetic analysis of light induced pH changes in Bacteriorhodopsin, containing particles from Halobacterium halobium, Biochemistry, 17:4691.

Fraley, R., Wilschut, J., Düzgunes, N., Smith, C., and Papahadjo-poulos, D., 1980, Studies on the mechanism of membrane fusion: Role of phosphate in promoting calcium ion-induced fusion of phospholipid vesicles, Biochemistry, 19:6021.

Happe, M., Teather, R. M., Overath, P., Knobling, A., and Oester-helt, D., 1977, Direction of proton translocation in proteo-liposomes formed from purple membrane and acidic lipids depends on the pH during reconstitution, Biochim. Biophys. Acta, 465:415.

Harold, F. M., Pavlasova, E., and Baarda, J. R., 1970, A transmem-brane pH gradient in Streptococcus faecalis: origin and dissi-pation by proton conductors and N,N'-dicyclohexyl carbodi-imide, Biochim. Biophys. Acta, 196:235.

Harold, F. M., 1977, Membranes and energy transduction in bacteria, Curr. Top. Bioenerg., 6:83.

Hellingwerf, K. J., 1979, "Structural and Functional Studies on Lipid Vesicles Containing Bacteriorhodopsin", Ph.D. Thesis, University of Amsterdam.

Hellingwerf, K. J., Arents, J. C., Scholte, B. J., and Westerhoff, H. V., 1979, Bacteriorhodopsin in liposomes. II. Experimental evidence in support of a theoretical model. Biochim. Biophys. Acta, 547:561.

Hellingwerf, K. J., and Konings, W. N., 1980, Kinetic and steady state investigations of solute accumulation in bacterial mem-branes by continuously monitoring the radioactivity in the effluent of flow-dialysis experiments, Eur. J. Biochem., 106:431.

Hellingwerf, K. J., and Konings, W.N., 1984, The energy flow in bacteria: the main free energy intermediates and their regula-tory role, Adv. Microb. Physiol., in press.

Kaback, H. R., 1974, Transport studies in bacterial membrane vesi-cles, Science, 186:882.

Kaback, H. R., 1982, Membrane vesicles, electrochemical ion gradients, and active transport, Curr. Top. Membr. Transp., 16:393.

Konings, W. N., and Michels, P. A. M., 1980, Electron Transfer Chain Solute Translocation across Bacterial Membranes, in: "Diversity of Bacterial Respiratory Systems", C. J. Knowles, ed., CRC Press Inc., Boca Raton.

Lanyi, J. K., 1979, The role of Na^+ in transport processes of bacterial membranes, Biochim. Biophys. Acta, 559:377.

Lolkema, J. S., Hellingwerf, K. J., and Konings, W. N., 1982, The effect of "probe binding" on the quantitative determination of the proton motive force in bacteria, Biochim. Biophys. Acta, 681:85.

Lowry, O. H., Rosebrough, N. J., Farr, A. J., and Randall, R. J., 1851, Protein measurements with the Folin phenol reagent, J. Biol. Chem., 193:265.

Maloney, P. C., 1982, Coupling between H^+ entry and ATP synthesis in bacteria, Curr. Top. Membr. Transp., 16:175.

Michels, P. A. M., Michels, J. P. J., Boonstra, J., and Konings, W. N., 1979, Generation of an electrochemical gradient in bacteria by the excretion of metabolic endproducts, FEMS Microbiol. Lett., 5:357.

Mitchell, P., 1966, "Chemiosmotic Coupling and Energy Transduction", Glynn Research Ltd., Bodmin.

Mitchell, P., 1972, Chemiosmotic coupling in energy transduction: a logical development of biochemical knowledge, Bioenergetics, 3:5.

Mukohata, Y., and Kaji, Y., 1981, Light-induced membrane potential increase, ATP synthesis, and proton uptake in Halobacterium halobium R1mR catalyzed by halorhodopsin: Effects of N,N'-dicyclohexylcarbodiimide, triphenyltin chloride and 3,5-ditert-butyl-4-hydroxybenzylidenemalononitrile (SF 6847), Arch. Biochem. Biophys., 290:72.

Oesterhelt, D., and Hess, B., 1973, Reversible photolysis of the purple complex in the purple membrane of Halobacterium halobium, Eur. J. Biochem., 37:316.

Otto, R., Sonnenberg, A. S. M., Veldkamp, H. and Konings, W. N., 1980, Generation of an electrochemical proton gradient in Streptococcus cremoris by lactate efflux, Proc. Natl. Acad. Sci. USA, 77:5502.

Otto, R., Lageveen, R. G., Veldkamp, H. and Konings, W. N., 1982, Lactate efflux induced electrical potential in membrane vesicles of Streptococcus cremoris, J. Bacteriol., 149:733.

Otto, R., 1981, "An Ecophysiological Study of Starter Streptococci", Ph.D. Thesis, University of Groningen.

Pennoyer, J. D., Kramer, H. J. M., van Grondelle, R., Westerhuis, W., Amesz, I., and Niederman, R.A., 1984, Excitation energy transfer in Rhodopseudomonas sphaeroides chromatophore membrane fused to liposomes, FEBS Lett., in press.

Racker, E., 1973, A new procedure for the reconstitution of biologically active phospholipid vesicles, Biochem. Biophys. Res. Commun., 55:224.

Ritchey, T. W., and Seeley, H. W., 1976, Distribution of cytochrome-like respiration in streptococci., J. Gen. Microbiol., 93:195.

Rosen, B. P., 1983, Bacterial Calcium Transport Systems, in: "Metal Ions in Biological Systems 17", H. Sigel, ed., Elsevier/North Holland, Amsterdam.

Schneider, H., Lemasters, J. J., Hochli, M., and Hackenbrock, C. R., 1980a, Fusion of liposomes with mitochondrial inner membranes. Proc. Natl. Acad. Sci. USA, 77:442.

Schneider, H., Lemasters, J. J., Hochli, M., and Hackenbrock, C. R. 1980b, Liposomes, mitochondrial inner membrane fusion, J. Biol. Chem., 255:3749.

Selwyn, M. J., Dawson, A. P., Stockdale, M., and Gains, N., 1980, Chloride-hydroxide exchange across mitochondrial, erythrocyte and artificial lipid membranes mediated by trialkyl- and triphenyl his compounds, Eur. J. Biochem., 14:120.

Sijpensteijn, A. K., 1970, Induction of cytochrome formation and stimulation of oxidative dissimilation of hemin in Strepto- coccus lactis and Leuconostoc mesenteroides, Ant. Leeuwenh., 36:335.

Shinbo, T., Kama, N., Kurihara, K. and Kobatake, Y., 1978, A PVC- based electrode sensitive to DDA$^+$ as a device for monitoring the membrane potential in biological systems, Arch. Biochem. Biophys., 187:414.

Skulachev, V. P., 1984, Membrane bioenergetics - Should we build the bridge across the river or alongside of it? Tr. Biochem. Sci., 9:182.

Snozzi, M., and Crofts, A. R., 1984, Electron transport in chroma- tophores from Rhodopseudomonas sphaeroides GA fused with lipo- somes, Biochim. Biophys. Acta, 766:451.

Stoeckenius, W., and Kunau, W. H., 1968, Further characterization of particulate fractions from lysed cell envelopes of Halobac- terium halobium and isolation of gas vacuole membranes, J. Cell. Biol., 38:337.

Struck, D. K., Hoekstra, D., and Pagano, R. E., 1981, Use of reso- nance energy transfer to monitor membrane fusion, Bioche- mistry, 20:4093.

Ten Brink, B., and Konings, W. N., 1980, Generation of an electro- chemical proton gradient by lactate efflux in Escherichia coli membrane vesicles, Eur. J. Biochem., 111:59.

Ten Brink, B, 1984, "The Generation of Metabolic Energy in Bacteria: The Energy Recycling Model", Ph.D. Thesis, University of Groningen.

Tokumaga, F., and Ebrey, T., 1978, The blue membrane: the 3-dehy- droretinal based artificial pigment of the purple membrane, Biochemistry, 17:1915.

CONTROL OF MITOCHONDRIAL RESPIRATION

Guy Brown

Department of Biochemistry
University of Cambridge
Cambridge, U.K.

INTRODUCTION

The control of the rate of respiration within cells is a complex process, central to metabolism, and exhibiting several different types of control at different levels of the overall process. The aim of this chapter is to outline some general principles of control in metabolic pathways and recent approaches to describing control; this is followed by a brief review of those steps, in the respiratory pathway of rat liver mitochondria, which are thought to be important in control; finally some experimental work is presented concerning the control of respiration by the proton motive force (Δp), and control of Δp by cation permeabilities.

PRINCIPLES OF CONTROL

The net flux through a metabolic pathway may be regulated by two separate means: (a) by altering the activity of the enzymes of the pathway, (b) by altering the concentration of metabolites in the pathway.

In general it only makes sense to alter the activity of certain enzymes in a pathway, that is, the rate determining enzymes. These are usually the enzymes with the lowest activity, and the reactions they catalyse are out of equilibrium. If a single enzyme in a pathway is rate determining it is said to be the rate limiting step, but often there is more than one rate determining step and it becomes necessary to quantify the amount of control exerted at each step. This is the aim of Control Theory, developed by Kacser & Burns (1973) and Heinrich and Rapoport (1974a). Here the extent to which alteration in the activity of a particular enzyme (v_i) alters the net flux through the whole pathway (J) is quantified in a parameter known as the control strength (C_i), where

$$C_i = dJ/dv_i$$

The sum of all control strengths in a pathway is 1. Where a single enzyme in a pathway is rate limiting its control strength approaches 1, while C_i for an enzyme close to equilibrium approaches 0. C_i is most conveniently determined by using an irreversible inhibitor of the enzyme. The interpretation of C_i becomes ambiguous with complex pathways, and should be used with some caution (Heinrich & Rapoport, 1974b). The application of this theory to control of respiration has been outlined by Tager et al. (1983).

The fact that an enzyme has significant control strength in a pathway, does not mean that this is where the pathway is physiologically regulated. It only means that it is a potential site for regulation. What we need to know is whether the enzyme has a physiological effector, whose concentration changes in vivo over a range which will significantly alter the enzyme's activity.

Control by metabolite concentrations can be regarded as control by pathways or processes external to the primary pathway. No metabolic pathway is a world unto itself. Some intermediates and cosubstrates and coproducts will be components of other pathways. Alterations in the concentration of intermediates through participation in other pathways may lead to changes in the rate of the primary pathway, but the quantitative effect on flux will depend on where the intermediate is located in the pathway. Many enzymes in a pathway use more than one substrate and product, and these reactants may not be regarded as intermediates in the primary pathway, but their concentration may still effect flux through the pathway. This will be particularly true of near equilibrium reactions in a pathway, where alteration in the concentration of cosubstrates and coproducts will alter the equilibrium concentration of intermediates and thus the flux through a pathway. For example, a pathway which includes an NAD-linked dehydrogenase which is close to equilibrium is likely to be responsive to changes in NAD/H redox level, or the electron flux through a redox linked proton pump will respond to changes in the electrochemical gradient of protons.

In order to quantify the effect of metabolite concentrations on the rate of pathways, it would be useful to be

464

able to model the kinetics of the whole pathway. Enzyme kinetics is of limited use for this purpose, because of the complexity of the equations describing reactions close to equilibrium, and reactions with more than one substrate or product. Also knowledge of each enzyme mechanism would often be required.

In order to give a simpler description of complex near-equilibrium reactions, Irreversible Thermodynamics was developed by Katchalsky & Curan (1967). This describes any system that is close to equilibrium by expressing the flows (reaction rates, fluxes) as linear functions of all corresponding forces (ΔG's of the reactions, free energy gradients). For example a redox coupled proton pump would be described as consisting of an electron flux (rate Je, force ΔEh) and a proton flux (rate Jh, force Δp), with the so called phenomenological equations:

$$Je = L_{ee}.\Delta Eh + L_{he}.\Delta p$$

$$J_{H+} = L_{eh}.\Delta Eh + L_{hh}.\Delta p$$

Here the L's are proportionality constants, for example $L(he)=d(Je)/d(\Delta p)$, and are analogous to conductivities in electrical circuits; this is because we have assumed a type of Ohm's law for chemical reactions. These equations can be further simplified as it can be derived that close to equilibrium $L(eh) = L(he)$, a relation known as Onsager's reciprocosity.

These relations have only been derived for reactions very close to equilibrium i.e. overall $\Delta G \ll RT$, which is of little use to describe real biological pathways. There is no theoretical justification for using these equations to describe reactions further from equilibrium. However they have been successfully been tested and applied to bioenergetic pathways in a relatively far from equilibrium domain (Pietrobon et al., 1982). Stucki et al. (1983) have demonstrated theoretically that linear flow/force relations can be far more efficient in coupling reactions than the non-linear relations we would expect from enzyme kinetics. He has argued that this may constitute a powerful evolutionary drive towards linear relations.

Control by enzyme activity and control by metabolite levels are interconnected. Stucki has shown that Control Theory and Irreversible Thermodynamics are complementary and the parameters interconvertible (Stucki, 1983)

Attempts to describe the control of a pathway are rarely finite, there are often many factors external to the pathway which affect it, and these factors may in turn be controlled by an ever receding system of controls.

In order to describe mitochondrial respiration we must widen our concepts of control to include transport processes and permeabilities, but in general the principles of control are the same. The main steps by which respiration in rat liver mitochondria is regulated are outlined below.

Substrate transport

All substrates of mitochondrial respiration have relatively specific coupled transport systems for transport across the inner mitochondrial membrane. Dicarboxylic acids such as malate, succinate, and 2-oxoglutarate may exchange for inorganic phosphate (Pi), or for each other. Citrate may enter in exchange for malate. Pyruvate, glutamate and Pi may enter in exchange for OH^- ions. Thus, potentially, the distribution of these organic acids is determined by the pH gradient across the inner mitochondrial membrane.

In isolated mitochondria, in the presence of excess uncoupler or in state 3 (excess ADP), the respiration rate does appear to be partially limited by substrate transport, (Palmieri et al., 1967) and (Harris et al., 1967). This limitation is overcome by high concentrations of substrate, or conditions that maintain a high ΔpH. The control strength of succinate transport in state 3 has been estimated as 0.33 (Tager et al., 1983). However, measurements of substrate distribution between the mitochondrial and cytosolic compartments of hepatocytes have indicated a substantial equilibration of the gradients of citrate, isocitrate, 2-oxoglutarate, glutamate, and malate with ΔpH, (Tischler et al., 1977). This would suggest that permeation of these compounds is not limiting in resting physiological conditions, however no study has been performed in high flux states. We do not know whether ΔpH may change significantly in physiological conditions and thus might affect substrate transport if this is in fact a near-equilibrium step.

Long chain fatty acids enter the mitochondria after esterification to fatty acyl-CoA, via the acylcarnitine carrier, which is physiologically inhibited by malonyl-CoA (an intermediate of fatty acid synthesis). However it is still controversial whether transfer into the mitochondria is limiting to oxidation. The subject has been reviewed by Sugden & Williamson (1982).

Dehydrogenases and the Tricarboxylate Cycle

Reviewed by Hansford (1980).

Pyruvate dehydrogenase, isocitrate dehydrogenase, and 2-oxoglutarate dehydrogenase all catalyse non-equilibrium reactions in pathways of substrate oxidation. All three appear to be physiologically stimulated by reduced ratios of NADH/NAD and ATP/ADP, (which occurs during state 4 to 3 transitions). In addition, pyruvate dehydrogenase is stimulated by a reduced ratio of acetyl-CoA/CoA, and 2-oxoglutarate dehydrogenase by a reduced ratio of succinyl-CoA/CoA. Calcium ions also stimulate all three as the intramitochondrial concentration is raised in the range 0.1 to 10μM (Denton & Hughes, 1978), and the calcium concentration may be under hormonal control (Williamson et al., 1981). However it is still uncertain what the intramitochondrial calcium concentration is.

Pyruvate dehydrogenase activity itself does not control respiration rates but rather regulates and is regulated by the ratio of acetyl-CoA/CoA. The other two dehydrogenases probabaly have some role in controlling respiration although it may be primarily a passive role, in responding to the NADH/NAD ratio.

Succinate dehydrogenase is inhibited by low levels of oxaloacetate, and this inhibition is relieved by a high ratio of ubiquinol/ubiquinone; ATP may also act as an activator (Gutman, 1981). The physiological relevance of this control is unknown.

The rate of TCA cycle appears to be partially limited by citrate synthetase activity, which is primarily regulated by one of its substrates oxaloacetate (Hansford, 1980). The level of oxaloacetate is determined by: (a) the activity of the glucogenic pathway from pyruvate, which crosses the TCA cycle at oxaloacetate, (b) the NADH/NAD ratio, which sets the near-equilibrium distribution of malate and oxaloacetate across malate dehydrogenase.

Respiratory Chain

Reviewed by Hansford (1980), and Ferguson and Sorgato (1982).

The respiratory chain oxidises NADH, succinate and fatty acyl-CoA and reduces O_2 to H_2O. The passage of electrons down this redox gradient, through three protein complexes, is used to pump protons across the inner mitochondrial membrane, generating an electrochemical gradient of protons, also refered to as the proton motive force ($\Delta p = \Delta\psi - 59\Delta pH$ at $25°C$). The activity of the chain is high and the redox drop (ΔEh) between NADH and cytochrome c appears to be close to equilibrium with ΔG_p , in state 4 (Owen & Wilson, 1974), while the reaction catalysed by cytochrome oxidase is probably far from equilibrium and is experimentally irreversible.

Anything that dissipates Δp will speed up respiration, presumably by reducing backpressure on the proton pumps, thus allowing the coupled electron flow to speed up. Δp may be dissipated experimentally with protonophores such as the uncouplers FCCP or DNP, or by combinations of ionophores that cause ion cycling coupled to proton transport, for example the combination of valinomycin which catalyses K^+ uniport, plus nigericin which catalyses K^+/H^+ exchange. Physiologically Δp may be diminished by H^+ coupled cation cycling of calcium, Na^+, or K^+. Free fatty acids may act as protonophores, and this has been suggested as the mechanism by which certain hormones stimulate respiration in white adipose tissue (Davis & Martin, 1982). The most important system diminishing the proton gradient is the ATP synthase, plus the adenine nucleotide carrier and phosphate carrier. The addition of ADP to respiring mitochondria results in a drop in Δp estimated at between 10-30mV. Thus the stimulation of respiration by ADP might be explained solely by its effect on Δp. If Δp is titrated with uncouplers, there is a linear increase in respiration rate with decline in Δp, up to the uncoupled rate (Azzone et al., 1978). However several groups have shown that for the decrease in Δp elicited by ADP, there is a much greater increase in respiration rate, than occurs with uncouplers (Azzone et al., 1978). This has been explained on the basis of localised chemiosmosis. This is a modification of the original chemiosmotic theory (recent example proposed by Westerhoff et al. (1984)), and suggests that the protons used by the respiratory chain and the ATP synthetase do not normally equilibrate with the bulk phases of the extramitochondrial medium and intramitochondrial matrix, but rather remain localised within or on the surface of the membrane. Thus there are microcircuits of protons between the H^+ pumps and the ATP synthase, and the bulk phase Δp underestimates the true Δp across a proton pump.

The implications of localised chemiosmosis for control are that processes producing or consuming bulk phase Δp, would be relatively independent of each other; the major control on the rate of a proton translocating process would be exerted by any other process to which it was linked by localised proton circuits. Thus respiration rate would be controlled more directly by ΔG_p , and would perhaps be relatively independent of other Δp consuming reactions such as substate transport or calcium cycling.

As yet no version of the localised theory has been

proposed in such a concrete form that it can itself be tested. The evidence for localised chemiosmosis consists of miscellaneous observations that appear anomalous when viewed in terms of bulk phase chemiosmosis. With regard to the anomalous relation between Δp and Jo during phosphorylation, Hansford has pointed out that respiratory control may depend not on Δp alone, but rather on the difference between $n.\Delta p$ and the redox potential difference (ΔEh) over a segment of the respiratory chain that translocates n protons per electron, thus

$$Je = f(\Delta Eh - n.\Delta p)$$

This, in common with irreversible thermodynamics, assumes that it is the thermodynamic forces that determine the rate. Note that there is no reason to expect this relation from enzyme kinetics.

This type of relation is tested in the experimental section.

Control by Adenine Nucleotides

Wilson and coworkers (Owen & Wilson, 1974; Wilson et al., 1977) have suggested a model of respiratory control that is independent of the chemiosmotic theory. They suggest that the rate of respiration is proportional to the amount of reduced cytochrome a_3 and Cu_B, these being the components in cytochrome oxidase that react directly with oxygen in an irreversible reaction. Unfortunately it is not yet possible to measure accurately the amount of reduced a_3 or intermediates in oxygen reduction. They further suggest that the concentration of reduced a_3 is determined by the redox state of NAD/H and the ΔG_P; this is because equilibrium between the redox drop from NAD/H to a_3 and the ΔG_P is assumed even for high flux states. This is supported by measurements of the redox drop from NAD/H to cytochrome c and ΔG_P in high flux states (Owen & Wilson, 1974).

However the evidence is contradicted by the finding that respiration responds to the ratio [ATP]/[ADP] rather than [ATP]/[ADP].[Pi] (Davis & Lumeng, 1975). This, together with a large amount of other evidence, (see Hansford (1980)), suggests that oxidative phosporylation and state 3 respiration may be partially rate limited by the adenine nucleotide translocase. The control strength of the translocase on respiration has been estimated to rise from 0 in state 4 to 0.3 in state 3 (Tager et al., 1983), in rat liver mitochondria. However it is close to zero in heart mitochondria, even in state 4 (Doussiere et al., 1984), presumably due to the higher content of the translocase in heart.

Fatty acyl-CoA inhibits the translocase (Vignais, 1976), and recent evidence (Soboll et al., 1984) suggests this may have a physiological role .

Control of Δp

The level of Δp depends on the rate at which protons are extruded by the respiratory chain and the rate at which they return to the matrix. Now

$$\text{Rate of } H^+ \text{ extrusion} = n. \text{ Jo}$$

where n is the stoichiometry of H^+ extruded per oxygen atom and Jo is the rate of oxygen consumption. And let us assume for the moment

$$\text{Rate of } H^+ \text{ return} = L.\Delta p$$

where L is the permeability of the inner mitochondrial membrane to protons. Now in a steady state, i.e. when

$$\text{Rate of } H^+ \text{ extrusion} = \text{rate of } H^+ \text{ return}$$

then $$n.Jo = L.\Delta p$$

or $$Jo = (L/n).\Delta p$$

Now, if we inhibit respiration with an inhibitor which does not change L, we might expect that a graph of Jo against Δp would give a straight line. In fact (see fig. 5) the graph is very non-linear, so that at high Δp respiration can be inhibited with very little effect on Δp. This implies either that L increases at high Δp, as Nicholls (1974) has suggested, or that n drops at high Δp as Pietrobon et al. (1981) have suggested. Thus there are internal controls on the level of Δp, effectively clamping it, so that Δp is buffered against any change.

The origin and possible function of this relation are discussed in the experimental section.

EXPERIMENTAL

Parallel measurements of ΔEh, Δp, & Jo

In order to assess whether $\Delta Eh-n.\Delta p$ correlates with the respiration rate, parallel measurements of the redox drop (ΔEh) across the bc_1 complex, the proton motive force (Δp) and the

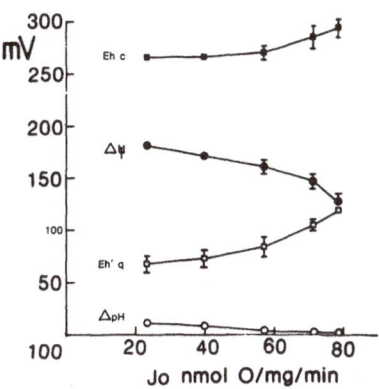

Fig.1 Redox states of ubiquinone-2 and cytochrome c as Δψ and Jo are titrated with FCCP.

1mg mitochondrial protein/ml was incubated in 120 KCl, 5mM (K$^+$)phosphate, 5mM (K$^+$)Hepes, 5mM mannitol, 1mM (K$^+$)EGTA, 100μM (K$^+$)acetate, 10μM ubiquinone-2, 5μM rotenone, 5μM TPMP bromide, 3.3μM cyt.c, 0.6μg oligomycin/ml, 0.3μg nigericin/ml, pH 7.00, at 25°C, for 2 minutes. Mitochondria were then energised with 5mM (K$^+$)succinate, and Δψ was reduced with successive additions of 16.7pmoles FCCP/mg protein, at 2 minute intervals when a steady state was reached, up to a final concentration of 66.7pmoles FCCP/mg protein. Parallel determinatios were made in each steady state of matrix volume, Δψ (●), ΔpH (O), Eh'(Q2) (□), Eh(cyt.c) (■), and rate of oxygen consumption.

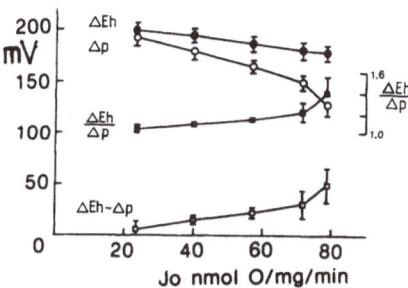

Fig.2 ΔEh, Δp, ΔEh/Δp, and ΔEh-Δp across the bc$_1$ complex, against respiration rate (Jo), calculated from the values in Fig.1.

ΔEh is Eh(cyt.c) - Eh'(Q2), (●); Δp is (Δψ - 59ΔpH), (O); ΔEh/Δp (■) is the force ratio; and ΔEh-Δp (□) is equal to (-ΔG/F) for electron flow through the bc$_1$ complex coupled to vectorial proton transport, assuming a constant stoichiometry of 1 H$^+$/e$^-$.

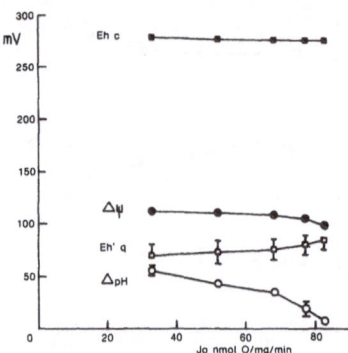

Fig.3 Redox states of ubiquinone-2 and cytochrome c, as ΔpH and Jo are titrated with FCCP.

1mg mitochondrial protein/ml was incubated in 200mM sucrose, 5mM KCl, 5mM (Li$^+$)Hepes, 5mM mannitol, 1mM (Li$^+$)EGTA, 100μM (K$^+$)acetate, 10μM Q2, 5μM rotenone, 5μM TPMP bromide, 3.3μM cytochrome c, 0.6μg oligomycin/ml , 320pmoles valinomycin/mg, pH 7.00, at 25°C, for 2 minutes. Mitochondria were then energized with 5mM (Li$^+$)succinate, and ΔpH was reduced with successive additions of 16.7pmoles FCCP/mg protein, at 2 minute intervals when a steady state was reached, up to a final concentration of 66.7pmoles/mg FCCP. Parallel determinations were made in each steady state as before. The points represent the redox state Eh'(Q2) (□), Eh(cyt.c) (■), the membrane potential Δψ (●), and 59ΔpH (O).

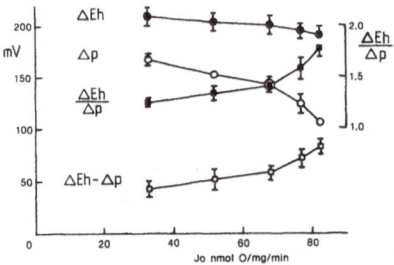

Fig.4 ΔEh, Δp, ΔEh/Δp, and ΔEh-Δp across the bc$_1$ complex, against respiration rate (Jo), calculated from the values in Fig.4.

ΔEh is Eh(cyt.c) - Eh'(Q2) (●); Δp is (Δψ - 59ΔpH) (O); ΔEh/Δp (■) is the force ratio; and ΔEh-Δp (□) is equal to (-ΔG/F) for electron flow through the bc$_1$ complex coupled to vectorial proton transport, assuming a constant stoichiometry of 1 H$^+$/e$^-$.

respiration rate (Jo) were made in intact rat liver mitochondria (Brown & Brand, 1984). The measurement of redox states was made possible by the addition of exogenous ubiquinone-2 and cytochrome c, the reduction states of which were monitored spectrally at 275-290nm and 550-540nm respectively. Ubiquinone-2 reacts rapidly on the reducing side of the bc_1 complex. The actual redox potential (Eh' q) was calculated assuming a midpoint potential of +100mV, and using the external pH. The midpoint potential of cytochrome c was measured as +254mV. $\Delta\psi$ was measured from the accumulation of the lipophilic cation $TPMP^+$, ΔpH from the accumulation of [3H]acetate, and internal volume from the difference between [14C]mannitol and ^3H_2O spaces. Jo was measured with an oxygen electrode. The Δp and Jo were titrated with the uncoupler FCCP, either in conditions where ΔpH was fixed low and roughly constant (nigericin, plus 120mM KCl) or in conditions were $\Delta\psi$ was fixed relatively constant (valinomycin, plus 5mM KCl).

Fig. 1 shows the changes in Eh and decline in $\Delta\psi$ plotted against Jo. Fig. 2 shows the values of Δp, ΔEh, $\Delta Eh/\Delta p$, and $\Delta Eh-\Delta p$ calculated from Fig. 1. We know that at equilibrium i.e. zero rate,

$$\Delta G/F = \Delta Eh - n.\Delta p = 0$$

thus $n=\Delta Eh/\Delta p$.

In fig. 2, $\Delta Eh/\Delta p$ extrapolates to a value close to 1; this agrees well with kinetic estimates of n (Di Viriglio et al, 1981). Now if n is taken as 1 for all rates, then $\Delta G/F = \Delta Eh - \Delta p$, for the bc_1 complex.

We find that Jo is approximately linearly related to $\Delta Eh-\Delta p$, up to the uncoupled rate. The same is true when ΔpH is varied rather than $\Delta\psi$, as in figs. 3 & 4. The plot of $\Delta Eh-\Delta p$ has a similar slope but does not pass through the origin, which may mean that ΔpH (or effectively the internal pH, as the internal pH is highly buffered) has some additional kinetic control on rate. The conclusion is that ΔEh is close to equilibrium with Δp in state 4 and $\Delta Eh-n.\Delta p$ correlates well with electron flow through the bc_1 complex, so that irreversible thermodynamics may provide a useful description of respiratory control.

Measurements of cation permeability as a functionof $\Delta\psi$

Fig. 5 shows the relationship between $\Delta\psi$ and Jo as respiration is inhibited with malonate (a competitive inhibitor of succinate dehydrogenase). As suggested before this may indicate an increased permeability to protons at high $\Delta\psi$. However it is rather difficult to directly measure proton

Fig.5 Malonate titration
of Jo and Δψ.

4mg mitochondrial protein
was incubated in 1ml 200mM
sucrose, 30mM (K$^+$)Hepes
(pH 7.2), 10mM KCl, 3mM
(K$^+$) EGTA, 5μM rotenone,
2μg oligomycin, plus 0
to 10 mM final concentration
of (K$^+$) malonate.
Mitochondria were energised
with 5mM (K$^+$) succinate.
ΔpH was determined to be
negligible.

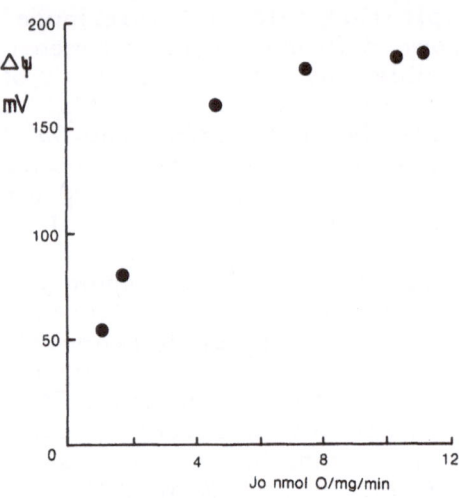

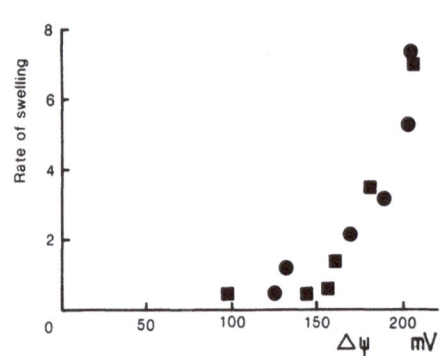

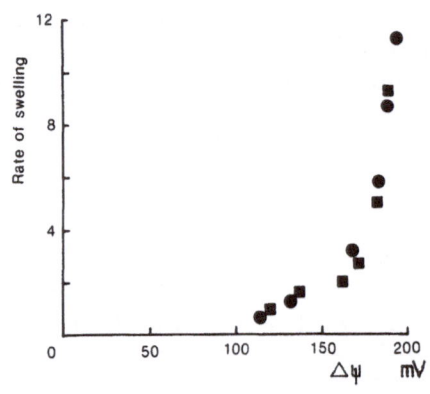

Fig.6   Fig.7

Fig.6 K$^+$ permeability as a function of Δψ.

2mg mitochondrial protein per ml was incubated in 120mM
(K$^+$)acetate, 5mM (K$^+$)Hepes (pH 7.1), 1mM (K$^+$)EGTA, 5μM rotenone,
1μg/ml oligomycin, plus 2.5mM (K$^+$)succinate, with either from 0
to 3mM (K$^+$)malonate (●) or from 0 to 100pmoles FCCP/mg protein
(■). The initial Δψ was determined with a TPMP$^+$-sensitive
electrode. The initial rate of swelling was taken as
proportional to the initial rate of absorbance decrease (at
540nm) on addition of succinate. Scale of swelling is in arbitary
units.

Fig.7 Tetramethylammonium permeability as a function of Δψ.

Same conditions as fig.6, except 120mM TMA acetate
replaced 120mM (K$^+$)acetate.

474

permeability as a function of $\Delta\psi$. On the other hand it is very easy to measure other cation permeabilities, simply by monitoring the rate of swelling of mitochondria suspended in phospate or acetate salts of the cation under study. Mitochondria scatter less light when they swell, and the initial rate of absorbance decrease is proportional to the rate of salt entry (Massari et al., 1972). Because of the very high rate of phosphate transport, swelling in these conditions will be rate limited by the cation permeability.

In fig. 6 the initial rate of swelling in K^+ phosphate is plotted against $\Delta\psi$ for mitochondria respiring on succinate. $\Delta\psi$ was measured with a $TPMP^+$-sensitive electrode, and was decreased either with the inhibittor malonate or with the uncoupler FCCP. With either method of decreasing $\Delta\psi$ it is evident that the permeability to K^+ increases dramatically above 150mv. This bears a striking similarity to what appears to be happening to the H^+ permeability in fig. 5.

In fig. 7 the experiment was repeated, with similar results, using an unphysiological ion, tetramethylammonium, replacing K^+. The same relationship also appears to occur with $Tris^+$ and $choline^+$ salts. This indicates a general unspecific increase in cation permeability at high $\Delta\psi$; this would fit in with certain evidence that it is a change in the structure of the mitochondrial lipids which is responsible for the change in permeability to protons at high $\Delta\psi$ (O'Shea et al, 1984).

A number of bacteria have also been shown to have large increases in general ion permeability at high $\Delta\psi$ (Clark et al., 1983).

What is the significance of this permeability change ?. Is it a general physical breakdown of bilayer permeability at high $\Delta\psi$?. Does it occur in cells ?. The answers are not yet known, but we do know that the permeability change is occuring over the "control" range of $\Delta\psi$, i.e. between state 3 and state 4; the range over which mitochondria in cells normally operate. The effect, if it occured in cells, would be to buffer $\Delta\psi$ against changes in substrate supply, substrate type, calcium cycling, limited uncoupling, and ATP turnover. The Δp of the cell would be effectively clamped and with it the cytosolic ΔG_p. A fixed ΔG_p might have many functional advantages, as has been pointed out (Stucki, 1980), however such teleological speculation is groundless until the permeability change can be shown to occur physiologically in cells.

CONCLUDING REMARKS

We have seen that respiration exhibits several types of control, which may be regarded either as allosteric control of far from equilibrium enzymes in the pathway, or as control of metabolite concentrations, often across close to equilibrium enzymes, by pathways or processess external to the primary pathway.

The rate of respiration is probably directly determined by the amount of reduced cytochrome a_3 or an intermediate in oxygen reduction. This in turn may well be determined by $\Delta Eh-n.\Delta p$ of the rest of the respiratory chain. ΔEh is affected by the activation state of the non-equilibrium dehydrogenases, and possibly also by transport of substrates. Δp is determined by the rate of ATP turnover, but is also probably internally controlled by ion permeabilities. Whether there are localised circuits of protons between the respiratory chain and the ATPase is as yet uncertain.

REFERENCES

Azzone, G.F., Pozzan, T., Massari, S. and Bragadin, M., 1978, Biochim. Biophys. Acta, 501:296.

Brown, G. and Brand, M., 1984, Biochem. J., In Press.

Clark, A.J., Cotton, N.P.J. and Jackson, J.B., 1983, Eur. J. Biochem., 130:575.

Davis, E.J. and Lumeng, L., 1975, J. Biol. Chem., 250:2275.

Davis, R.J. and Martin, B.R., 1982, Biochem. J., 206:611.

Di Virgilio, F., Pozzan, M. and Azzone, G.F., 1981, Eur. J. Biochem., 177:225.

Doussiere, E.J., Ligeti, E., Brandolin, G. and Vignais, P.V., 1984 Biochim. Biophys. Acta, 776:492.

Denton, R.M. and Hughes, W.A., 1978, Int. J. Biochem., 9:545.

Gutman, M., 1980, Biochim. Biophys. Acta, 594:53.

Hansford, R.G., 1980, Curr. Topic. Bioenerg., 10:217.

Harris, E.J., Hofer, M.P. and Pressman, B.C., 1967, Biochemistry, 6:1348.

Heinrich, R. and Rapoport, T.A., 1974a, Eur. J. Biochem., 42:89.

Heinrich, R. and Rapoport, T.A., 1974b, Eur. J. Biochem., 42:97.

Kacser, H. and Burns, J.A., 1973, in "Rate Control of Biological Processes", P.D. Davies, ed., Cambridge University Press, London.

Katchalsky, A. and Curran, P.K., 1967, in "Non-Equilibrium Thermodynamics in Biophysics, Harvard University Press, Cambridge, Mass.

Massari, S., Frigeri, L. and Azzone, G.F., 1972, J. Membrane Biol., 9:71.

Nicholls, D.G., 1974, Eur. J. Biochem., 50:305.

O'Shea, P.S., Thelen, S., Petrone, G. and Azzi, A., 1984, FEBS Lett., 172:103.

Owen, C.S. and Wilson, D.F., 1974, Arch. Biochem. Biophys., 161:581.

Palmieri, F., Cisternino, M. and Quagliariello, E., 1967, Biochim. Biophys. Acta, 143:625.

Pietrobon, D., Azzone, G.F. and Walz, D., 1981, Eur. J. Biochem., 117:389.

Pietrobon, D., Zoratti, M., Azzone, G.F., Stucki, J.W. and Walz, D., 1982, Eur. J. Biochem., 127:483.

Soboll, S., Seitz, H.J., Seis, H., Ziegler, B. and Socholz, R., 1984, Biochem. J., 220:371.

Stucki, J.W., 1980, Eur. J. Biochem., 109:257.

Stucki, J.W., 1983, Biophys. Chem., 18:111.

Stucki, J.W., Compiani, M. and Caplan, S.R., 1983, Biophys. Chem., 18:101.

Sugdena M.C. and Williomson, D.H., 1982, in "Metabolic Compartmentation", H. Sies, ed., Academic Press.

Tager, J.M., Wanders, R.J.A., Groen, A.K., Kunz, W., Bohnensack, R., Kurter, U., Letko, G., Bohme, G., Duszynski, J. and Wojtczak, L., 1983, FEBS Lett., 151:1.

Tischler, M.E., Friedrichs, D., Coll, K. and Williamson, J.R., 1977, Arch. Biochem. Biophys., 184:222.

Vignais, P.V., 1976, Biochim. Biophys. Acta, 456:1.

Westerhoff, H.V., Melandri, B.A., Venturoli, G., Azzone, G.F. and Kell, D.B., 1984, FEBS Lett., 165:1.

Williamson, J.R., Cooper, R.H. and Hoek, J.B., 1981, Biochim. Biophys. Acta, 639:242.

Wilson, D.F., Owen, C.S. and Holian, A., 1977, Arch. Biochem. Biophys., 182:749.

MOLECULAR STUDY ON SOLUTE CARRIERS IN BIOMEMBRANES, EXEMPLIFIED
BY TWO WELL-CHARACTERIZED CARRIERS FROM MITOCHONDRIA, THE ADP/ATP
CARRIER AND THE UNCOUPLING PROTEIN

Martin Klingenberg

Institut für Physikalische Biochemie
Universität München
München, BRD

INTRODUCTION

Specific solute transport is a central function of biomembranes for facilitating the translocation of specific solutes
such as metabolites and concomittant ions. For this purpose
biomembranes are equipped with specific catalysts, biomembrane
carriers or translocators which catalyze the transport of these
solutes. This process is not yet well understood and one of the
most important problems of membrane functions waited to be solved.
Since the central issue in the transport is the catalytic translocation across the membrane, it is advantageous to concentrate
on those systems where the catalyst has only this function.

In active transport systems to the actual translocation an
energy supply mostly from ATP is superimposed. Thus in the ATP-
driven cation pumps an energy transduction machinery complicates
the analysis of the translocational section in the carrier. On
the other hand, simple transport systems for solutes such as for
glucose at the plasma membrane and the solute carriers in the
inner mitochondrial membrane provide only the catalytic section
to the solute transport. In these instances concentration
gradients can be also generated by superimposition of a membrane
potential to the electrically charged solute translocation. This
is not obligatory and we can use these transport systems for
concentrating on the catalytic mechanism of translocation by dis-
regarding "primary or secondary" active transport.

Analogy and Difference Between Carrier and Enzyme Catalysis

Solute transport through biomembranes involves "carriers" or translocators which are intrinsic membrane proteins. These proteins recognize and select specifically their substrates for the translocation within a catalytic cycle. In many respects it is useful to compare the function of these protein catalysts with enzymes. For this purpose their function can be described in terms of a catalytic cycle as shown in Fig. 1. A substrate A from the outside binds to the carrier forming a carrier substrate complex similar to the enzyme substrate complex. The actual catalytic reaction with the carrier is the vectorial transposition of the substrate such that it faces the other side of the membrane where it can dissociate into the internal space. In contrast, in enzymes the catalytic step is the chemical interconversion of the substrate. The catalytic step of the carrier is a vectorial one, the enzymatic step a scalar one.

Whereas the enzyme has completed the catalytic cycle after dissociation of the product and is ready for the next reaction, the free carrier has still to reconvert its binding site from the internal to the external phase of the membrane before commencing the next transport cycle. In summary an important feature in carriers and enzymes is the recognition of binding of the substrates. Also in both cases, it is the interaction between protein and substrate which facilitates the catalysis, i.e. the decrease of the activation barrier. Differently, the substrate in carrier catalysis retains the identity but is transformed by the enzyme. As we shall discuss below, carrier catalysis involves large conformation changes of the protein whereas the enzyme catalysis concentrates more on the distortion of the substrate in preparation for the chemical transformation.

In many cases transport catalysis involves an exchange system where for the uptake of one solute molecule another molecule is exported. In this case the carrier does not return unloaded but only when it binds solutes also from the inside to the outside. This obligatory counter-exchange can have important advantages for the catalytic process since the return of the carrier molecules is again facilitated by substrate protein interaction as we shall discuss further below.

For understanding and analysis of the biomembrane transport catalysis, we shall now concentrate on two solute transport systems occurring in the inner mitochondrial membrane. These are relatively elementary transport catalysts for either the unidirectional (uncoupling protein) or bidirectional exchange (ADP/ATP carrier) transport. Both consist of a relatively small membrane protein which has been isolated and well characterized. As we shall see further, a comparison of these two systems shows several common but also several constrasting features which are particularly revealing for understanding the molecular characteristics of solute transport

Catalytic Cycle

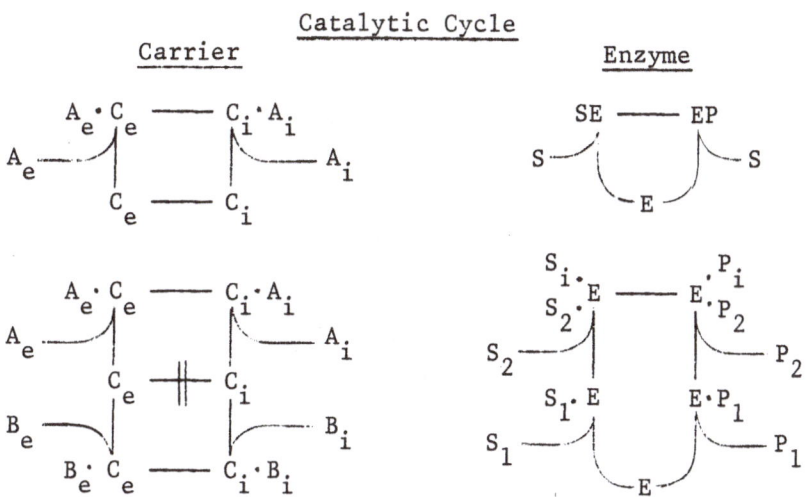

Fig. 1.   Comparison between the catalytic cycles of biomembrane carrier and enzymes. Unidirectional and exchange carriers. One and two substrate enzymes.

General Features of the Two Translocators

All mitochondria contain the ADP/ATP carrier (AAC), even those which are deficient in respiratory components or in coupled ATP-synthase, for example mitochondria from yeast mutants.[1] Thus the ADP/ATP carrier has become the most universal indicator of mitochondria or, more specifically, of the inner mitochondrial membrane (Fig. 2). In contrast, the uncoupling protein (UCP), which we regard as a H^{+}-carrier,[2] is expressed specifically only in mitochondria from brown adipose tissue and has not been found so far in any other cell. Furthermore, whereas the ADP/ATP carrier functions as a supplier of free energy to the cytosol in the form of ATP, the uncoupling protein dissipates the free energy generated by the respiratory chain in the form of heat. Already here we have two major contrasts between the two systems and it will be interesting to see what factors or intrinsic properties of the translocators are responsible for these contrasts.

Both proteins bind ADP and ATP and this common denominator was actually the original reason for our interest in the uncoupling protein (Table 1).[3,4] Moreover, the common ligands justify a comparison between the two proteins. Most important for further analysis is the contrast between these interactions: whereas ADP

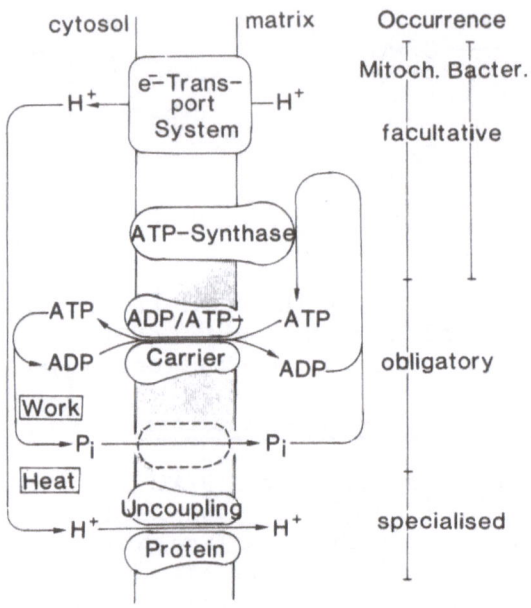

Fig. 2.   The occurrence and function of two translocators in the
          inner membrane of mitochondria, the ADP/ATP carrier (AAC)
          and the uncoupling protein (UCP).

Table 1. Introductory Properties

	ADP/ATP Carrier (AAC)	Uncoupling Protein (UCP)	Ref.
	Similarity		
M.W. in SDS-gel k_D	30	33	
Nucleotide ligand	ADP/ATP ——— ADP/ATP		
Localization	inner mitochondrial membrane		
Abundance, $\dfrac{Mol}{Cyt\ aa_3}$	2 to 3		5,6
% of mitochondrial protein	14 (BHM)	18 (HBAT)	6,7
	Contrast		
Function	providing free energy to cytosol	converting free energy to heat	
Occurrence	universal to mitochondria	only in BAT-mitochondria	
Substrate	ADP/ATP-largest transport solute	H^+-smallest solute	

BHM = beef heart mitochondria, HBAT, hamster brown adipose tissue mitochondria

and ATP are substrates for AAC, they are inhibitors for UCP. For UCP, ADP and ATP are probably endogenous regulators, whereas for AAC exogenous antibiotics are developed in plants and prokaryotes, such as carboxyatractylate and bongkrekate. The most important common feature of the two proteins is the similar size of their peptides of about 33 kD, which originally encouraged us to isolate UCP according to similar methods as those used for AAC. The similarly high abundance of the two proteins in the respective mitochondria bears on the similar isolation procedures. Both AAC and UCP are the most abundant membrane proteins in these mitochondria.

The Protein Detergent Micelles

Being translocators, both AAC and UCP can be expected to traverse the membrane and, because of their relatively low

Table 2. Isolation

	ADP/ATP Carrier (AAC)	Uncoupling Protein (UCP)	Comment
Source of mito-chondria	beef heart	hamster brown adipose tissue	
	preloading with atractylate	---	
	removal of loose protein with Brij 38-detergent		
	sediment		
	solubilization with <u>Triton X-100</u> + Na_2SO_4		
TX/protein g/g	2.8	1.8	UCP easier solubilized than AAC
Na_2SO_4 mM	100	25	
	supernatant		
	<u>hydroxylapatite</u>		unprotected AAC denatured and retained by HTP
	at $4^{\circ}C$	at 25°	
	breakthrough		
	60 to 70% yield		
	Agarose (e.g. ACA 34)	Sephadex (G 150)	for UCP only removal of lipids and detergents
(Optional)	sucrose gradient centrifugation		
Purity	95%	90%	

molecular weight, to exist as deeply membrane-imbedded proteins which do not markedly protrude from the membrane. As a result both proteins are bound tightly to membranes. Like the ADP/ATP carrier the uncoupling protein also requires strong detergents for solubilization. In fact, as expected, the time-proven procedure originally developed for purification of the ADP/ATP carrier could be applied also to isolate the uncoupling protein, with only minor modifications. This method has in the meantime been so successful that it has made possible the isolation of several other membrane proteins in our laboratory.[8-11] Again Triton X-100 proved to be the most useful detergent. Fortunately - quite differently from the situation with AAC - in UCP the binding of nucleotide was retained after solubilization, indicating that the protein remained intact. Thus the convenient assay with ^{14}C-GDP was utilized to trace and quantitate UCP.[4]

Table 3. Characteristics of Carrier-Detergent Micelle

	AAC	UCP
Gel filtration		
Stokes' radius $\overset{\circ}{A}$	65	61
Triton X-100 g/g	1.47	1.9
mol/mol	150	180
Phospholipid mol/mol	16	2
Analytical UC		
Sedimentation coefficient (10^{-13}sec)	3.93	4.0
Mr sedimentation equilibrium	172	180
Mr protein	64	62

Ref. 12, 13

A comparison of the isolation of both membrane proteins is given in Table 2. UCP appears in the breakthrough of hydroxylapatite which is indicative of a large envelope of Triton around the protein. Because of the high original concentration of hydroxylapatite, this step already gives an 80% pure preparation which can be used for many applications.[4,6] Further purification steps may include the sucrose gradient or gel filtrations,yielding finally 95% pure UCP.

Whereas AAC is isolated as carboxyatractylate-complex, UCP can be isolated unprotected without a ligand. Both the isolated AAC and UCP exist as mixed protein-Triton X-100 micelles of a surprisingly similar overall structure (Table 3) and contain a high amount of bound Triton. This indicates that both proteins are deeply imbedded in the membrane and have a large hydrophobic surface normally covered by the acyl-chains and now replaced by the hydrophobic section of the detergent. In gel filtration both proteins migrate with an unusually high Stokes' radius of 62 to 65 Å. Obviously the large detergent shell with the sponge-like polyoxethelene chain forms a large hydrophilic belt. Sedimentation runs give a relatively high sedimentation coefficient of S = 3.9 to 4.2 x 10^{-13}sec, again indicative of similarities between the two protein micelles. Sedimentation equilibrium gives nearly the same molecular weight for the two protein-detergent micelles. Subtracting from this value the amount of Triton bound, one arrives at 64 kD for AAC and 62 kD for UCP. This indicates that both proteins are dimers, made up of two subunits each with similar Mr values, as inferred from the SDS gels. This overall structural similarity has been recently further substantiated by a comparison of the amino acid sequences.

Dimer Structure and Half-Site Reactivity

A striking feature of the AAC is the binding of only one molecule of inhibitor or substrate to apparently two subunits. This was derived from the functional Mr as calculated from the binding per protein.[14] Only later the hydrodynamic studies showed that in fact the AAC exists as a dimer.[12] It is of great significance that also in the UCP a functional Mr as determined from the binding of ADP or GDP to the purified protein corresponds to 60 kD and thus to a dimer.[4,6] Again this is confirmed by the hydrodynamic data.[13] Thus obviously both proteins have the peculiar property that they exhibit only one binding site in the dimeric molecule although each of the subunits might be expected to carry the information for one binding domain. For this reason we had speculated that the binding site is along the two-fold axis in the center of the protein where binding of one ligand sterically prevents the attachment of a second ligand on the opposite site.[15,16] As a result we suggested that the translocation path in these molecules also follows the central two-fold axis.

Nucleotide-Protein Interaction

Since both proteins bind nucleotides, a comparison of nucleotide interaction with the two carriers is imperative and highly rewarding. Whereas in the AAC, ADP or ATP "activate" their own transport,[17] in UCP they function as inhibitors of the H^+-transport (Table 4). Cocomitantly in AAC the affinity for ADP or ATP is relatively low,[5] while conversely in UCP it is high.[6] This is understandable in view of their different function in the two proteins. Rapid dissociation is required for transporting ADP and ATP and slow dissociation for the inhibition.

However, there is much more to this affinity difference which tells us about the activation energy in the translocation mechanism. It may seem paradoxical that in AAC loose binding is associated with higher specificity only for adenine nucleotides whereas the much higher affinity in UCP is less specific, also allowing for guanine or inosine nucleotides. The high specificity indicates higher intrinsic binding energy at the AAC than at the UCP. However, in AAC considerable catalytic energy is withdrawn from the intrinsic binding energy for lowering the activation energy. This energy is invested into the strong conformation changes required in the AAC structure for translocation (see below). As a result there is a strong "debinding" of ADP and ATP and, in balance, a low affinity to the AAC. In contrast, on binding of ADP or ATP

Table 4. Nucleotide Interaction

	ADP/ATP Carrier	Uncoupling Protein	
ADP-ATP influence	activating transport	inhibiting transport	c
Specificity	only adenosine-DP, TP	also guanosine-DP, TP	c/s
pH-dependence of binding	low	strong	c
Sidedness of binding	c- and m-side	only c-side	c
Affinity	low ($<10^5 \ M^{-1}$)	high ($>10^6 \ M^{-1}$)	c
Binding to purified protein	$\frac{\mu Mol}{g \ prot}$ 17 (CAT*)	15 (GTP,ATP,etc.)	s
Equivalent (functional)Mr,kD	≥ 59	≥ 62	s
Functional dimer →	- "half-site" reactivity -		
	"DAN"- nucleotide**		
Specificity	DAN-AMP,-ADP,-ATP	DAN-ADP, -ATP	s
Fluorescence	strongly fluorescent (E_m 520 nm)		s
Sidedness	only m-side	only c-side	c
Affinity	$10^6 \ M^{-1}$	10^5 to $10^6 \ M^{-1}$	s
Transport	inhibiting	non-inhibiting	c

*CAT - carboxyatractylate,   c = contrast, s = similarity
**DAN-nucleotide = 1,5-dimethylamino-naphthoyl-3-0'-AXP

as inhibitors to the UCP the conformation change of the protein can be expected to be smaller. Even if the intrinsic binding energy may be smaller than in the AAC, as reflected in the lower specificity, the only small energy investment for the protein conformation change still leaves a high external binding affinity.

This conclusion receives further strong support from the influence of ADP/ATP binding on the protein structure. In the AAC the protein becomes more labile when a number of amino acid residues are exposed.[7,18] For example lysine and cysteine become more accessible to the respective amino acid reagents. Moreover, the protein becomes more susceptible to proteolytic attacks. In contrast, in UCP the ADP/ATP binding causes a stabilization of the protein structure and provides protection against trypsin attack (unpublished observation). We may visualize that nucleotide interaction in the AAC "fluidizes" the protein, in line with its activation of the conformational transition, whereas in the UCP the protein becomes more rigid.

Fluorescence analogues of adenine nucleotides such as in dimethylamino-naphthoyl-AMP, -ADP, and -ATP have proven to be very useful ligands for studying the AAC.[19,20] Since these three ligands exhibit strong fluorescence only on binding to the m-side they permitted to directly record ligand binding and transition from c- to m-state as well as the m-state distributions. Binding affinity is higher than with ADP and ATP (Table 4). Furthermore, the DAN-nucleotide binding inhibits ADP/ATP exchange. This agrees well with our concept of inhibitor interaction with the AAC in that they do not require conformational energy investment which causes a debinding of the DAN-ADP.

It was very gratifying that DAN-ADP and DAN-ATP bind to the UCP also by exhibiting a strong fluorescence.[21] This binding is competitive with ADP or ATP and therefore obviously occurs at the same inhibitory site and thus is directed only to the c-side of UCP. Also differently from AAC, the affinity of DAN-ADP/ATP, although still relatively high, is lower than that of ADP and ATP. Further in contrast to AAC, DAN-ADP/ATP binding does not inhibit markedly the H^+-transport of UCP. This complements other kinetic observations which made us assume that the binding of DAN-AMP cannot induce the same conformation change in UCP as ADP or ATP and which causes the inhibition of transport (unpublished results).

In conclusion, the fact that the binding of identical ligands to similar proteins causes opposite effects on the protein structure strongly enhances the significance of each individual case and increases our understanding of the interaction of ligands with carriers in general.

Inhibitor-Carrier Interaction

For UCP the nucleotides are inhibitors of the transport action, i.e., of H^+-translocation. For AAC strong and specific inhibitors are produced as antibodies by particular plants or bacteria. The glucoside carboxyatractylate is the most potent inhibitor of the ADP/ATP transport[22,23] It has a structure quite different from ADP and ATP, but also carries four negative charges. Another effective inhibitor, bongkrekate, has a complete different structure, being a polyisoprenoid tricarboxylic acid, but is again equipped with a common denominator, namely three anionic charges. These two inhibitors have been instrumental in developing the single-site-gated-pore mechanism of the molecular events in the translocation process, construed about 10 years ago for the ADP/ATP carrier.[24] It is implied that these inhibitor ligands overlap essentially with the free binding site. Most importantly, CAT binds only when the binding center faces to the outer side, while BKA binds preferentially when it faces the inner side of the membrane. In the case of the AAC this has been called the "c" (cytosolic) state or the "m" (matrix) state of the carrier center. By binding these inhibitors the carrier can be fixed either in the c-state or in the m-state. Only one type of ligand, substrate or inhibitor at a time can bind to a carrier molecule.

In both cases, in the AAC and the UCP, the inhibitors are "side-specific", i.e. they interact with the carrier only from one side of the membrane (Fig. 3). First of all, this is in line with the asymmetric localization of the carrier with the two-fold axis traversing the membrane plane.[16] Accordingly, the binding center should have a different conformation when looking to the external or internal side respectively. Whereas in substrate binding this difference is not apparent, it is manifest with inhibitors which fix either one of the two states.

Here a comparison to enzyme catalysis is particularly useful. In enzymes, "transition state" analogues are often powerful inhibitors.[25] The structure of these inhibitors imitates the structure of the substrate molecule, being distorted in the transition state enzyme complex. This enzyme-inhibitor-complex reflects a central state in the catalytic cycle. In contrast, we propose that in carriers the inhibitor complements the ground state of the protein which strongly differs from the catalytic transition state.[17] Here the inhibitor seems to pull the ground states further away from the transition state towards a still higher side specificity. Earlier on we called this an "abortive ground state inhibition".[17] The sidedness of these inhibitors for AAC and UCP is typical also for all other known membrane carriers.

We assume that the binding of these inhibitors to the carrier site reflects the great conformational flexibility of the binding center.[15] This flexibility is a necessary result of the translocation process, associated with the "squeezing" of the substrate through the carrier. For a further understanding it is very useful to consider activation energy profiles (Figure 4). The catalytic energy ΔG_{cat} corresponds to the difference in activation energy ΔG between the uncatalyzed and catalyzed reactions. In enzyme catalysis this energy corresponds mainly to the distortion of substrate in preparation of the chemical bond changes, i.e. $\Delta G_{cat} \simeq \Delta G_{S**} - \Delta G_{ES**}$.[26] We propose that in contrast in carrier catalysis the energy is mainly invested into the protein conformation change, i.e. $\Delta G_{cat} \simeq \Delta G_{C_{ie}**} - \Delta G_{C_{ie}** \cdot S}$. Here the preparative

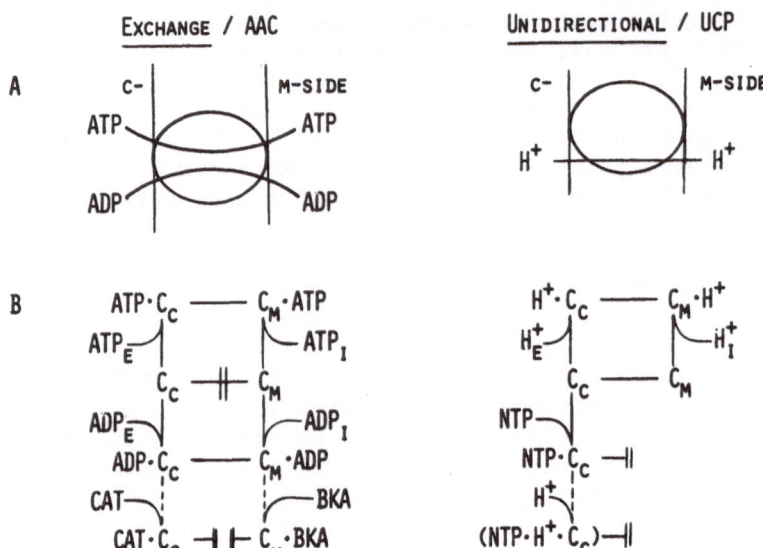

Fig. 3. The comparison of the catalytic transport cycle for the AAC and UCP and the point of interference of inhibitor ligands. CAT = carboxyatractylate, BKA = bongkrekate, NPT = purine-TP, c = cytosolic side, m = matrix side.

chemical bond distortion is unnecessary and there exists no substrate transition state. In both cases this catalytic energy is withdrawn from the intrinsic binding energy of the substrate with the protein.

In carriers, therefore, no "transition analogue" exists and no binding of inhibitor in the central transition state can be expected. On the other hand, starting from the ground state, intrinsic binding energy can be well utilized and, since it is not consumed by protein conformational changes, this is expressed in the high affinity of an inhibitor. In fact, the structure of inhibitors may more faithfully complement the structure of the active centers in the ground states than the substrates. Accordingly great differences of the binding center which are to be expected when it opens to the internal or external side, are complemented in the large structural differences for the AAC between the inhibitor ligands atractylate and bongkrekate. Thus by pointing out the sidedness of inhibitors in binding to the carrier, in contrast to enzymes, and by developing its significance for the carrier binding site, important insight into carrier catalysis is obtained.

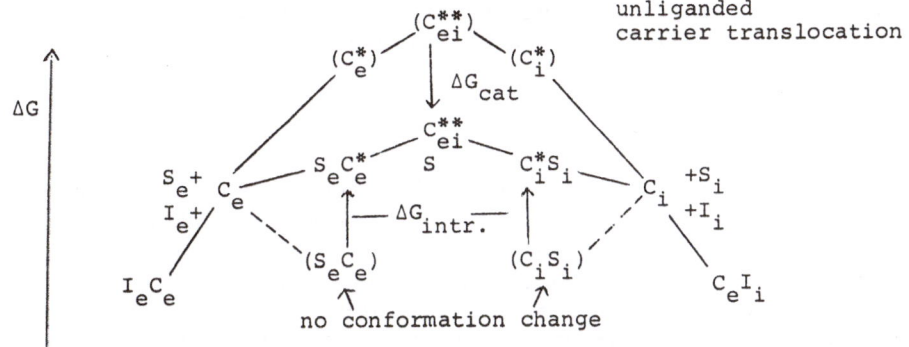

Fig. 4. The activation energy profile of the carrier catalyzed substrate translocation and the energies of the interaction of the carrier with substrate and inhibitor. Energy gradients between both sides are suppressed for simplicity.

C_e = carrier binding center facing the external face,
C_i = the internal face of the membrane,
C_{ie} = transfer state of binding center.

Chemical Characteristics

The amino acid sequence of the ADP/ATP carrier and thus the complete chemical description are now known.[27] It was the first amino acid sequence of a biomembrane carrier protein established according to conventional sequencing methods. The total molecular weight was found to be 32 870 daltons. The externally largely hydrophobic character of the protein can obviously not derive from the primary structure since many polar amino acids are distributed along the chain. There are 6 glutamic and 15 aspartic acids, 23 lysines and 17 arginines. Since the N-terminus is blocked by acetyl, a net difference of 18 positive charges remains. This is in agreement with the unusually high isoelectric point of more than 10.5. There are remarkably frequent clusters of polar amino acids forming triplets or doublets and ion pairs. An unusual feature is the occurrence of the trimethyl lysine in position 51.

Predictions of the secondary structure of membrane proteins from the primary structure are now often made, featuring trans-membrane hydrophobic α-helical segments comprising about 18 to 22 residues. Only two such segments are found in this case between Tyr 111 and Leu 133 and between Ile 209 and Phe 230. Another candidate for hydrophilic segments is present from Gly 171 to Tyr 194, interrupted only by one polar residue Arg 187.

With computerized "hydroplots" using an averaging alogarithm of hydrophobic increments,[28] it has been claimed that five possible α-helical segments can be discerned in the primary structure.[29] The α-helical content based on circular dichroism (41%) could accommodate five transmembrane α-helical segments. Whether all of this content is in the form of transmembrane α-helices seems doubtful since a similar α-helical content is found in other nucleotide-binding soluble enzymes with α-helical stretches of varying length. Thus the secondary structure has to provide first of all the nucleotide binding domain as known from other proteins.

The secondary structure folding across the membrane was probed with the lysine reagent pyridoxal phosphate (PLP) taking advantage of the abundant distribution of lysines across the chain. Comparing the PLP incorporation to inner mitochondria and with sonic particles, the localization of the labeled lysines in the c- or m-state has been deduced.[30] Together with the potential transmembrane helical segments a tentative folding pattern has been described. By comparing the incorporation of PLP into the c- and m-state those lysines sensitive to conformational changes could be detected. These are possibly close to the translocation channel where the major changes take place (see below). Unmasking of lysine by removal of ligands such as atractylate should reveal lysines involved at the binding site. The appearance of several additional lysines on isolating the carrier in the nonionic detergent micelle has been attributed to those groups which are at the "neckline" of the protein and which are, in the membrane, masked and compensated by the phosphatidic, negatively charged headgroups. The tentative assignments of these structural features to the folding and distribution of the primary structure are summarized in Fig. 5.[31]

Conformational Change and Carrier Mechanism

A fundamental conclusion from the binding studies on the ADP/ ATP carrier in mitochondria was the existence of the reorienting carrier mechanism, also reclassified as single-site gated pore mechanism of carrier action as will be discussed below. This mechanism implies a drastic conformation change of the carrier protein associated with the translocation step.[15] Probably the most direct evidence is a drastic change of the mitochondrial membrane associated with the transition between the c- and m-state as reviewed above. At high density of membrane molecules in the membrane, the opening of the binding site to the c- or m-side respectively could conceivably be responsible for the drastic configuration changes of the internal cristae.[32] In fact these changes were early interpreted as spreading of c-surfaces and

492

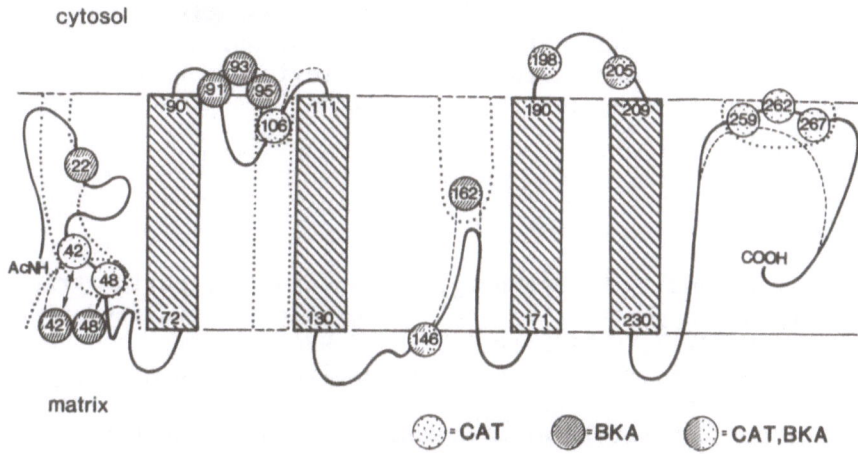

Fig. 5. Possible spanning of the ADP/ATP carrier across the
membrane as derived from lysine labeling in the c- and
m-state in sonic particles and mitochondria. The dotted
circles represent lysines, labeled in c-state, the
hatched circles indicate the newly or more exposed
positions of the m-state. Dotted lines show possible
pore regions, implicated by reactivity from both sides of
the membrane. Dashed lines link exclusively m-state
reactive lysines to the sequence.[31]

decreasing of the m-surface in transition to the c-state as
compared to the reversed situation when the carrier is brought
into the m-state.

The high sensitivity to proteases of the carrier in the m-
state on the one hand and its resistance in the c-state on the
other hand also represent early evidence for large conformational
differences.[7,18] Obviously the carrier is more open "in the m-
state", easier unfolded and denatured. Immunological evidence for
specific conformation of the c-state was early obtained when the
antibodies raised against the CAT-protein complex did not react
very well with the BKA-protein complex or denatured protein.[33]
Although antibodies against the BKA-protein are raised, these had
Also considerable cross reactivity with the CAT-protein.[34]

Particularly significant is the unmasking of SH-groups on
transition from the c- to the m-state (see above). This can be
readily observed both in membrane and isolated protein. Using
reagents with absorption or fluorescent changes, the kinetics of
the transition can be directly recorded.[35] Arginine groups
possible involved in the binding center are also more accessible

to phenylglyoxal in the m-state.[36,37] Difference in lysine incorporation between c- and m-state, the sites of which were precisely localized in the primary structure, were reviewed above.[31] Also inactivation of binding due to UV-radiation is stronger in the m- than in the c-state.[38]

REFERENCES

1. J. Kolarov and M. Klingenberg, The adenine nucleotide translocator in genetically and physiologically modified yeast mitochondria, FEBS Lett. 45:320 (1974).
2. D. G. Nicholls, The influence of respiration and ATP hydrolysis on the proton-electrochemical gradient across the inner membrane of rat-liver mitochondria as determined by ion distribution, Eur. J. Biochem. 50:305 (1974).
3. M. Klingenberg, H. Hackenberg, R. Krämer, C. S. Lin and H. Aquila, Two transport proteins from mitochondria: I. Mechanistic aspects of asymmetry of the ATP/ATP translocator. II. The uncoupling protein of brown adipose tissue mitochondria, Ann. N.Y. Acad. Sci. 358:83 (1980).
4. C. S. Lin and M. Klingenberg, Isolation of the uncoupling protein from brown adipose tissue mitochondria, FEBS Lett. 113:299 (1970).
5. M. J. Weidemann, H. Erdelt and M. Klingenberg, Adenine nucleotide translocation of mitochondria. Identification of carrier sites, Eur. J. Biochem. 15:313 (1970).
6. C. S. Lin and M. Klingenberg, Characteristics of the isolated purine nucleotide binding protein from brown fat mitochondria, Biochemistry 21:2950 (1982).
7. M. Klingenberg, P. Riccio and H. Aquila, Isolation of the ADP/ATP carrier as the carboxyatractylate-protein complex from mitochondria, Biochim. Biophys. Acta 503:193 (1978).
8. P. Riccio, H. Schägger, W.D. Engel and G. von Jagow, bc₁-Complex from beef heart: One-step purification by hydroxyapatite chromatography in Triton X-100, polypeptide pattern, Biochim. Biophys. Acta 459:250 (1977).
9. U. I. Flügge and H. W. Heldt, The phosphate translocator of the chloroplast envelope, isolation of the carrier protein and reconstitution of transport, Biochim. Biophys. Acta 638:296 (1981).
10. G. Unden, H. Hackenberg and A. Kröger, Isolation and functional aspects of the fumarate reductase involved in the phosphorylative electron transport of Vibrio succinogenes, Biochim. Biophys. Acta 591:275 (1980).
11. A. Kröger, E. Winkler, A. Innerhofer, H. Hackenberg and H. Schägger, The formate dehydrogenase involved in electron transport from formate to fumarate in Vibrio succinogenes, Eur. J. Biochem. 94:465 (1979).

12. H. Hackenberg and M. Klingenberg, Molecular weight and hydro-dynamic parameters of the adenosine 5'-diphosphate-adenosine 5'-triphosphate carrier in Triton X-100, Biochemistry 19:548 (1980).

13. C. S. Lin, H. Hackenberg and M. Klingenberg, The uncoupling protein from brown adipose tissue mitochondria is a dimer. A hydrodynamic study, FEBS Lett. 113:304 (1980).

14. P. Riccio, H. Aquila and M. Klingenberg, Purification of the carboxy-atractylate binding protein from mitochondria, FEBS Lett. 56:133 (1975).

15. M. Klingenberg, P. Riccio, H. Aquila, B.B. Buchanan and K. Grebe, Mechanism of carrier transport and the ADP/ATP carrier, in: "The Structural Basis of Membrane Functions," Y. Hatefi and L. Djavadi-Ohaniance, eds., Academic Press, New York, pp. 293-311 (1975).

16. M. Klingenberg, Membrane protein oligomeric structure and transport function, Nature 290:449 (1981).

17. M. Klingenberg, The mechanism of the mitochondrial ADP/ATP carrier as studied by the kinetics of ligand binding, in:"Dynamics of Energy-Transducing Membranes," L.Ernster et al., eds., Elsevier, Amsterdam, pp. 511-528 (1974).

18. H. Aquila, W. Eiermann, W. Babel and M. Klingenberg, Isolation of the ADP/ATP translocator from beef heart mitochondria as the bongkrekate-protein complex, Eur. J. Biochem. 85:549 (1978).

19. I. Mayer, A. S. Dahms, W. Riezler and M. Klingenberg, Inter-action of fluorescent adenine nucleotide derivatives with the ADP/ATP carrier in mitochondria. 1. Comparison of various 3'-O-ester adenine nucleotide derivatives, Biochemistry 23:2436 (1984).

20. M. Klingenberg, I. Mayer and A. S. Dahms, Interaction of fluorescent adenine nucleotide derivatives with the ADP/ATP carrier in mitochondria. 2. (5-(dimethyl-amino)-1-naphthoyl)adenine nucleotides as probes for the transition between c- and m-states of the ADP/ATP carrier, Biochemistry 23:2442 (1984).

21. M. Klingenberg, Characteristics of the uncoupling protein from brown fat mitochondria, Biochem. Soc. Transactions 12:390 (1984).

22. P. V. Vignais, P. M. Vignais and G. Defaye, Adenosine di-phosphate translocation in mitochondria. Nature of the receptor site for carboxyatractyloside, Biochemistry 12:1508 (1973).

23. M. Klingenberg, K. Grebe and B. Scherer, The binding of atractylate and carboxy-atractylate to mitochondria, Eur. J. Biochem. 52:351 (1975).

24. M. Klingenerg, H. Buchholz, H. Erdelt, G. Falkner, K. Grebe, H. Kadner, B. Scherer, L. Stengel-Rutkowski and M. J. Weidemann, The adenine nucleotide carrier: study of its

translocating mechanism by binding with adenosine di-
phosphate, atractyloside and bongkrekic acid, in:"Biochemistry and Biophysics of Mitochondrial Membranes,"
G. F. Azzone et al., eds., Academic Press, New York/
London, pp. 465-486 (1972).

25. G. E. Lienhard, Transition analogue inhibitors, Science 188:
149 (1973).

26. W. P. Jencks, Binding energy, specificity, and enzymic
catalysis: the circe effect, in: "Advances in Enzymology,
Vol. 43," A. Meister, ed., John Wiley & Sons, New York,
pp. 219-410 (1975).

27. H. Aquila, D. Misra, M. Eulitz and M. Klingenberg, Complete
amino acid sequence of the ADP/ATP carrier from beef heart
mitochondria, Hoppe Seyler's Z. Physiol.Chem. 363:345
(1982)

28. J. Keyte and R. F. Doolittle, A simple method for displaying
hydropathis character of a protein, J. Mol. Biol. 157:105
(1982).

29. M. Saraste and J. E. Walker, Internal sequence repeats and
the path of polypeptide in mitochondrial ADP/ATP translocase, FEBS Lett. 144:250 (1982).

30. W. Bogner, H. Aquila and M. Klingenberg, Surface labeling of
membrane-bound ADP/ATP carrier by pyridoxal phosphate,
FEBS Lett. 146:259 (1982).

31. W. Bogner, H. Aquila and M. Klingenberg, Probing the
structure of the ADP/ATP carrier with pyridoxal phosphate,
in: "Structure and Function of Membrane Proteins,"
E. Qualiariello and F. Palmieri, eds., Elsevier Science
Publ., Amsterdam, pp. 145-156 (1983).

32. B. Scherer and M. Klingenberg, Demonstration of the relationship between the adenine nucleotie carrier and the
structural changes of mitochondria as induced by adenosine
5'-diphosphate, Biochemistry 13:161 (1974).

33. B. B. Buchanan, W. Eiermann, P. Riccio, H. Aquila and M.
Klingenberg, Antibody evidence for different conformational
states of ADP/ATP translocator protein isolated from
mitochondria, Proc. Natl. Acad. Sci. 73:2280 (1976).

34. W. Eiermann, H. Aquila and M. Klingenberg, Immunological
characterization of the ADP/ATP translocator protein
isolated from mitochondria of liver, heart and other
organs, evidence for an organ specificity, FEBS Lett.
74:209 (1977).

35. H. Aquila and M. Klingenberg, The reactivity of -SH groups
in the ADP/ATP carrier isolated from beef heart mitochondria, Eur. J. Biochem. 122:141 (1982).

36. M. Klingenberg and M. Appel, Is there a common binding center in the ADP/ATP carrier for substrate and inhibitors? Amino acid reagents and the mechanism of the ADP/ATP translocator, FEBS Lett. 119:195 (1980).

37. M. R. Block, G.J.M. Lauquin and P.V.C. Vignais, Chemical modification of atractyloside and bongkrekic acid binding sites of the mitochondrial adenine nucleotide carrier. Are there distinct binding sites? Biochemistry 20:2692 (1981).

38. M. R. Block, J. M. Lauquin and P. V. Vignais, Differential inactivation of atractyloside and bongkrekic acid binding sites on the adenine nucleotide carrier by ultraviolet light. Its implication for the carrier mechanism, FEBS Lett. 104:425 (1979).

THE MITOCHONDRIAL F_0-F_1 ADENOSINE TRIPHOSPHATASE:

ISOLATION BY SEPHAROSE HEXYLAMMONIUM CHROMATOGRAPHY

Georges Dreyfus

Departamento de Bioenergética
Centro de Investigaciones en Fisiología Celular
Universidad Nacional Autónoma de México
Ap. Postal 70-600; 04510
México, D.F.

INTRODUCTION

Most ATP is synthesized by oxidative and photophosphorylation in mitochondria and chloroplasts respectively. Peter Mitchell proposed a theory which formulates that ATP is synthesized by the ATPase complex utilizing an electrochemical gradient generated by substrate oxidation or by light[1]. The energy stored in this electrochemical potential is also utilized as a driving force in other processes such as active transport and motility.

After nearly two decades of controversy the general features of this hypothesis now have firm experimental support. It is now widely accepted that proton translocation and the role of $\Delta\mu H^+$ are crucial for ATP synthesis although arguments persist over the details of the coupling pathway and mechanisms[2]. It has been shown that the coupling factor F_0-F_1 is responsible for the synthesis of ATP after it was reconstituted in liposomes in which this complex was the only protein component present[3-6]. The generation of the $\Delta\mu H^+$ to drive ATP synthesis can be achieved artificially by changing the external pH or generation of a K^+-diffusion potential with valinomycin.

The F_0-F_1 ATPase complex is composed of two sectors the proton translocating portion (F_0) which is an intrinsic membrane component and the soluble F_1 sector. They can be separated and their functions studied as autonomous entities. These H^+-ATPases are found in mitochondria, chloroplasts and many bacteria[7] and are remarkably similar in structure.

499

The sector of the complex catalyzing ATP synthesis which is the F_1 portion can be easily dissociated from the membrane and is soluble in aqueous buffers lacking detergents. The portion of the molecule remaining in the membrane after removal of F_1 is F_0, often termed proton channel.

Two models have been very helpful in the study of this complex. The F_0-F_1 ATPase from thermophilic bacterium[3] and the one from *Escherichia coli*[8] these two complexes have been thoroughly studied.

On the one hand the reconstitution studies which were performed in PS3 thermophilic bacterium were indeed very encouraging for further research on the same line with other F_0-F_1 systems. On the other, knowledge of the structure and function of the ATPase complex from *E. coli* derived from genetic studies developed very fast. The first mutant affecting the F_1 ATPase described by Butlin et al[9] was the beginning of this challenging field.

The F_1 portion of the complex is composed in most systems of five subunits termed α, β, γ, δ, ϵ in order of decreasing molecular weight[7]. In mammalian systems and endogenous low molecular weight polypeptide also known as the natural ATPase inhibitor[10] has been described. This protein is believed to regulate the catalytic activity of the enzyme[11-13].

The hydrophobic membrane sector of the ATPase complex in eukariotic systems is not well known. It catalyzes the conduction of protons from one side of the membrane to the other in a passive manner. Proton conducting abilities of proteolipids extracted with organic solvents have been demonstrated[14-15] although the resulting preparations show different sensitivities to proton channel inhibitors depending apparently on the organic solvent used for the extraction.

In bacteria it is composed of three subunits commonly known as a, b and c[16,17]. Subunit c binds Dicyclohexylcarbodiimide (DCCD) in a stoichiometric fashion resulting in a complete inhibition of proton conductance.

The link between F_0 and F_1 in mitochondria is believed to occur by means of three polypeptides these are the oligomycin sensitivity conferring protein (OSCP), F_6 and F_B which have been well characterized[18-20]. The approach now is to determine the interactions between these polypeptides and other subunits of the ATPase complex.

The study of the mitochondrial F_0-F_1 complex has been attempted by several groups[21-25] by solubilizing the complex with various detergents, the purification procedures involve mainly molecular sieve chromatography and differential centrifugation.

These preparations vary in the number of polypeptide components revealed by their SDS-gel patterns, they range from 10 to 18 poly-peptides. This is a main concern in the definition of the structure of the ATPase complex which is still uncertain. The wide range in the values of the enzymatic activities found in these studies, could also be attributed to variation in the number of components present in each preparation. Nevertheless these preparations have retained the properties of the membrane bound enzyme such as P_i-ATP exchange and sensitivity to proton channel inhibitors.

It is our interest to further characterize the composition of the complex by the use of a previously described ability of the F_1 sector to bind to a Sepharose-Hexylammonium (AH) column[26]. We tested for the first time in mitochondrial systems Lauryl dimethyl-amino oxide (LDAO) a zwitterionic detergent (Fig. 1) for the solu-bilization of the ATPase complex.

We believe that this new preparation could help in the clari-fication of the overall structure of the F_0-F_1 complex by adding information on the number of components and catalytic activities of the complex.

EXPERIMENTAL RESULTS

Submitochondrial particles were used as starting material, they were prepared according to[27], solubilization with LDAO was carried out at room temperature and centrifugation of the solubilized material was carried out for 30 min at 100 000 x g. For further ex-perimental details see[31]. Figure 2 shows how increasing amounts of LDAO in the medium increase the ATPase hydrolytic activity liberated to the supernatant while it decreases in the pellet. It can also be seen that the optimal concentration of LDAO for solubilization is 7.5 mM which was the concentration used throughout this study. LDAO has a very low critical micellar concentration (cmc) 125 μM this fact makes the dialysis of this detergent almost impossible. One of the advantages of this method is that the ATPase complex is retained in the Sepharose-AH column. This allows the exchange of LDAO for cholate which is dialyzable due to its high cmc.

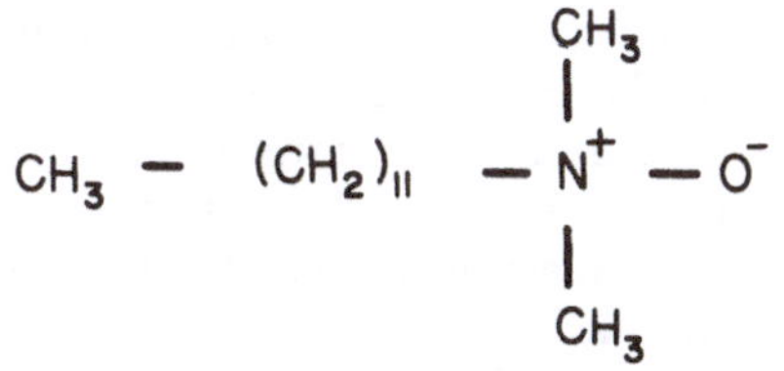

$$CH_3 - (CH_2)_{11} - \overset{\displaystyle CH_3}{\underset{\displaystyle CH_3}{\overset{|}{\underset{|}{N^+}}}} - O^-$$

Fig. 1. Lauryl Dimethylamino Oxide

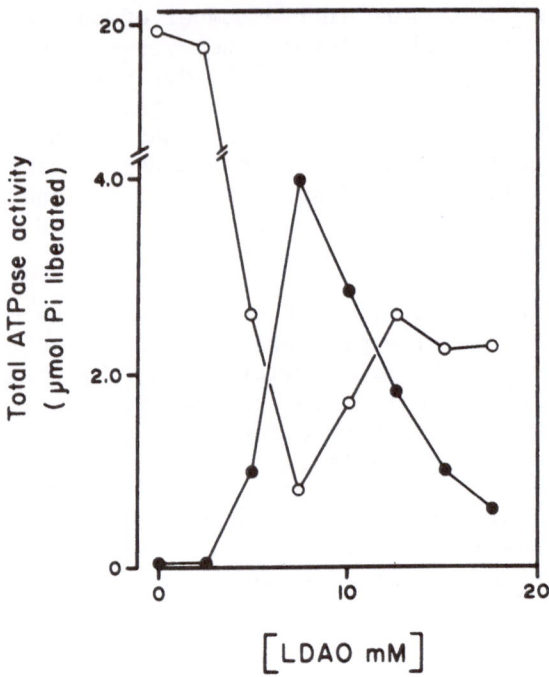

Fig. 2. Solubilization of the ATPase Complex by Varying Amounts of
LDAO. The indicated concentrations of LDAO were employed to
solubilize submitochondrial particles. ATP hydrolysis was
measured in the supernatant (●—●) and in the pellet (o—o).
After solubilization the samples were centrifuged 1 hour at
100 000 x g and the supernatant and pellet separated for
enzymatic activity determination. Hydrolysis of ATP was
measured in a medium containing 100 mM Tris-SO$_4$ pH 7.8, 10
mM MgCl$_2$, 5 mM ATP, 5 mM phosphoenolpyruvate and 32 µg of
pyruvate kinase. Incubation was carried out in 350 µl at
30°C for 10 minutes and the reaction was stopped by the addition of trichloroacetic acid 7% final concentration. The
amount of liberated phosphate was determined according to[32].

The supernatant resulting from the solubilized submitochondrial
particles is passed through a Sepharose-AH column (3x1cm). The
column is thoroughly washed and the retained ATPase complex is released by the addition of 1M KCl (Figure 3). The active fractions
are precipitated with 50% saturated ammonium sulfate centrifuged
and the resulting pellet is resuspended in a buffer containing
soybean phospholipids. The sample is then dialyzed to achieve the
reconstitution of the enzyme.

The purification scheme is shown in Table 1 the purification

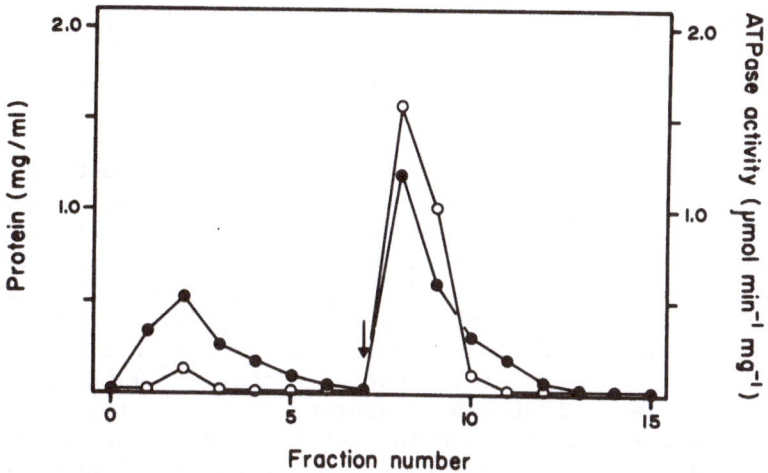

Fig. 3. Sepharose Hexylammonium Chromatography. Elution patterns
of protein (●—●) and ATPase hydrolytic activity (o—o)
of the solubilization supernatant. The arrow indicates
when 1M KCl was applied. The elution buffer consisted of
20 mM Tris-Cl pH 7.5, 2 mM EDTA, 10 mM sodium cholate and
3.0 mg/ml of sonicated phospholipids.

factor was 7.5 and the yield 18%. The purification efficiency is
quite high considering that 10% of the mitochondrial membrane protein
corresponds to the ATPase complex.

The polypeptide composition of the preparation obtained from

Table 1. Purification of the F_1-F_0 ATPase
by Sepharose-AH Chromatography.

	Total Protein (mg)	Total activity[a]	Specific activity[b]	Yield (%)
Submitochondrial particles	22.3	30.14	1.35	100
LDAO extract	5.6	18.2	3.25	60.4
Sepharose-AH	0.58	5.51	9.25	18.3

[a]Total activity expressed as μmol of Pi liberated by the
amount of protein indicated.
[b]Specific activity expressed as μmol Pi liberated $min^{-1}mg^{-1}$.

the Sepharose-AH column was analyzed in SDS-gel electrophoresis. The number of components present was approximately 18 (Table 2) 5 correspond to subunits of the F_1 sector (α, β, γ, δ, ε). We were also able to recognize OSCP and the DCCD binding proteolipid.

The participation in the ATPase complex of the rest of the components remains to be determined, specially the high molecular weight polypeptides. This will be done by adding further purification steps to this procedure i.e. additional ammonium sulfate fractionation. To further confirm the presence of the DCCD binding proteolipid we labelled the preparation with [14C]Dicyclohexyl-carbodiimide (DCCD) which is known to bind to a proteolipid in the F_0-sector. After 12 hours of incubation excess [14C]DCCD was washed and the binding was estimated by measuring radioactivity in the sliced gel. Another gel was run in parallel and stained to obtain the densitometric trace shown in Fig. 4. It can be observed that [14C]DCCD binds to one component which corresponds in molecular

Table 2. Polypeptide Composition of the
Sepharose-AH Eluate.

	BAND	F_1-F_0
	1	72000
	2	65500
	3	63100
	4	61000
	5	53700
α	6	52300
β	7	51000
	8	46000
	9	44000
	10	41000
	11	39250
	12	37900
γ	13	29300
OSCP	14	26700
δ	15	17400
DCCD	16	13800
	17	9750
ε	18	7300

The molecular weights used were bovine serum albumin (68,000) egg albumin (45,000), pepsin (34,700), trypsin (24,000), β-lacto-globulin (18,400), lysosyme (14,300), which were obtained from Sigma Chemical Company, using also the soluble F_1 ATPase purified as in[26].

504

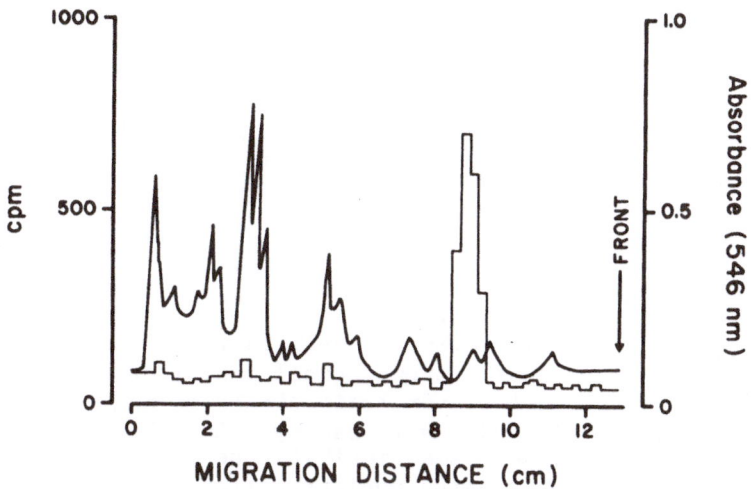

Fig. 4. Affinity Labeling of the Sepharose-AH eluate with
[14C]Dyciclohexylcarbodiimide. Bars show total
radioactivity in counts per minute and the continuous
line the densitometric trace of 546 nm of a gel run
in parallel and stained with coomassie blue.

weight to the DCCD binding proteolipid. Confirming the presence of
this polypeptide in this preparation.

The molecular composition of our preparation resembles very
much the composition reported in previous studies[21-25], for the F_0-
F_1 complex from mitochondria isolated by different procedures. We
reconstituted the fraction obtained from the Sepharose-AH column
into phospholipid vesicles by the cholate dialysis method[28]. The
resulting proteoliposomes were tested for their ability to carry
out membrane coupled functions i.e. P_i-ATP exchange and ATP hydrol-
ysis-dependent membrane energization of the F_0-F_1 containing
liposomes.

The P_i-ATP exchange relates the ability of the ATPase complex
to carry out the incorporation of [32P] into ATP. This is done by
hydrolyzing ATP which results in the movement of protons to the
interior of the liposomes. These delocalized protons then support
the inverted sense of the reaction (scheme 1) giving [γ-32P]-ATP.

This reaction is strictly dependent on the integrity of the
ATPase complex and the permeability of the membrane. In other words,
the reaction will not take place if inhibitors of the ATPase complex
are present or if protons are made permeable by the addition of a
ionophore. Table 3 shows that the proteoliposomes are able to carry

Scheme 1

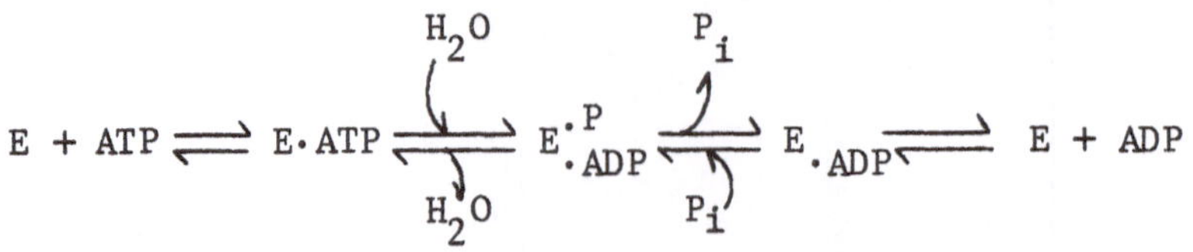

$$E + ATP \rightleftharpoons E \cdot ATP \xrightleftharpoons[H_2O]{H_2O} E \cdot {}^P_{\cdot ADP} \xrightleftharpoons[P_i']{P_i} E \cdot ADP \rightleftharpoons E + ADP$$

1	*2*	*3*	*4*

out the P_i-ATP exchange reaction as well as the hydrolysis of ATP (Table 3). The presence of proton channel inhibitors inhibit the exchange reaction while the hydrolysis of ATP is only partially inhibited by Oligomycin and unaffected by DCCD. The presence of a proton ionophore like FCCP, also inhibits the P_i-ATP exchange while the hydrolytic reaction remains unaltered (Table 3).

One observation must be made given that the P_i-ATP exchange is rather low if it is compared to other F_0-F_1 reconstituted preparations (Table 4). Specifically those of Penin et al[24] and Hughes et al[23] who reported exchange activities of the order of 1.6 µmol $min^{-1}mg^{-1}$. Less active but still high are the preparations of Serrano et al[22]; Stigall et al[21], which range between 0.15-0.4 µmol $min^{-1}mg^{-1}$. On the other hand if we compare our results to those obtained by McEnery et al[25] and Weber and Schäfer[29] who prepared F_0-F_1 with a zwitterionic detergent, we find that the three preparations show similar P_i-ATP activities. They range between 25 and 87 nmol $min^{-1}mg^{-1}$, this low exchange activity could be related to the detergent employed to solubilize the complex.

Table 3. Effect of Oligomycin, DCCD and FCCP
on the Reconstituted F_0-F_1 complex.

	ATP-Pi exchange (nmol $min^{-1}mg^{-1}$)	Inhibition (%)	ATPase activity (µmol $min^{-1}mg^{-1}$)	Inhibition (%)
Control	70	0	9.5	0
Oligomycin	0.0	100	3.6	62
DCCD	0.9	98	9.5	0
FCCP	0.0	100	9.6	0

Table 4. The P_i-ATP Exchange Reaction in Various F_0-F_1 Reconstituted Systems.

P_i-ATP exchange μmol min^{-1}mg^{-1}	Detergent used	Charge	Author
1.6	Lysolecithin	anionic	Hughes et al., 1982
1.0	"	"	Penin et al., 1982
0.4	Cholate	cationic	Stigall et al., 1978
0.15	"	"	Serrano et al., 1976
0.087	CHAPS*	zwitterionic	McEnery et al.,1984
0.07	LDAO	"	This study, 1984.
0.025	CHAPSO	"	Weber and Schäfer, 1984

*CHAPS, 3-(3-cholamydopropyl) Dimethylammonio-1-propanesulfonate.

Another valuable tool often used to determine the reconstitution of the ATPase complex is the ATP-dependent fluorescence quenching. This procedure reflects the capacity of the reconstituted enzyme to generate a pH gradient acid inside upon the hydrolysis of ATP. Figure 5 shows how the fluorophore 9 aminoacridine responds to changes in the interior pH of the proteoliposomes (Fig. 5a).

However if a proton channel inhibitor is added the effect is reverted (Fig. 5 b,c) and the same phenomenon is observed if submitochondrial particles are employed (Fig. 5 d).

Finally, we have visualized the reconstitution of the F_0-F_1 complex by electron microscopy of the samples stained with phosphotungstic acid (PTA), as it is shown in figure 6.

It can be seen that particles are protruding from the membrane the diameter of these structures was determined to be of about 90 Å very similar to that reported for the F_1 ATPase[30].

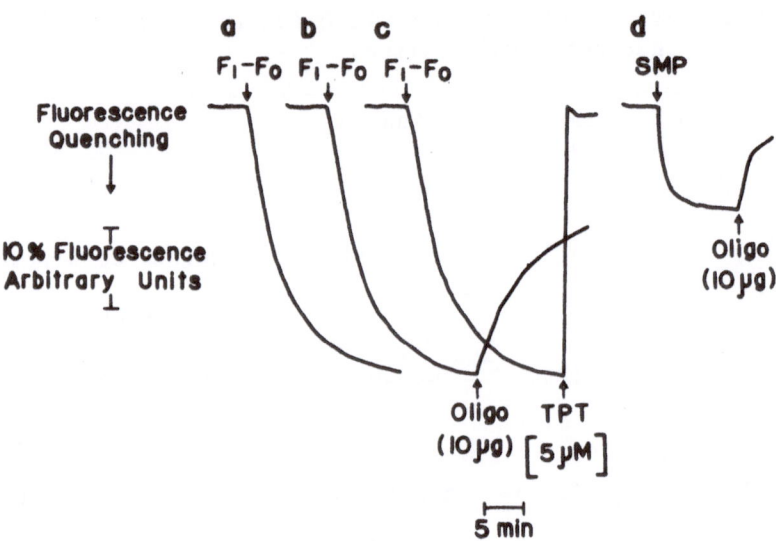

Fig. 5. 9-Aminoacridine Fluorescence quenching during ATP-
dependent Energization of the F_0-F_1 Containing
Liposomes and Submitochondrial Particles.
a,b,c, - The reaction was started with 20 μl (20 μg)
of proteoliposomes, d - The reaction was started
with 20 μl (200 μg) of submitochondrial particles. Additions:
Oligomycin 10 μg, triphenyltin 5 μM final concentration.
The reaction mixture contained 0.1 M sucrose, 40 mM Tris-
Cl, pH 8.0, 0.1 mM EDTA, 0.2% defatted bovine serum
albumin, 2 mM ATP, 5 mM $MgCl_2$, 10 μM 9-amino acridine,
the reaction volume was 1.5 ml.

CONCLUSIONS

The set of experiments described in this paper demonstrate the
obtainment of an F_0-F_1 preparation based on the adsorption of the
enzyme to hexylammonium covalently bound to Sepharose beads and the
specific release by K^+ ions[31]. An effort must still be done in
order to further clarify the structure of the ATPase complex and
determine to what extent a minimal function composition can be
obtained.

On the other hand, the solubilization of the ATPase complex
by different detergents could also be helpful in the study of the
relationship between structure and function. Apparently the exchange
activity of the resulting preparation depends on the detergent used
for the extraction, probably due to the presence or absence of components important for the enzymatic activity. We believe that

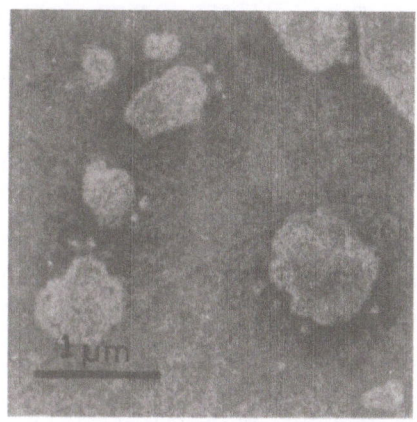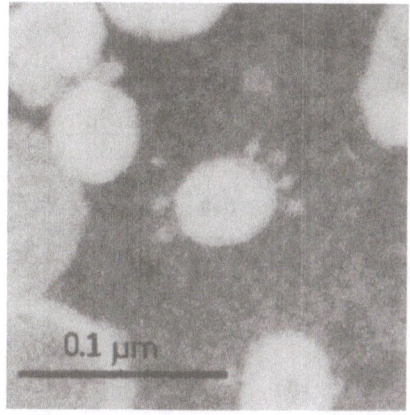

Fig. 6 Electron micrographs of the phosphotungstic acid stained liposomes containing the F_0-F_1 complex.

reconstituted preparations will still be considered useful in the study of the regulation of the catalytic functions of the ATP synthase.

ACKNOWLEDGEMENTS

I gratefully acknowledge J. Sepúlveda and R. Paredes for the electron microscopy and Mrs. Guadalupe Ramírez for excellent typing assistance. Partially supported by CONACYT, México.

REFERENCES

1. P. Mitchell, Nature 191:144 (1961).
2. R.L. Cross, Ann. Rev. Biochem. 50:681 (1981).
3. Y. Kagawa, Biochim. Biophys. Acta 505:45 (1978).
4. I.J. Ryrie and P.F. Blackmore, Arch. Biochem. Biophys. 176:127 (1976).
5. U. Pick and E. Racker, J. Biol. Chem. 254:2793 (1979).
6. N. Sone, M. Yoshida, H. Hirata and Y. Kagawa, J. Biol. Chem. 252:2956 (1977).
7. R.H. Fillingame, Ann. Rev. Biochem. 49:1079 (1980).
8. A.E. Senior and J.G. Wise, J. Membr. Biol. 73:105 (1983).
9. J.D. Butlin, G.B. Cox and F. Gibson, Biochem. J. 124:75 (1971)
10. M.E. Pullman and G.C. Monroy, J. Biol. Chem.238:3762 (1963).
11. D.A. Harris and A.R. Crofts, Biochim. Biophys. Acta 502:87 (1978)
12. A. Gómez Puyou, M. Tuena de Gómez-Puyou and L. Ernster. Biochim Biophys. Acta 547:252 (1979).

13. G. Dreyfus, A. Gómez-Puyou, and M. Tuena de Gómez-Puyou, Biochem. Biophys. Res. Commun. 100:400 (1981).
14. R. S. Criddle, L. Packer and P. Shieh, Proc. Nat. Acad. Sci. USA 7:4306 (1977).
15. H. Célis, Biochem. Biophys. Res. Commun. 92:26 (1980).
16. J. Hoppe and W. Sebald, Biochim. Biophys. Acta 1:27 (1984).
17. A. E. Senior, Biochim. Biophys. Acta 726:81 (1983).
18. T. Hundal, B. Norling and L. Ernster, 3rd EBEC Hannover pp. 315-317, ICSU Press (1984).
19. A. Dupuis, M. Satre and P.V. Vignais, FEBS Lett. 156:99 (1983).
20. S. Joshi, J.B. Hughes, R. L. Houghton and D.R. Sanadi, Biochim. Biophys. Acta 637:504 (1981).
21. D. L. Stigall, Y.M. Galante and Y. Hatefi, J. Biol. Chem. 253:956 (1978).
22. R. Serrano, B.I. Kanner and E. Racker, J. Biol. Chem. 251:2453 (1976).
23. J. Hughes, S. Joshi, K. Torok and R.D. Sanadi, J. Bioenerget. Biomembr. 14:287 (1982).
24. F. Penin, C. Godinot, J. Comte and D.C. Gautheron, Biochim. Biophys. Acta 679:198 (1982).
25. M. W. McEnery, E.L. Buhle , U. Aebi and P.L. Pedersen, J. Biol. Chem. 259:4642 (1984).
26. M. Tuena de Gómez-Puyou and A. Gómez-Puyou, Arch. Biochem. Biophys. 182:82 (1977).
27. G. Klein, M. Satre and P.V. Vignais, Biochim. Biophys. Acta 681:226 (1982).
28. Y. Kagawa, Methods Enzymol. 55:711 (1979).
29. J. Weber and G. Schäfer, 3rd EBEC Hannover pp. 377-378, ICSU Press (1984).
30. L. M. Amzel, M. McKinney, P. Narayanan and P.L. Pedersen, Proc. Natl. Acad. Sci. USA. 79:5852 (1982).
31. G. Dreyfus, H. Celis and J. Ramírez, Anal. Biochem. 142 (1984) In Press.
32. C. A. Fiske and Y. Subbarow, J. Biol. Chem. 177:751 (1925).

THE UNFOLDING OF A CATALYTIC MECHANISM

FOR THE REMARKABLE ATP SYNTHASE

Teri Mélèse

Department of Chemistry and Biochemistry
University of California
Los Angeles, California   90024

INTRODUCTION

In the past two decades, investigators in the field of bio-energetics have shown that a particular multisubunit protein complex (the ATP synthase) on the inner membrane of mitochondria, and on plant and bacterial membranes uses the energy supplied by an electrochemical or pH gradient to drive the net synthesis of ATP (see Figure 1). This energy could exert its effect either indirectly through a physical change in the structure of the protein itself (i.e. a conformational change) which would facilitate changes at the catalytic site, or directly, by the active donation of protons into the catalytic site which would be used in the covalent bond step when ATP is formed from ADP and P_i. Mitchell (1974, 1978) has suggested that effective removal of an oxygen group from P_i could be accomplished by such protonations at the catalytic site. However, over the past 10 years, increasing evidence has accrued showing that the energy provided by electron transport down the respiratory chain is used indirectly for the phosphorylation of ADP by P_i.

How energy is used to drive ATP synthesis

In 1953, Mildred Cohn demonstrated that mitochondria catalyzed a rapid exchange of phosphate oxygens with water oxygens. We now know that this oxygen exchange results from the reversal of the overall reaction of oxidative phosphorylation (ADP + P_i $\leftrightarrows$ ATP + HOH). An unusual property of this phosphate-oxygen exchange in submitochondrial particles is its resistance to the uncoupling agent, 2,4-dinitrophenol (Mitchell et al. 1967).

511

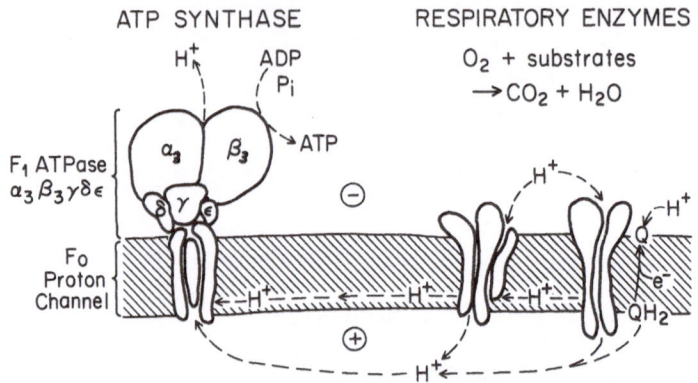

Fig. 1. Schematic illustration of oxidative phosphorylation.

        Later research showed that the uncoupler insensitive por-
tion of the overall sequence was the readily reversible covalent
bond-forming step as shown in Figure 2 (Rosing et al., 1977).
Because this step is relatively insensitive to uncouplers, and
because these uncouplers dissipate transmembrane proton motive
force, it is unlikely that such a force could be used to drive
ATP formation by the direct protonation of P_i oxygens.

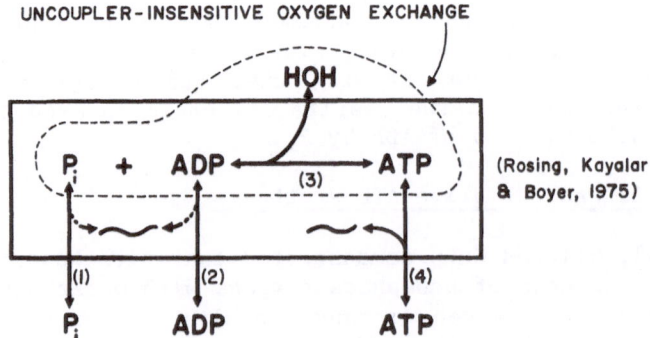

Fig. 2. Schematic illustration of the active site of the ATP
        synthase.

Even though it was not yet apparent that the uncoupler insensitive step was the formation of the covalent bond, in 1973 Boyer suggested that a prominent role of energy in oxidative phosphorylation was not manifested in direct effects on the covalent bond formation of ATP, but rather on the release of ATP from the catalytic site (Boyer et al., 1973). It was suggested later by other evidence that ADP and P_i binding are also associated with energy input from oxidation (Boyer et al., 1975; Jagendorf, 1975; Rosing et al. 1976; Kayalar et al., 1977).

The reason that the point of energy input during oxidative phosphorylation for ATP synthesis was commonly thought to be the formation of the anhydride bond was the assumption that this reaction was as highly unfavourable on the surface of the enzyme as it was in solution. However, since the energy-linked binding change mechanism was first proposed considerable additional evidence has accumulated for a number of protein kinases showing that this is not the case (Cohn, 1979). Most of these studies were done using ^{31}P NMR techniques instead of ^{18}O-exchange. These methods can be used to determine the K_{eq} on the enzyme surface, and for many enzymes the value is nearly equal to 1, indicating that this reaction is freely reversible. Recently the K_{eq} was calculated for the chloroplast ATPase and was also found to be close to 1 (Feldman and Sigman, 1982). However, the overall free energy for the synthesis of ATP on the enzyme cannot be different from that for the reaction occurring free in solution. The energy-linked binding change model accounts for this apparent difference by proposing differences in the binding affinities of the enzyme for substrates and products.

Catalytic Cooperativity

Another related facet of this work was the observation by Kayalar, Rosing, and Boyer (1977) of substrate modulations of the ^{18}O exchange reactions which lead to the proposal that there was a cooperative interaction of the subunits of the ATP synthase during catalysis. Just as the presence of exchange, as discussed above, serves as an indirect measure of the reversibility of the catalytic bond-forming step in ATP synthesis, monitoring the number of ^{18}O oxygens exchanged into one mole of product formed, provides information as to how many reversals of the catalytic step take place before the product is released. Therefore measuring the effect of substrate concentration on the number of oxygens incorporated into the product can yield information about whether the rate of substrate binding can influence the rate of product release. This type of experiment showed that the binding of ADP and P_i promotes the release of ATP during net synthesis (Kayalar et al., 1977), and that the binding of ATP promotes the release of ADP and P_i during net hydrolysis (Hackney and Boyer, 1978; Choate et al., 1979).

A more direct test for subunit cooperativity is to measure the effect of substrate binding on the rate of product release. In his 1981 review, Cross sites the work of Hutton and Boyer (1979), where they show that the addition of adenine nucleotides accelerates the release of tightly bound P_i, and also the work of Penefsky (1979), Chernyak and Kozlov (1979), and Nalin and Cross (1980) showing that the rate of binding and release of AMPPNP, a nucleotide analogue, is strongly influenced by the binding of adenine nucleotides at additional sites.

Returning briefly to the problem of energy input during oxidative phosphorylation, and the observation that ADP and P_i binding as well as ATP release required energy, if there were cooperativity between subunits it would allow this energy to be split physically between substrate binding and product release; thus it would be easy to envision one conformational change being able to drive both ADP and Pi binding at one catalytic site and ATP release from another. As discussed by Boyer (1982) this concept is appealing because it utilizes the energy from substrate oxidation or light transduction in a most efficient manner.

Whether or not the idea of cooperative interaction between catalytic subunits is feasible as a mechanism for this enzyme depends on the structure of the ATP synthase. Investigators have shown that the enzyme is a multisubunit complex composed of five different subunits with the following stoichiometry, $\alpha_3 \beta_3 \gamma, \delta, \epsilon$, having molecular weights of 60,000, 57,000, 36,000, 12,500 and 7,500, respectively, with the catalytic sites located on the β subunits (Baird and Hammes, 1979; Merchant et al., 1983; Moroney et al., 1983). Since all the catalytic β subunits are genetically coded for by one gene (Senior et al., 1983, and Senior and Wise, 1983) it appears likely that all three are identical in structure, and therefore would be expected to behave similiarly during the catalytic cycle.

A schematic drawing of the suggested mechanism of the ATP synthase demonstrating both the energy-linked binding change and the idea of cooperativity of the catalytic subunits is shown in Figure 3 (Boyer 1983; and Gresser et al., 1983). The drawing depicts the conformational changes that the three catalytic sites would undergo during one catalytic cycle. Although each site is regarded as having identical functions they participate in sequence; therefore, during catalysis, the subunits are not functionally equivalent at any one time. There are essentially four steps: substrate binding; energy-linked binding changes for ATP, ADP, and P_i; chemical interconversion of substrates; and product release. ATP synthesis is depicted in the clockwise direction and ATP hydrolysis in the counterclockwise direction. When it is removed from the membrane the synthase complex is incapable of doing net ATP synthesis but can catalyze ATP hydrolysis, and has

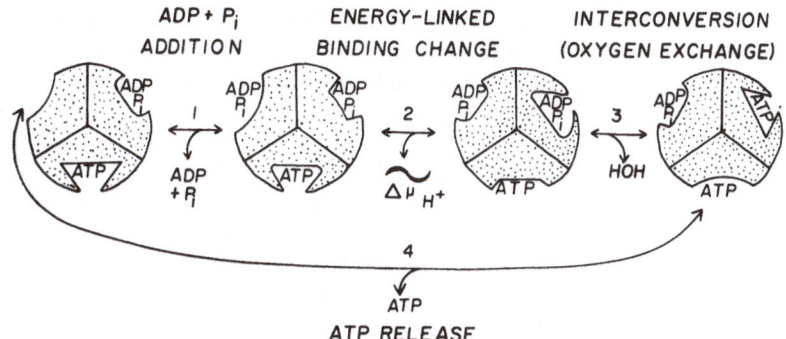

ADP + Pᵢ ADDITION   ENERGY-LINKED BINDING CHANGE   INTERCONVERSION (OXYGEN EXCHANGE)

ATP RELEASE

Fig. 3. The energy-linked binding change mechanism with three catalytic sites participating.

thus been commonly referred to as an ATPase. There is consider-able evidence that the hydrolysis reaction is simply a reversal of the synthesis reaction, and researchers now agree that they probably occur by the same general mechanism.

Experimental evidence for cooperativity between subunits

One of the important pieces of data for cooperativity be-tween subunits concerns the retention at the catalytic site of 1 mole of ADP per mole of enzyme when ATP is being hydrolyzed at concentrations far below that required for half-maximal veloc-ity. This would not be expected for independent catalytic sites which should follow Michaelis-Menten Kinetics, i.e., the amount of product on the enzyme should drop linearly with substrate. This retention of bound intermediates has been demonstrated dur-ing both oxidative and photophosphorylation (Gresser, et al., 1979 and Rosen et al., 1979), and indicates that they will remain bound until substrate binds to another catalytic site. As stated earlier, the results of Kayalar, Rosing, and Boyer (1977) using ^{18}O techniques had shown this retention of product by demonstrat-ing that the number of reversals of formation of product ATP on the enzyme increased when the substrate concentration was low-ered. In other words, the enzyme was "stuck" synthesizing and hydrolyzing ATP continously because it could not release its pro-duct until more substrate bound to another catalytic site.

Gresser, Meyers, and Boyer (1982), showed the retention of a tightly bound ADP during hydrolysis of ATP by the soluble ATP-ase. During steady state hydrolysis in the presence of unlabeled ATP, the bound ADP present acquires a ^{3}H label rapidly if a trace of [^{3}H]ATP is added to the medium, indicating that hydrolysis of the labeled ATP occurred. Furthermore, the bound [^{3}H]ADP is chased off the enzyme if excess unlabeled ATP is added. These experiments support the idea that the bound ADP represents a catalytic intermediate, and that release of the product ADP occurs when ATP binds to a second site.

Along these same lines, Feldman and Sigman (1982) demonstrated the synthesis of ATP from tightly bound ADP and high concentrations of medium $^{32}P_i$ on the isolated ATPase from chloroplasts. This evidence is compelling not only because it shows that the ADP is located at a catalytic site, but also because it dramatically demonstrates that energy is not used in the covalent bond-forming step of ATP synthesis. Finally, recent work reported by Abbot, Czarnecki, and Selman (1984) using the photoaffinity label 2-azido ADP, which binds to the catalytic subunit with the same K_d as ADP and can also be covalently attached to it, showed that the tightly bound ADP present during steady state photophosphorylation is located on the β subunit of the chloroplast ATP synthase.

Another form of evidence that the catalytic sites work in a cooperative fashion is the fact that a single site derivatization at one catalytic subunit will inhibit the whole enzyme. Ferguson, Lloyd, and Radda (1975) showed that derivatization of a single tyrosine residue totally inhibited the activity of the ATPase. This type of one-third site reactivity has been shown with a number of different derivatizing reagents (Kohlbrenner and Cross, 1978; Cross and Kohlbrenner, 1978; Satre et al., 1979).

Do the catalytic subunits rotate?

At this point in the study of the ATP synthase it is interesting to ask how the structure and function of the molecule are related. In other words we have a rudimentary architecture for this enzyme and we also have defined how the catalytic subunits function by the study of the kinetics of ATP synthesis and hydrolysis; but what will the architecture allow in terms of how the enzyme carries out this function?

We know that all the catalytic subunits are made from one gene and that they are essentially structurally identical. Therefore, in order for these subunits to function independently during catalysis they must form unique associations with other subunits in the complex. In other words since the β subunits are not intrinsically asymmetric this asymmetry is created by unique associations accompanying catalysis such that at any one time each subunit is in a different conformational environment based on attachments to other subunits (see Fig. 4). Furthermore, since all the subunits are also required to go through the same series of conformational changes during catalysis these associations must be made and broken in a cyclical manner such that

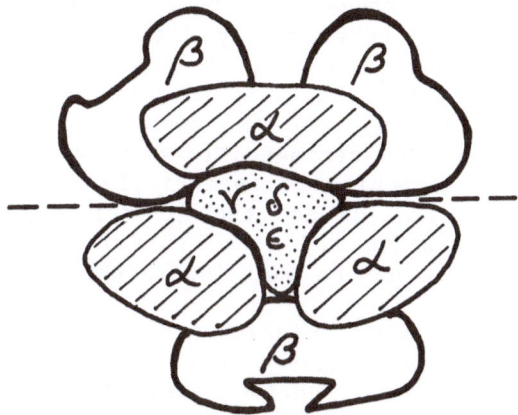

Fig. 4. Illustration of possible asymmetry in the subunit
structure of the ATP synthase.

each subunit will have different connections at different periods
of time during the catalytic cycle.   The best candidates for
these types of specific interactions are the single copy subunits
that form the "core" of the multisubunit complex. These subunits
also form a structural bridge between the oxidation and phosphorylation reactions and ATP synthesis.   The γ subunit of the ATP
synthase is crucial to the ability of the enzyme to carry out
catalysis and if modified appears to alter the interaction of the
proton gradient with the catalytic site (McCarty, 1979).   The
catalytic subunits may be able to rotate relative to the core
subunit such that each β subunit sees the identical unique association with the γ subunit one third of the time, or the γ subunit could act through the α subunits that in turn make contact
with the β subunits. This type of rotary movement has been proposed as the mechanism for the movement of several bacteria
(Doetsch and Sjoblad, 1980).   Motile bacteria have structures
known as flagella that generate bacterial movement by rotating in
a whip-like motion. This motion of the flagella is coupled to an
ionic gradient usually composed of H^+ ions, but Na^+ can be substituted (Hirota, and Imae, 1983).   The mechanism of bacterial
flageller motion was used by Boyer as an analogy for the catalytic mechanism of the ATP synthase when he first proposed in 1981
that catalytic subunits of the ATP synthase might rotate about
the central core of minor subunits (as discussed above).

Experimental evidence for rotation

Testing whether the flageller subunits actually rotated
about a core was ingeniously accomplished by tethering the flagella and watching the bacteria rotate in opposition (Silverman

517

and Simon, 1974). Unfortunately such an elegant observation for the ATP synthase would be extremely difficult, although we have considered such an idea. Finding an approach to study the positional interchange of the catalytic subunits is not easy, but our current investigations are yielding some interesting preliminary data. Basically we have attempted to develop a system for tagging the catalytic subunits differentially during catalysis by using chemical modification and then identifying the individual native catalytic subunits and those that have been modified by visualization in a two dimensional isoelectric focusing system.

A prediction of the cooperative sequential site mechanism for the ATP synthase as discussed in this manuscript is that when the enzyme is catalytically active, rapid interconversion of the subunits from one conformation to another should render each subunit equal in chemical reactivity towards appropriate chemical modifying reagents. The reason for this is that the subunits are proposed to go through the same series of conformational changes during one catalytic cycle and as long as the reaction time of the derivatizing reagent is slow compared to the turnover of the enzyme (i.e., many turnovers will occur before reaction is complete) all the subunits will have the same opportunity to bind the modifying reagent. On the other hand, when the enzyme is catalytically inactive no such interconversion of the subunit conformations would occur, and therefore, the catalytic subunits should show unequal chemical reactivity towards the same modifying reagents.

The first requirement for these studies was finding an enzyme preparation that yielded catalytic subunits that were homogeneous with respect to their isoelectric points. Specific derivatizing reagents were used that would alter the net overall charge, but in order to visualize these changes one must start from a known population of unmodified subunits. When separated on two-dimensional gels the chloroplast ATPase prepared by the method of Binder et al.,(1978), or the method of Younis et al., (1977) but with the addition of protease inhibitors to the buffers, showed about 85% one major component and another single component of about 15% as shown in the bar graph (Figure 5A).

To assess the effect of catalysis on the conformation of the β subunits we exposed the enzyme to either succinic anhydride or iodoacetate under conditions where catalytic activity was retained during hydrolysis and when the enzyme was not turning over. The distributions of the migrating isoelectric species of β subunits following exposure of chloroplast ATPase to succinic anhydride are shown in the bar graphs in Figure 5B and C. When catalysis is not occurring, two subunits react readily but one remains unmodified (Figure 5B). This can be attributed to nonequivalent conformational changes induced by the tight binding of

518

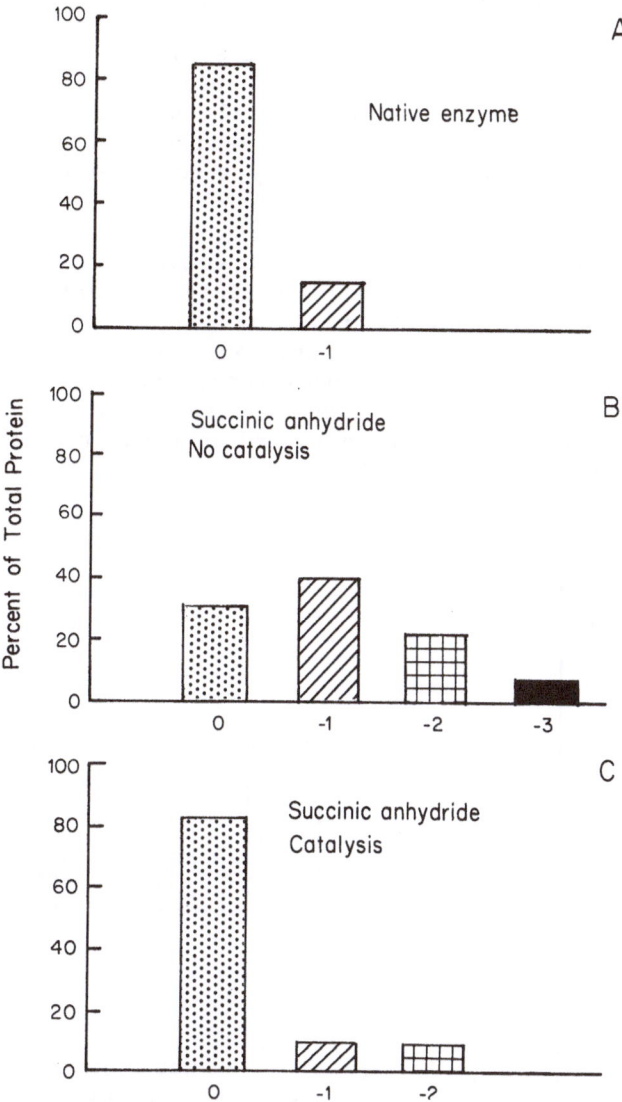

Fig. 5. Protein distribution profiles from two-dimensional electrophoresis of the β subunits of chloroplast ATPase. The designations 0, -1, -2 and -3 indicate apparent change in charge due to derivatization with succinic anhydride.

nucleotides to one β subunit. Under otherwise identical exposure conditions but with catalysis occurring, all β subunits appear to be protected from succinic anhydride as there is very little change from the pattern of the native enzyme (Figure 5C). If all β subunits pass rapidly through the protected "tight" conformation during catalysis this would be the predicted pattern. This points to equivalent participation of each of the three subunits in the catalytic process, as if they were required to proceed in sequence.

The reaction of the chloroplast enzyme with iodoacetate is similar to that observed with succinic anhydride in that the subunits show heterogeneity in their isoelectric mobility when catalysis is not occurring (Figure 6A). However, when catalysis is initiated under the same conditions, the total amount of reaction with iodoacetate decreases but the distribution is now as would be expected if all subunits had an equal reactivity towards the reagent (Figure 6B).

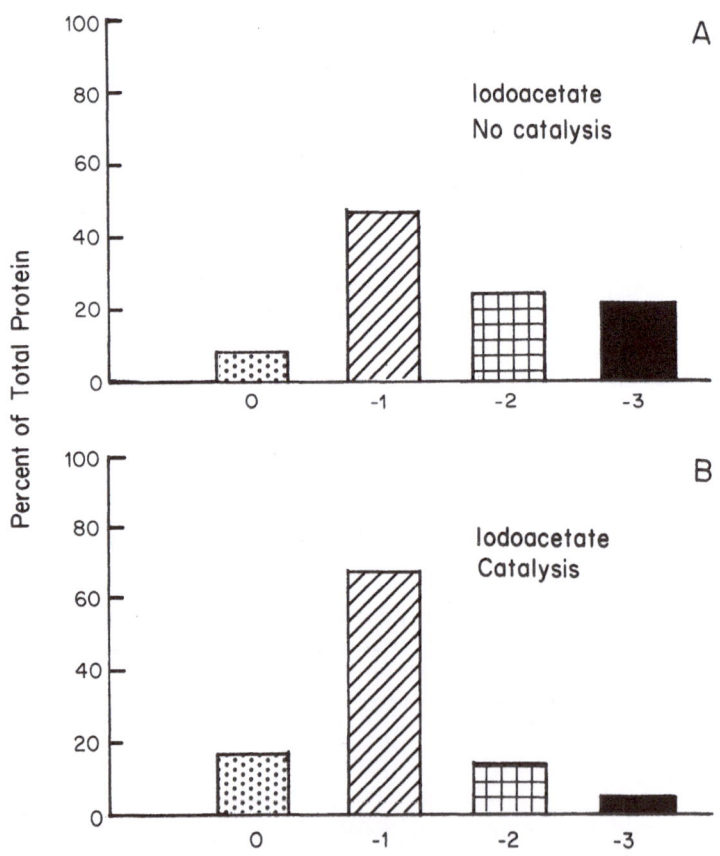

Fig. 6. Protein distribution profiles after derivatization with iodoacetate.

Another demonstration of sequential site participation during catalysis is the following type of experiment: Attach a modifying reagent, such as succinic anhydride, that reacts completely with specific residues on the catalytic subunits and then turn the enzyme over. If the subunits change conformation then new reactive surface groups may be exposed and additional labeling could occur with the same or different derivatizing reagents. Preliminary data shows that this is indeed the case for the reaction of succinic anhydride under these conditions (Mélèse and Boyer, 1984, manuscript in preparation).

CONCLUSION

The modification techniques discussed above coupled with the visualization of the resulting patterns using isoelectric focusing should prove useful for observing the type of positional interchange of the subunits suggested by our studies. Our initial evidence indicates that the catalytic subunits are indeed going through a series of different conformational changes as demonstrated by the marked difference in their reactivity toward modifying reagents depending on whether or not the enzyme is undergoing catalysis.

ACKNOWLEDGEMENTS

My research was carried out with the aid of US Public Health Service Grant GM 11094, and Department of Energy Contract DE-AT03-76ER70102, P.D. Boyer principal investigator. I am also indebted to the National Science Foundation for a travel grant that allowed attendance at the Spetsai Summer School. Special thanks to Lester Packer for submitting me for this fellowship, and to Kerstin Stempel for her help in carefully editing and revising the manuscript as well as for doing the illustrations.

REFERENCES

Abbot, M.S., Czarnecki, J.J., and Selman, B.R. 1984, Localization of the high affinity binding site for ATP on the membrane bound chloroplast ATP synthase, J. Biol. Chem. 259: 12271.

Baird, B.A., and Hammes, G.G. 1979, Structure of oxidative and photophosphorylation coupling factor complexes, Biochim. Biophys. Acta 549: 31.

Binder A., Jagendorf, A., and Ngo E. 1978, Isolation and composition of the subunits of spinach chloroplast coupling factor protein, J. Biol. Chem. 253: 3094.

Boyer, P.D. 1983, How cells make ATP, In: <u>Biochemistry of Meta-bolic Processes</u>, ed. F.W. Stratman et al., pp 465-477, Elsevier, New York.

Boyer, P.D., Cross, R.L. and Momsen, W. 1973, A new concept for energy coupling in oxidative phosphorylation based on a molecular explanation of the oxygen exchange reactions, Proc. Nat'l. Acad. Sci. USA 70: 2837-2839.

Boyer, P.D., Smith, D.J., Rosing, J., and Kayalar, C. 1975, Bound nucleotides and conformational changes in oxidative and photophosphorylation, In: <u>Electron Transfer Chains and Oxidative Phosphorylation</u>, ed. E. Quagliariello, et. al., pp. 361-72, Amsterdam: North Holland.

Chernyak, B.C., and Kozlov, I.A. 1979, Adenylylimidodiphosphate release from the active site of submitochondrial particles ATPase, <u>FEBS Lett</u>. 104: 215.

Choate, G.L., Hutton, R.L., and Boyer, P.D. 1979, Occurrence and significance of oxygen exchange reactions catalyzed by mitochondrial adenosine triphosphatase preparations, <u>J. Biol. Chem</u>. 254: 286.

Cohn, M. 1953, A study of oxidative phosphorylation with O^{18}-labeled inorganic phosphate. <u>J. Bio. Chem</u>. 201: 735.

Cross, R.L., 1981, The mechanism and regulation of ATP synthesis by F_1-ATPases, <u>Ann. Rev. Biochem</u>. 50: 681.

Cross, R.L., and Kohlbrenner, W.E. 1978, The mode of inhibition of oxidative phosphorylation by efrapeptin (A23871), <u>J. Biol. Chem</u>. 253: 4865.

Doetsch, R.N., and Sjoblad, R.D. 1980, Flagellar structure and function in Eubacteria, <u>Ann. Rev. Microbiol</u>. 34: 69.

Feldman, R.I., and Sigman, D. 1982, The synthesis of enzyme-bound ATP by soluble chloroplast coupling Factor 1, <u>J. Biol. Chem</u>. 257: 1676.

Ferguson, S.J., Lloyd, W.J., and Radda, G.K. 1975, The mito-chondrial ATPase: Selective modification of a nitrogen residue on the   subunit, <u>Eur. J. Biochem</u>. 54: 127.

Gresser, M.J., Cardon, J., Rosen, G., and Boyer, P.D. 1979, Demonstration and quantitation of catalytic and noncatalytic bound ATP in submitochondrial particles during oxidative phosphorylation, <u>J. Biol. Chem</u>. 254: 10649.

Gresser, M.J., Meyers, J.A., and Boyer, P.D. 1982, Catalytic site cooperativity of beef heart mitochondrial F_1 adenosine triphosphatase: Correlations of initial velocity, bound intermediates and oxygen exchange measurements with an alternating three-site model, <u>J. Biol. Chem</u>.

Hackney, D.D., and Boyer, P.D. 1978, Evaluation of the partition-ing of bound P_i during medium and intermediate $P_i \leftrightharpoons$ HOH oxygen exchange reactions of yeast inorganic pyrophospha-tase, <u>J. Biol. Chem</u>. 254: 3164.

Hirota, N., and Imae, Y. 1983, Na$^+$-driven flagellar motors of an alkalophilic bacillus strain YN-1, J. Biol. Chem. 258: 10577.

Hutton, R.L., and Boyer, P.D. 1979, Subunit interaction during catalysis: Alternating site cooperativity of mitochondrial adenosine triphosphatase, J. Biol. Chem. 254: 9990.

Jagendorf, A.T. 1975, Chloroplast membranes and coupling factor conformations, Fed. Proc. 34: 1718.

Kayalar, C., Rosing, J., and Boyer, P.D. 1977, An alternating site sequence for oxidative phosphorylation suggested by measurement of substrate binding patterns and exchange reaction inhibitions, J. Biol. Chem. 252: 2486.

Kohlbrenner, W.E., and Cross, R.L. 1978, Efrapeptin prevents modification by phenylglyoxal of an essential arginyl residue in mitochondrial adenosine triphosphatase, J. Biol. Chem. 253: 7609.

Merchant S., Shaner, S. and Selman, B.A. 1983, Molecular weight and subunit stoichiometry of the chloroplast coupling Factor 1 from Chamydomonas reinhardi, J. Biol. Chem. 258: 1026.

Mitchell, P. 1974, A chemiosmotic molecular mechanism for proton translocating adenosine triphosphatases, FEBS Lett. 43: 189.

Mitchell, P. 1979, Keilin's respiratory chain concept and its chemiosmotic consequences, Science 206: 1148.

Mitchell, R.A., Hill, R.D. and Boyer, P.D. 1967, Mechanistic implications of Mg^{++} and adenine nucleotide requirements for energy-linked reations catalyzed by mitochondrial particles, J. Biol. Chem. 242: 1793.

Moroney, J.V., Lopresti, L., McEwen, B.F., McCarty, R.E., and Hammes, G.G. 1983, The M_r-value of chloroplast coupling factor 1, FEBS Lett. 158:58.

Nageswara Rao, B.D., Kayne, F., and Cohn, M. 1979, ^{31}P NMR studies of enzyme-bound substrates of rabbit muscle, J. Biol. Chem. 254: 2689.

Nalin, C.M., and Cross, R.L. 1980, Cooperativity between adenine nucleotide binding sites on mitochondrial F_1 ATPase, Fed. Proc. 39: 1843.

O'Farrell, P.H. 1975, High resolution two-dimensional electrophoresis of proteins, J. Biol. Chem. 250: 4007.

Penefsky, H.S. 1979, Mitochondrial ATPase, Adv. Enzymol. 49: 223.

Rosen, G., Gresser, M.J., Vinkler, C., and Boyer, P.D. 1979, Assessment of total catalytic sites and the nature of bound nucleotide participation in photophosphorylation, J. Biol. Chem. 254: 10654.

Rosing, J., Kayalar, C. and Boyer, P.D. 1977, Evidence for energy-dependent change in phosphate binding for mitochondrial oxidative phosphorylation based on measurement of medium and intermediate phosphate-water exchanges, J. Biol. Chem. 252: 2478.

Satre, M., Lunardi, J., Pougeois, R., and Vignais, P.V. 1979, Inactivation of Escherichia coli BF$_1$-ATPase by dicyclohexylcarbodiimide chemical modification of the subunit, Biochem. 18: 3134.

Senior, A.E., Langman, L., Cox, G.B., and Gibson, F. 1983, Oxidative phosphorylation in Escherichia coli, Biochem. J. 210: 395.

Senior, A.E., and Wise, J.G. 1983, Proton ATPase of E. coli and mitochondria, J. Memb. Biol. 73: 105.

Silverman, M., and Simon, M. 1974, Flagellar rotation and the mechanism of bacterial motility, Nature 249: 73.

Younis, H.M., Winget, G.D., and Racker, E. 1977, Requirement of the δ subunit of chloroplast coupling factor 1 for photophosphorylation, J. Biol. Chem. 252: 1814.

MEMBRANE-OSCILLATOR HYPOTHESIS OF METABOLIC CONTROL IN
PHOTOPERIODIC TIME MEASUREMENT AND THE TEMPORAL
ORGANIZATION OF DEVELOPMENT AND BEHAVIOUR IN PLANTS

E. Wagner+, M. Bonzon++, and H. Greppin++

+Department of Biology II, University of Freiburg
 D-7800 Freiburg
++Department of Plant Physiology, University of Geneva
 CH-1211 Genève 4

SUMMARY

    As a working hypothesis we proposed that the oscillatory feed-
back system of cellular energy metabolism should be the basis for
the endogenous timing of growth, development and behaviour in eukaryotic systems. The interaction of environmental signals with an endogenous physiological rhythm or clock was assumed to occur at membrane-organized receptors which modulate membrane-bound energy
transduction. The energy-dependent state of membranes was in turn
considered to determine the sensitivity of membrane-bound receptors.
The structural and functional principles for the physiological oscillators were supposed to be the same as those underlaying the
theory of membrane-bound energy transduction (1,2). The circadian
system is genetically fixed and provides the temporal frame for
physiological and behavioural patterns that are necessary for survival of organisms and populations. In photoperiodic acclimation of
organisms photoredox systems most likely function in signal transduction as modulators of vectorial metabolism in general and of co-
translational and post-translational protein translocation in particular (1,3,4).

INTRODUCTION

    During evolution living systems have been subjected to continu-
ous variations of light and darkness, heat and cold and other rhythmic parameters of the physical environment. In response to these external periodicities the cells and organisms show periodic behaviour.
In contrast to such exogenously determined responses, living systems

525

Table 1. Chronobiology and Photoperiodism

TEMPORAL ORGANIZATION OF LIVING SYSTEMS	CHRONOBIOLOGY
	CHRONOPATHOLOGY
and their	CHRONOPHARMACOLOGY
	CHRONOTHERAPY

Adaptation to cyclic environmental signals from distinct regions of the electromagnetic spectrum

PHOTOPERIODISM – Signal perception, – transduction, – amplification and – realization

Photoreceptors : Phytochrome, Flavoproteine
Timer   : Circadian Rhythm
                  a) Period length in constant conditions : 22 – 28 h
                  b) Temperature compensated, $Q_{10} \approx 1$
                  c) Entrainment by light and temperature signals, phase shifting
                  d) Genetically determined period length
                  e) Insensitive to chemical manipulation
                  f) Sensitive to D_2O (period lengthening)
Mechanisms   : Biochemical – biophysical oscillations and feedback
                  Ions
                  Membranes and transport
                  Mitchell – Hypothesis of energy transduction

also display circadian (~ 24 h) rhythms which persist in constant environment and are believed to be the manifestation of an endogenous, physiological clock (5,6). In this context, homeostasis of living systems can be understood as the constancy of temporal relations among rhythmic biochemical and biophysical processes on all levels of biological organization. Under natural conditions, the physiological clock is entrained by environmental zeitgebers or synchronizers such as light-dark and temperature cycles. These environmental signals are perceived by photo- and thermoreceptors and are transduced into the modulation of metabolism, thus allowing adaptation of developing systems to cope with environmental constraints by appropriate photo- and thermoperiodic responses.

The experimental evidence indicates that the spatio-temporal organization of living systems might ultimately depend on the proper functioning of endogenous physiological clocks and their interaction with receptors for internal and environmental signals.

Investigations on the temporal organization at the subcellular, cellular and organismic level are the topic of an interdisciplinary science – Chronobiology – with subdivisions including Chronopathology, Chronopharmacology and Chronotherapy. In Table 1, the main aspects of Chronobiology are summarized with special emphasis on Photoperiodism in plants.

526

The comprehensive and far-reaching character of this field of research is due to the ubiquity of rhythms in physiological and behavioural variables. There are rhythms with cycles/second to cycles/year or even more (1,5). Oscillatory phenomena can be observed on all levels of biological organization. Rhythmicity seems to be a function of the molecular make up of protein molecules in their interaction with their solvent environment (8,9). Rhythmic functions have been demonstrated with metabolic sequences in vitro and in vivo (10,11,12) with isolated cell organelles (13), with single cells (14), with organisms and with populations of organisms (5,15).

The eukaryotic cells obviously internalized, i.e. fixed in their genetic program the circadian periodicity and use it as an endogenous physiological clock with the characteristics summarized in Table 1.

By programming responses in relation to the seasonal changes in the daily photoperiod, plants and animals adapt to predictable variations in the environment. By timing dormancy and initiation of flowering at a particular season, the continuity of species is assured. In plants, photoperiodic behaviour has been used very successfully to analyse the main photoreceptor involved in the perception of photoperiodic signals - the phytochrome system -, a photochromic pigment which distinguishes between light and dark through photomodulation and the action of the active form of the pigment on metabolism as outlined in Fig. 1. Metabolic activities of plants change drastically at dusk from dominant photosynthesis to utilization of reserves in the dark and the reverse occurs at dawn. The change in the active form of the pigment and the timing in the availability of substrates on which it acts are probably the essentials of photoperiodic control in plants as shall be discussed later.

In the case of the circadian frequency range, it seems very probable that the network of interacting subsystems of the eukaryotic cell has evolved to be synchronizable to external 24 h geophysical periodicities. The network of cellular metabolism seems to be a prerequisite for the free running circadian rhythm of cell functions in constant conditions since isolated metabolic sequences or cell organelles can only display non-circadian frequencies of metabolic activity (1).

COMPARTMENTATION AND CIRCADIAN RHYTHMICITY

The existence of circadian rhythmicity only at the eukaryotic organisational level (prokaryotes show no circadian rhythms) suggests that metabolic compartmentation may be essential. Furthermore, circadian rhythms in photosynthesis, respiration and chloroplast shape suggest precise interactions between organelles during acclimation of a photosynthetic cell to the lighting environment (Fig. 2).

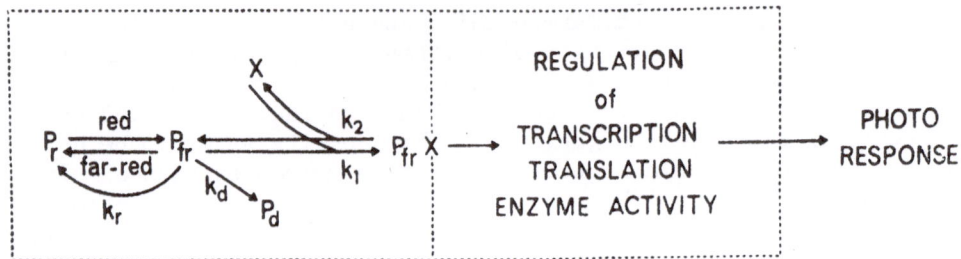

Fig. 1. Kinetic model of the phytochrome system and potential ways for signal transduction. In the scheme, K_d is the reaction rate of P_{fr} destruction, leading to P_d which is inactive biologically and photochemically. K_r is the reaction rate of dark reversion and P_{fr}. X is the hypothetical reaction partner of P_{fr}, and K_1 and K_2 are the specific rates for the formation and decay respectively of the P_{fr}X complex. The photoresponse is then a function of the concentration of P_{fr}X. X might be membrane-bound (from 7).

Interrelations between cellular compartments are evident from reciprocal changes in the ultra-structure of chloroplasts and mito-chondria in daily light-dark cycles (16). Within the same cell mito-chondria swell and chloroplasts contract upon illumination and the reverse occurs in the dark (17). These metabolic interactions are mediated through the structurally integrated multi-enzyme systems of the oxidative phosphorylation reaction chains in the inner membranes of mitochondria and the vectorial electron transport chains of both light reactions across the thylakoids of chloroplasts. In these multi-enzyme systems the well-defined interaction of macromolecules provides the basis for sequential or concerted reactions.

The importance of the entire metabolic network for the display of circadian oscillations is underlined by the fact that, in con-trast to the temperature-compensated circadian oscillations of the intact system, isolated organelles display high-frequency oscilla-tions (13,18) which are temperature-dependent.

On these and other experimental grounds, we suggested that cir-cadian rhythmicity is based on a circadian rhythm in energy meta-bolism (2,7). This rhythm would be the result of a compensatory con-trol oscillation between glycolysis and oxidative phosphorylation, coupled in plants to photophosphorylation (1,2). This mechanism of circadian rhythmicity could involve energy control of transport processes at the membranes of cells and organelles (19).

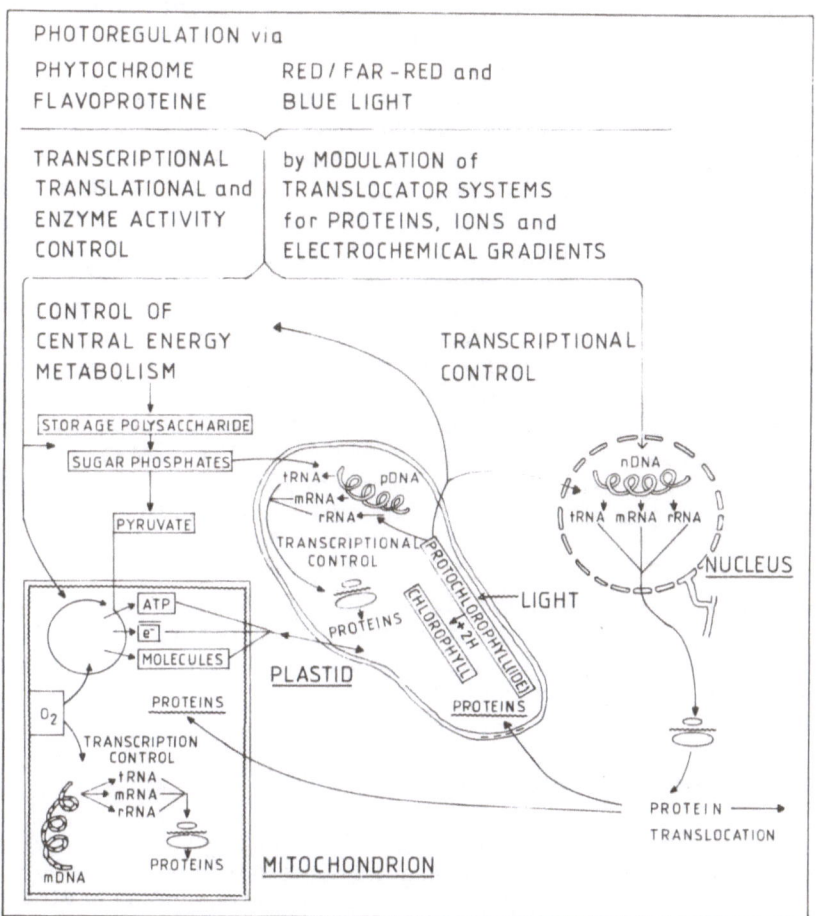

Fig. 2. Flow diagram of nutritional and informational interactions between compartments of a eukaryotic cell during photocontrol of development (from 3).

     Taking into account the symplastic organisation of higher plants the concept of compartmental feedback becomes even more attractive in view of bioelectric phenomena. The symplast might be the basis for transduction of electric, photoperiodic and photomorphogenetic stimuli. The energetic integration of the entire system could be based on the same symplastic organisation of the organism, so that proton translocation or a flow of "proticity" and concomitant ion movements would give rise to a circadian rhythm in electric potential (1). The circadian rhythm in transcellular current of a single cell, as observed in Acetabularia (20), probably arises from the compartmental feedback between mitochondria chloroplasts and glycolysis, as suggested in our concept of circadian rhythmicity. The vacuole could be involved in this control net by acting as a reservoir for metabolites (acids and sugars) (21).

As mentioned above, the symplast of higher plants is probably not only the network for rapid electrical integration of metabolic activities but also the route for the translocation of sucrose and the transfer of the flowering stimulus. An observation that may have some bearing on the significance of changes in membranes during signal transduction is the detection of alterations in the distribution of the endoplasmic reticulum in cells of the shoot apex of Chenopodium album after photoperiodic stimulation (22). The endoplasmic reticulum may have particular significance in cell coordination since it is continuous with the nuclear membrane and probably with the intercellular connections as well and might be the network for protein translocation. In the case of flower induction, the temporal organisation of development at the apical meristem might involve rhythmic symplastic transport of metabolites and the interaction of rhythmic (bioelectric) signals originating in the leaves (23).

Bearing in mind the circadian rhythm in transcellular current reported in Acetabularia (20) it seems possible that a circadian rhythm in proton flow may be of great significance in the circadian rhythmic coordination of the whole plant.

CIRCADIAN RHYTHMS AND PHOTOPERIODIC CONTROL

In his theory of the physiological clock, Bünning suggested (5,6) that endogenous physiological rhythms acted as photoperiodic timers in the control of growth, differentiation and behaviour of living systems. In an attempt to analyse the mechanism of timing in photoperiodism we have suggested that circadian rhythmicity is an evolutionary adaptation of eukaryotic energy metabolism in order to optimise energy conservation, with respect to daily environmental cycles of energy supply (2). The mechanism would be the result of evolutionary refinement of energy conservation which probably occurred in parallel to the evolution from prokaryotic to eukaryotic organisms.

From fermentation in the primaeval anoxygenic soup, energy conservation progressed to anaerobic photosynthetic processes and then to CO_2 fixation with concomitant acceptance of electrons by water and evolution of O_2. Thereafter, in a progressively oxygenic environment, respiration developed with O_2 as a terminal electron acceptor. Eventually, light became the ultimate source of energy to sustain life. The evolution of energy metabolism was accompanied by a progressive refinement in compartmentation of cellular metabolic sequences. In plants, photosynthesis may be controlled through the interaction of circadian rhythms and phytochrome and/or flavoproteins as the main components of photoperiodic control (1).

530

In a review on rhythmic processes in plants (7) we introduced the discussion on the mechanism of photoperiodic control by quoting Nachmansohn (24), concerning the role of acetylcholine in the permeability cycle of excitable membranes. His ideas on acetylcholine action seemed to provide an analogous basis for phytochrome action. He said "The theory assumes that the reaction of acetylcholine with its receptor induces conformational changes. Since proteins are polyelectrolytes with a number of positive and negative charges, it appears likely that conformational changes, even limited ones, will lead to a shift of charge, thereby inducing a series of processes which may markedly affect ionic permeability of a membrane." We also drew attention to similarities between bioelectric effects in plants and animals. For example, an intense flash of light illuminating a leaf of Phaseolus vulgaris can evoke a fast electrical response with a time course in the millisecond range, the voltage-generating source, a pigment, being fixed longitudinally in the leaf (25). Diurnal leaf movements of Canavalia ensiformis have also been correlated with alternations in electric potential (26). In addition, circadian leaf movements of Phaseolus vulgaris are paralleled by a circadian rhythm in electric potential (27). The advances in acetylcholine receptor research as reviewed in (32) shall therefore be briefly summarized as far as they seem to be pertinent for research in the phytochrome system.

In animals and humans physiology and behaviour display circadian rhythms. Circadian rhythmic activity in the nervous system seems to be essential for normal human performance (28). Disturbances in the circadian organization might be responsible for mental disorder (29). Specific phase-relationships between rhythms in metabolic subsystems are apparently crucial for the normal functioning of plants, animals and humans (3,28,29,65). From the existence of circadian rhythmic changes in mental and behavioural activities one may conclude that circadian rhythms possibly modulate signal transduction at synaptic junctions in animals and humans in a similar way as they modulate phytochrome-dependent signal transduction in plants. Circadian rhythmic modulation of signal transduction in eukaryotes and the advances in acetylcholine receptor research might offer a similar conceptual frame for the analysis of the molecular mechanisms involved in synaptic and phytochrome-mediated signal transduction. Especially the evolution of the synaptic complexes during embryogenesis might serve as a model for the development of the phytochrome photoreceptor system during de-etiolation and greening in plants. The following features of the phytochrome system (c.f. Table 1, Fig. 1) should be kept in mind when discussing molecular aspects of the acetylcholine receptor:

- In dark-grown seedlings phytochrome is evenly distributed in high amounts throughout the cell; most of it is degraded upon photostimulation.
- During acclimation to light phytochrome is becoming localized in distinct areas of the cell and is then stable.

- In the case of phytochrome-controlled chloroplast orientation, the pigment seems to be localized close to the plasma membrane in a dichroic structure and most likely acts on an actomyosin system to turn the chloroplast (c.f. Fig. 5).
- In several systems there are changes in bioelectric potential upon phytochrome stimulation.

The pertinent literature for the above-mentioned phenomena is reviewed in references 30 and 31.

Acetylcholine Receptor Function

On the conceptual level a comparison can be made between embryo-genesis in neuronal systems and the establishment of the phytochrome system during de-etiolation of plants.

The information about the molecular properties of the acetyl-choline receptor as discussed by J.-P. Changeux may serve as a guide-line and an analogy for the analysis of the photoreceptor phyto-chrome-mediating plant photoperiodism and photomorphogenesis. Some essential features of the molecular mechanisms of the acetylcholine receptor are therefore briefly summarized from Changeux's Harvey lecture (32).

The electrogenic action of acetylcholine in the membrane is considered to involve two structures: the receptor sensu stricto that carries the acetylcholine receptor site (the regulatory "allo-steric" site) and the ionophore involved in the selective transfer of ions (the biologically "active" site). Both of these would be coupled by an allosteric transition within the same macromolecular complex. Since the receptor protein is membrane-bound, the possibi-lity that cooperative interactions may take place between a large number of neighboring receptor-ionophore complexes organized in a two-dimensional lattice was also considered. The acetylcholine re-ceptor complexes which are condensed in the motor endplate of the neuromuscular junction seem to be anchored to a complex network of filamentous structures. These receptor-rich membranes appear as a dense coaggregate of lipids with two main proteins present in close to equimolar amounts, the membrane-integral acetylcholine receptor and a peripheral 43,000 dalton protein which possesses a definite structural role in the organization of the subsynaptic membrane.

During embryogenesis the receptor protein and acetylcholine-esterase are uniformly distributed on the cell surface. In the adult, the acetylcholine receptor is almost exclusively present in the sub-synaptic membrane in a densely pooled form. The development of the adult structure was proposed as follows:

"At the early stages of synapse formation, the nerve terminal already releases the neurotransmitter, elicits electrical phenomena,

and even transmits nerve impulses. This 'activity', which originates
from the spontaneous oscillatory behaviour of embryonic nerve cells
and is subsequently relayed by evoked activity, may regulate some
critical steps of endplate formation and possibly the pattern of
endplates."

During synapse formation the following changes have been observed: The receptor protein is mobile and labile in the plasma membrane of the embryonic muscle cell and becomes immobile and metabolically more stable in the adult subsynaptic membrane. In the adult
synapse, the metabolic degradation of the receptor protein takes
place with a half-life of up to 10 days or more, while that of the
nonjunctional receptor is much more rapid: the half-life is 17-22
hours. Most likely the 43,000 dalton polypeptide is involved in the
localization process. The nonjunctional receptor thus disappears
during development. This elimination results from the regulation of
receptor biosynthesis.

The intracellular chemical signals engaged in this regulation,
which most likely takes place at the level of gene transcription,
have not yet been identified. The electrical activity of the muscle
triggers the shut-off of receptor biosynthesis presumably via a
change of intracellular ionic concentration (Ca^{2+} or Na^+, K^+) consecutive to the opening of the voltage-sensitive ionic channels.
The ions, however, may not act directly at the gene level, and other
intermediate "internal" signals have been considered such as the
cyclic nucleotides cAMP and cGMP. An involvement of cyclic nucleotides as "second messengers" in the information transfer between the
cell membrane and the genome therefore appears plausible but deserves
further investigation. A very similar situation seems to emerge for
autoregulation of the photoreceptor phytochrome by photostimulation
and feedback control of its own mRNA (33).

In the native membrane, the acetylcholine receptor is strongly
immobilized by the peripheral 43,000 dalton protein through protein-
protein interactions. This protein bound to the inner face of the
membrane cross-links several receptor molecules into a stable "two-
protein" network under the nerve terminal. The 43,000 dalton polypeptide might then serve as an "intermediate piece" between receptor
and cytoskeleton. These structural interactions should be kept in
mind when discussing phytochrome control of chloroplast orientation
(c.f. Fig. 5).

ENERGY METABOLISM - PHOTORECEPTOR INTERACTION

The observation of a circadian rhythm in "leakage" of betacyanin indicated a circadian rhythm in membrane stability (2) and
prompted us to suggest that phytochrome could act as a "membrane
operator" in a system where overall energy metabolism should oscil-

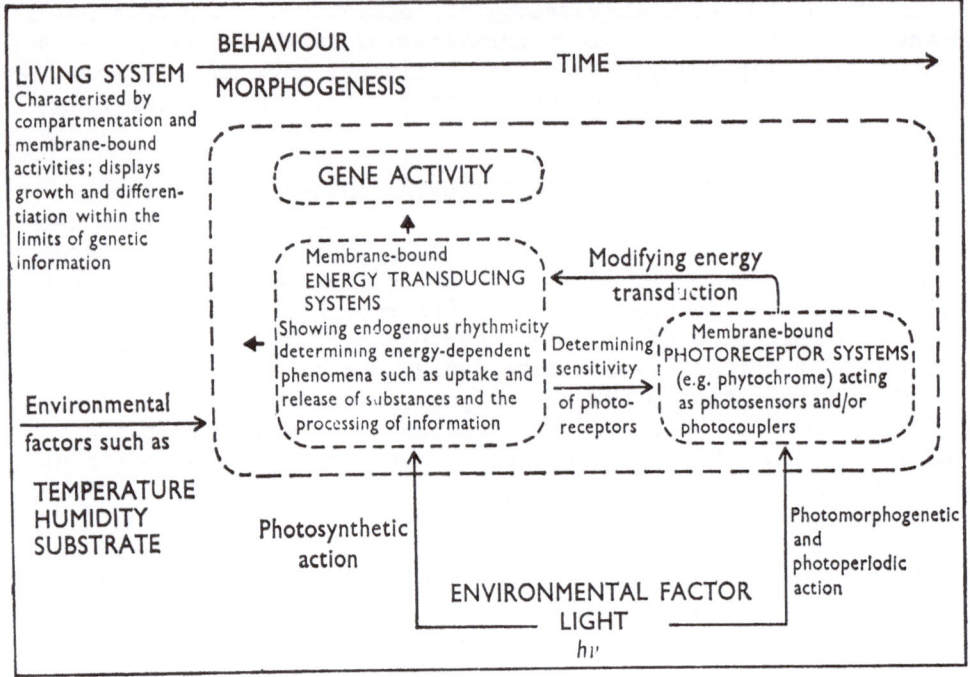

Fig. 3. Flow diagram of postulated regulatory loops involved in endogenous rhythmicity and phytochrome action (from 2).

late in a circadian manner. More recently a circadian rhythm in the kinetics of acid denaturation of cell membranes has been found in Euglena gracilis (34). Analysis of the effect of potassium and ethyl alcohol on light-induced phase shifts of circadian leaf movements in Phaseolus indicated that phase advances could be caused by accelerated membrane depolarisation and delays could be caused by stabilising effects on membranes (35). As a working hypothesis, we put forward the interaction between circadian energy transduction and membrane-bound photoreceptors as shown in Fig. 3. We considered conformational changes in membrane proteins to be essential for the mechanism of these membrane changes within the framework of Mitchell's chemiosmotic hypothesis (36).

With more specific information on a circadian rhythm in energy metabolism, in energy charge and in redox potential, we advanced a membrane oscillator hypothesis of photoperiodic control (37,38). Other membrane models of circadian rhythms have also been proposed (39,40).

Membrane Oscillator Hypothesis of Photoperiodic Control

In an attempt to elucidate the mechanism of endogenous rhythmicity and phytochrome action, we emphasized the importance of cellular energy metabolism in relation to phytochrome action.

Photoperiodic synchronization and switching of metabolism was conceived of as follows: that phytochrome is a membrane-bound photoreceptor which upon illumination undergoes photoisomerization, this is followed by conformational changes in the proteinaceous part of the holochrome which could lead to a change in the structure of the membrane, thus modifying vectorial energy transduction and also altering the activity of membrane-bound enzymes. On the other hand, alterations in energy transduction will be accompanied by alterations in membrane structure. Thus, alterations in membrane potential and concomitant conformational changes are due to alterations in energy transduction and may be the cause of alterations in sensitivity of the membrane-bound photoreceptor(s) on a circadian scale.

It was therefore suggested (2) that the basis of circadian rhythmicity should be energy transduction based on alterations in the conformation of flexible lipoprotein ion-exchange membranes, with phytochrome acting as a membrane operator, capable of phase shifting rhythmic energy transduction. Experimental evidence on phytochrome action on energy metabolism has been discussed in some detail (1,2,4). It was suggested that signal transduction in photoperiodism on a molecular level should involve a membrane oscillator which could display co-operative structural changes in membrane-bound activities conditioned by the (rhythmic) energy state and modulated through photochromic ligands.

Experimental evidence was obtained later, demonstrating circadian organization of energy metabolism in Chenopodium rubrum L. (1). In addition, phase dependent phytochrome modulation of enzymes involved in energy metabolism was shown (41). Rapid, phytochrome-mediated changes in pyridine-nucleotide pool size levels observed in Chenopodium rubrum and Sinapis alba seemed good candidates for transmitting the environmental signal (1,42).

The energy transducing sequences of the cell would also be the information transducers of environmental conditions through adenine and/or pyridine nucleotide modulation of metabolic activity. Fine control of this type of information network could be achieved through activity modulation of compartmentalized adenylate-kinase isozymes as discussed elsewhere (4).

Phytochrome: A Photochromic Receptor and Membrane Operator

In green plants, phytochrome is very stable in contrast to the case of destruction in etiolated plants after phototransformation to the active form (43). Under natural conditions phytochrome is capable of sensing light quality changes such as those at dawn and dusk, and of transducing this information into metabolic signals.

Table 2. Power Transmission by Proticity and Potential Phytochrome Action (from 1 after 36).

POWER TRANSMISSION BY PROTICITY (AFTER MITCHELL, 1976)

PROTICITY PRODUCERS	PROTICITY CONSUMERS
PHOTOREDOX CHAINS	REDOX CHAIN (IN REVERSE)
REDOX CHAINS	⟶ H^+-ATPASE (IN REVERSE)
⟶ H^+-ATPASE	⟶ H^+-PYROPHOSPHATASE (IN REVERSE)
⟶ H^+-PYROPHOSPHATASE	PROTON-COUPLED SOLUTE TRANSPORT SYSTEMS
BACTERIORHODOPSIN	PROTIC HEATING
PHOTORECEPTOR RHODOPSIN (?)	BACTERIAL FLAGELLA (?)

SUGGESTED PHYTOCHROME ACTION

PHYTOCHROME AS PHOTOCOUPLER IN PHOTOREDOX CHAINS UNDER CONDITIONS OF PROLONGED IRRADIATION	PHYTOCHROME AS PHOTOSWITCH IN MEMBRANE-ORGANISED REDOX CHAINS OR ⟶H^+-ATPASE IN RESPONSE TO BRIEF SPECTRAL QUALITY CHANGES AT DAWN AND DUSK

Signal perception and/or transduction could possibly be modulated by a circadian rhythm in redox state of the whole system. Phytochrome could even be directly involved in manipulating photosynthetic electron flow by controlling the level of NADP available for photoreduction (44) by modulating NAD-kinase (45). Under natural conditions, it is very likely that phytochrome control in photoperiodism is depending on the light quality changes at dawn and dusk. This suggestion is supported by data demonstrating that an endogenous rhythm controlling fluctuation in enzyme activity is acting like an on/off switch which determines the potential amplitude modulation of enzyme activity from dusk through the first hours of darkness (41).

These data suggest a dual control of the expression in enzyme activity: a) the endogenous oscillation, which predetermines the course of activity changes, and b) phytochrome, which acts as a phase-dependent amplitude modulator.

It is the phasing of the endogenous rhythm which qualitatively predetermines the amount of phytochrome control at the light-dark transition in daily photoperiodic cycles. There are a number of phytochrome effects which could be rationalized when considered in terms of energy transduction by proton flow. There is, for instance,

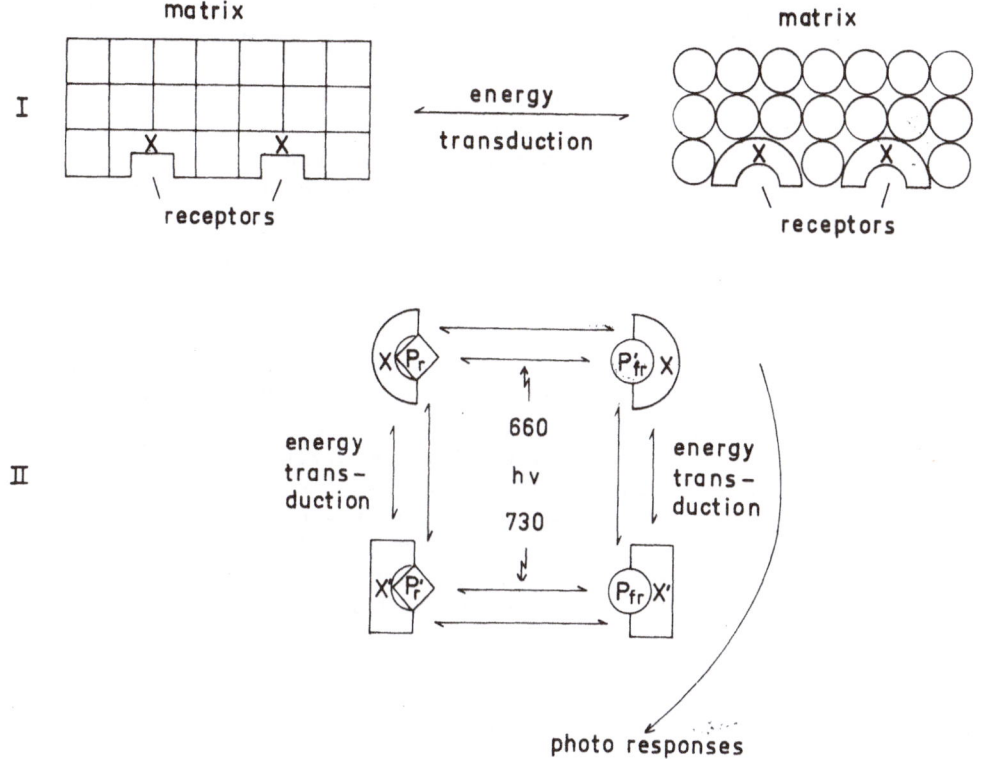

Fig. 4. Structural representation of a membrane oscillator model of photoperiodic control. For explanation, see text (from 38).

phytochrome control of pyridine nucleotide levels after light pulses and under continuous irradiation (42). In the first case the pigment could act as a (conformational) switch while in the latter case it would function as an energy transducer (Table 2). Phytochrome modulation of pyridine nucleotide pools and transitory phytochrome control of ATP levels in C. rubrum and other systems provide supporting evidence for the concept that phytochrome might be involved primarily in redox changes, as was recently discussed in relation to dormancy and germination (46) or might regulate electron flow (47,48).

Our hypothesis of phytochrome-membrane oscillator interaction (38) suggests that the receptor sites for phytochrome are incorporated in a matrix or membrane. The matrix oscillates between two stable configurations via a metastable transitory state in the course of rhythmic energy transduction. Timing of ligand and receptor could either lead to a signal transduction or not, as in the phase-dependent amplitude modulation of enzyme activity. With the structural representation of a membrane-oscillator interacting with phytochrome in Fig. 4, we can distinguish two possibilities, (i) in which phytochrome may reversibly interact with receptor sites, and (ii) where phytochrome is permanently receptor-bound as may be realised in light-grown plants. The receptor site changes its conformation during the cycle of energy transduction and also undergoes structural changes during photoconversion. The photoconversion may result in signal transduction with a subsequent regeneration of the state of the receptor site. With short-term irradiations (switching function), the structural changes are probably accompanied by transitory changes in reduction and/or energy charge.

Phytochrome, Flavoproteins and Interactions

The involvement of different photoreceptors in photoperiodic control was deduced from experiments on photoperiodic control of mating behaviour in Chlamydomonas (49). There was a change in the action spectrum of light controlled sexual activity at the minimum and maximum of a diurnal rhythm in readiness to agglutination of opposite sexes. It was concluded that depending on the time of day there might be a change in the relative contribution to the response from phytochrome and from a blue light receptor. Experiments on the "low irradiance movement" of Mougeotia chloroplasts were also indicative of an interaction between phytochrome and a blue light receptor (50). Blue light induced absorbance changes in membrane fractions of Zea mays and Neurospora crassa implicate a flavin-mediated reduction of a b-type cytochrome with a flavoprotein as the actual photoreceptor (51).

Blue light dependent and flavoprotein mediated changes of membrane-organized redox systems of the cell could quite well transduce environmental signals into cellular energy metabolism. As membrane-bound photoreceptors, both, phytochrome and flavoproteins, very likely could display complex interactions depending on the metabolic (redox) state of their micro-environment as discussed recently for plants acclimating to their natural environment (3).

INTEGRATION OF METABOLIC CONTROLS AND THE TRANSDUCTION OF ENVIRONMENTAL SIGNALS

The internal coupling of metabolic pathways as well as the transduction of environmental signals is probably achieved through nucleotide and ionic (e.g. Ca^{2+}/Mg^{2+}) ratios via redox shuttles and transport systems in the different energy transducing biomembranes. Thus, nucleotides and ions could function as secondary messengers in transduction and amplification of signals. This regulatory concept is supported by observations of photocontrol of NAD-kinase (45) and adenylate kinase (41,52). In the case of NAD-kinase, the Ca^{2+}-calmodulin amplifier system is involved (53),

SURFACE RECEPTOR ACTIN- NETWORK ORGANELLE INTERACTION	PHYTOCHROME CONTROL OF CHLOROPLAST ORIENTATION

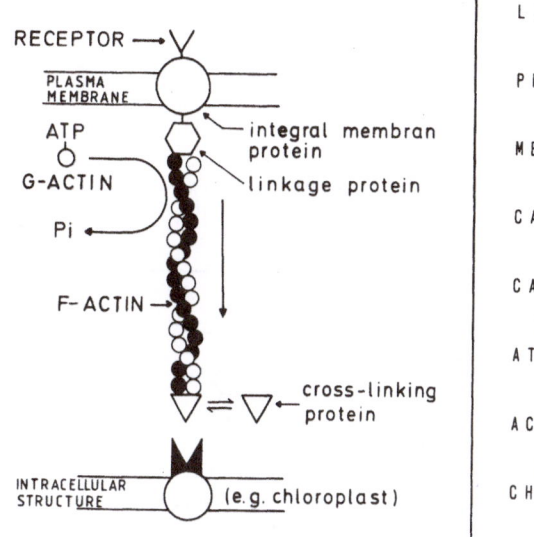

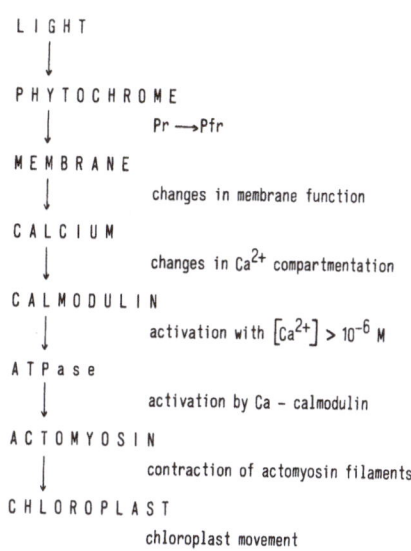

RECEPTOR →

PLASMA MEMBRANE

ATP

G-ACTIN

Pi

F-ACTIN →

integral membran protein

linkage protein

cross-linking protein

INTRACELLULAR STRUCTURE — (e.g. chloroplast)

LIGHT
↓
PHYTOCHROME
↓ $Pr \longrightarrow Pfr$
MEMBRANE
↓   changes in membrane function
CALCIUM
↓   changes in Ca^{2+} compartmentation
CALMODULIN
↓   activation with $[Ca^{2+}] > 10^{-6}$ M
ATPase
↓   activation by Ca - calmodulin
ACTOMYOSIN
↓   contraction of actomyosin filaments
CHLOROPLAST
    chloroplast movement

Fig. 5. Molecular aspects of cytoskeleton dynamics and movement of
cell organelles in response to environmental signals.
Left: Model for transmembrane modulation of actin-organelle
interactions. Right: Potential reaction sequence involved
in ATP and Ca^{2+} dependent orientation of Mougeotia chloroplasts (from 65).

In this context, the importance of ionic ratios in the control
of metabolism has to be stressed (54). Of great significance for
this regulatory network is the participation of ambiquitous enzymes
(55) which can change their micro-compartmentation depending on the
metabolic state of the cell. This might also be the case for soluble
cytoplasmic enzymes (e.g. 56) which could reversibly bind to the
microtrabecular lattice which seems to interconnect all cell structures (57). In this way the network of energy transduction, during
perception and transmission of signals, could involve changes in
membrane potential and surface charge densities as general regulators of membrane protein activities. Changes in membrane structure
could directly be involved in information processing as suggested
for signal transduction in cellular functions on cytoskeletal proteins (58) or in phytochrome control of chloroplast orientation
(Fig. 5) (50). An interplay between oscillating enzymatic reactions
and contractile elements of the structural proteins of the cell has
recently even been used to design a mechano-chemical model of the
biological clock (59).

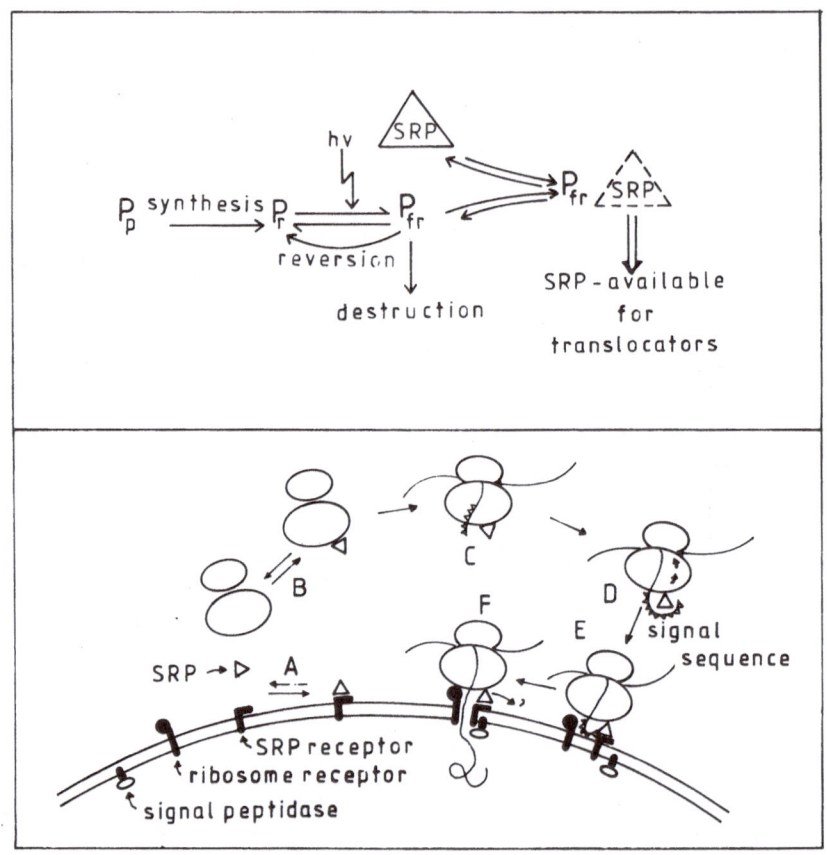

Fig. 6.   Upper half: Kinetic model of the phytochrome system. Lower
          half: Model for the function of SRP in co-translational
          translocation of proteins (from 4).

Spatio Temporal Organisation of Biological Systems

     The genetic determination of membrane components of the eukary-
otic cell could be the basis for the genetically fixed period lengths
of circadian rhythms.

     The participation of the nucleus in circadian rhythms of photo-
synthetic capacity was shown with Acetabularia by exchanging nuclei
between plants kept under opposite L:D regimes. The periodicity of
oxygen evolution was changed in the plants after nuclear transplan-
tation, in accordance with the L:D rhythm to which the nucleus was
previously exposed. These and other manipulations showed that the
nucleus is capable of shifting in the phase of a cytoplasmic circadian rhythm, probably by transfer of material from the nucleus to
the cytoplasm (60).

540

The involvement of phytochrome in membrane transport of β-fructosidase (61,62) could be suggestive of phytochrome involvement in co-translational and past-translational translocation of proteins which might be essential for the cooperation of cell organelles as previously discussed (4) and shown in Fig. 6.

It was proposed that phytochrome ribonucleoprotein interactions, as evidenced by pelletability experiments, might be significant for phytochrome action in vivo on membrane-organized translocator systems (4). As shown in the upper half of Fig. 6 the hypothetical reaction partner "X" for phytochrome action (c.f. Fig. 1) has been replaced by SRP, the "signal recognition protein" (63), the functional unit being a 7s RNA-protein-complex. Based on findings on co-translational translocation of proteins, Walter and Blobel (64) concluded that an SRP should interact with the signal peptide of the nascent chain (emerged from the large ribosomal subunit; D in Fig. 6) that modulates translation and thereby causes an arrest in chain elongation. This arrest is released upon SRP-mediated binding of the ribosome to the microsomal membrane, resulting in chain completion and translocation into the microsomal vesicle. In Fig. 6 an equilibrium between a membrane-bound and a free form of SRP is suggested (A in Fig. 6). Further, an equilibrium is also proposed for free SRP and SRP bound to monomeric ribosomes (B in Fig. 6). Upon translation of a mRNA (C in Fig. 6) the expression of the nascent protein's signal sequence causes a dramatic enhancement in the apparent affinity of SRP for polysomes. Concomitantly, SRP arrests the synthesis of the protein (D in Fig. 6). Only when membranes with "translocation competent sites"are offered to the arrested polysome, does it attach to the membranes (E in Fig. 6), resulting in the assembly of the functional translocation machinery, with the restart of protein synthesis and concomitant translocation of the protein across the membrane (F in Fig. 6). In the case of the existence of signal-sequence-specific SRP analogs for post-translational translocation systems, a similar model was considered feasible (63). It was suggested (4) that phytochrome might manipulate SRP availability or affinity for binding to ribosomes and membranes. The binding to membranes could be controlled by a circadian rhythm in the competence of a membrane-bound SRP receptor.

CONCLUSIONS

The structural and functional principles for physiological oscillators or clocks are supposed to be the same as those determining vectorial metabolism in energy transforming biomembranes. Temporal organization of development and behaviour is dependent on a network of physiological oscillations. There are subsystems engaged in signal perception, transduction and amplification involving a sequence of primary and secondary messengers. The basic endogenous timer probably is a genetically determined circadian rhythm in energy transduction which sets the temporal order for modulation of transcription translation and enzyme activity by environmental signals.

ACKNOWLEDGEMENTS

The authors are grateful for the careful typing of the manuscript by Ms. L. Lohmann and Ms. C. Nolte and the critical discussion by Drs. Ch. Beggs and S.M. De Looze.

REFERENCES

1. E. Wagner, Molecular basis of physiological rhythms, in "Society for Experimental Biology Symposium 31", University Press, Cambridge (1977).
2. E. Wagner and B. G. Cumming, Betacyanin accumulation, chlorophyll content and flower initiation in Chenopodium rubrum as related to endogenous rhythmicity and phytochrome action, Can. J. Bot. 48:1 (1970).
3. E. Wagner, S. Bissbort, M. Schwall, A. Lecharny, R. Bergfeld, I. Kossmann, M. Bonzon, and H. Greppin, Circadian rhythmicity in energy metabolism: A prerequisite for the cooperation between the organelles of eukaryotic cells, Int. J. on Endocytobiosis and Cell Research 1, in press (1984).
4. E. Wagner, U. Haertle, I. Kossmann, and S. Frosch, Metabolic and developmental adaptation of eukaryotic cells as related to endogenous and exogenous control of translocators between subcellular compartments, in "Endocytobiology II", H. Schenk and W. Schwemmler, eds., Walter de Gruyter & Co., Berlin, New York (1983).
5. E. Bünning, "Die physiologische Uhr", Springer Verlag, Berlin (1977).
6. E. Bünning, Die endogene Tagesrhythmik als Grundlage der photoperiodischen Reaktion, Ber. Deutsch. Bot. Ges. 54:590 (1936).
7. B. G. Cumming and E. Wagner, Rhythmic processes in plants, Ann. Rev. Plant Physiol. 19:381 (1968).
8. A. J. Mandell, Redundant mechanisms regulating brain tyrosine and tryptophan hydroxylases, Ann. Rev. Pharmacol. Toxicol. 18:461 (1978).
9. A. J. Mandell and P. V. Russo, Striatal tyrosine hydroxylase activity: multiple conformational kinetic oscillators and product concentration frequencies, J. Neurosci. 1:380 (1981).
10. A. Goldbeter and S. R. Caplan, Oscillatory enzymes, Ann. Rev. Biophys. Bioengineering 5:449 (1976).
11. S. Jerebzoff, Systèmes métaboliques oscillants chez les végétaux inférieurs, Bull. Soc. Bot. Fr. 126:23 (1979).
12. A. Boiteux and B. Hess, Design of glycolysis, Phil. Trans. R. Soc. London B 293:5 (1981).

13. D. Gooch and L. Packer, Oscillatory states of mitochondria. Studies on the oscillatory mechanism of liver and heart mitochondria, Arch. Biochem. Biophys. 163:759 (1974).

14. T. Vanden Driessche, K. J. Doege, C. Minder, and W. L. Cairns, Circadian rhythm in cyclic AMP content in Acetabularia, in "Chronopharmacology", A. Reinberg and F. Halberg, eds., Pergamon Press, Oxford and New York (1979).

15. A. Boiteux, B. Hess, and E. E. Sel'Kov, Creative functions of of instability and oscillations in metabolic systems, in "Current Topics in Cellular Regulation, Vol. 17", Academic Press, New York (1980).

16. W. Könitz, Elektronenmikroskopische Untersuchungen an Euglena gracilis im tagesperiodischen Licht-Dunkel-Wechsel, Planta (Berlin) 66:345 (1965).

17. S. Murakami and L. Packer, Light-induced changes in the conformation and configuration of the thylakoid membrane of Ulva and Porphyra chloroplasts in vivo, Plant Physiol. 45:289 (1970).

18. A. V. Gylkhandanyan, Yu. V. Evtodienko, A. M. Zhabotinsky, and M. N. Kondrashova, Continuous Sr^{2+}-induced oscillations of the ionic fluxes in mitochondria, FEBS Lett. 66:44 (1976).

19. W. J. Vredenberg, Energy control of ion transport processes at the membranes of cells and organelles, Ber. Deutsch. Bot. Ges. 87:473 (1974).

20. B. Novak and C. Sironval, Circadian rhythms of the transcellular current in regenerating enucleated posterior stalk segments of Acetabularia mediterranea, Plant Sci. Lett. 6:273 (1976).

21. U. Lüttge and C. K. Pallaghy, Light triggered transient changes of membrane potentials in green cells in relation to photosynthetic electron transport, Zeitschr. Pflanzenphysiol. 61:58 (1969).

22. E. M. Gifford and K. D. Steward, Ultrastructure of vegetative and reproductive apices of Chenopodium album, Science 149:75 (1965).

23. R. W. King, Multiple circadian rhythms regulate photoperiodic flowering responses in Chenopodium rubrum, Can. J. Bot. 53:2631 (1975).

24. D. Nachmansohn in "Proc. Intern. Symp. Impact Basic Sciences Medicine, Jerusalem 1965", B. Shapiro and M. Prywes, eds., Academic Press, New York (1966).

25. T. Ebrey, Fast light-evoked potential from leaves, Science 155:1556 (1967).

26. Y. Yamaguti, Über elektrische Potentialveränderungen an periodisch sich bewegenden Primärblättern von Canavalia ensiformis, D.C., Botanical Magazine (Tokyo) 46:216 (1932).

27. R. Aimi and S. Shibasaki, Diurnal change in bioelectric potential of Phaseolus plant in relation to the leaf movement and light condition, Plant and Cell Physiol. 16:1157 (1975).

28. J. M. D. Rutenfranz and W. P. Colquhoun, Circadian rhythms in human performance, Scand. J. Work, Environ. & Health 5:167 (1979).

29. T. A. Wehr, D. Sack, N. Rosenthal, W. Duncan, and J. C. Gillin, Circadian rhythm disturbances in manic-depressive illness, Fed. Proc. 42:2809 (1983).

30. L. H. Pratt, Molecular properties of phytochrome, Photochem. Photobiol. 27:81 (1978).

31. D. Marmé, Phytochrome: Membranes as possible sites of primary action, Ann. Rev. Plant Physiol. 28:173 (1977).

32. J.-P. Changeux, The acetylcholine receptor an "allosteric" membrane protein, in "The Harvey Lectures Series 75", Academic Press, London (1981).

33. P. H. Quail, Phytochrome: a regulatory photoreceptor that controls the expression of its own gene, TIBS 9:450 (1984).

34. K. Brinkmann, Circadian rhythm in the kinetics of acid denaturation of cell membranes of Euglena gracilis, Planta (Berlin) 129:221 (1976).

35. E. Bünning and I. Moser, Light-induced phase shifts of circadian leaf movements of Phaseolus: comparison with the effects of potassium and of ethyl alcohol, Proc. Natl. Acad. Sci. USA 70:3387 (1973).

36. P. Mitchell, Vectorial chemistry and the molecular mechanics of chemiosmotic coupling: power transmission by proticity, Biochem. Soc. Transactions 4:399 (1976).

37. E. Wagner, S. Frosch, and G. F. Deitzer, Metabolic control of photoperiodic time measurement, J. Interdisciplinary Cycle Res. 5:240 (1974a).

38. E. Wagner, S. Frosch, and G. F. Deitzer, Membrane oscillator hypothesis of photoperiodic control, in "Proceedings of the Annual European Symposium on Plant Photomorphogenesis", J. A. De Greff, ed., Campus of the State University Center, Antwerp (1974b).

39. L. N. Edmunds Jr. and V. P. Cirillo, On the interplay among cell cycle, biological clock and membrane transport control systems, International J. Chronobiol. 2:233 (1974).

40. D. Njus, F. M. Sulzman, and J. W. Hastings, Membrane model for the circadian clock, Nature 248:116 (1974).

41. S. Frosch and E. Wagner, Endogenous rhythmicity and energy transduction. II. Phytochrome action and the conditioning of rhythmicity of adenylate kinase. NAD- and NADP-linked glyceraldehyde-3-phosphate dehydrogenase in Chenopodium rubrum by temperature and light intensity cycles during germination, Can. J. Bot. 51:1521 (1973a).

42. S. Frosch, E. Wagner, and H. Mohr, Control by phytochrome of the level of nicotinamide nucleotides in the cotyledons of the mustard seedling, Z. Naturforsch. 290:392 (1974).

43. M. Jabben and M. G. Holmes, Phytochrome in light-grown plants, in "Encyclopedia of Plant Physiology, New Series 16B, Photomorphogenesis", W. Shropshire, Jr. and H. Mohr, eds., Springer-Verlag, Berlin, Heidelberg, New York, Tokyo (1983).

44. J. F. Allen, Oxygen reduction and optimum production of ATP in photosynthesis, Nature 256:599 (1975).

45. T. Tezuka and Y. Yamamoto, Kinetics of activation of nicotin-amide adenine dinucleotide kinase by phytochrome-far-red-absorbing form, Plant Physiol. 53:717 (1974).

46. R. B. Taylorson and S. B. Hendricks, Aspects of dormancy in vascular plants, Bioscience 26:95 (1976).

47. H. Greppin, B. A. Horwitz, and L. P. Horwitz, Light-stimulated bioelectric response of spinach leaves and photoperiodic induction, Z. Pflanzenphysiol. 68:336 (1973).

48. H. Greppin and B. Horwitz, Floral induction and the effect of red and far-red preillumination on the light-stimulated bio-electric response of spinach leaves, Z. Pflanzenphysiol. 75:243 (1975).

49. K. M. Hartmann, "Die Regulationder Gametogenese von Chlamydo-monas eugametos und Chlamydomonas moewusii durch exogene und endogene Faktoren. Vergleichende morphologische, physiologi-sche und biophysikalische Untersuchungen. Dissertation der Eberhard-Karls-Universität Tübingen (1962).

50. W. Haupt, Schwachlichtbewegung des Mougeotia-Chloroplasten im Blaulicht, Z. Pflanzenphysiol. 65:248 (1971).

51. R. D. Brain, J. A. Freeberg, C. V. Weiss, and W. R. Briggs, Blue light-induced absorbance changes in membrane fractions from corn and Neurospora, Plant Physiol. 59:948 (1977).

52. S. Frosch and E. Wagner, Endogenous rhythmicity and energy transduction. III. Time course of phytochrome action in ade-nylate kinase, NAD- and NADP-linked glyceraldehyde-3-phos-phate dehydrogenase in Chenopodium rubrum, Can. J. Bot. 51: 1529 (1973).

53. J. M. Anderson, H. Charbonneau, H. P. Jones, R. O. McCann, and M. J. Cormier, Characterization of the plant nicotinamide adenine dinucleotide kinase activator protein and its iden-tification as calmodulin, Biochemistry 19:3113 (1980).

54. F. L. Bygrave, The ionic environment and metabolic control, Nature 214:667 (1967).

55. J. E. Wilson, Ambiquitous enzymes: Variation in intracellular distribution as a regulatory mechanism, Trends Biochem. Sci. 3:124 (1978).

56. S. De Looze, In vitro and in vivo regulation of chloroplast gly-ceraldehyde-3-phosphate dehydrogenase isozymes from Cheno-podium rubrum. III. The molecular basis of the aggregation phenomenon: chloroplast glyceraldehyde-3-phosphate dehydro-genase as an ambiquitous enzyme. Physiol. Plant. 57:243 (1983).

57. K. R. Porter and J. B. Tucker, The ground substance of the living cell, Scientific American 244:41 (1981).

58. S. R. Hameroff and R. C. Watt, "Computer-like" information pro-cessing in cytoskeletal proteins, Biophys. J. B7:A348 (1982).

59. T. S. Sørensen and J. L. Castillo, Spherical drop of cytoplasm with an effective surface tension influenced by oscillating enzymatic reactions, J. Colloid Interface Sci. 76:399 (1980).

60. H. G. Schweiger and E. Schweiger, The role of the nucleus in a cytoplasmic diurnal rhythm, in "Circadian Clocks", J. Aschoff, ed., North-Holland, Amsterdam (1965).

61. M. Zouaghi, D. Klein-Eude, and P. Rollin, Phytochrome regulated transfer of fructosidase from cytoplasm to cell wall in Raphanus sativus L. hypocotyls, Planta 147:7 (1979).

62. A. Ghorbel, B. Mouatassim, and L. Faye, Studies on ß-fructosidase from radish seedlings. V. Immunochemical evidence for an enzyme photoregulated transfer from cytoplasm to cell wall, Plant Sci. Lett. 35:35 (1984).

63. G. Blobel, Regulation of intracellular protein traffic, Cold Spring Harbor Symp. Quant. Biol. 46:7 (1982).

64. P. Walter and G. Blobel, Translocation of proteins across the endoplasmic reticulum. III. Signal recognition protein (SRP) causes signal sequence-dependent and site-specific arrest of chain elongation that is released by microsomal membranes, J. Cell Biol. 91:557 (1981).

65. E. Wagner and L. Fukshansky, Die zeitliche Organisation des eukaryotischen Energiestoffwechsels als Grundlage für die Signalverarbeitung im Photo- und Thermoperiodismus, Ber. Deutsch. Bot. Ges., in press (1985).

INDEX

Glutathione (GSH) (continued)
  in cellular defense, 287
  in drug biotransformation, 287
  in mammalian cells, 287
  oxidation, 288
  peroxidase, 361, 365
  reductase, 362-365
  system, 286-288
  transferase, 287
Glycogen, 149
Glycogenolysis, 149
Glycogen phosphorylase, 149-150
  ligand interactions,
    conformational changes, 157
    dissociation constants, 153,
      155
    microenvironmental proper-
      ties, 155
    polarity of binding sites,
      150-153
    structural changes, 153-160
Glycolipids, 360
Glycophorin, 22, 25
Gramicidin, 126, 396
Guinier approximation, 4-5

Half-site reactivity, 486
Halobacteria, 393
  ion transport, 393
Halobacterium halobium, 75, 394,
  396
  purple membranes, 442, 453
Halorhodopsin, 87, 393
  anion binding sites, 400-402
  as chloride pump, 396-397
  chloride transport model,
    402-404
  functional reconstitution,
    399-400
  light-dependent molecular
    events, 402
  light-driven chloride trans-
    port, 394-397
  purification/identification,
    398-400
Harmothoe lunalata, see Scale
  worms
Hemolysis, 69-70
Hepatocytes,
  drug transformation, 286

quinone metabolism, 292-297
Higher plant mitochondria,
  autotropic metabolism, 241-255
  DNA, 252-255
  features, 237
    anion transports, 246
    cyanide-resistant electron
      pathway, 243-244
    electron transports, 241-243
    NAD^+ transport, 247-249,
      254-255
    oxaloacetate transport, 247
    rotenone-resistant electron
      pathway, 244-246
  genome, 253-254
  malate oxidation mechanism,
    250-252
  preparation, 238-241
  TCA cycle, 252
Hill plots, 156
Hydrophobicity, 32
1-hydroxy-6,8-naphthalenedisulfonic
  acid, 124
Hyperfine splitting, 59-61, 64

Inducible isozymes, 287
Influenza virus, 65
Inhibitor-carrier interactions,
  488-491
Interferon, 71-72
Intracellular second messenger
    system,
  role of phosphatidylinositol
    kinase, 46-52
Invasive techniques, 321-325
Ionic double layer theory, 68-69
Ion transport, 425
Isocitrate dehydrogenase, 467
Isoelectric focusing, 518

Jablonski diagram, 100

Kornberg-Pricer pathway, 188

Lauryl dimethylamino oxide, 501
Ligand-receptor interactions, 47
Light harvesting pigment protein,
  371
Lipid bilayers, 7-0, 104
  virus-membrane interactions,
    65-68